Instructor's Management System

Engineering Drawing and Design

Fifth Edition

Cecil Jensen

Former Technical Director
R. S. McLaughlin Collegiate and
   Vocational Institute
Oshawa, Ontario, Canada

Jay D. Helsel

Professor and Chairman
Department of Industry
   and Technology
California University of Pennsylvania
California, Pennsylvania

Dennis Short

Associate Professor
Department of Technical Graphics
Purdue University
West Lafayette, Indiana

GLENCOE

McGraw-Hill

New York, New York   Columbus, Ohio   Mission Hills, California   Peoria, Illinois

CONTENTS

Preface iii
Tech Prep. . . . . v–xiv
Information and Instructions for CD-ROM . . . . xv–xx
Solutions. 1–258
Chapter Tests and Answers 259–319
Transparency Masters 320–571

P/N G01872.50   Part of ISBN 0–02–801872–9

Send all inquiries to:
GLENCOE/McGRAW-HILL
936 Eastwind Drive
Westerville, Ohio 43081

For technical support, contact:
HyperGraphics Corporation
308 North Carroll Drive
Denton, TX 76201
(800) 369–0002

Printed in the United States of America

Introduction

This *Instructor's Management System* (IMS) written for both *Engineering Drawing and Design* and *Fundamentals of Engineering Drawing* is organized to provide the instructors with those components that will improve their ability to deliver course material in an efficient and organized manner. Components for laboratory instruction and for lecture presentation are included for the traditional manual drafting class, the CAD-based class or for a class that is transitioning between traditional manual skills and CAD. The goal of the IMS is allow the instructor to focus on teaching and not on materials preparation.

Components

The IMS consists of several components:

- The CD-ROM
- Solution keys to all assignments
- Transparency masters
- Chapter tests and keys

CD-ROM

One of the major innovations of the IMS is the CD-ROM. The CD-ROM contains

- Solution keys to assignments
- Step-by-step solutions to selected problems
- Additional specialized material

The CD-ROM contains images of the solution keys for the assignments found at the end of each chapter in *Engineering Drawing and Design*. These images may be displayed on a computer screen for reference while grading a student's drawing or they can be displayed during lecture for discussion with the entire class.

The CD-ROM also contains examples of the step-by-step procedures required to solve selected problems as indicated in the Instructor's Edition of the text. You can use these step-by-step solutions as the basis for a lecture review, as verification of your own procedural solution, or to provide an independent walk-through for a student having difficult with a drawing type. These step-by-step solution examples can also be used to provide an independent learning environment for students who are interested in advanced problems or topics not covered in your course.

Two advanced topics not covered in the test are included in the CD-ROM: applied mechanics and strength of materials. Information is updated on these two topics that had previously been covered in the text and not included in the CD-ROM. Instructors who need to include these topics will find the text, illustrations, reference material, and assignments available.

Solution Keys

The solution keys for all of the end-of-chapter problems are included in the IMS. These solutions are also available on the CD-ROM.

Transparency Masters

A major concern of instructors of engineering graphics courses is the availability of good visuals for lectures that are appropriate to the topics being presented. The preparation of original artwork can be very time-consuming and students do not have copies to reference and study unless distributed by the instructor. One option many instructors use is to copy and enlarge illustrations from the text that support the lecture topic. The problem with this approach is that the quality of the visuals suffers considerable degradation when enlarged and projected.

The IMS includes a significant number of transparency masters, carefully selected with your lecture in mind, that have been generated from the original artwork for the text. Transparencies can be produced from these masters directly and the quality of the projected image is greatly improved. Those illustrations that are available as transparency masters are identified in the Instructor's Edition of the text; they are identical to the illustrations in the text.

PREFACE

Chapter Tests

Five-item multiple choice tests of twelve to fourteen questions each are available in the IMS for each chapter in the text. These questions are knowledge based and are intended to supplement the traditional graphics based testing that occurs in engineering graphics classes. These tests can be administered on scan sheets for automated grading or they can be graded manually. The keys for the tests are grouped to correspond to the five parts of the text and are located after the test duplication masters in the *Instructor's Management System*.

Appreciation

Everyone involved in the production of the the the fifth edition of *Engineering Drawing and Design* and *Fundamentals of Engineering Drawing* would like to thank you for selecting our text for your course. We hope you find that the Instructor's Management System and the new technology represented by the CD-ROM prove helpful in the administration and management of your course.

Cecil Jensen

Jay Helsel

Dennis Short

Introduction

The Tech Prep initiative is directed toward establishing a constructive bridge between high school and post-secondary education to provide better direction for students in terms of their future careers. Some high schools have changed from general education programs to Tech Prep/Associate Degree (TPAD) applied-academics programs to parallel the college prep and vocational educations programs. TPAD programs are directed toward the middle range of occupations that require some postsecondary education and training for which students receive certificates or associate degrees. They are carefully structured and coordinated programs where high school and community college or technical institute teachers, instructors, and administrators work closely with the area business community.*

A goal of TPAD programs is to reduce the high school dropout rate. Most dropouts leave school between the tenth and eleventh grade. The Tech Prep program represents a viable alternative for students who are contemplating an end to their high school years. Students can select the TPAD major in their junior year and continue for two years of postsecondary education and training. This common form results in these programs being referred to as 2 + 2 Tech Prep programs.

Studies and reports support the need for improving the school-to-work transition.**
- Of the high school graduates who go directly to the workforce, more than 60% do so without training in specific skills.
- Approximately 50% of high school graduates go to a four-year postsecondary institution, and only 50% of those students graduate within a ten-year period.
- Nearly 80% of all future jobs will require education beyond high school.

The need for qualitative and quantitative technical knowledge is increasing exponentially. The nation's economic transformation from an industrial-labor base to information-technology-service base has created two dramatic changes within the workplace:

- Employers are demanding higher performance that requires advanced technical skills and an ability to understand complex theories and ideas.
- Businesses are requiring increased skills in problem solving, computer applications, critical thinking, teamwork, and higher-level skills in math, the sciences, and communications.

Research suggests the following statistics to the workforce and education:
- Graduates from a technical college have higher employment rates, higher average pay, and a greater likelihood of success in life-long learning than those from other educational institutions.
- Approximately 11% of the students attending the technical/community colleges do so directly from high school. This low figure suggests a "floundering" period during which the majority of youth are likely to find themselves in minimum-wage, dead-end, high-turnover jobs.
- Relevant education and training in high school will enable students to make an easier transition to advanced training in postsecondary institutions.

Many students today are not being prepared for a current and rapidly expanding technological, global economy. Youth need access to a full range of education-based, future-oriented life-choice options. *College prep, tech prep,* and *work prep* are three terms used to describe postsecondary choices students can make. School districts and community/technical college systems must work together to enable students to make a successful transition from high school to postsecondary education or work.

Federal Provision

The federal government has devoted significant funds to support Tech Prep. Increased expectations for integration of occupational and academic learning accompany this support. The Carl Perkins Vocational and Applied Technology Education Act Amendments of 1990 devoted an entire part to Tech Prep (Title III, PART E—TECH-PREP EDUCATION, Sections 342-347, September 25, 1990).

---

* Portions of the information contained in this article have appeared in other Glencoe/McGraw-Hill publications, including *Strategies for Implementing Tech Prep,* 1995, by Loock and Voiers; *Strategies for Implementing Tech Prep into Communication,* 1994, by Loock and Voiers; and *Tech Prep Strategies in Accounting,* 1995, by Ross, Swinehart, Hogan, Morrison, Hoyt, and Haber. While footnotes are used sparingly, a complete list of references does appear at the end of this Tech Prep segment.
**See items 2, 7, and 14 of the References and Suggested Readings listing.

TECH PREP

Criteria

The Tech Prep part of the Perkins Act includes these criteria:

- Provides planning and demonstration grants to consortia of local education agencies and postsecondary education institutions that will develop and operate four-year Tech Prep education programs. Strong links will exist between secondary and postsecondary educational institutions.
- Programs will consist of two years of secondary and two years of postsecondary education or two years of apprenticeship. The program leads to a two-year associate degree or a two-year certificate.
- Programs will combine nontraditional school-to-work technical education, will use state-of-the-art equipment, and will use appropriate technologies.
- Programs have a common core of required proficiency in mathematics, science, communication, and technologies.
- Programs must include strong teacher and counselor in-service and training to effectively implement Tech Prep education curricula, as well as recruit students and ensure that students not only successfully complete such programs but also are placed in appropriate employment.

For long-term success, however, states and local school districts must continue to support the restructuring established by Tech Prep. Thus, many states and local school districts have passed legislation and allocated funds to supplement federal monies and continue this endeavor.

Objectives

The long-term objectives for Tech Prep in many states are as follows:

- An increase in the percentage of students entering a community/technical college program within one semester after high school graduation.
- An increase in the percentage of students successfully completing a community/technical college program within three years of initial entry.
- An increase in the percentage of students qualifying for advanced standing in a community/technical college program after high school graduation.
- A reduction in the percentage of recent high school graduates needing remedial course work to be admitted to, or to receive a diploma/degree from the community/technical college.
- A reduction in the percentage of recent high school graduates who seek entrance to a community/technical college uncertain about what program they want to pursue.

What is Tech Prep?

Tech Prep is a competency-based program of combined secondary and postsecondary educational and occupational experience that includes a common core of required proficiency in mathematics, science, communication, and technologies designed to lead to an associate degree or certificate in specific career fields. The Tech Prep process is intended to help students select a sequence of courses and experiences that will enable them to achieve the academic and technical competence they will need to successfully pursue a specific associate degree program in a community/technical college after high school graduation.

Student Goals

Avoiding overlap and duplication between secondary and postsecondary experiences is a major feature of Tech Prep. Tech Prep emphasizes the advanced skills rather than time-shortened education. Students who have successfully completed Tech Prep will

- Identify and capitalize on their interests, aptitudes, abilities, values, and preferences.
- Recognize the broad range of postsecondary options that capitalize on their identified strengths.
- Establish tentative postsecondary education or employment goals and plans to achieve these goals.
- Select learning experiences designed to help them achieve their postsecondary educational or employment goals.
- Earn advanced standing, dual credit, and the like in a technical/community college for high school courses that result in the same student competencies.

Common Denominators of Tech Prep

All Tech Prep programs have five things in common:
- Continuity in learning
- Content-based teaching (applied academics)
- Competency-based teaching
- Communication between learning institutions (high school and postsecondary technical institutes and community colleges and the local business community)
- Completion of the program with an associate degree (or certificate)

The Tech Prep/Associate Degree (TPAD) approach assists students in developing broad-based competencies in a career field of their choice and in identifying long-term education and training goals. It provides a chance for many students who might otherwise leave high school with few opportunities to reach for promising long-term, higher-paying employment fields. By having the business community involved with TPAD programs, both the students and future employers have a greater sense of the skills and talents these students bring to the workplace.

Developing Core Personal Abilities

Based upon the SCANS (Secretary's Commission on Achieving Necessary Skills, United States Department of Labor), the underlying Tech Prep goals include "soft" general competencies, and the "hard" occupationally specific competencies. These general competencies are valued assets for all managers and employees. Two national reports, "Workplace Basics: The Skills Employers Want" and "What Work Requires of Schools, A SCANS Report for America 2000," identify the skills, competencies, and abilities students need to live full lives. Every Tech Prep program should incorporate these skills, competencies, and abilities. A discussion of these two reports follows.

Workplace Basics: The Skills Employers Want.

Although Workplace Basics focuses primarily on skills needed for productive employment, it does not diminish the important of skills people need to participate in their communities, to raise families, and to enjoy their leisure time. In fact, these new "basics" are the keys to greater opportunity and a better quality of life.

Under a grant from the Employment and Training Administration, a two-year research project was conducted by the American Society for Training and Development and the U.S. Department of Labor. The portion of research that summarized findings on workplace skills found that employers want good basic academic skills. Specifically, the summary identified these seven skill groups:

1. *Learning to learn.* The most basic of all skills is knowing how to learn. The skill is the key to all other skills for learning and is now a fact of life in the workplace. Employers are putting a premium on the ability to absorb, process, and apply new information quickly and effectively.

2. *Reading, writing, and computation.* Today's workforce is called on to use these basic skills in ways totally different from the ways workers have been traditionally taught. Traditionally students have been taught reading in isolation, writing for creativity, and mathematics in concept only. Educators know now, through their research and through implementation of applied academics, that these basics are learned more effectively when taught in context, using or adapting materials, and concepts that are job-based.

   Students must learn the process of reading to locate information and of using higher-level thinking strategies to solve problems. Writing should rely on analysis, conceptualization, synthesis, and dissemination of information in a clear, concise, and correct manner. Building on the learner's prior math knowledge and emphasizing problem identification, reasoning, estimation, and problem solving, the instruction can simulate specific job tasks, bringing computation into a "real world" context.

3. *Communication: Listening and oral communication.* Students need listening and speaking skills to ensure good communication in the workplace.

4. *Creative thinking and problem solving.* Creative thinking should focus on expanding the thinking process of students. Students need to escape from logical and sequential thought patterns and solve problems creatively.

5. *Self-esteem, goal setting/motivation, personal/career development.* This skill area can be accomplished by assisting students in

   a. Recognizing their current skills and knowledge

   b. Being aware of their impact on others

   c. Understanding their personal emotional set points to cope with varying situations.

6. *Interpersonal skills, negotiation, and teamwork.* This ongoing self-development required first an awareness and then an ability to deal effectively with inappropriate behavior to provide inspiration to others, to share responsibility, and to interact confidently.

7. *Organizational effectiveness and leadership.* Students must first understand structures of organizations in general: what they are, why they exist, and how an individual moves within the structure.

TECH PREP

What Work Requires of Schools: A SCANS Report for America 2000. To examine the changes for learning, the Secretary of Labor organized the Secretary's Commission on Achieving Necessary Skills (SCANS). The Commission spent 12 months talking to the working public, union officials, and workers at all levels. Their consensus was that all workers need to be able to put knowledge to work and to be creative and responsible problem solvers with the skills and attitudes on which employers can build.

The SCANS report listed three findings:

1. All American high school students must develop a new set of competencies and foundation skills if they are to enjoy a productive, full, and satisfying life.

2. The qualities of high performance that today characterize our most competitive companies must become the standard for the vast majority of our companies, large and small, local and global.

3. The nation's schools must be transformed into high-performance organizations in their own right.

Although these conclusions may appear abstract in isolation, the Commission asks that they be considered, instead, in the context of traditional education, current business needs, and the standards of our schools. The SCANS report then addresses what these convictions mean in practice on the job and in the school.

The report identifies five competencies that, in conjunction with a three-part foundation of skills and personal qualities, lie at the heart of job performance today. The eight areas identified are not only consistent with the purposes of a Tech Prep program but also complement and add credence to the efforts of Tech Prep proponents. These eight areas represent essential preparation for all students, both those going directly to work and those planning further education. The Commission maintains that all eight areas must be an integral part of each student's education and should be taught and understood in a way that reflects the workplace contexts in which they are applied.

The five competencies differ from a person's technical knowledge. An effective worker should be able to productively use resources, interpersonal skills, information, systems, and technology.

Workplace Know-How

According to SCANS, workplace know-how consists of the five competencies and a three-part foundation of skills and personal qualities. The five competencies necessary for effective job performance are the following:

1. **Resources:** allocating time, money, materials, space, and staff

2. **Interpersonal skills:** working on teams, teaching others, serving customers, leading, negotiating, and working well with people from culturally diverse backgrounds

3. **Information:** acquiring and evaluating data, organizing and maintaining files, interpreting and communicating, and using computers to process information

4. **Systems:** understanding social, organizational, and technological systems, monitoring and correcting performance, and designing or improving systems

5. **Technology:** selecting equipment and tools, applying technology to specific tasks, and maintaining and troubleshooting technologies

The three-part foundation includes
- **Basic skills:** reading, writing, arithmetic and mathematics, speaking and listening
- **Thinking skills:** thinking creatively, making decisions, solving problems, seeing things in the mind's eye, knowing how to learn, and reasoning
- **Personal qualities:** individual responsibility, self-esteem, sociability, self-management, and integrity

The Commission defines each of these competencies as generic across industries and at many steps of the career ladder. The five competencies and the resulting behaviors are defined as follows:

1. **Resources:** Identifies, organizes, plans, and allocates resources
   a. *Time*—selects goal-relevant activities, ranks them, allocates time, and prepares and follows schedules
   b. *Money*—uses or prepares budgets, makes forecasts, keeps records, and makes adjustments to meet objectives
   c. *Materials and facilities*—acquires, stores, allocates, and uses materials or space efficiently
   d. *Human resources*—assesses skills and distributes work accordingly, evaluates performance and provides feedback

2. **Interpersonal Skills:** Works with others
   a. *Participates as member of a team*—contributes to group effort
   b. *Teaches others new skills*
   c. *Serves clients/customers*—works to satisfy customer's expectations
   d. *Exercises leadership*—communicates ideas to justify position, persuades and convinces others, assumes responsibility, challenges existing procedures and policies
   e. *Negotiates*—works toward agreements involving exchanges of resources, resolves divergent interests
   f. *Works with diversity*—works well with people from diverse backgrounds

3. **Information:** Acquires and uses information
   a. *Acquires and evaluates information*
   b. *Organizes and maintains information*
   c. *Interprets and communicates information*
   d. *Uses computers to process information*

4. **Systems:** Understands complex interrelationships
   a. *Understands systems*—knows how social, organization, and technical systems work and operates effectively with them
   b. *Monitors and corrects performance*—distinguishes trends, predicts impacts on system operations, diagnoses deviations in system's performance and corrects malfunctions
   c. *Improves or designs systems*—suggests modifications in existing systems and develops a new or alternative design to improve performance

5. **Technology:** Works with a variety of technologies
   a. *Selects technology*—chooses procedures, tools, or equipment, including computers and related technologies
   b. *Applies technology to task*—understands overall intent and proper procedures for setup and operation of equipment
   c. *Maintains and troubleshoots equipment*—prevents, identifies, or solves problems with equipment, including computers and other technologies.

The Commission's work represents a further evolution of the recommendations in *America's Choice.** More importantly, it develops a framework for a skills structure that schools and employers can use.

Incorporating the findings from both reports can serve as the core of personal abilities from which further content integration can occur in the redesign of curriculum and the restructure of education. However, for any curriculum to be as effective as possible, it needs the support and acceptance from more than just the educational community. A redesign of curriculum, such as suggested by Tech Prep, needs the support of students, parents, business and industry, and the community at large. The first step in gaining this support is through knowledge and understanding, which can be achieved through marketing of the program.

*See items 2 and 16 in References and Suggested Readings.

TECH PREP

Developing Learning Strategies

To dramatically change all schools—from elementary through high school, technical training or community college, and four-year colleges and university—educators must begin by placing learning objectives within real environments so that students no longer first learn in the abstract what they will then be expected to apply. When instructors teach skills in the context of competencies, students will learn the skill more rapidly and will be more likely to apply that skill to real situations.

Any teacher or instructor can support the goals of the Tech Prep program that is obligated to help students learn the subjects, as well as to achieve these skills:

- Think logically
- Write well
- Read critically
- Listen perceptively
- Speak clearly
- Use imagination creatively
- Develop interpersonal skills

Many vocational educators have, by the nature of their subject matter, traditionally taught skills in context of real situations but have not emphasized the inherent "academic skills" use in their application. In contrast, instructors in the academic wing have taught "the basics" in isolation without demonstrating the link to the real world. Though both of these statements are generalities, neither group can maintain its status quo. Academic and vocational instructors will have to collaborate in the transformation needed for a Tech Prep program to be one of cooperation, quality, equity, and learning.

Tech Prep Sequence of Courses

No single sequence of courses exists that all Tech Prep students will pursue. To the contrary, students will select a sequence of elective courses depending on their tentative postsecondary educational or employment goals. In some instances, a student planning to pursue a specific bachelor's degree program will choose the same course work as a student planning a related associate degree major in a technical college. One students may choose to "prep," or prepare for a four-year college (College Prep), whereas another student may play to pursue a technical college degree (Tech Prep).

Tech Prep Components

- **Common Core outcomes (for K-12).** Each student must demonstrate an ability to perform complex tasks—the use of mathematics, languages, and artistic expression in everyday application.
- **Specialty courses and experiences (primarily in grades 11 and 12).** Some specialty courses will be academically based whereas others will be vocationally based; still others will be based in the arts, foreign languages, or other areas of special interest. In some instances, these specialty courses may be formally articulated to offer advanced placement opportunities in two-year or four-year institutions.

Applied Academics

Applied academics combines thinking and doing. Applied academics is the integration of a particular academic discipline (such as mathematics, science, or English) with personal workforce applications (hands-on laboratories dealing with practical equipment and devices). It is imperative that applied academics be understood as a competence-orientated curriculum with high expectations rather than a traditional curriculum with lower expectations. Applied learning can motivate students who currently fail to make the connection between classroom study and real-life application. Applied academic programs provide mathematics, science, communication, and social science courses which offer activities that are relevant to experiences students will have as adults. The courses are developed in a competency-based curriculum that provides concepts and skills related to real-world applications.

Tech Prep in Middle/Junior High Schools

Tech prep also has a significant role in the middle/junior high school. More applied curriculum and instruction needs to take place in the middle/junior high school, along with greater opportunities to explore future options. The competencies identified in the SCANS reports, such as learning to learn, critical thinking, and problem solving, need to be integrated components of the middle school curriculum.

Designing the Curriculum

Curriculum Clusters. Many states have chosen to cluster student competencies and curriculum topics into five major occupational areas:
- Health and medical services
- Business and marketing
- Agribusiness/agriscience
- Family/consumer services
- Technical/industrial

Curriculum Models. The basic design of most Tech Prep curriculum models is focused on applied academics. The integration of applied math, applied science, applied communication, and the principles of technology is the foundation for each cluster and its specific career paths. The actual outcomes from these models encompass a broad range of competencies that prepare Tech Prep students to be productive workers and good citizens. Following is a curriculum model in the area of physics.

TECH PREP

COMPETENCIES FOR ENGINEERING TECHNOLOGY: DRAFTING, DESIGN, AND CAD

1. <u>Resources:</u> Identifies, organizes, plans, and allocates resources. This includes time, money, human resources, material, and facilities.

2. <u>Interpersonal:</u> Includes skills for working with others. This includes participating as a member of a team, teaching, serving clients, exhibiting leadership, negotiation, and working with people from diverse backgrounds.

3. <u>Information:</u> Acquires and uses information. This includes evaluation, organization, maintenance, interpretation, communication, and using computers to process information.

4. <u>Systems:</u> Understands complex interrelationships. This includes design or redesign and monitoring or correcting performance.

5. <u>Technology:</u> Works with technologies including selection, application to tasks, maintenance, and troubleshooting equipment.

6. <u>Basic Laboratory Skills:</u> Uses drawing equipment and supplies for drawing reproductions, lettering, and line weights.

7. <u>Geometric Construction:</u> Draws straight lines, angles, plane figures, circles and arcs, and irregular figures.

8. <u>Orthographic and Auxiliary Projection:</u> Understands projection principles and ability for sketching, drawing multiview, and auxiliaries.

9. <u>Sectioning:</u> Understands standard sections, special sections, and conventional breaks.

10. <u>Pictorials:</u> Applies drawing principles to axonometric projection, oblique drawings, presentation drawings, and exploded assemblies.

FEATURES THAT MEET COMPETENCIES

1. Identifies material and drafting/design facilities. Organizes technical equipment, including drawing materials and computer-assisted drawing. Provides introduction to working within engineering team and planning for the use of costly materials and equipment.

2. Students should be assigned team projects, where they can review manufacturers' designs and prepare team reports that they present to the class. Student demonstrations of technical applications to textbook theory are encouraged.

3. Shows how to acquire and use information for interpretation. Provides introduction to using CAD systems. Demonstrates the importance of geometric dimensioning and tolerancing for understanding designs, interpretation of dimensions, and communicating required specifications. Uses reference information.

4. Includes design and redesign. System designs can be presented in more detail via student projects and demonstrations.

5. Explains application to technical tasks. Plan for completion of final parts based upon technical designs, manufacturing processes, and with United States or international specifications.

6. Teaches how to use manual and CAD equipment and supplies. Reinforces basic drawing principles and requirements based on national standards.

7. Illustrates all facets of both plane and applied geometry.

8. Devotes a chapter to each orthographic projection and auxiliary.

9. Illustrates all types of sections and conventions.

10. Examines isometric, oblique, perspective (one-, two-, three-point) technical illustration, exploded assemblies.

COMPETENCIES FOR ENGINEERING TECHNOLOGY: DRAFTING, DESIGN, AND CAD

11. Intersection and Developments: Developments, transitions, pieces, and intersections.

12. Dimensioning: Develops techniques for tolerances, fits, and geometric tolerances.

13. Computer-Aided Design: Develops basic skills, CAD skills, drawing, editing, display, options, save/plot, symbols/libraries, layers/line types, and attributes.

14. Machine Drawings: Understand and prepare details, assemblies, fasteners, fluid power, cam/gear, and design.

15. Architectural: Provides instruction on how to draw floor plans, plot plans, foundations, elevations, schedules, sections, details, roof plans, electrical, plumbing, and how to design residential structures.

16. Structural: Prepares drawings of shapes, connections, erection plans, and details.

17. Civil and Surveying: Gathers information, set up and operate transit, boundary lines, and survey drawing.

18. Electrical and Electronic: Prepares drawings of electrical assemblies, wiring diagrams, electronic assemblies, and schematics.

19. ADVANCED CAD: Develops understanding of text editors, macros/LISP, and modeling.

20. Employability Skills: Career development, decision making, problem-solving work ethic, job-seeking skills, retention, advancement, lifelong learning, economics, work and family, citizenship, leadership, and entrepreneurship.

FEATURES THAT MEET COMPETENCIES

11. Shows development, transition pieces, and intersections.

12. Examines dimensioning, tolerances, and fits, including an extensive geometric tolerancing chapter.

13. Includes a chapter on Computer-Aided Drafting. Highlights appropriate CAD concepts throughout many chapters.

14. Extensive machine drawing chapters include detail assembly, fluid power, cam/gear design, plus two complete fastener chapters.

15. Provides information about basic drawing principles that can be applied to separate courses on architecture. Glencoe/ McGraw-Hill offers separate textbook programs for architectural drawing and design.

16. Discusses shapes, beam connections, sectioning, and details.

17. Provides all necessary background about basic drawing principles. Civil engineering technology topics are treated in separate textbooks and courses.

18. Includes electrical pictorial and connection (wiring) diagrams plus electronic schematic, printed circuit, and block diagrams.

19. Provides general introduction to CAD. Advanced CAD, including LISP and modeling applications, and CAD icons are treated in separate CAD book by the same authors.

20. Discusses career development, problem solving, and job-seeking skills. Provides photographic inserts representing various career opportunities.

References and Suggested Readings

1. America 200: An Education Strategy Sourcebook (April 1991). Washington, D.C.: United States Department of Education.

2. *America's Choice: High Skills or Low Wages* (June 1990). National Center on Education and the Economy, P.O. Box 10670, Rochester, New York 14610.

3. *The AVA Guide to the Carl Perkins Vocational and Applied Technology Education Act of 1990*. AVA, 1410 King Street, Alexandria, Virginia 22314.

4. Bureau for Vocational Education, Wisconsin Department of Public Instruction, 125 South Webster Street, Madison, Wisconsin 53702.

5. Center on Education and Training for Employment, The Ohio State University, 1900 Kenny Road, Columbus, Ohio 43210.

6. Center for Occupational Research and Development, 601 Lake Air Drive, Waco, Texas 76710.

7. *The Forgotten Half: Non-College Youth in America*. Washington, D.C.: Youth and America's Future: The William T. Grant Foundation Commission on Work, Family and Citizenship, 1988.

8. Grubb, N.W., Davis, G., Lum, J., Phihal, J., and Morgaine, C. (July 1991). "The Cunning Hand, The Cultured Mind": Models for integrating vocational and academic education. Berkeley: National Center for Research in Vocational Education, University of California.

9. Hull, D., and Parnell, D. "Tech-Prep Associate Degree: A Win-Win Experience." 1991. Center for Occupational Research and Development, 601 Lake Air Drive, Waco, Texas 76710.

10. Loock, J.W., and Voiers, J.S. *Strategies for Implementing Tech Prep*. 1995, Glencoe/McGraw-Hill.

11. Loock, J.W., and Voiers, J.S. *Strategies for Implementing Tech Prep into Communication*. ©1994, Glencoe/McGraw-Hill.

12. *A Nation at Risk*. Washington, D.C.: U.S. Department of Education, 1983.

13. *National Tech-Prep Network Newsletter*. Membership available from the Center for Occupational Research and Development, 601 Lake Air Drive, Waco, Texas 76710.

14. Parnell, D. *The Neglected Majority*. 1985. Center for Occupational Research and Development, 601 Lake Air Drive, Waco, Texas 76710.

15. Ross, Maryanne; Swinehart, Carole P.; Hogan, Diane; Morrison, Connie; Hoyt, Bill; and Haber, Barry. *Tech Prep Strategies in Accounting*. ©1995, Glencoe/McGraw-Hill.

16. *SCANS: Secretary's Commission on Achieving Necessary Skills*. SCANS Office, Room C-2318, United States Department of Labor, 200 Constitution Ave. N.W., Washington, D.C. 20210.

17. *Training Strategies: Preparing Non-College Youth for Employment in U.S. and Foreign Countries*. 1990. United States General Accounting Office, PO Box 6015, Gaithersburg, Maryland, 20877.

18. United States Statutes at Large, 1990 and Proclamations, Volume 104 in Six Parts, Public Law 101-392, September 25, 1990, "Carl D. Perkins Vocational and Applied Technology Education Act Amendments of 1990," United States Government Printing Office, Washington, D.C.

19. *Workplace Basics: The Essential Skills Employers Want* (October 1988). American Society for Training and Development, 1630 Duke Street, Box 1443, Alexandria, Virginia 22313

INFORMATION AND INSTRUCTIONS FOR CD-ROM

Instructor's CD-ROM
Information and Instructions

Welcome to *Engineering Drawing and Design on CD-ROM*. On this disk, you can explore four programs designed to enhance classroom experiences for students and instructors. The programs are designed for use on a multimedia PC and can be projected on a screen for classroom use.

Graphical Solutions. The first program contains all of the solutions for the drawing assignments in *Engineering Drawing and Design, Fifth Edition* by Jensen and Helsel. From any chapter instructors can select a specific problem assignment. The solution will then appear on the PC screen, or it can be projected to a larger screen for classroom use. This time-saving feature allows instructors to deal collectively with common assignment problems, or spend more time with an individual student. The solutions can also be used for classroom discussion and chapter review prior to an examination. Students can raise questions regarding their individual or common concerns; instructors can then project solutions which best answer their questions.

Tutorials. Preparing students for an assignment and providing them with detailed instructions has always been a time-consuming activity. This CD-ROM contains twenty tutorials. Two to four tutorials are included for Chapters 6, 7, 8, 9, 10, 14, 15, and 16, with one tutorial provided for Chapter 18. The tutorials are designed to give step-by-step procedures students should follow in creating a solution to the given assignment. Each tutorial contains three to six "layers" that begin with the assignment and evolve to the solution. These tutorials are excellent instructional devices for explaining how students should proceed in solving related assignments.

Advanced Topics. The CD-ROM also contains two advanced topics: *(a) applied mechanics*; and *(b) strength of materials*. Both topics are designed for use in a lecture-presentation mode where "frames" are projected on the screen. The sequence of the frames carefully takes students through these advanced concepts. Vectors are added in a progressive format so students can easily visualize the resultant forces. Many practical examples are included. The example frames provide a statement of the problem and then carefully provide the solution. These advanced topics are designed to supplement *Engineering Drawing and Design*.

ANSI, ISO, and ISO 9000. This program enhances the book's coverage of technical standards and provides a comparison and contrast between ANSI and ISO drawing standards. It also includes sources and references these standards. Actual drawings for *Engineering Drawing and Design, Fifth Edition* are included so students can refer to specific pages and figure numbers in their textbook. Some coverage of ISO 9000 is also included to aid students in understanding the need for designing and manufacturing high-quality products.

INSTALLING SOFTWARE

Installation and Operation

The *Hyper*Graphics system is an instructional delivery system utilizing a Windows 3.1 or higher platform. It is coordinated with the textbook, *Engineering Drawing and Design, Fifth Edition* by Jensen and Helsel. The system is designed to provide you as an instructor with a teaching tool that is customized to support your view of the materials and your teaching style.

Minimum Hardware Requirements

IBM - PC or 100% compatible

- multimedia PC, with a 386SX or higher processor, and a mouse
- 4 MB RAM
- hard drive with 20 to 30 MB available for Windows
- double speed CD-ROM drive
- Windows 3.1

Instructions for Projection Device Setup

1. Place the LCD panel on top of an overhead projector or your projection device. The panel should be on a flat surface with the lens pointing towards a projection screen.

2. Make sure the computer is turned **off.**

3. Plug the "LCD" end of the furnished cable into the device.

4. Unplug the monitor cable from the computer and plug it into the "Monitor" end of the furnished cable.

5. Plug the "VGA" end of the furnished cable into the monitor connector on the computer.

6. Plug the power cord into the device, then into the electrical outlet.

7. Turn on the projection device, the monitor, and then the computer.

8. A function key may be required to switch over from the monitor to the projection device. This information will be found in the computer's user manual.

Setup Instructions

To set up the *Hyper*Graphics software:

1. Place the *Jensen-Helsel Solutions CD* into the CD-ROM drive.

2. Start Windows.

3. From Windows Program Manager, choose **File, Run.**

4. In the Command Line box, type **D:\SETUP. (D** is the drive letter of your CD-ROM drive.)

5. Follow instructions on the screen. **SETUP** creates a program group called **J***ensen-Helsel Solutions CD.*

> After completing the installation process, your desktop manager will include a new program group called *Jensen-Helsel Solutions CD.* The program group includes icons representing each chapter and additional icons for *Advanced Topics* and *ANSI/ISO.*

Moving within the Software

*Hyper*Graphics provides you with three modes of movement within the software. You may use any of the following devices to navigate through a chapter:

- keyboard
- mouse
- remote control

Executing the Software

Highlight the chapter icon of your choice and press Enter (keyboard) or double click the chapter icon of your choice (mouse).

Note that you can use the *Hyper*Graphics Software System with an instructor's remote control. If it is not yet installed, exit the software, attach the device, and restart the software. If you have not installed the remote control, you will receive a message indicating the absence of the device. **Ignore** this message if you want to use the system without the remote control.

Title Screen

The title screen of the software will display the title of the book, *Engineering Drawing and Design, Fifth Edition,* by Jensen and Helsel.

To advance to the Main Chapter Menu, do the following:
- Click the **CONTINUE** button (mouse).
OR
- Press **Spacebar** (keyboard).
OR
- Press **up arrow** (remote).

Main Chapter Menu

The main chapter menu is the control center for the solutions. From this screen you have the option to view solutions by Assignment or by Unit. Tutorials are available in selected chapters.

The main chapter menu consists of four components:
1. Solutions by Assignment
2. Solutions by Unit
3. Tutorials (chapters 6, 7, 8, 9, 10, 14, 15, 16, and 18)
4. Exit

1. Solutions by Assignment. You can pick **Solutions by Assignment** by doing one of the following:
- Click the left mouse button on the Assignment Icon (mouse).
OR
- Press key 1 and press **Enter** (keyboard).
OR
- Press key 1 and press **Enter** (remote).

Upon entering **Solutions by Assignment,** you will arrive at the Assignment Submenu. You can select one of the assignments listed by using the same conventions described above.

2. Solutions by Unit. You can pick **Solutions by Unit** by doing one of the following:
- Click the left mouse button on the Unit Icon (mouse).
OR
- Press key 2 and press **Enter** (keyboard).
OR
- Press key 2 and press **Enter** (remote).

Upon entering **Solutions by Unit,** you will arrive at the Unit Submenu. You can select one of the units listed by using the same conventions.

After entering one of these units, you will see a menu with all figures for the chapter.

EXECUTING SOFTWARE

3. Tutorials. The **Tutorials** are available in chapters 6, 7, 8, 9, 10, 14, 15, 16, and 18. You can select Tutorials by doing one of the following:
- Click the left mouse button on the Tutorial Icon (mouse).
  OR
- Press key 3 and press Enter (keyboard).
  OR
- Press key 3 and press Enter (remote).

Upon entering **Tutorials,** you will arrive at the Tutorial Submenu. You can select the Tutorial listed by using the same conventions.

4. Exit. **Exit** allows you to quit the chapter. To quit the program immediately, press **Q** (keyboard) or press down arrow and hold it down for two seconds (remote). To quit from the Main Chapter Menu, do the following:
- Click the left mouse button on the Exit Icon (mouse).
  OR
- Press key 3 or 4 (pending Tutorials) and press **Enter** (keyboard).
  OR
- Press key 3 or 4 (pending Tutorials) and press **Enter** (remote).

Moving through Solutions. Each page of solutions will have a navigation icon (a square within a diamond) to be used in coordination with the mouse.

Continuing Through an Instructional Sequence.
- Click **up arrow** in the navigation icon or click on any other non-icon position on the screen (mouse).
  OR
- Press **Spacebar** (keyboard).
  OR
- Press **up arrow** (remote).

If you continue to move forward, you will eventually move through all the solutions and return to the **unit/assignment submenu.** This is the normal mode you will use as you move through the material.

Repeating an Instructional Sequence. To repeat an instructional sequence from the beginning:
- Click **down arrow** in the navigation icon (mouse).
  OR
- Press **R** (keyboard).
  OR
- Press **down arrow** (remote).

Use the continue function to proceed with the instruction.

Backing Up to a Previous Instructional Sequence. To move to the beginning of the **previous** instructional sequence:
- Click **left arrow** in the navigation icon (mouse).
  OR
- Press **P** (keyboard).
  OR
- Press **left arrow** (remote).

Use the continue function to proceed with the instruction.

Skipping Forward to the Next Instructional Sequence. To move to the beginning of the **next** instructional sequence:
- Click **right arrow** in the navigation icon (mouse).
  OR
- Press **N** (keyboard).
  OR
- Press **right arrow** (remote).

Use the continue function to proceed with the instruction.

Moving Back to the Previous Menu. To move to the most recently used menu:

- Click **M** in the **center of the navigation icon** (mouse).
  OR
- Press **M** (keyboard).
  OR
- Press **Menu** button (remote).

Use the continue function to proceed with the instruction.

Help. To obtain a **Help** screen pertaining to the use of the instructor's remote control, the keyboard, and the mouse, do one of the following:

- Click the **Question Mark** in the **upper right hand corner of the screen** (mouse).
  OR
- Press **1** and **Enter** (keyboard).
  OR
- Press **1** and **Enter button** (remote).

The **Help** screen is not available from the Main chapter Menu. Once you have entered **Help,** follow the on-screen directions. To return, press **M** (keyboard), click the **M** on the screen (mouse), or press the **Menu button** (remote).

CONTENTS

Solutions

Basic Drawing and Design

Chapter 1 Engineering Graphics as a Language 2

2 Computer-Aided Drafting (CAD) (No problems for this chapter)

3 Drawing Media, Filing, Storage, and Reproduction
(No problems for this chapter)

4 Basic Drafting Skills 2

5 Applied Geometry 3

6 Theory of Shape Description 4

7 Auxiliary Views and Revolutions 30

8 Basic Dimensioning 46

9 Sections 58

Fasteners, Materials, and Forming Processes

Chapter 10 Threaded Fasteners 70

11 Miscellaneous Types of Fasteners 78

12 Manufacturing Materials 86

13 Forming Processes 89

Working Drawings and Design

Chapter 14 Detail and Assembly Drawings 95

15 Pictorial Drawings 134

16 Geometric Dimensioning and Tolerancing 147

17 Drawings for Numerical Control 172

18 Welding Drawings 174

19 Design Concepts 182

Power Transmissions

Chapter 20 Belts, Chains, and Gears 186

21 Couplings, Bearings, and Seals 199

22 Cams, Linkages, and Actuators 207

Special Fields of Drafting

Chapter 23 Developments and Intersections 216

24 Pipe Drawings 229

25 Structural Drafting 232

26 Jigs and Fixtures 241

27 Electrical and Electronics Drawing 246

Chapter 1 Engineering Graphics as a Language
Unit 1–4,   1. 1–4–A

1. A = 2.36   F = 9'-9
   B = 4.00   G = 58
   C = 2 3/16   H = 108
   D = 3 3/4   J = 270
   E = 1'-10   K = 1500

2.–4. 1–4–B

	ASSIGNMENT		
	2	3	4
A	8.80	111.5	4'-4 3/4
B	5.80	73.5	2'-10 3/4
C	2.20	28	1'-1 1/4
D	2.70	30	1'-2
E	.60	7.5	0'-3 1/2
F	.80	20	0'-4 3/4
G	1 13/16	46	0'-3 5/8
H	4 5/16	110	0'-8 5/8
J	5 1/2	140	0'-11
K	3 7/16	87	0'-6 7/8
L	2 9/16	165	0'-5 3/16
M	3/4	50	0'-1 1/2
N	.68	85	2'-9
O	4.92	622	19'-8
P	4.24	538	16'-11
Q	1.56	395	6'-3
R	1.18	300	5'-9
S	.32	80	1'-3
T	.12	30	0'-6
U	2 25/32	706	7'-5
V	1 9/16	2000	4'-2
W	31/32	1230	2'-7
X	15/16	1200	2'-6 1/2
Y	1 1/32	1300	2'-9
Z	4 31/32	6320	13'-3

Chapter 4 Basic Drafting Skills
Unit 4–1
10. 4–1–L

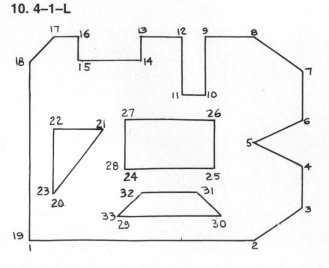

10. 4–1–M

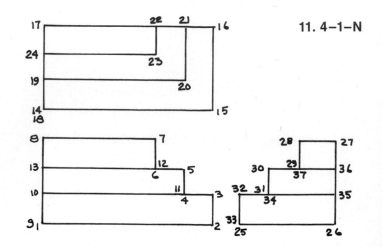

11. 4–1–N

12. 4–1–P

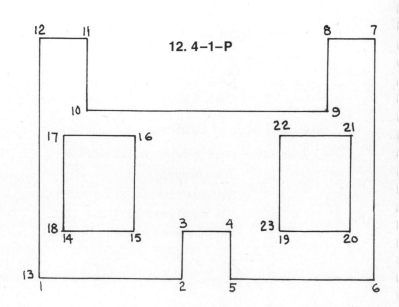

Unit 4–1
14. 4–1–R

FIG. 4-1-R			
START AT A			
FROM	TO	DEG	DIST
A	B	180	3.75
B	C	150	2.38
C	D	45	1.75
D	E	135	2.19
E	F	15	3.62
F	G	285	1.50
G	H	15	2.06
H	J	120	1.12
J	K	330	2.75
K	L	270	3.62
L	M	150	4.12
M	A	CLOSE	

4–1–B

FIG. 4-1-B				
START AT A				
FROM	TO	DEG	DIST in.	mm
A	B	210	2.50	60
B	C	120	3.00	75
C	D	90	2.40	60
D	E	0	8.00	190
E	F	270	1.75	45
F	G	225	2.83	70
G	A	CLOSE		

4–1–C

FIG. 4-1-C				
START AT A				
FROM	TO	DEG	DIST in.	mm
A	B	120	1.40	35
B	C	240	1.40	35
C	D	180	.90	22
D	E	90	3.00	75
E	F	0	1.00	25
F	G	90	1.50	40
G	H	0	1.50	40
H	J	270	.40	10
J	K	0	1.80	40
K	L	90	.40	10
L	M	0	1.50	40
M	N	270	1.50	40
N	P	0	1.00	25
P	R	270	3.00	75
R	S	180	75	18
S	T	135	1.27	31
T	U	225	1.27	31
U	A	CLOSE		
START AT V				
V	W	90	1.20	30
W	X	0	1.20	30
X	Y	270	1.20	30
Y	V	CLOSE		

Chapter 5 Applied Geometry
Unit 5–2
7. 5–2–E

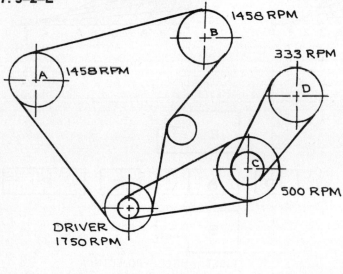

Unit 5–3

10. 5–3–B

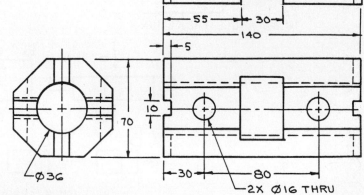

Chapter 5

Unit 5–3
10. CONTD 5–3–D

11. 5–3–E

Unit 5–5
15. 5–5–B

Chapter 6 Theory of Shape Description
Unit 6–1
1. 6–1–A

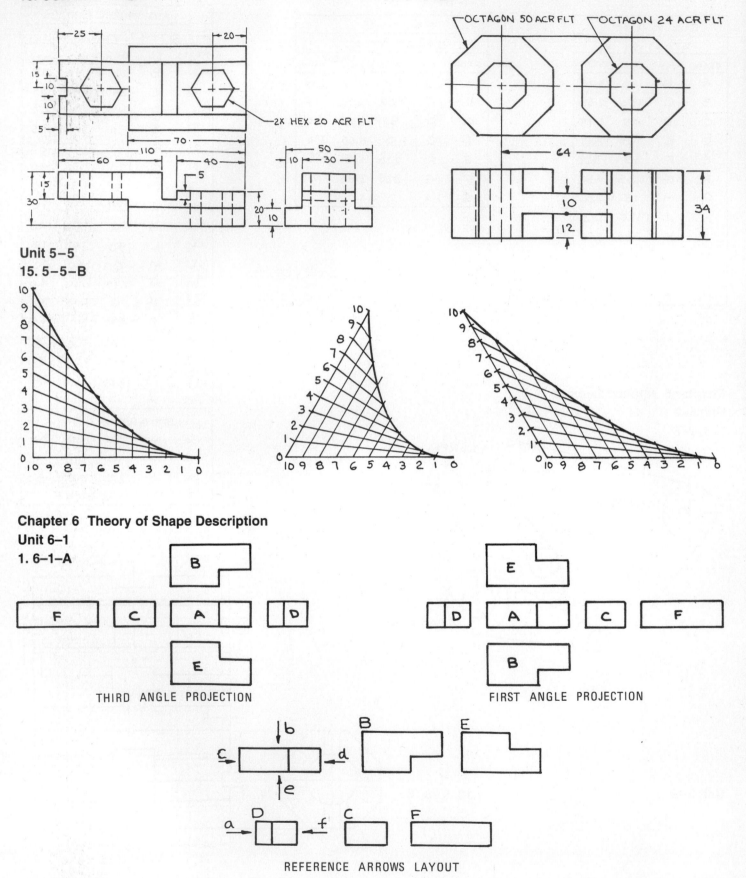

THIRD ANGLE PROJECTION

FIRST ANGLE PROJECTION

REFERENCE ARROWS LAYOUT

4 Solutions

Copyright © Glencoe/McGraw-Hill

Unit 6-1

1. CONTD 6-1-B

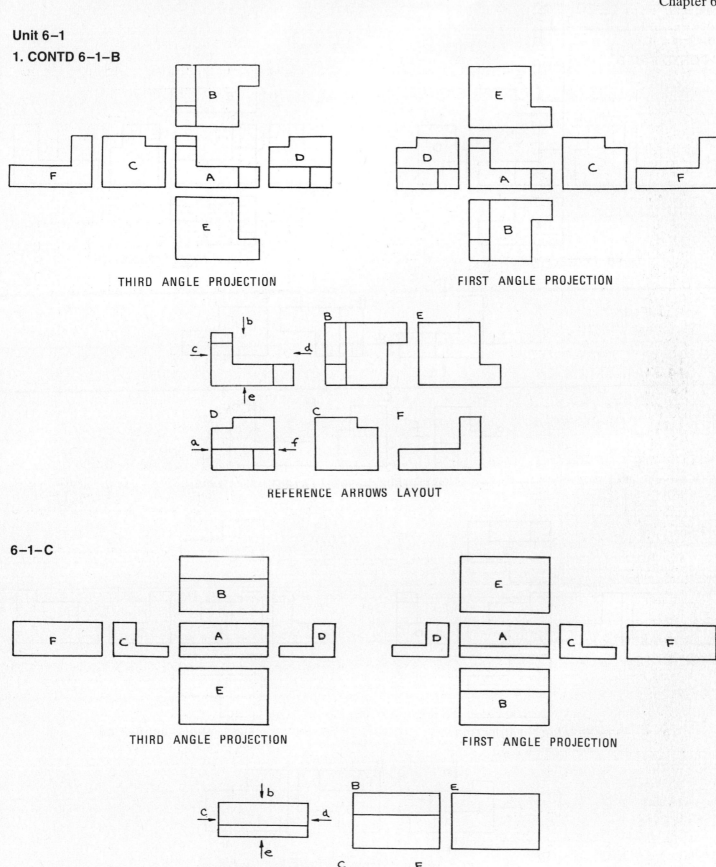

THIRD ANGLE PROJECTION

FIRST ANGLE PROJECTION

REFERENCE ARROWS LAYOUT

6-1-C

THIRD ANGLE PROJECTION

FIRST ANGLE PROJECTION

REFERENCE ARROWS LAYOUT

Unit 6–1
1. CONTD 6–1–D

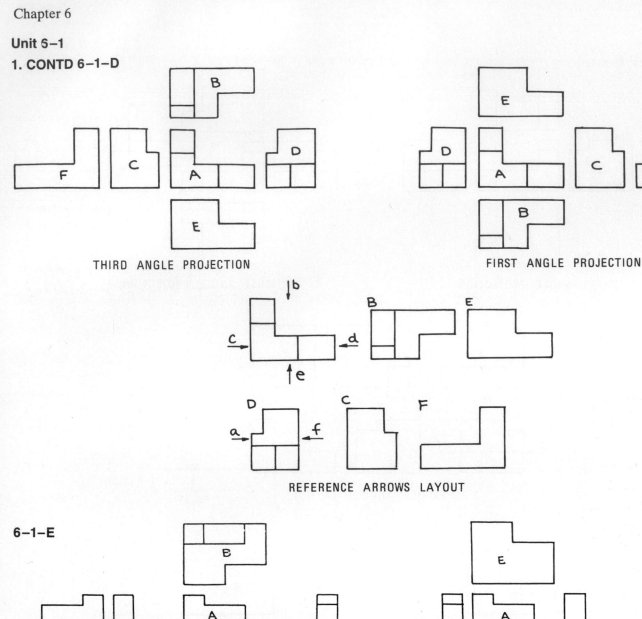

THIRD ANGLE PROJECTION

FIRST ANGLE PROJECTION

REFERENCE ARROWS LAYOUT

6–1–E

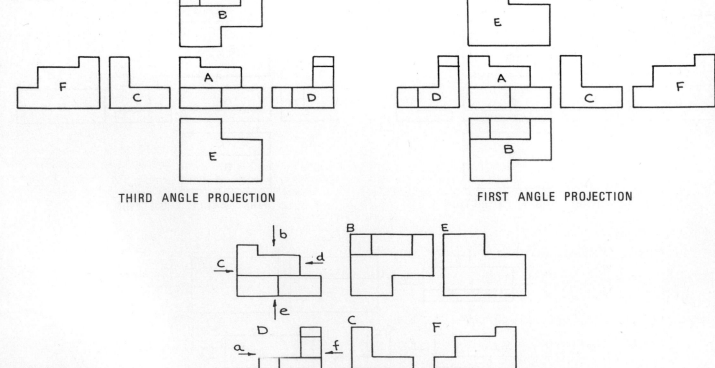

THIRD ANGLE PROJECTION

FIRST ANGLE PROJECTION

REFERENCE ARROWS LAYOUT

Unit 6–1

2. 6–1–F

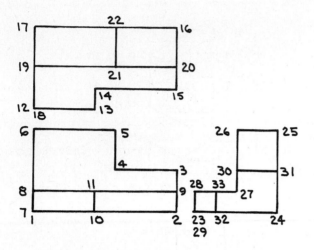

3. 6–1–J

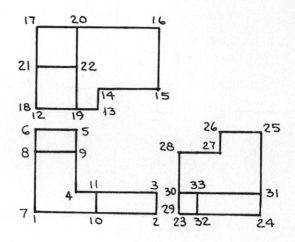

6–1–G

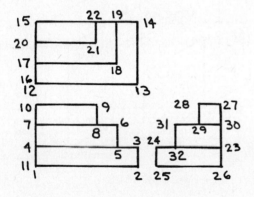

6–1–K

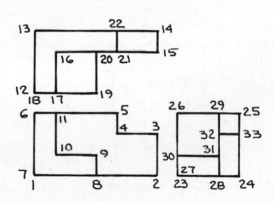

6–1–H

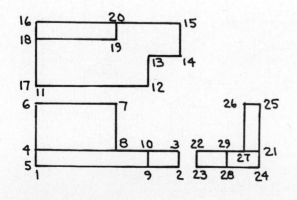

6–1–L

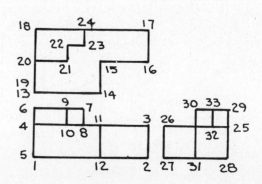

Unit 6-1

4. 6-1-M

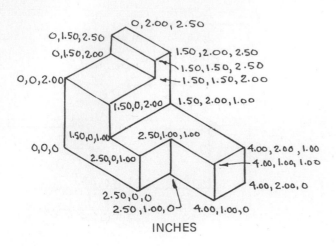

INCHES

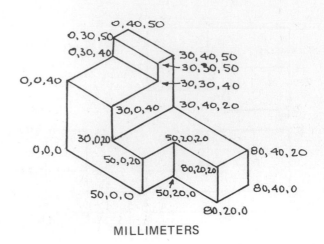

MILLIMETERS

6-1-N

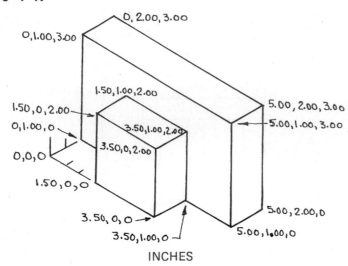

INCHES

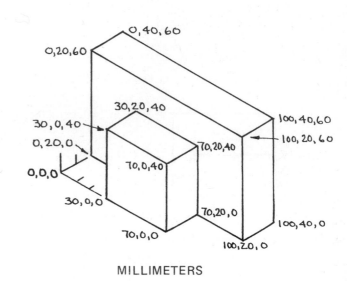

MILLIMETERS

6-1-P

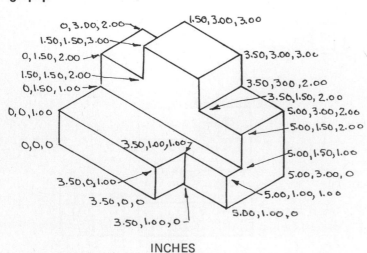

INCHES

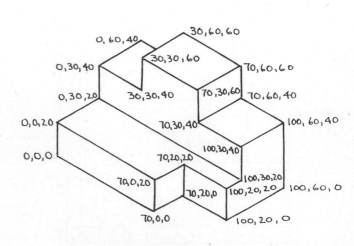

MILLIMETERS

Unit 6–1

4. CONTD 6–1–R

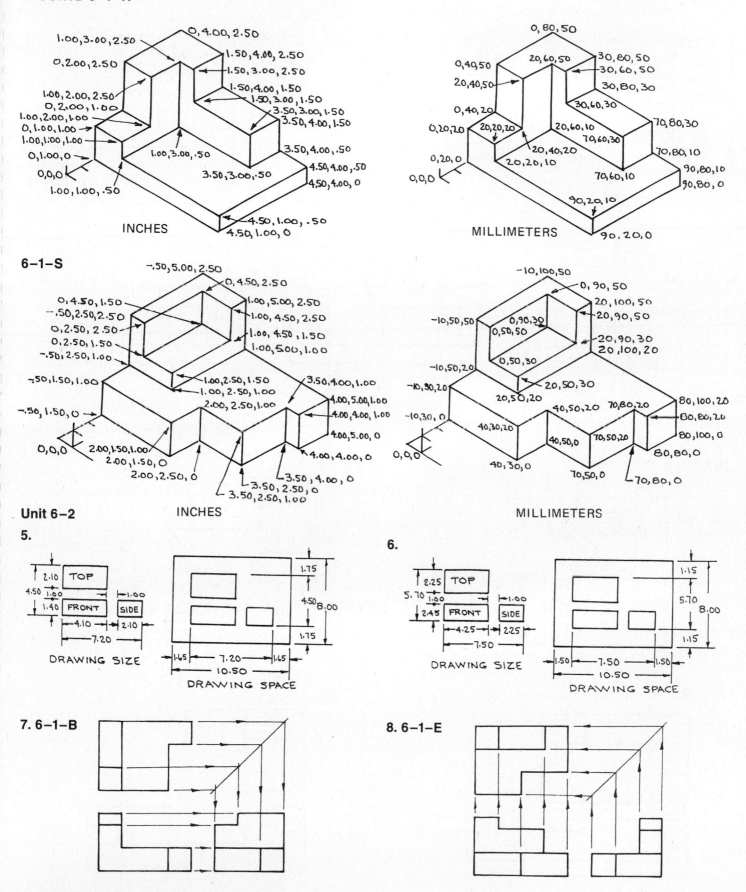

Unit 6–2

5.

6.

7. 6–1–B

8. 6–1–E

Unit 6-3

9. 6-3-A

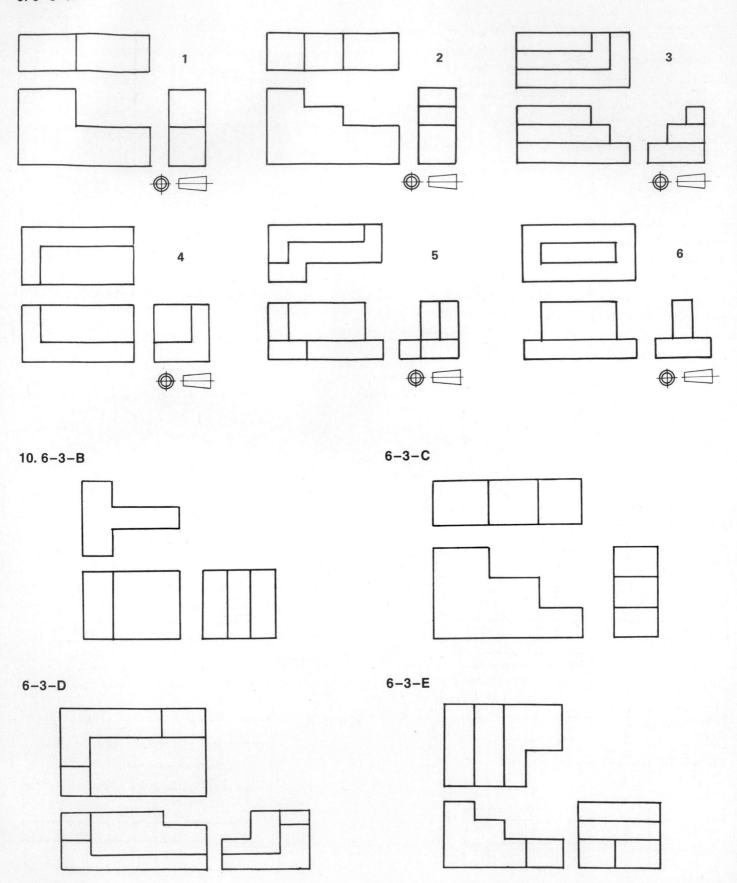

10. 6-3-B

6-3-C

6-3-D

6-3-E

Unit 6–4

11. 6–4–A

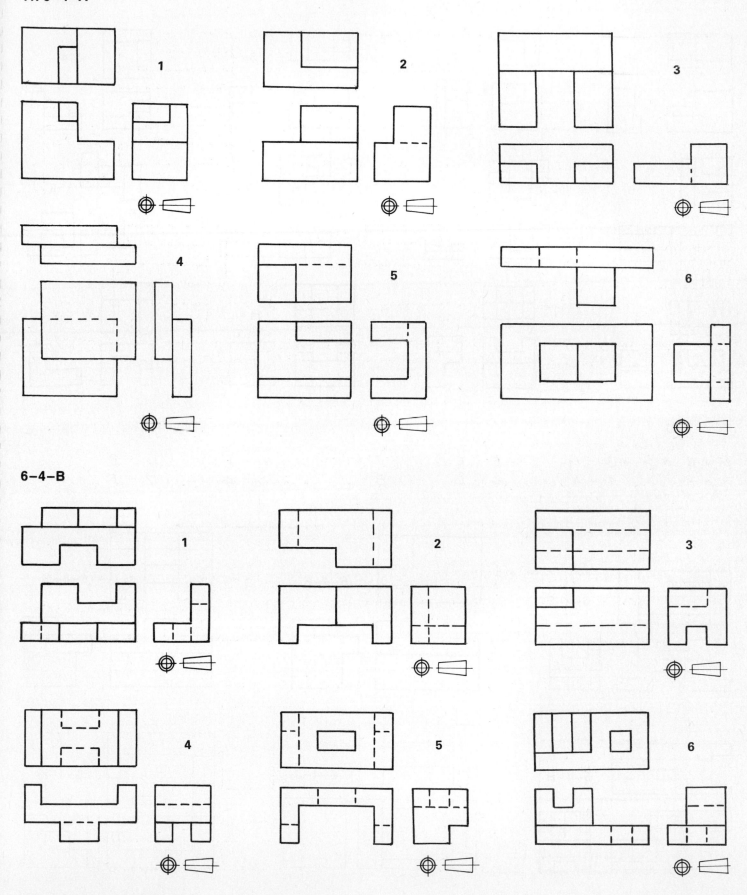

6–4–B

Unit 6–4

12. 6–4–C

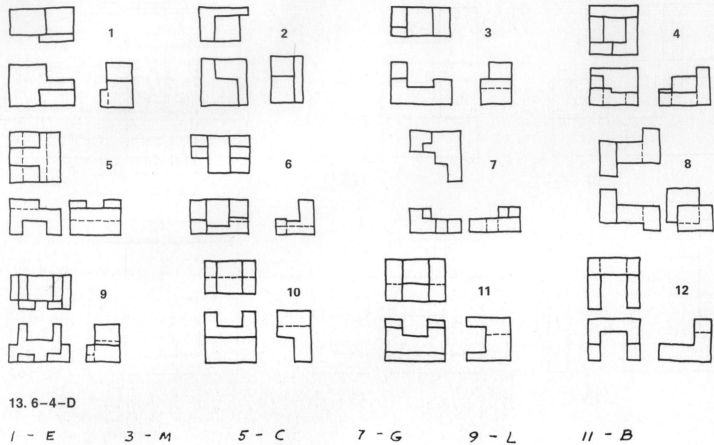

13. 6–4–D

1 – E	3 – M	5 – C	7 – G	9 – L	11 – B
2 – H	4 – K	6 – D	8 – J	10 – A	12 – F

14.

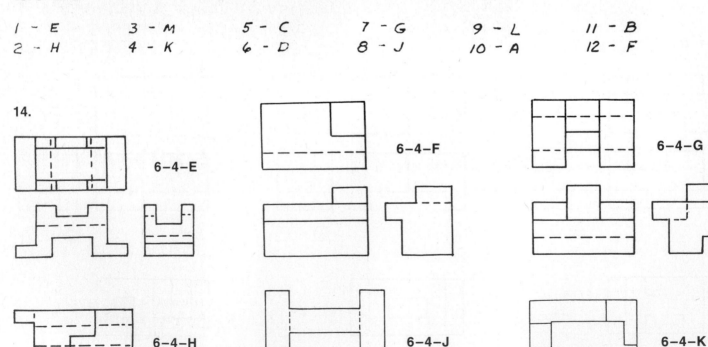

6–4–E
6–4–F
6–4–G
6–4–H
6–4–J
6–4–K

Unit 6–4

15.

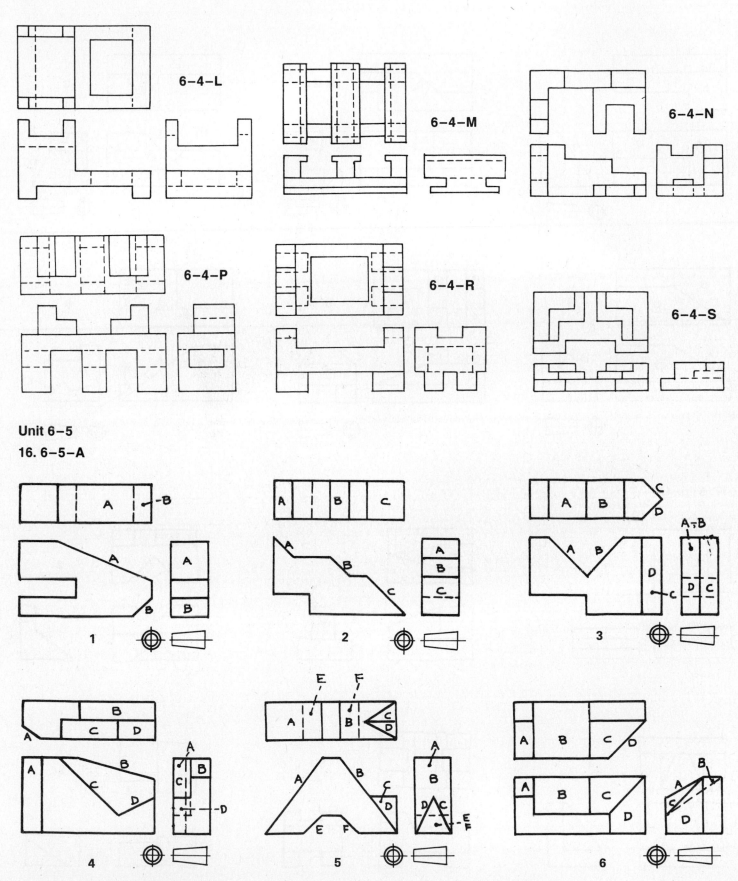

6–4–L

6–4–M

6–4–N

6–4–P

6–4–R

6–4–S

Unit 6–5

16. 6–5–A

1

2

3

4

5

6

Unit 6-5
16. CONTD 6-5-B

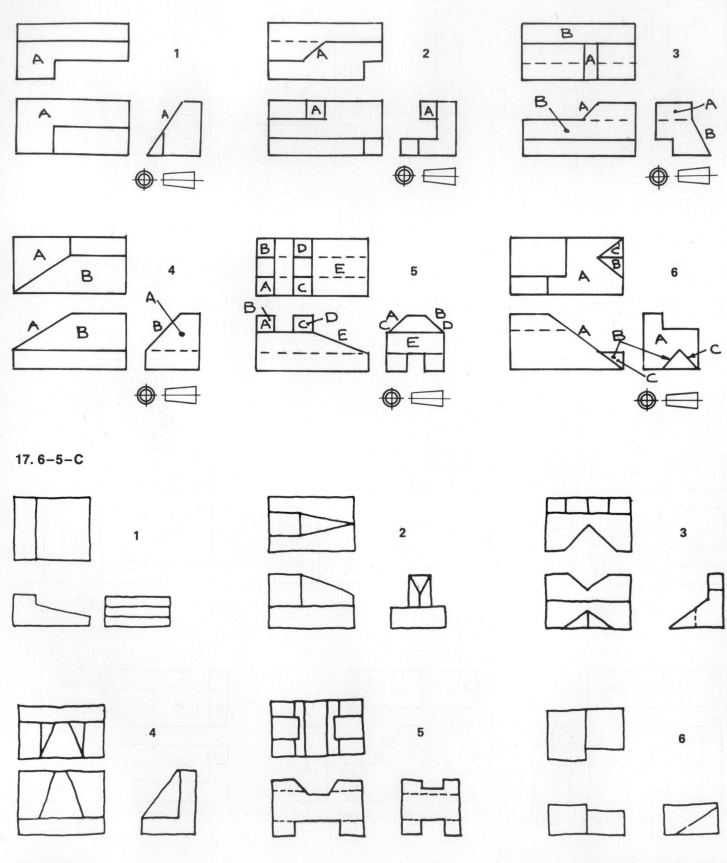

17. 6-5-C

Unit 6–5
17. 6–5–C CONTD

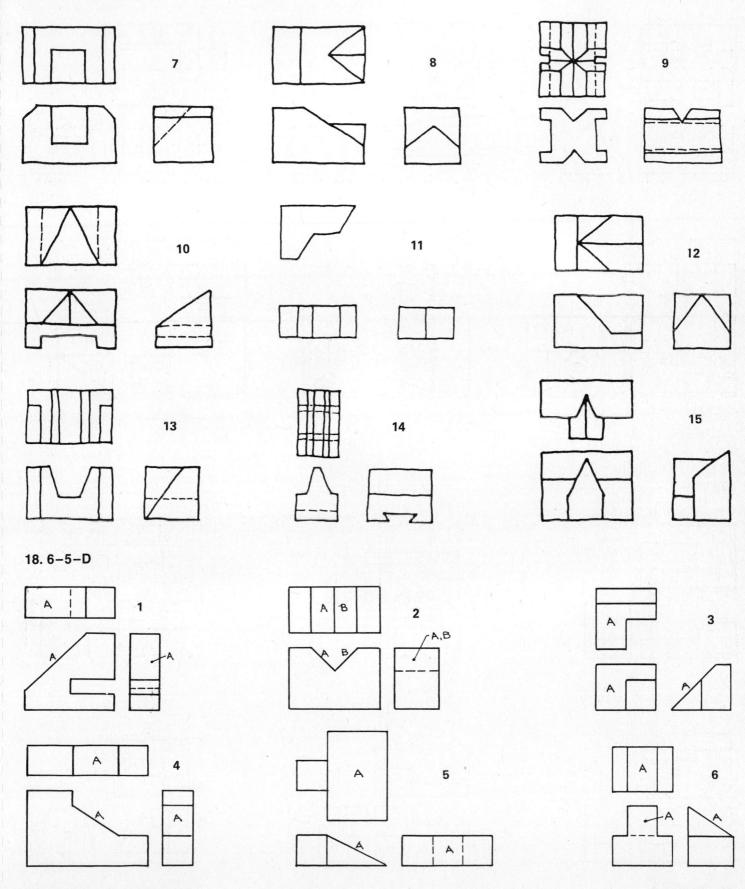

18. 6–5–D

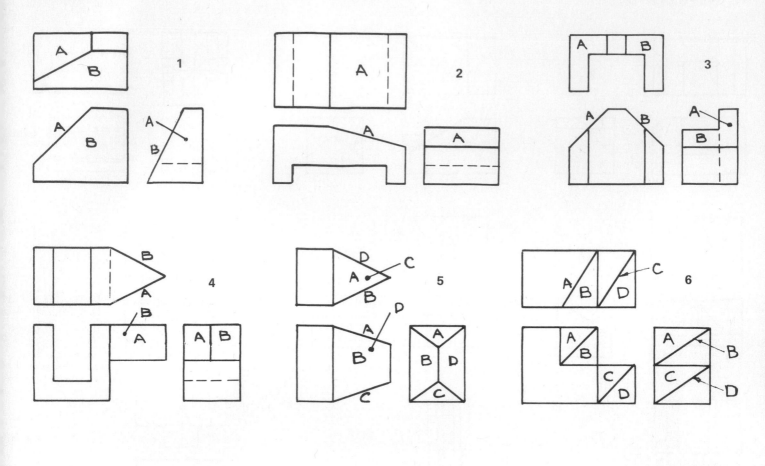

6-5-F

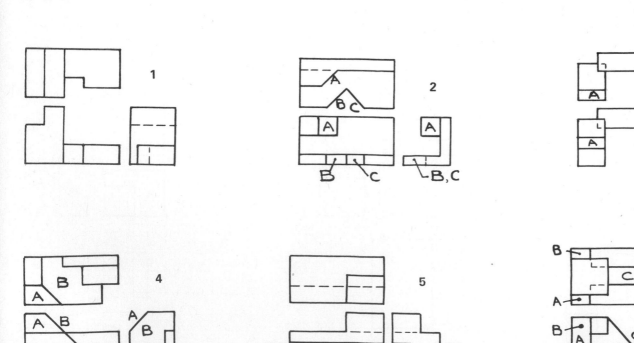

Unit 6–5
18. CONTD 6–5–G

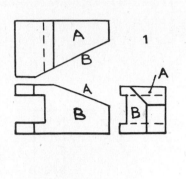
1

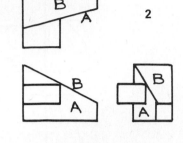
2

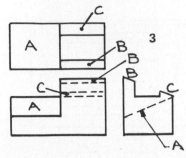
3

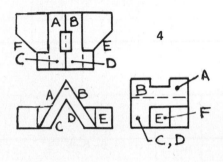
4

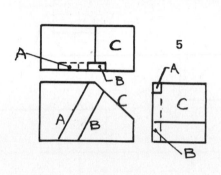
5

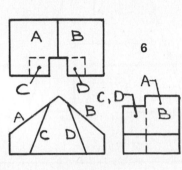
6

19. 6–5–H

| 13 – R | 15 – Q | 17 – S | 19 – N | 21 – W | 23 – X |
| 14 – T | 16 – Y | 18 – O | 20 – P | 22 – U | 24 – V |

20.

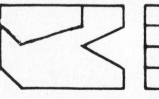
6–5–J

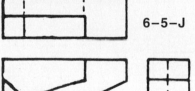

6–5–K

6–5–L

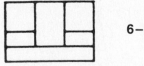

6–5–M

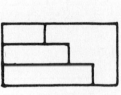
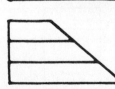

6–5–N

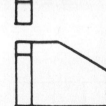
6–5–P

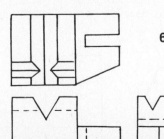

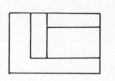
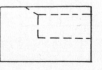

Unit 6–5

21.

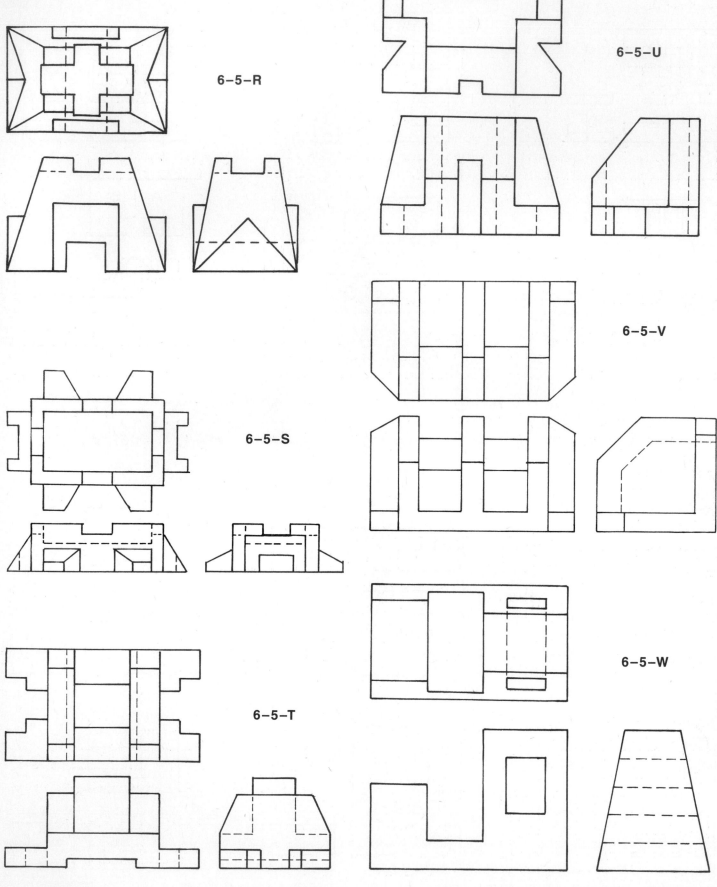

6–5–R

6–5–S

6–5–T

6–5–U

6–5–V

6–5–W

Unit 6−6

22. 6−6−A

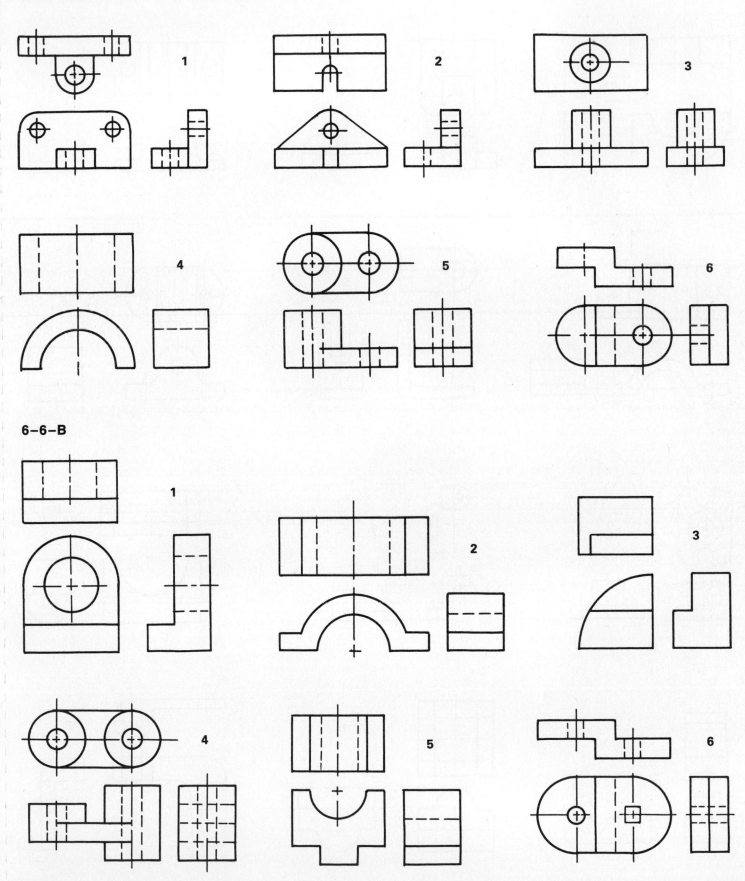

6−6−B

Unit 6–6
23. 6–6–C

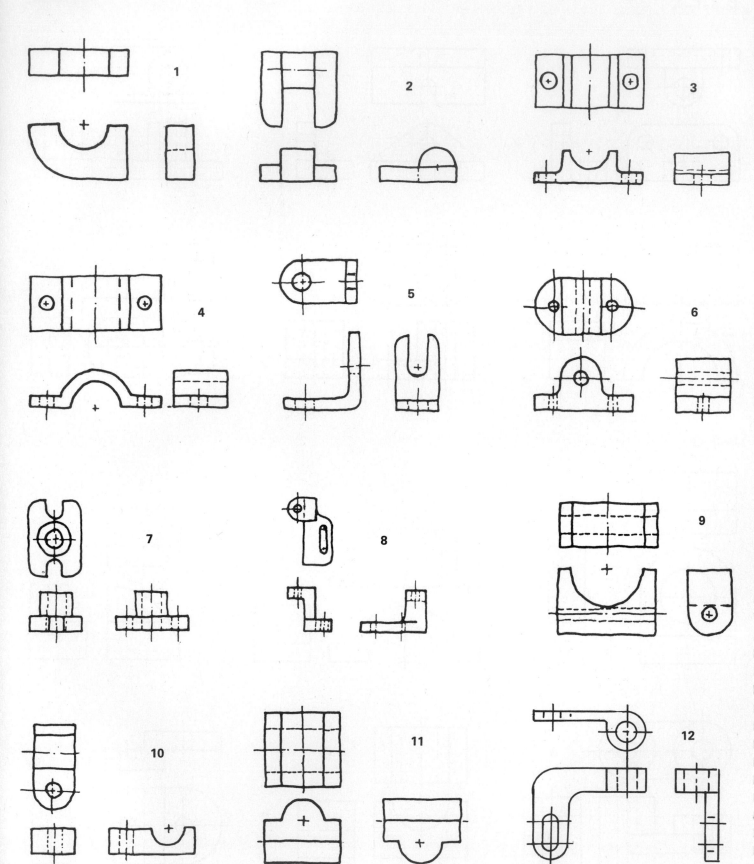

Unit 6–6
24. 6–6–D

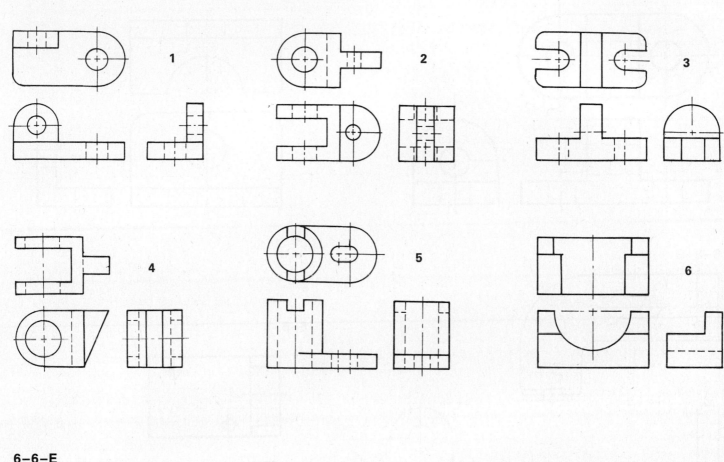

6–6–E

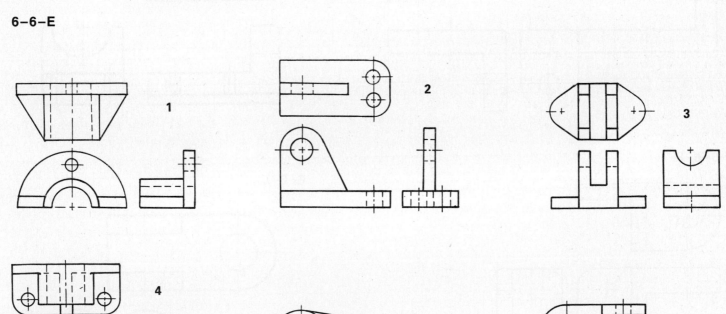

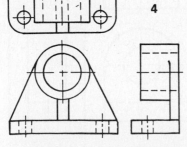

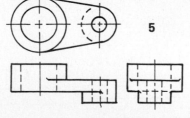

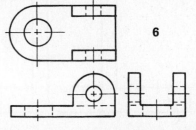

Unit 6–6

25.

6–6–F

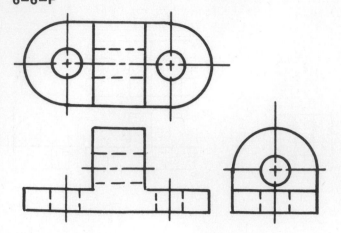

6–6–J

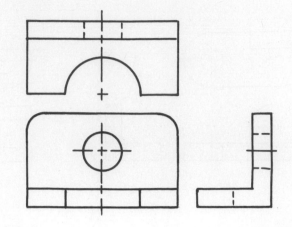

6–6–G

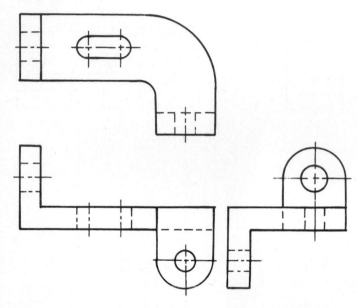

6–6–K

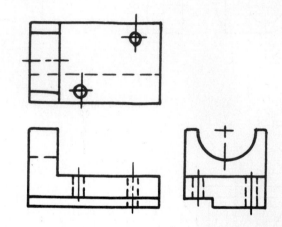

6–6–H

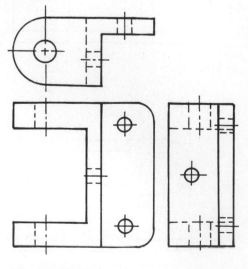

6–6–L

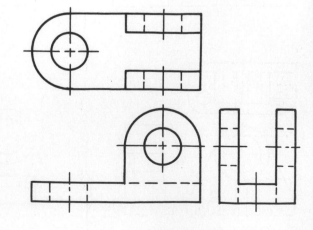

Unit 6–6

26. 6–6–M

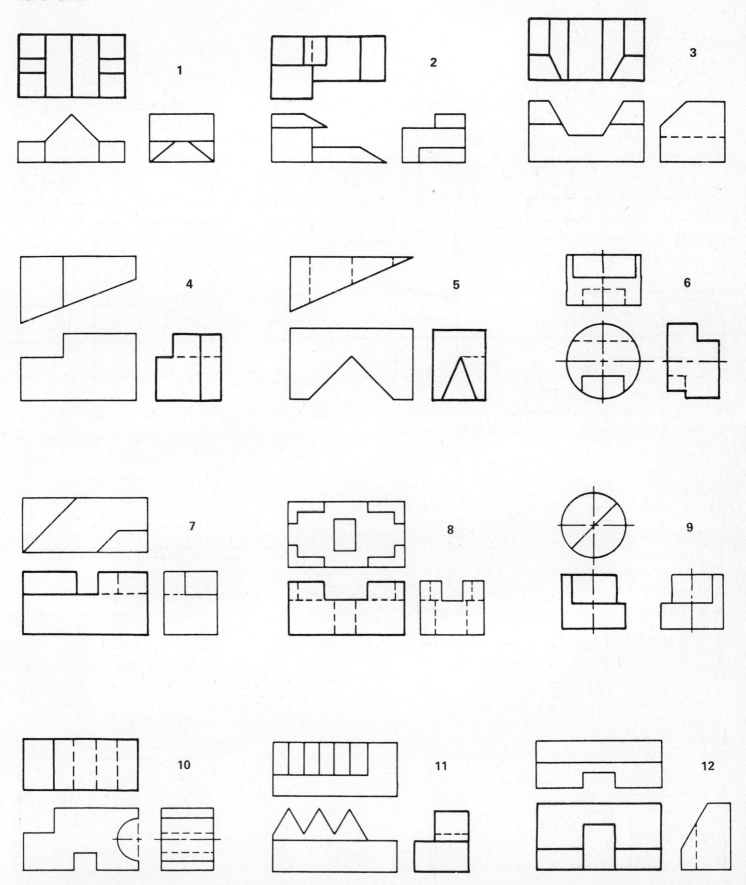

Unit 6-7

27. 6-7-A

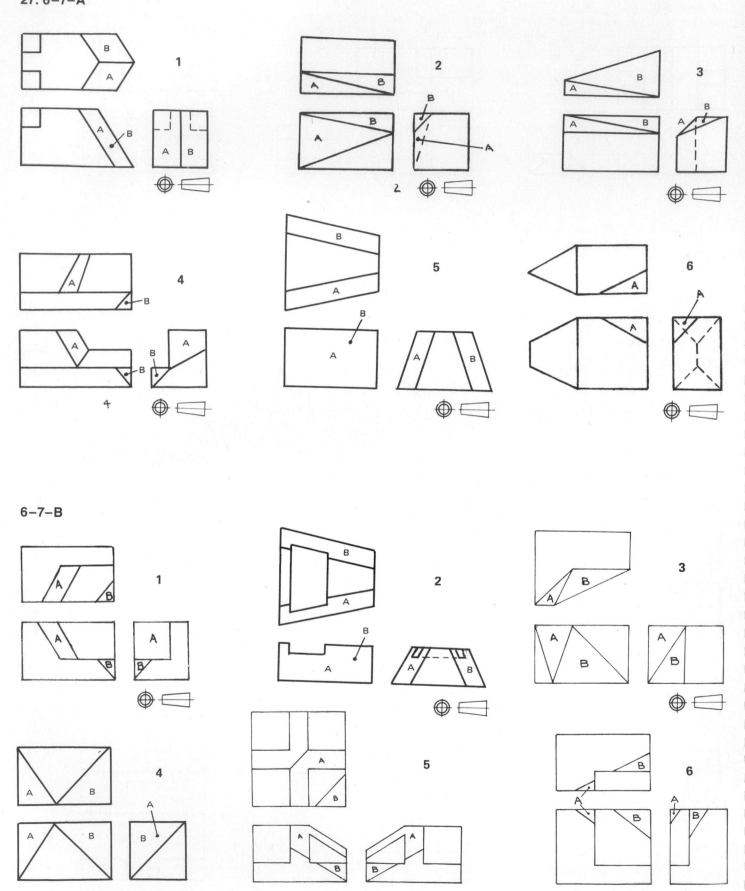

6-7-B

Unit 6–7

28.

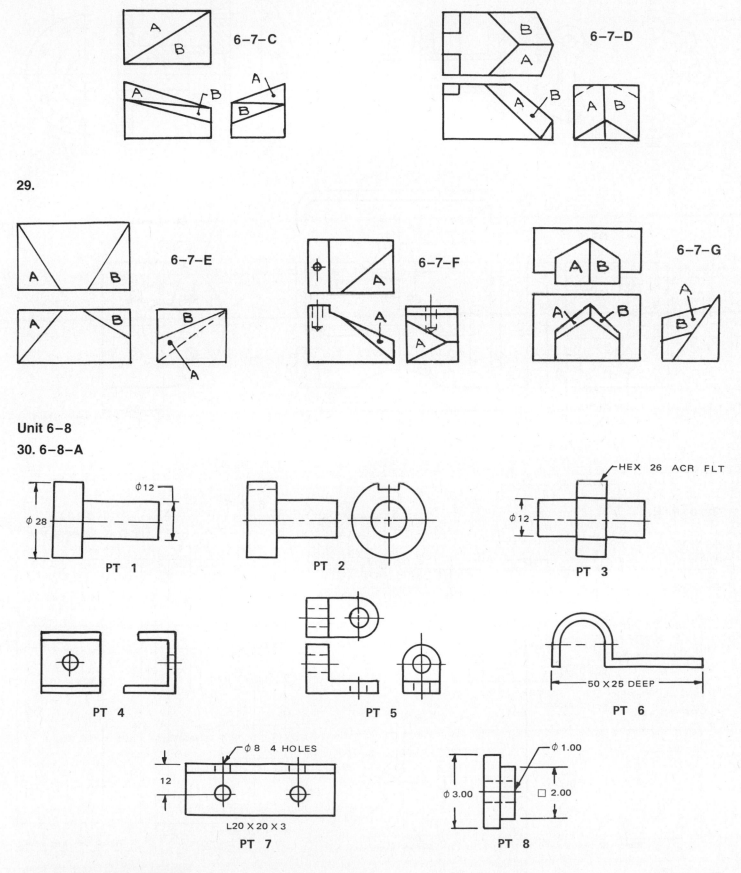

29.

Unit 6–8

30. 6–8–A

Unit 6–9

31. 6–9–A   6–9–B   6–9–C

6–9–D

32. 6–9–E   6–9–F

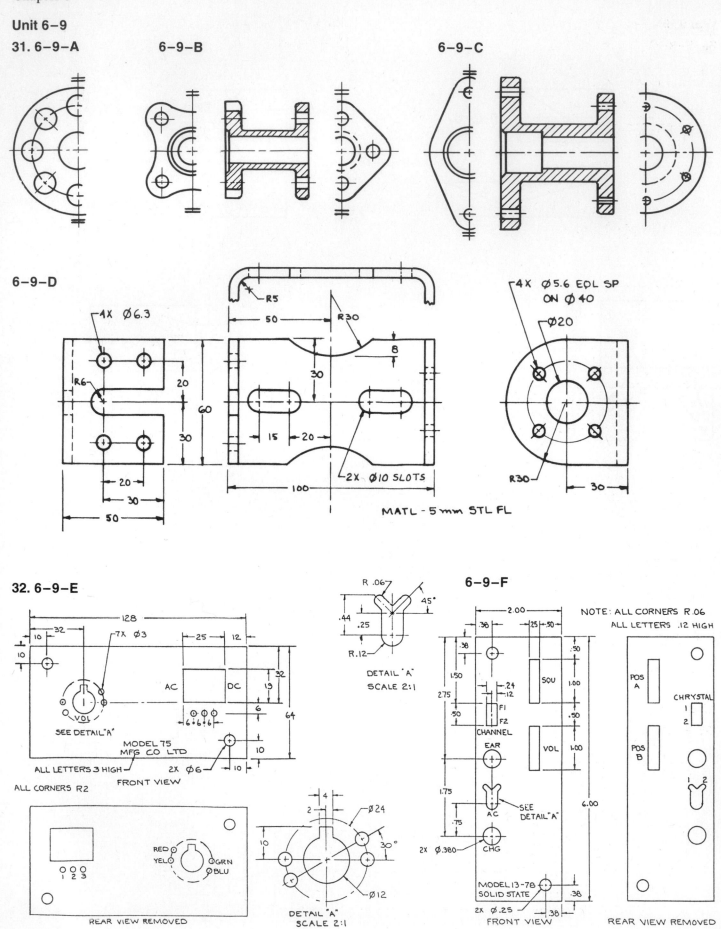

Unit 6-9
33. 6-9-G

GUSSET DETAIL A

GUSSET DETAIL B

GUSSET DETAIL C

GUSSET DETAIL D

Unit 6-10
34. 6-10-A

18X EQL SP   12X ØXX EQL SP   PD 33 DIAMOND KNURL

ØXX   XX   ØXX

MATL-SAE 1050

6-10-B

R SPHER   ØXX   XXX   XXX

18 TEETH EQ. SPACED

Ø XXX 12 HOLES EQ. SPACED ON ØXX B.C.

PXX DIAMOND KNURL

Unit 6-11
35. 6-11-A

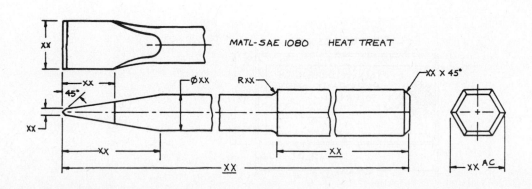

MATL-SAE 1080   HEAT TREAT

ØXX   RXX   XX X 45°   45°   XX AC

Unit 6–11
35. CONTD 6–11–B

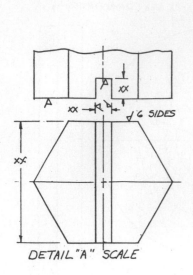

DETAIL "A" SCALE

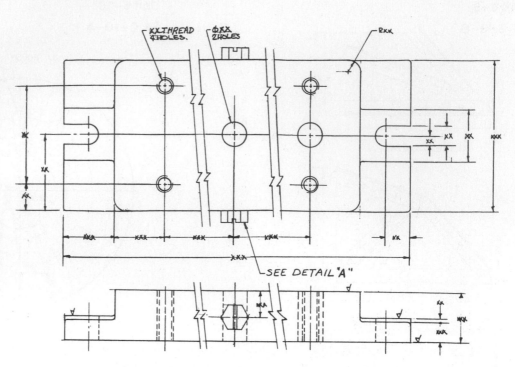

SEE DETAIL "A"

Unit 6–12
36. 6–12–A

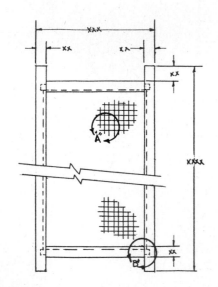

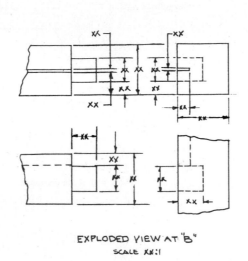

EXPLODED VIEW AT "B"
SCALE XX:1

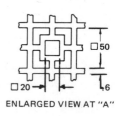

□ 50

□ 20   6

ENLARGED VIEW AT "A"

6–12–B

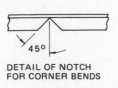

45°

DETAIL OF NOTCH
FOR CORNER BENDS

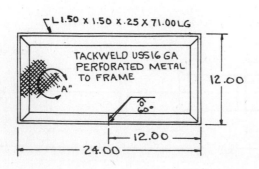

L 1.50 X 1.50 X .25 X 71.00 LG

TACKWELD USS 16 GA
PERFORATED METAL
TO FRAME

12.00

12.00

24.00

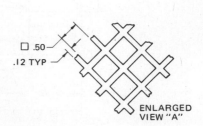

□ .50
.12 TYP

ENLARGED
VIEW "A"

Unit 6–13

37. 6–13–A

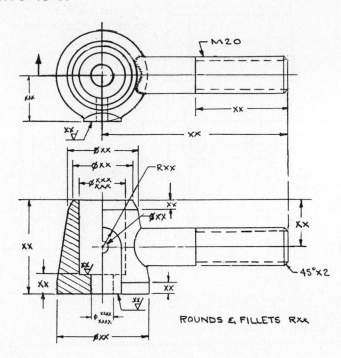

ROUNDS & FILLETS RXX

6–13–B

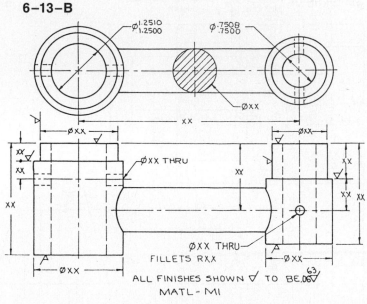

ALL FINISHES SHOWN ▽ TO BE.06▽
MATL – MI

6–14–B

28.12

φ25

VIEW 'A'
SCALE 1:1

Unit 6–14

38. 6–14–A

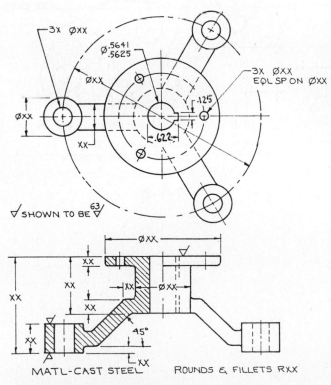

MATL–CAST STEEL   ROUNDS & FILLETS RXX

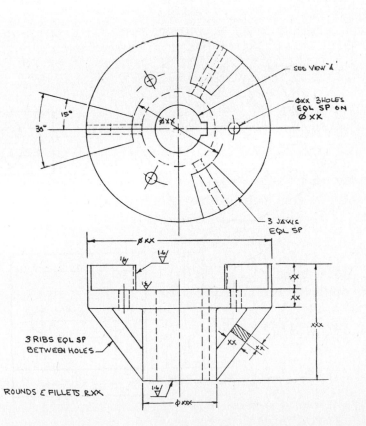

ROUNDS & FILLETS RXX

Unit 6–15
39. 6–15–A

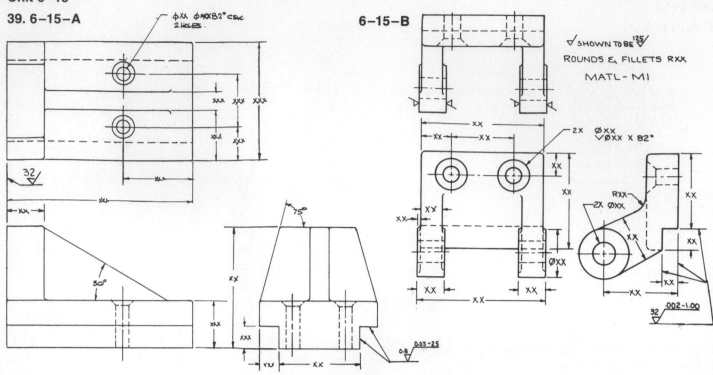

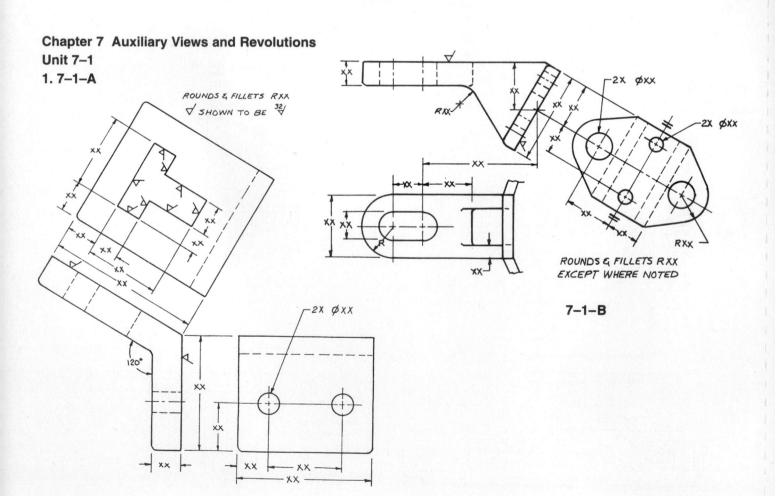

Chapter 7 Auxiliary Views and Revolutions
Unit 7–1
1. 7–1–A

Unit 7-1

1. CONTD 7-1-C

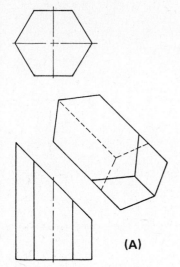

(A)

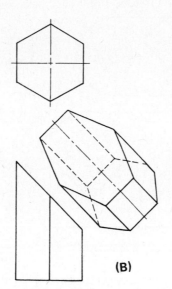

(B)

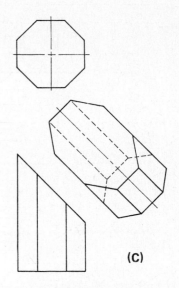

(C)

7-1-E

(A)

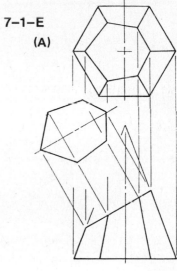

7-1-D

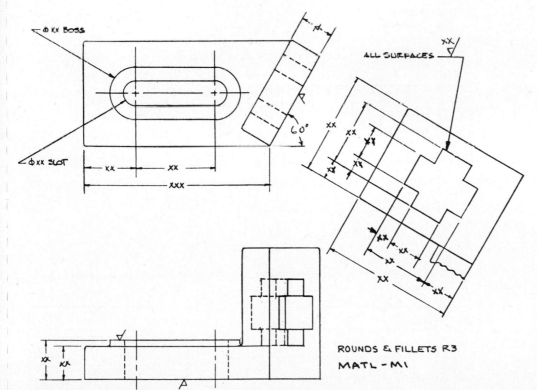

(B)

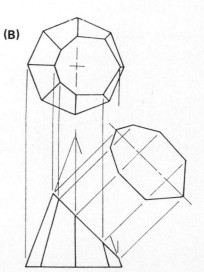

Unit 7–2

2. 7–2–A

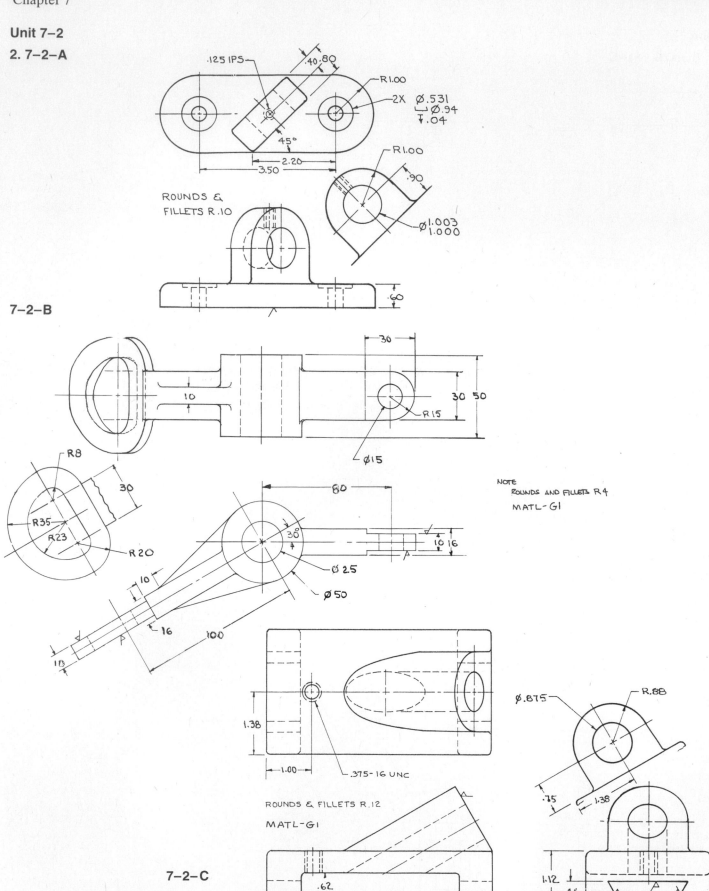

7–2–B

7–2–C

Unit 7–2

2. CONTD 7–2–D

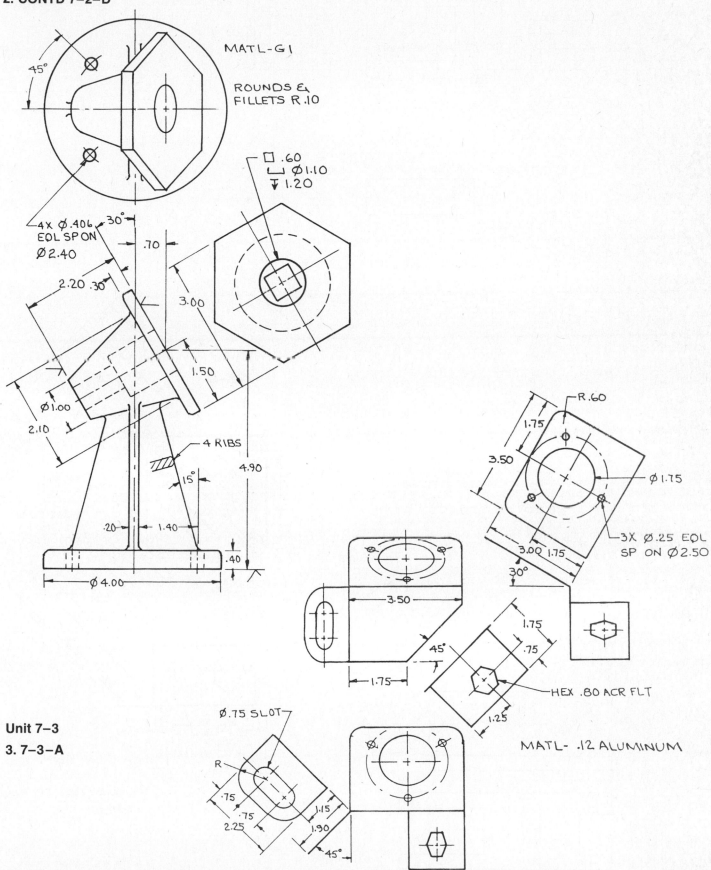

MATL-G1

ROUNDS &
FILLETS R .10

Unit 7–3

3. 7–3–A

MATL- .12 ALUMINUM

Unit 7–3
3. CONTD 7–3–B

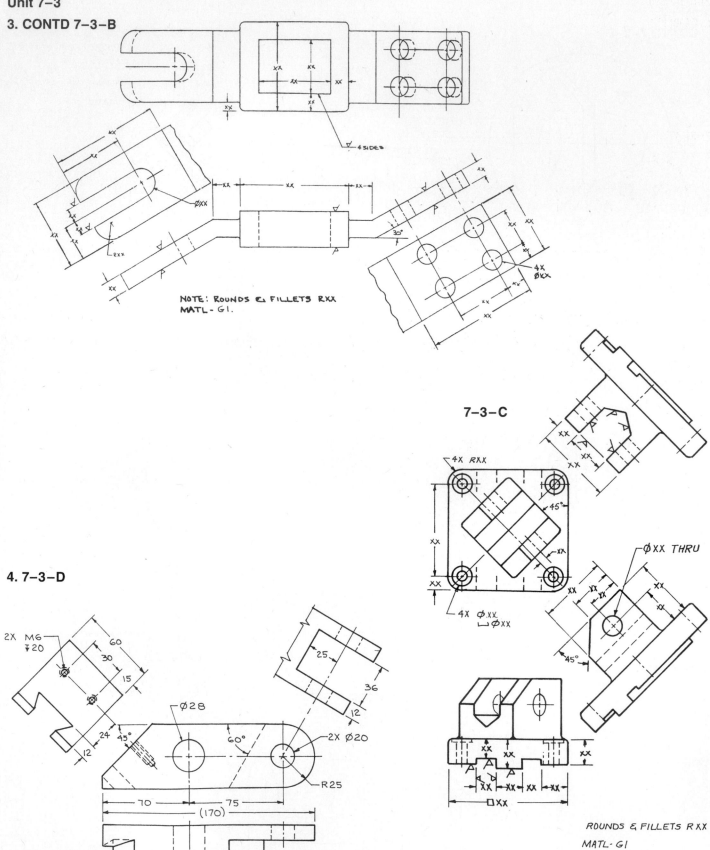

NOTE: ROUNDS & FILLETS RXX
MATL- GI.

7–3–C

4. 7–3–D

ROUNDS & FILLETS R XX

MATL- GI

Unit 7–3

4. CONTD

7–3–E

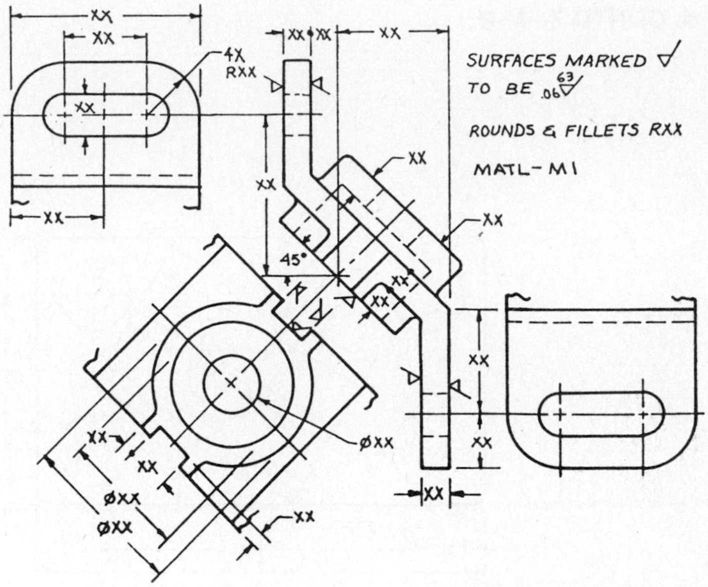

SURFACES MARKED ✓
TO BE .06 $\frac{63}{\sqrt{}}$

ROUNDS & FILLETS RXX

MATL– M1

4X RXX

45°

ØXX

7–3–F

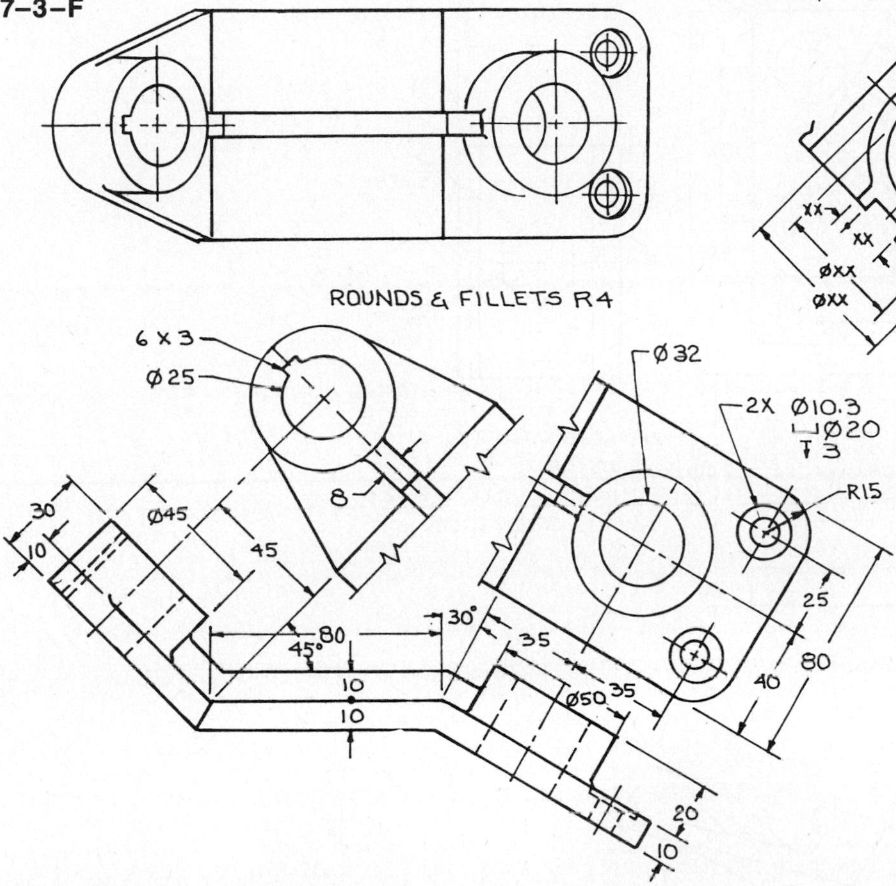

ROUNDS & FILLETS R4

6 X 3
Ø25
8
Ø32
2X Ø10.3
⌴Ø20
↧3
R15
Ø45
30
10
45
45° 80
30°
35
10
10
35
Ø50
25
80
40
20
10

15
30°
15
50

Unit 7–4

5. 7–4–A

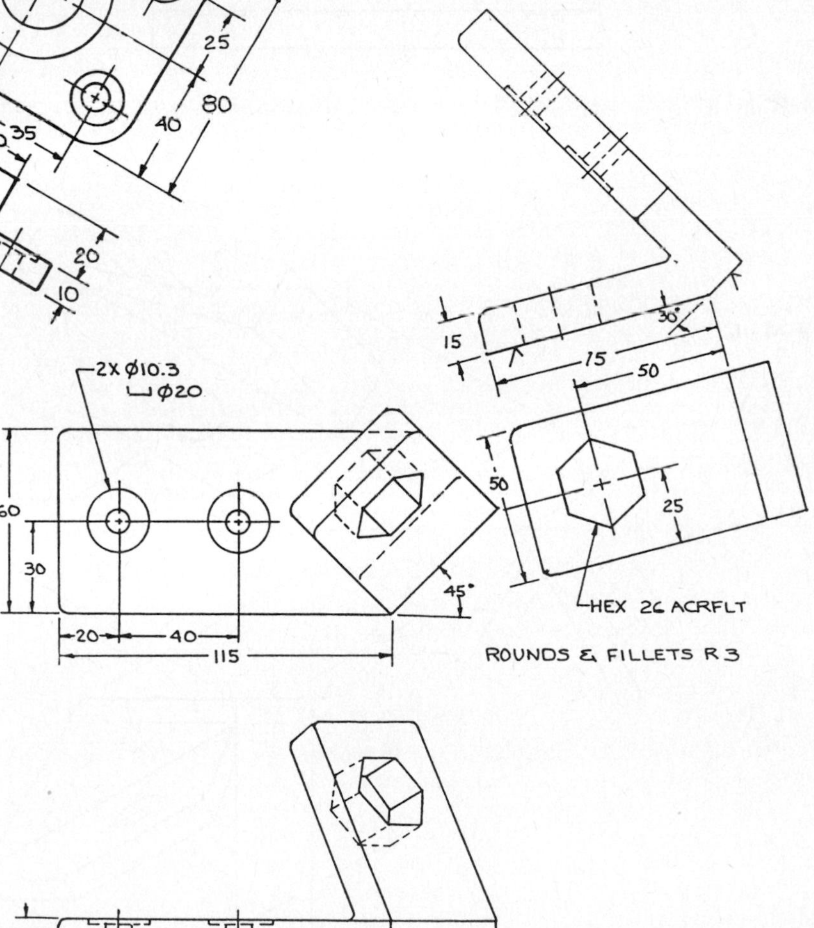

2X Ø10.3
⌴Ø20
60
30
20
40
115

45°

50
50
25

HEX 26 ACRFLT

ROUNDS & FILLETS R 3

15

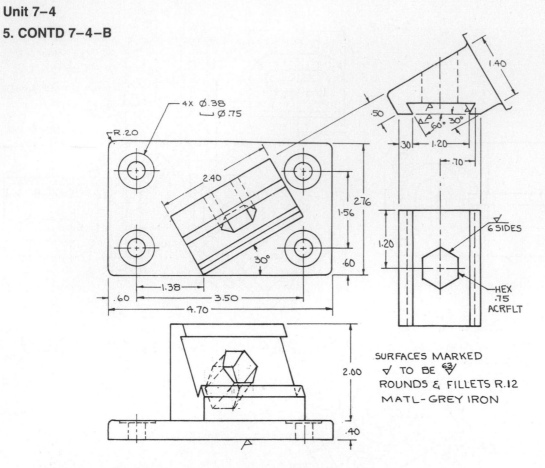

SURFACES MARKED
√ TO BE 63/
ROUNDS & FILLETS R.12
MATL–GREY IRON

7–4–C

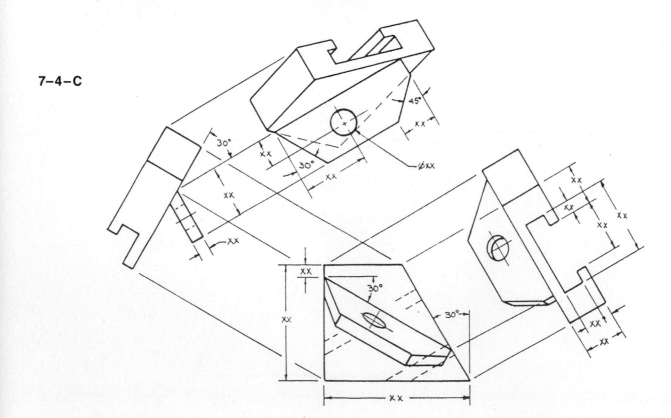

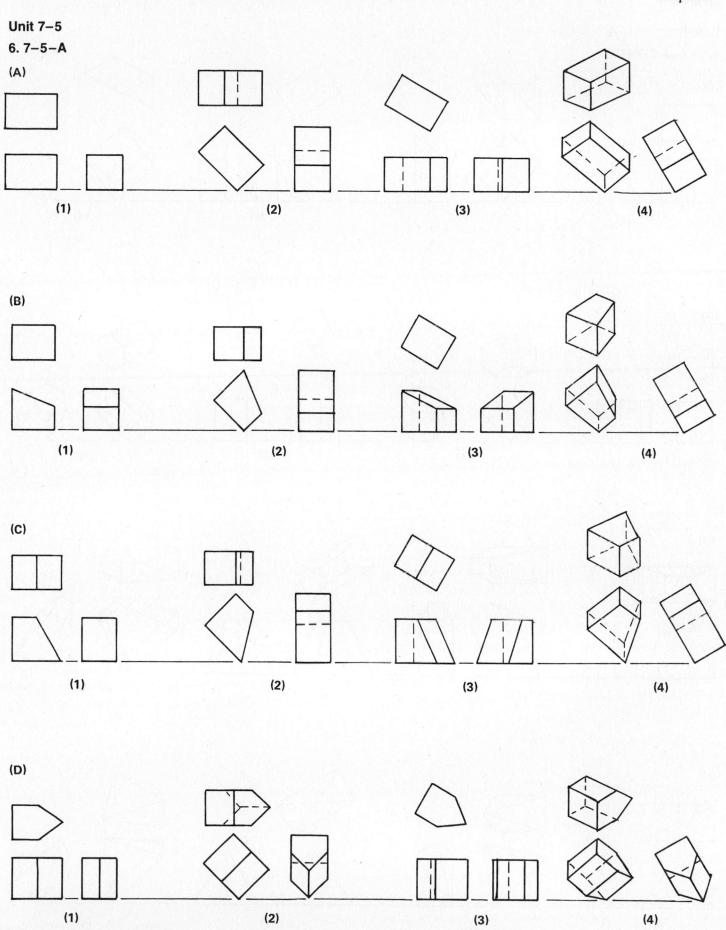

Unit 7–5

6. 7–5–A

(A)

(1)   (2)   (3)   (4)

(B)

(1)   (2)   (3)   (4)

(C)

(1)   (2)   (3)   (4)

(D)

(1)   (2)   (3)   (4)

Unit 7–5

6. 7–5–A CONTD

(E)

(1)   (2)   (3)   (4)

(F)

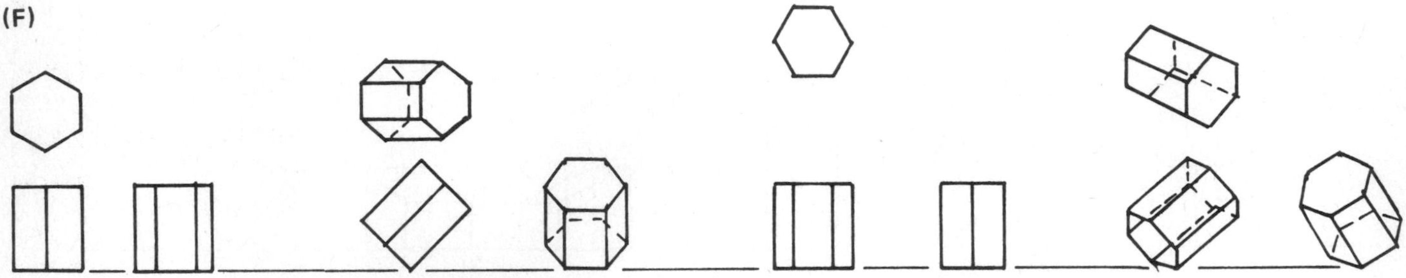

(1)   (2)   (3)   (4)

(G)

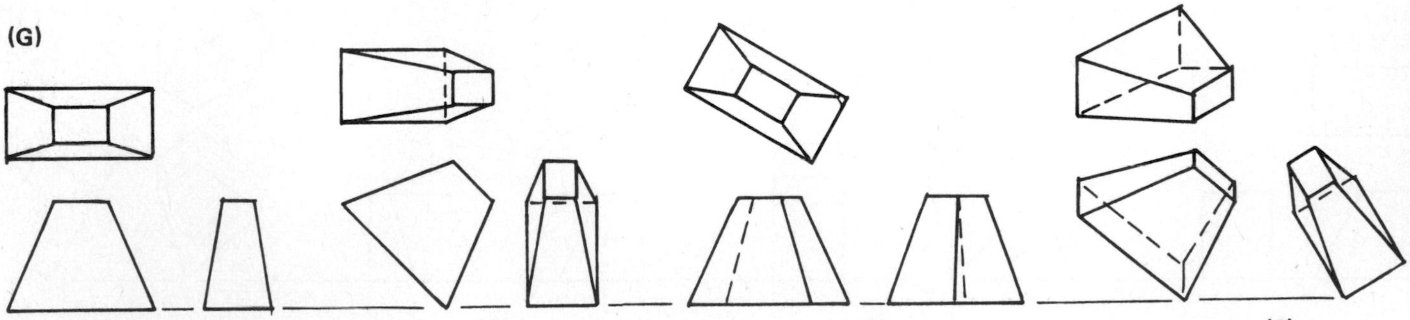

(1)   (2)   (3)   (4)

(H)

(1)   (2)   (3)   (4)

Unit 7–5

6. 7–5–A CONTD

(J)

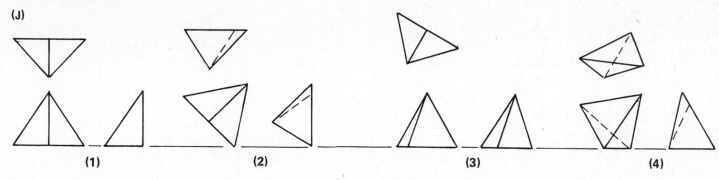

(1) (2) (3) (4)

(K)

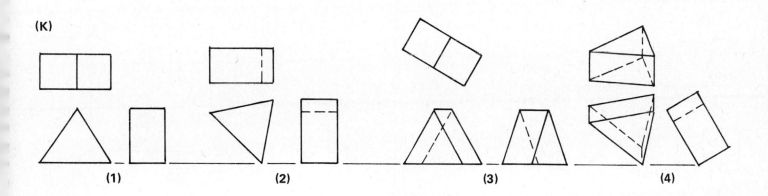

(1) (2) (3) (4)

(L)

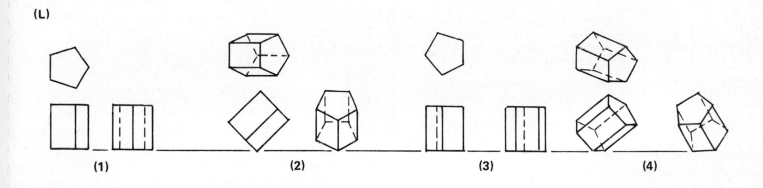

(1) (2) (3) (4)

(M)

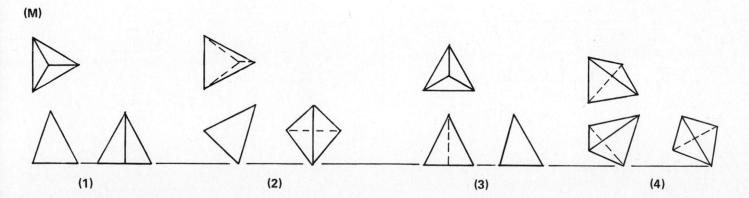

(1) (2) (3) (4)

Unit 7-6

7. 7-6-A

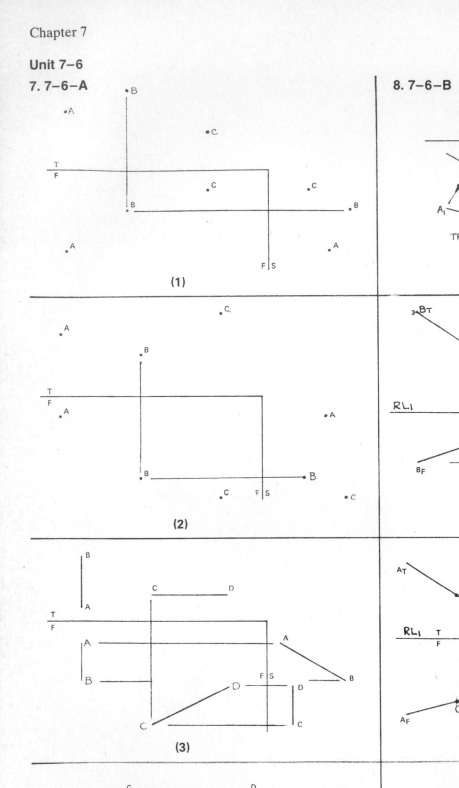

(1)

(2)

(3)

(4)

8. 7-6-B

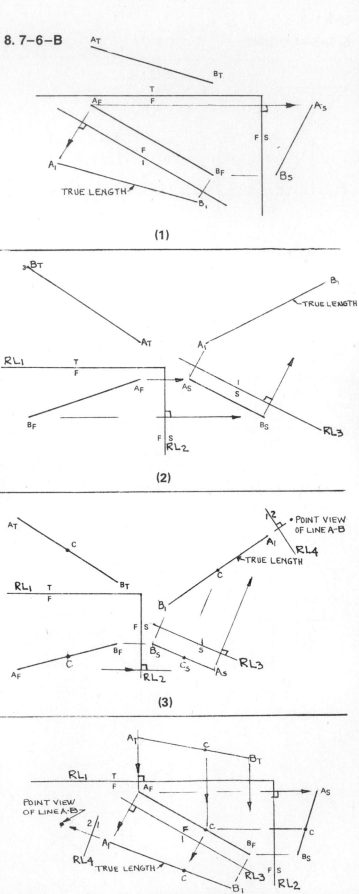

(1)

(2)

(3)

(4)

Unit 7–7

9. 7–7–A

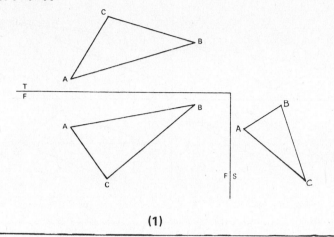

(1)

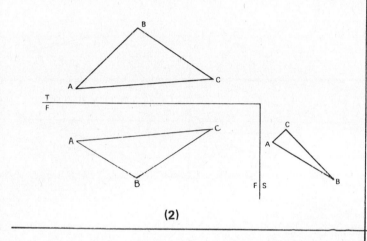

(2)

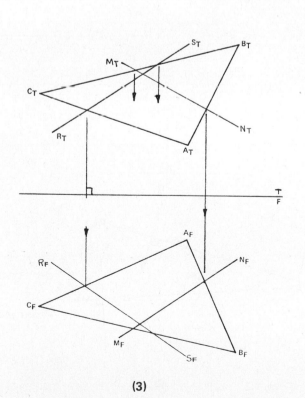

(3)

10. 7–7–B

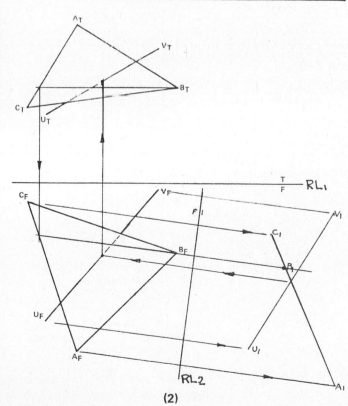

(1)

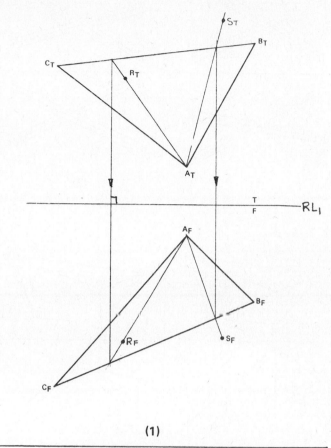

(2)

Unit 7–8
11. 7–8–A

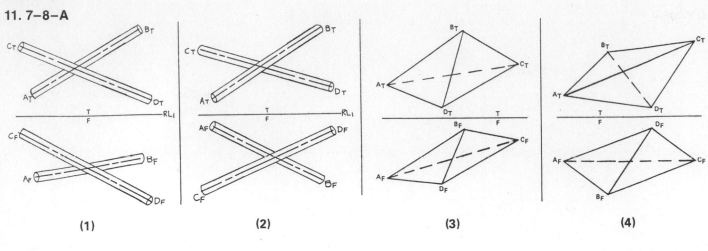

(1) (2) (3) (4)

12. 7–8–B

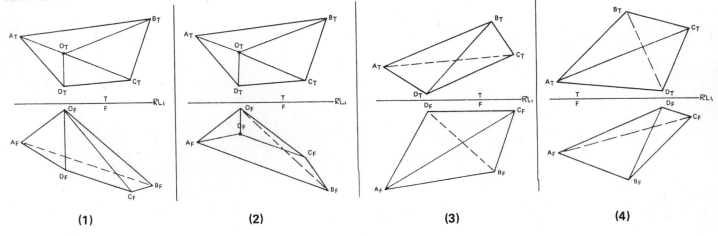

(1) (2) (3) (4)

Unit 7–9
13. 7–9–A

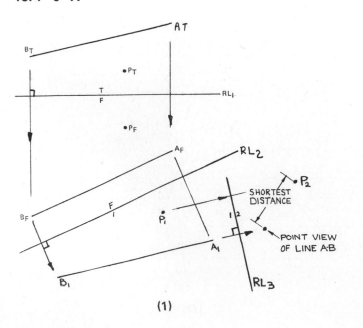

(1)

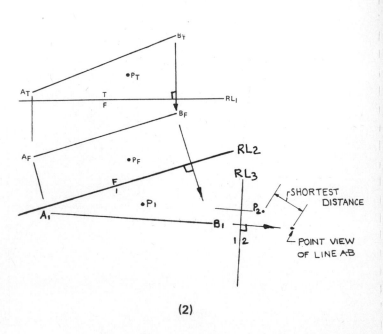

(2)

Unit 7–9

14. 7–9–B

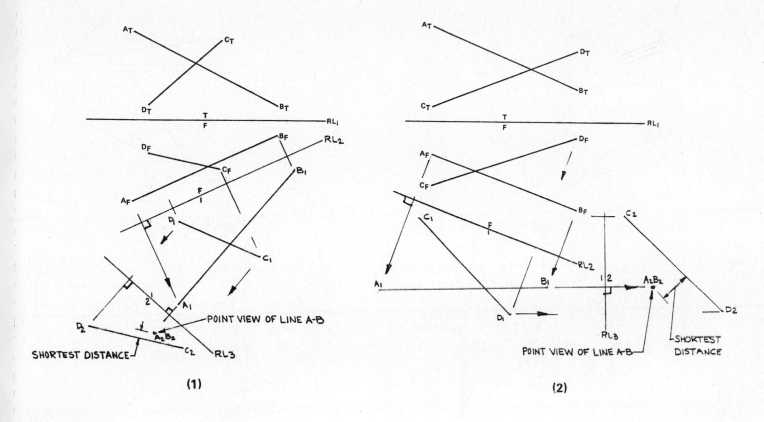

(1)

(2)

Unit 7–10

15. 7–10–A

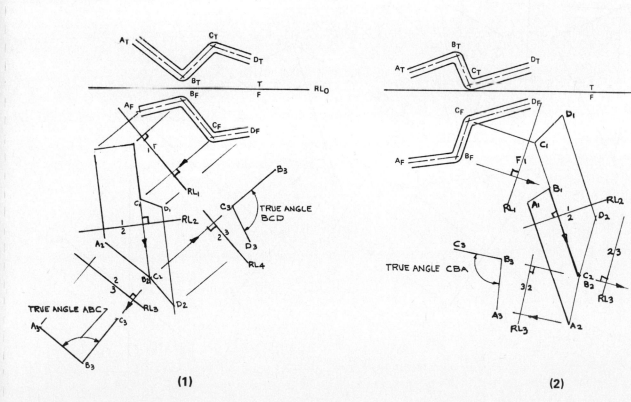

(1)

(2)

Unit 7–10

16. 7–10–B

(1)

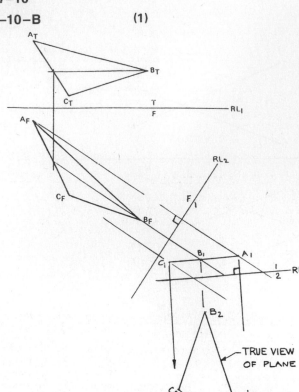

(2)

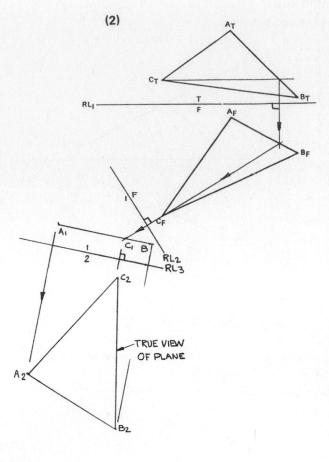

TRUE VIEW OF PLANE

TRUE VIEW OF PLANE

Unit 7–11

17. 7–11–A

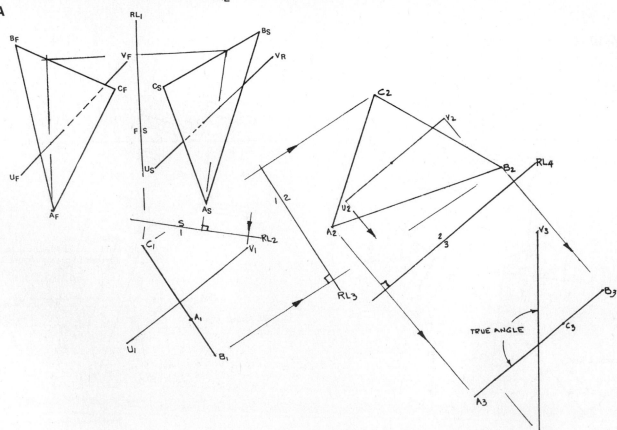

TRUE ANGLE

Unit 7–11

18. 7–11–B

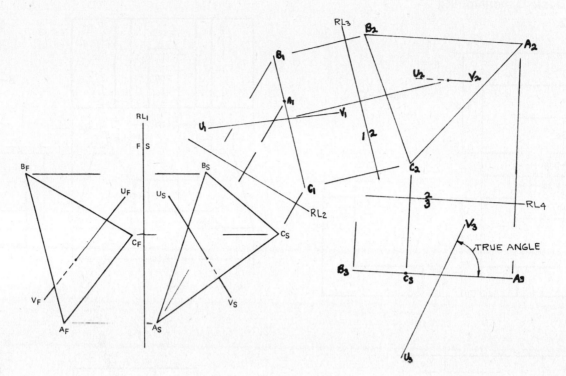

19. 7–11–C

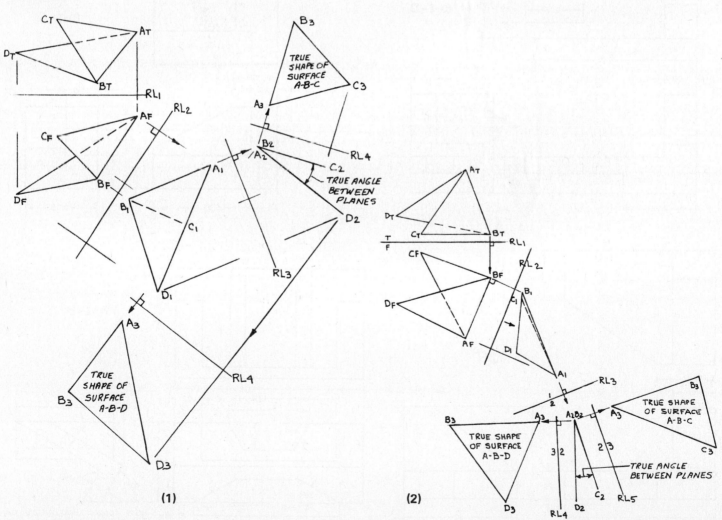

(1)

(2)

Chapter 8 Basic Dimensioning
Unit 8–1, 2.

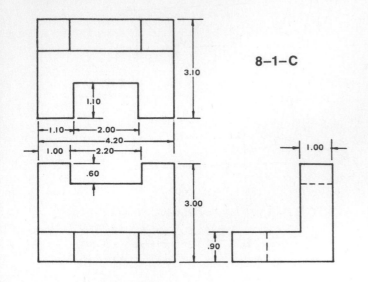

8–1–C

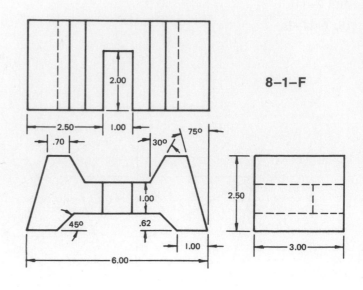

8–1–F

3.

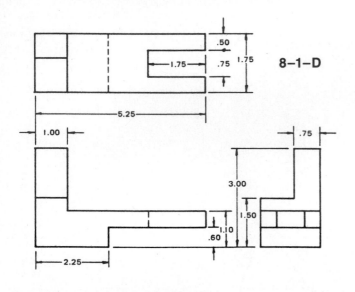

8–1–D

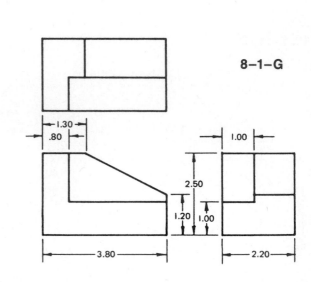

8–1–G

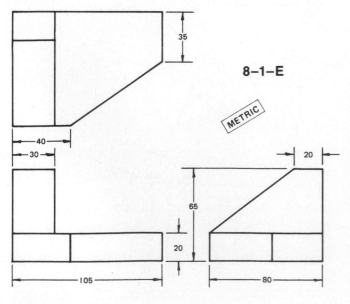

8–1–E

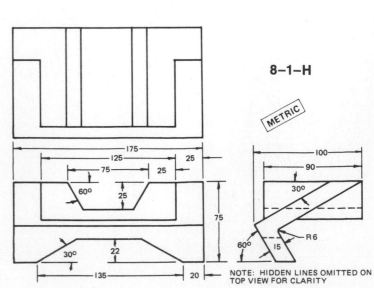

8–1–H

NOTE: HIDDEN LINES OMITTED ON TOP VIEW FOR CLARITY

Unit 8–1

3. CONTD

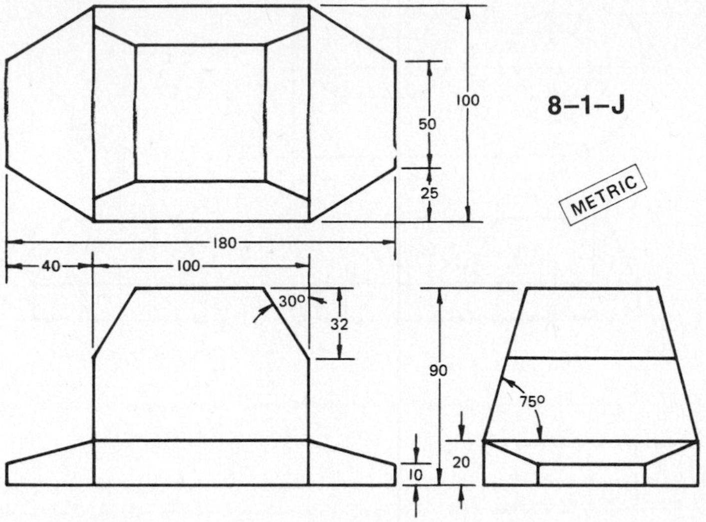

8–1–J

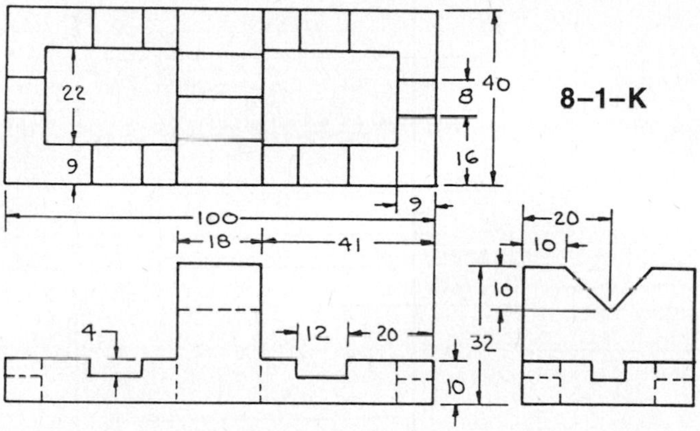

METRIC

8–1–K

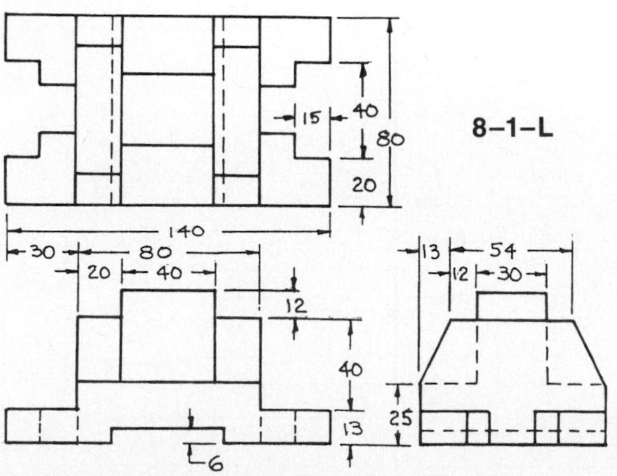

8–1–L

Unit 8–2

5.

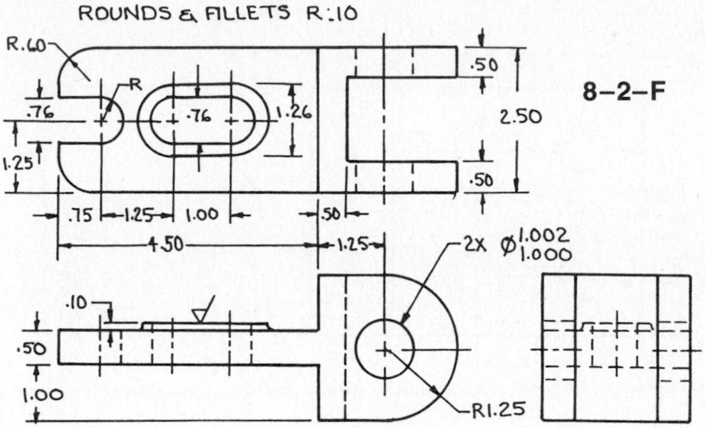

ROUNDS & FILLETS R .10

8–2–F

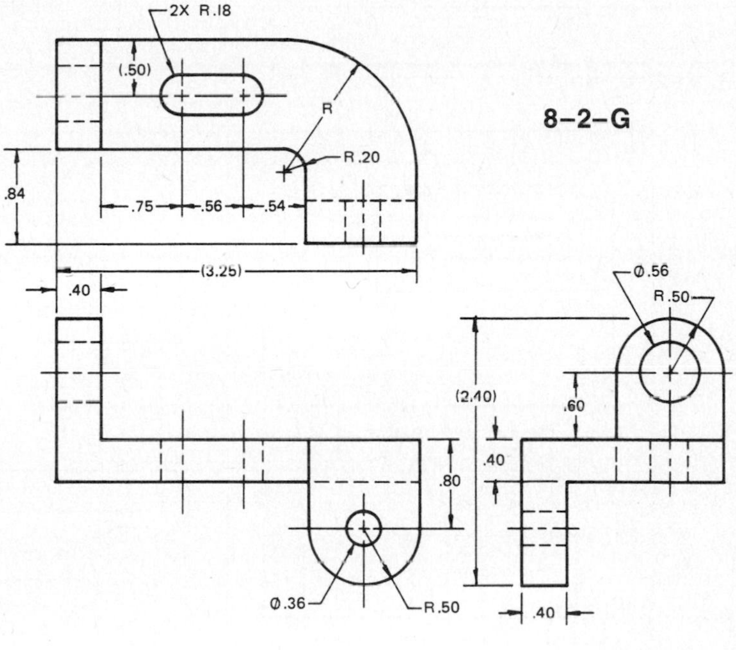

8–2–G

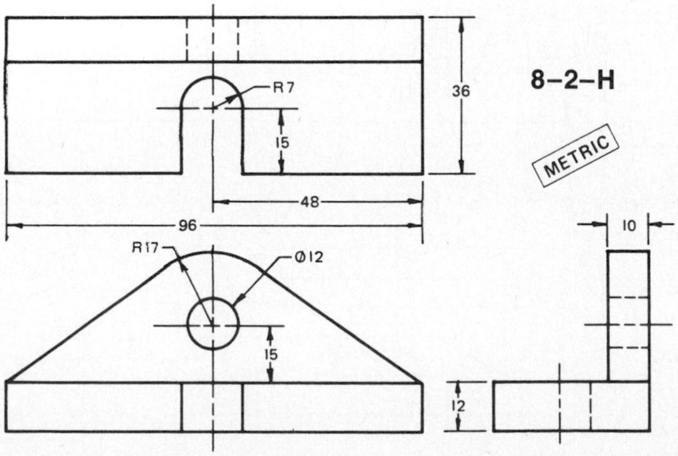

8–2–H

METRIC

Unit 8-2

5. CONTD

8-2-K

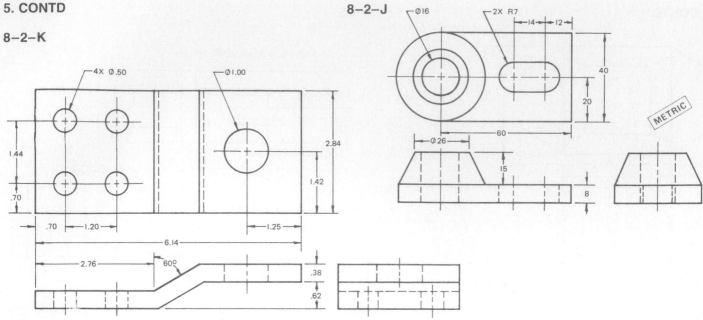

8-2-J

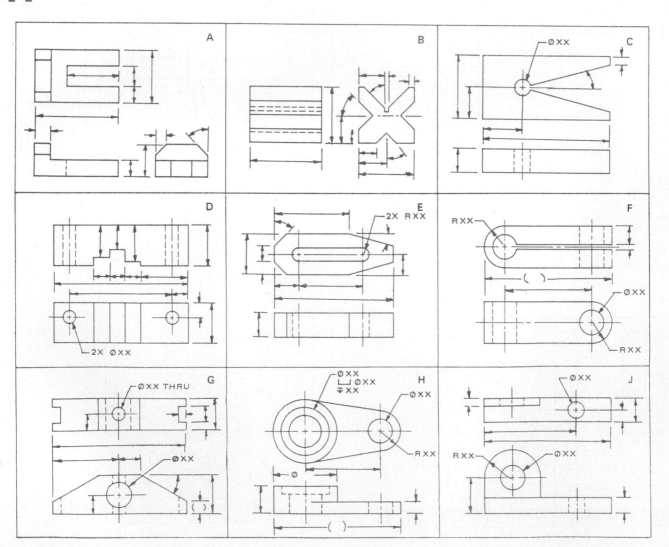

6. 8-2-L

Unit 8–2
6. 8–2–L CONTD

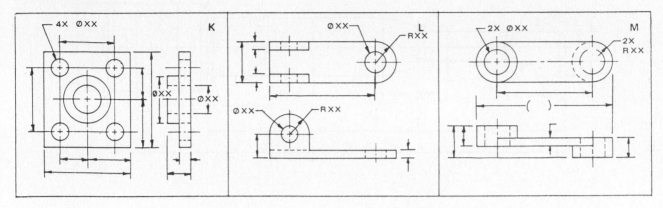

Unit 8–3
7. 8–3–A

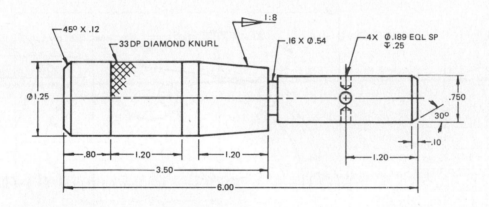

8. 8–3–B

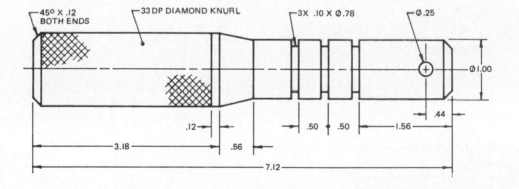

9. 8–3–C

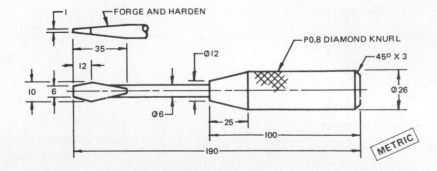

Unit 8-3

10. 8-3-D

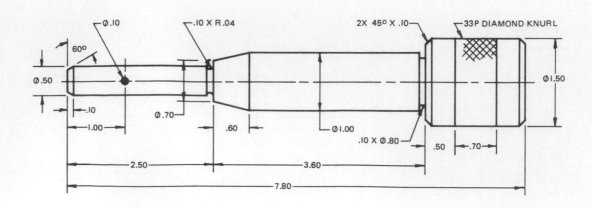

8-3-G

11. 8-3-E

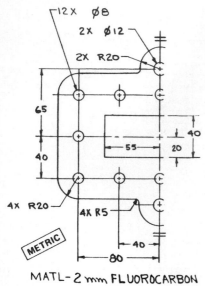

MATL-2 mm FLUOROCARBON
PLASTIC

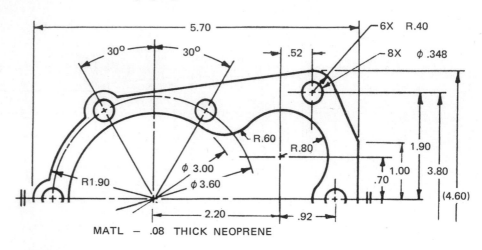

MATL — .08 THICK NEOPRENE

8-3-F

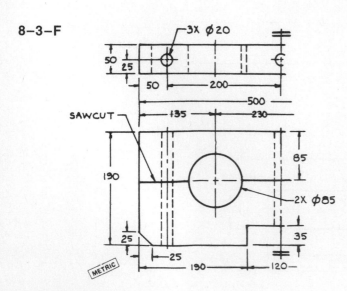

12. 8-3-H

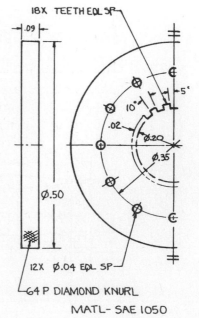

12X Ø0.04 EQL SP

64P DIAMOND KNURL

MATL- SAE 1050

Unit 8-4

13. 8-4-A

HOLE	DIA
A	8
B	4
C	5
D	76
E	12

METRIC

MATL- SAE 1006
3mm THK

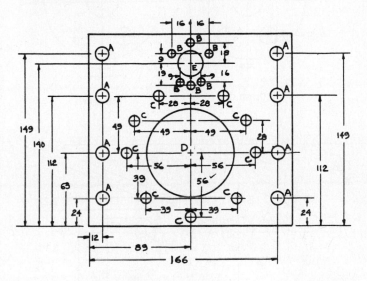

8-4-B

MATL-SAE 1008
.12 THK

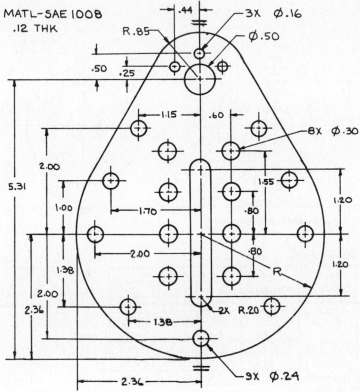

14. 8-4-C

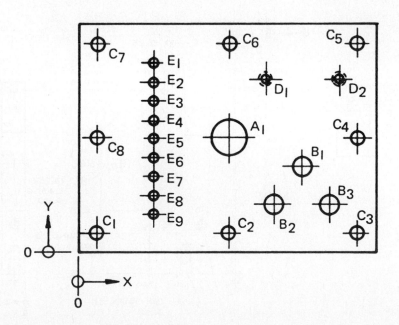

HOLE DIA	HOLE SYMBOL	LOCATION	
		X	Y
.40	A₁	2.00	1.50
.25	B₁	3.00	1.10
	B₂	2.60	.60
	B₃	3.40	.60
.20	C₁	.25	.25
	C₂	2.00	.25
	C₃	3.75	.25
	C₄	3.75	1.50
	C₅	3.75	2.75
	C₆	2.00	2.75
	C₇	.25	2.75
	C₈	.25	1.50
10-32 UNC	D₁	2.50	2.20
	D₂	3.50	2.20
.1015	E₁	1.00	2.50
	E₂	1.00	2.25
	E₃	1.00	2.00
	E₄	1.00	1.75
	E₅	1.00	1.50
	E₆	1.00	1.25
	E₇	1.00	1.00
	E₈	1.00	.75
	E₉	1.00	.50

Unit 8–4

15. 8–4–D

HOLE SYMBOL	HOLE SIZE
A	.50
B	.40

HOLE DIA	HOLE SYMBOL	LOCATION X	Y
.50	A₁	—	1.17
	A₂	58.5	-101.32
	A₃	-58.5	-101.32
.40	B₁	109.94	40.01
	B₂	-109.94	40.01

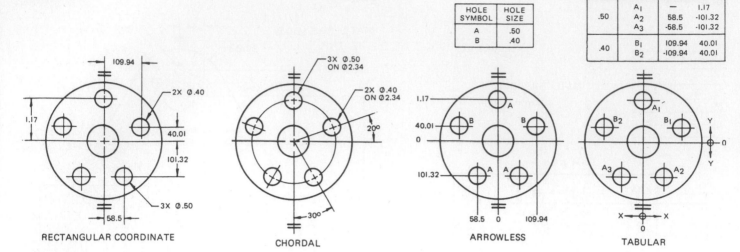

RECTANGULAR COORDINATE   CHORDAL   ARROWLESS   TABULAR

16. 8–4–E

HOLE DIA	HOLE SYMBOL	LOCATION X	Y	Z
Ø.406	A1	.50	5.45	THRU
	A2	.50	3.85	THRU
	A3	.50	2.25	THRU
	A4	.50	.65	THRU
Ø.401 ⊔Ø.88 ⊤.40	B1	8.75	5.30	THRU
	B2	11.25	5.30	THRU
Ø.438	C1	12.20	2.25	THRU
	C2	13.70	2.25	THRU
Ø.468	D1	3.50	5.00	THRU
	D2	11.00	3.49	THRU
Ø.689 .687	E1	12.65	5.30	THRU
Ø1.436 1.432	F1	8.00	2.25	THRU
.250-20 UNC-2B	G1	6.46	2.45	THRU
	G2	7.06	.94	THRU
.375-16 UNC-2B	H1	12.20	.65	THRU
	H2	13.70	.65	THRU

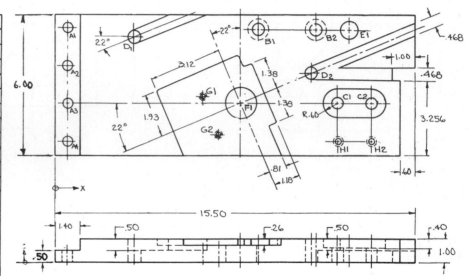

8–4–G

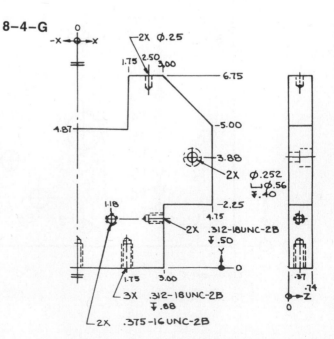

17. 8–4–F

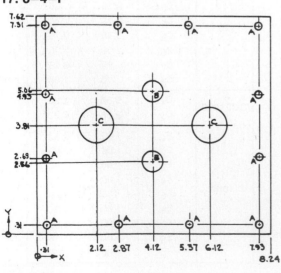

HOLE SYMBOL	HOLE SIZE
A	.28
B	.56
C	1.12

Unit 8–4

18. 8–4–H

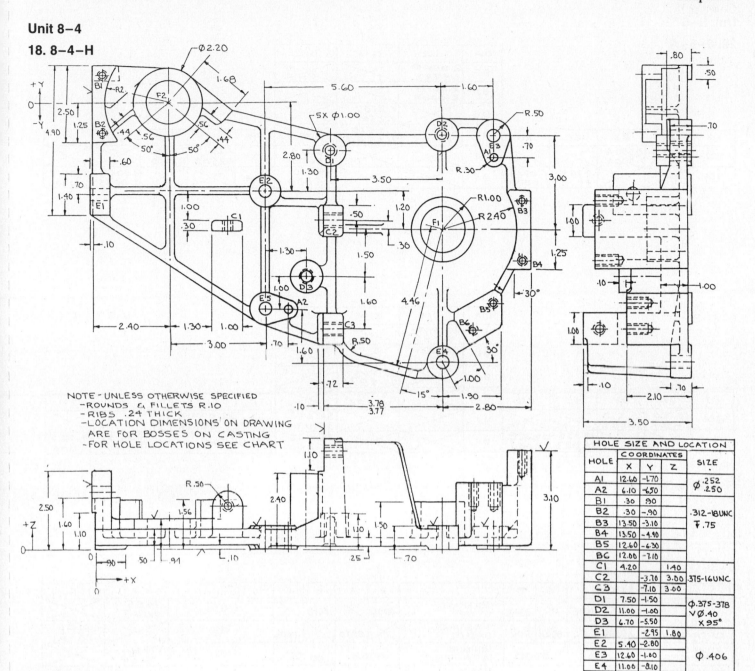

NOTE – UNLESS OTHERWISE SPECIFIED
- ROUNDS & FILLETS R.10
- RIBS .24 THICK
- LOCATION DIMENSIONS ON DRAWING
  ARE FOR BOSSES ON CASTING
- FOR HOLE LOCATIONS SEE CHART

HOLE SIZE AND LOCATION

HOLE	COORDINATES X	COORDINATES Y	COORDINATES Z	SIZE
A1	12.60	-1.70		Φ .252 .250
A2	6.10	-6.50		
B1	.30	.90		.312-18UNC ▼.75
B2	.30	-.90		
B3	13.50	-3.10		
B4	13.50	-4.90		
B5	12.60	-6.30		
B6	12.00	-7.10		
C1	4.20		1.40	.375-16UNC
C2		-3.70	3.00	
C3		-7.10	3.00	
D1	7.50	-1.50		Φ.375-.378 ∨Φ.40 X 95°
D2	11.00	-1.00		
D3	6.70	-5.50		
E1		-2.95	1.80	Φ .406
E2	5.40	-2.80		
E3	12.60	-1.00		
E4	11.00	-8.10		
E5	5.40	-6.50		
F1	11.00	-4.00		Φ 1.3760 1.3746
F2	2.40	0		

Unit 8–5

19. 8–5–A

		A	B	C	D	E			F	G	H	J	K			L	M	N	P	R
BASIC SIZE		3.50	1.00	2.25	.24	✕	BASIC SIZE		3.00	2.00	.50	.75	✕	BASIC SIZE		3.44	1.75	1.00	.32	✕
TOLERANCE		.02	.01	.002	.002	✕	TOLERANCE		.03	.04	.004	.001	✕	TOLERANCE		.12	.02	.02	.001	✕
LIMITS OF SIZE	MAX	3.50	1.01	2.251	.241	1.28	LIMITS OF SIZE	MAX	3.00	2.02	.502	.750	1.02	LIMITS OF SIZE	MAX	3.50	1.76	1.00	.321	1.02
	MIN	3.48	1.00	2.249	.239	1.19		MIN	2.97	1.98	.498	.749	.95		MIN	3.38	1.74	.98	.320	.98

		A	B	C	D	E			F	G	H	J	K			L	M	N	P	R
BASIC SIZE		90	25	60	6	✕	BASIC SIZE		75	50	12.5	20	✕	BASIC SIZE		90	50	25	10	✕
TOLERANCE		0.5	0.3	0.04	0.04	✕	TOLERANCE		0.76	1.0	0.04	0.02	✕	TOLERANCE		3	0.5	0.05	0.02	✕
LIMITS OF SIZE	MAX	90	25.3	60.02	6.02	33.1	LIMITS OF SIZE	MAX	75	50.5	12.50	20.00	25.50	LIMITS OF SIZE	MAX	91.5	50.25	25.00	10.02	30.5
	MIN	89.5	25.0	59.98	5.98	32.1		MIN	74.24	49.5	12.46	19.98	23.74		MIN	88.5	49.75	24.95	10.00	29.5

Unit 8–6

20. 8–6–A

DIA	LC2	LT3	LN2
Ø E	2.0000 / 1.9993		
Ø F	2.0018 / 2.0000		
Ø G		1.2507 / 1.2501	
Ø H		1.2510 / 1.2500	

FIT	Ø A	Ø B	Ø C	Ø D	DIA	LC2	LT3	LN2	FIT	Ø L	Ø M	Ø N	Ø P
RC2	1.4996 / 1.4992	1.5006 / 1.5000			Ø J		1.5016 / 1.5010		FN1	1.2513 / 1.2509	1.2506 / 1.2500		
RC5			.9984 / .9976	1.0012 / 1.0000	Ø K		1.5010 / 1.5000		FN4			2.0042 / 2.0035	2.0012 / 2.0000

8–6–B

DIA	H7/h6	K7/h6	H7/p6
Ø E	20.000 / 19.987		
Ø F	20.021 / 20.000		
Ø G		30.000 / 29.987	
Ø H		30.000 / 29.985	

FIT	Ø A	Ø B	Ø C	Ø D	DIA	H7/h6	K7/h6	H7/p6	FIT	Ø L	Ø M	Ø N	Ø P
G7/h6	25.000 / 24.987	25.028 / 25.007			Ø J			35.042 / 35.026	U7/h6	32.000 / 31.987	31.960 / 31.939		
H9/d9			18.935 / 18.883	19.052 / 19.000	Ø K			35.025 / 35.000	U7/s6			50.059 / 50.043	50.025 / 50.000

8–6–C

	A	B	C	D		E	F	G	H		J	K
TOLERANCE ON HOLE		.0005		.0012	TOLERANCE ON PART	.0006		.0010		Q1	.7455 / .7443	
TOLERANCE ON SHAFT	.0004		.0008		TOLERANCE ON SLOT		.0010		.0006	Q2		1.2526 / 1.2516
MINIMUM CLEARANCE		.0003	.0008		MINIMUM INTERFERENCE	.0008	.0025			Q3	.7466 / .7454	
MAXIMUM CLEARANCE		.0012	.0028		MAXIMUM INTERFERENCE	.0024	.0041			Q4		1.2522 / 1.2514

8–6–D

	A	B	C	D		E	F	G	H		J	K
TOLERANCE ON HOLE		0.05		0.02	TOLERANCE ON PART	0.26		0.24		Q1	19.01 / 18.96	
TOLERANCE ON SHAFT	0.05		0.5		TOLERANCE ON SLOT		0.25		0.12	Q2		31.84 / 31.77
MINIMUM CLEARANCE		0.02	0		MINIMUM INTERFERENCE	0.25	0.13			Q3	18.98 / 18.96	
MAXIMUM CLEARANCE		0.12	0.07		MAXIMUM INTERFERENCE	0.76	0.49			Q4		31.87 / 31.80

Unit 8–6
21. 8–6–E

DESIGN SKETCH	BASIC DIAMETER SIZE	SYMBOL	BASIS	FEATURE	LIMITS OF SIZE		CLEARANCE OR INTERFERENCE	
					MAX	MIN	MAX	MIN
A	.375	LN3	HOLE	HOLE	.3756	.3750	-.0012	-.0002
				SHAFT	.3762	.3758		
A	.250	RC4	HOLE	HOLE	.2509	.2500	.0020	.0005
				SHAFT	.2495	.2489		
B	.500	LT1	HOLE	HOLE	.5007	.5000	INT. .0002	CL. .0009
				SHAFT	.5002	.4998		
B	.625	RC4	HOLE	HOLE	.6260	.6250	.0023	.0006
				SHAFT	.6244	.6237		
B	.750	LN3	HOLE	HOLE	.7508	.7500	-.0004	-.0017
				SHAFT	.7517	.7512		
C	.312	LC6	SHAFT	HOLE	.3139	.3125	.0028	.0005
				SHAFT	.3120	.3111		
D	.188	LC3	HOLE	HOLE	.1887	.1880	.0012	.0000
				SHAFT	.1880	.1875		
D	.3120	RC7	SHAFT	HOLE	.3150	.3136	.0039	.0016
				SHAFT	.3120	.3111		
E	.812	FN2	HOLE	HOLE	.8128	.8120	INT. .0019	INT. .0006
				SHAFT	.8139	.8134		
DESIGN SKETCH	BASIC DIAMETER SIZE	SYMBOL	BASIS	FEATURE	LIMITS OF SIZE		CLEARANCE OR INTERFERENCE	
					MAX	MIN	MAX	MIN
A	10	H7/P6	HOLE	HOLE	10.015	10.000	-0.024	0.000
				SHAFT	10.024	10.015		
A	6	H8/f7	HOLE	HOLE	6.018	6.000	0.040	0.010
				SHAFT	5.990	5.978		
B	12	H7/k6	HOLE	HOLE	12.018	12.000	CL 0.017	INT -0.012
				SHAFT	12.012	12.001		
B	16	H8/f7	HOLE	HOLE	16.027	16.000	0.061	0.016
				SHAFT	15.984	15.966		
B	20	H7/P6	HOLE	HOLE	20.021	20.000	-0.001	-0.035
				SHAFT	20.035	20.022		
C	8	H7/h6	SHAFT	HOLE	8.015	8.000	0.024	0.000
				SHAFT	8.000	7.991		
D	5	H7/h6	HOLE	HOLE	5.012	5.000	0.020	0.000
				SHAFT	5.000	4.992		
D	8	F8/h7	SHAFT	HOLE	8.035	8.013	0.050	0.013
				SHAFT	8.000	7.985		
E	18	H7/u6	HOLE	HOLE	18.021	18.000	0.017	0.049
				SHAFT	18.054	18.041		

Unit 8–6
22. 8–6–F

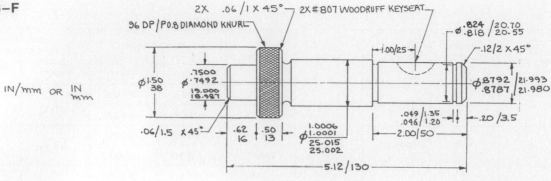

IN/mm OR $\frac{IN}{mm}$

23. 8–6–G

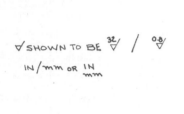

∇ SHOWN TO BE $\frac{32}{\nabla}$ / $\frac{0.8}{\nabla}$

IN/mm OR $\frac{IN}{mm}$

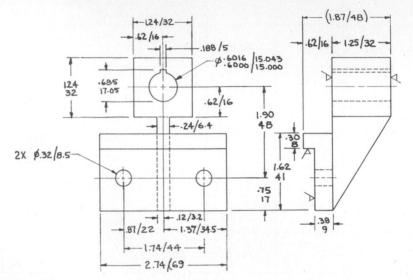

Unit 8–7
24. 8–7–A

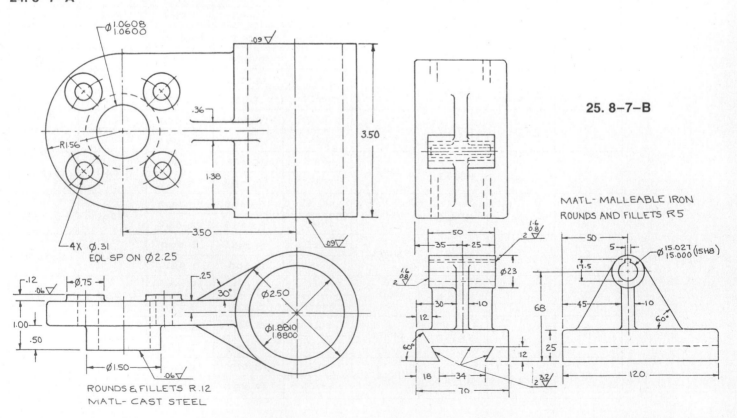

25. 8–7–B

MATL– MALLEABLE IRON
ROUNDS AND FILLETS R5

ROUNDS & FILLETS R.12
MATL– CAST STEEL

Unit 8-7

26. 8-7-C

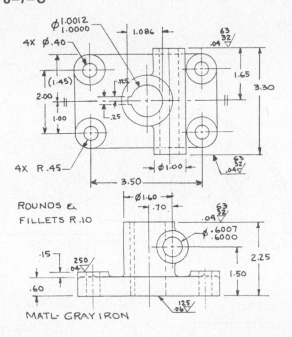

ROUNDS &
FILLETS R.10

MATL - GRAY IRON

28. 8-7-E

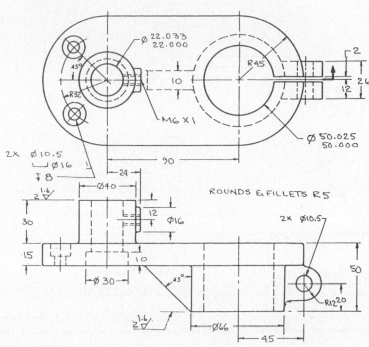

ROUNDS & FILLETS R5

27. 8-7-D

ROUNDS AND FILLETS R 2.5

MATL - GRAY IRON

8-7-F

MATL - GRAY IRON

8-7-G

ROUNDS &
FILLETS R5

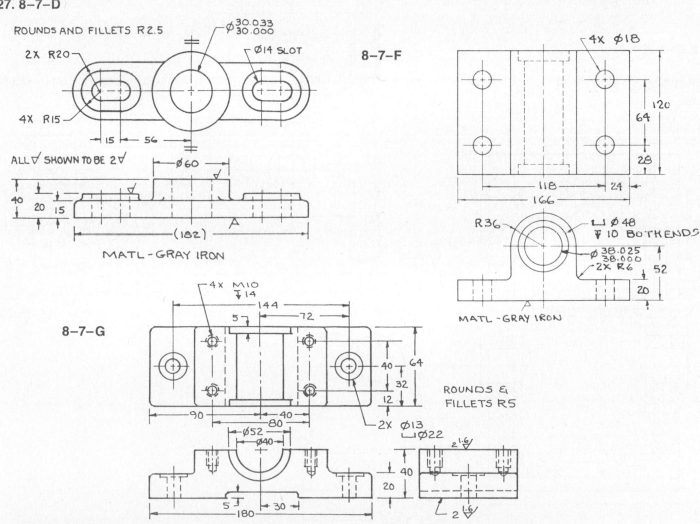

Chapter 9 Sections
Unit 9–1,   1. 9–1–A

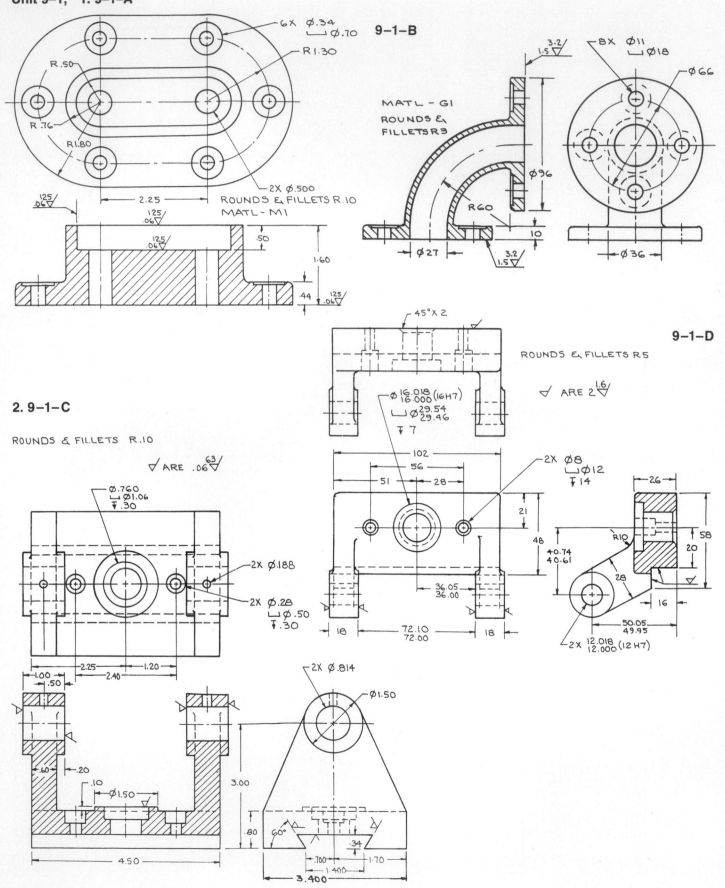

9–1–B

2. 9–1–C

9–1–D

Unit 9–2

3. 9–2–A

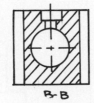  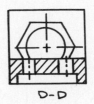

A-A   B-B   C-C   D-D

9–2–B

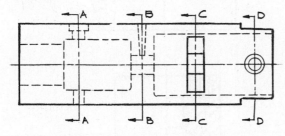  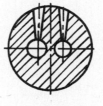  

SECTION A-A   SECTION B-B   SECTION C-C   SECTION D-D

4. 9–2–C

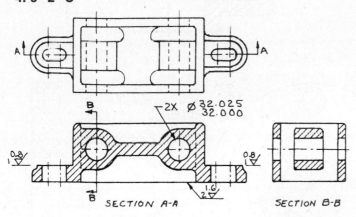

2X Ø 32.025 / 32.000

SECTION A-A   SECTION B-B

9–3–B

MATL-MI   ROUNDS & FILLETS R.16

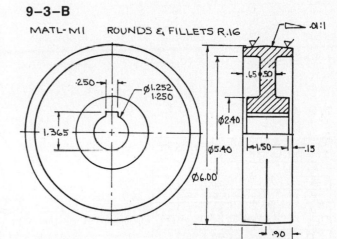

.250

Ø1.252 / 1.250

1.365

.65 Ø.50

Ø2.40

Ø5.40

1.50   .15

Ø6.00

.90

1.80

.01:1

Unit 9–3

5. 9–3–A

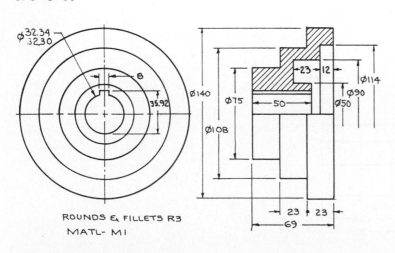

Ø 32.34 / 32.30

8

35.92

Ø140

Ø108

Ø75

Ø114

Ø90

Ø50

23   12

50

23   23

69

ROUNDS & FILLETS R3

MATL- MI

9–3–C

ROUNDS & FILLETS RXX

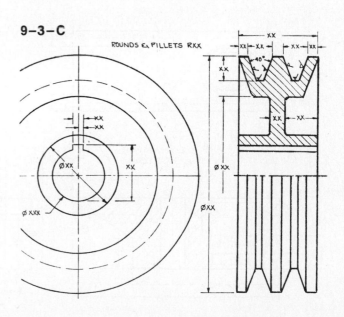

40°

XX XX   XX XX

XX

XX XX

Ø XX

Ø XXX

XX

Ø XX

Ø XX

Unit 9–3
5. CONTD 9–3–D

ROUNDS & FILLETS Rλ

Unit 9–4
6. 9–4–A

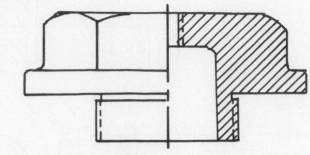

9–4–B

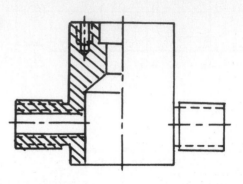

9–4–C

4X ⌀10
⌴⌀14
▼8

UNDERCUT

M64 X 4

M42 X 45

⌀98   ⌀118

⌀73

ROUNDS & FILLETS R5
MATL- M1

16
28
70

7. 9–4–D

HOLE SYMBOL	HOLE SIZE	LOCATION		
		X	Y	Z
A1	⌀18	16	15	
A2		20	15	
B1	⌀14	16	15	
B2		20	15	
C	⌀13.5	18	15	
D	M12 x 1.25	40	15	
E	⌀6.5-6.6	70		45

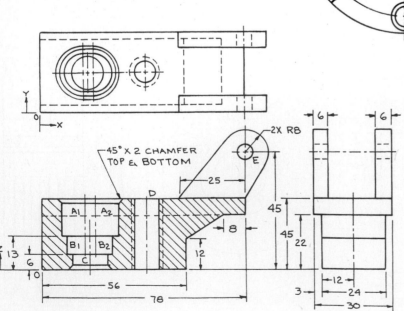

45° X 2 CHAMFER
TOP & BOTTOM

2X R8

25

8

45

45

22

12

12

24

30

3

6   6

13

6

56

78

Unit 9−5
8. 9−5−A

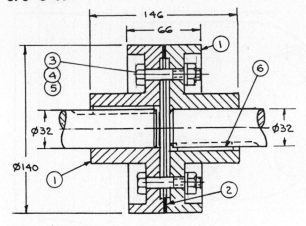

QTY	ITEM		MATL	DESCRIPTION	PT NO.
2	FLANGE		C 1	PATTERN No. XXXX	1
1	GASKET		NEOPRENE	Ø140 X 2	2
4	BOLT	HEX HD	STL	M12 X 1.75 X 50 LG	3
4	NUT	EX-REG	STL	M12	4
4	LOCKWASHER		STL	M12	5
2	KEY	SQ	STL	8 X 8 X 70	6

9−5−B

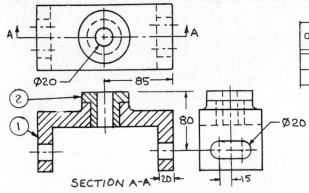

SECTION A-A

QTY	ITEM	MATL	DESCRIPTION	PT NO.
1	BASE	G 1	PATTERN XXXX	1
1	BUSHING −DRILL	XXXX	20 ID X 36 OD X 40	2

9−5−C

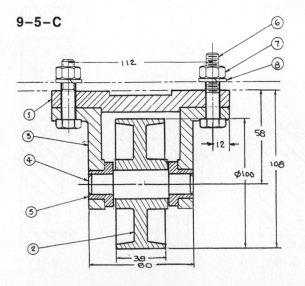

QTY	ITEM		MATL	DESCRIPTION	PT NO.
1	TOP PLATE		G1	PATTERN XXXX	1
1	WHEEL		G1	PATTERN XXXX	2
2	AXLE SUPPORT		G1	PATTERN XXX	3
1	AXLE		STL	Ø22 X 80 LG	4
2	BUSHING-HEAD T.		BRONZE	16 ID X 22 OD X 12	5
4	BOLTS	HEX HD	STL	M10 X 50 LG	6
4	NUT	HEX HD	STL	M10 REG.	7
4	LOCKWASHER		STL	M10	8

Unit 9–5

10. 9–5–E

QTY	ITEM	MATL	DESCRIPTION	PT
1	LINK	MI	PATTERN NO. XXXX	1
1	BUSHING	BRONZE	BOSTON GEAR # XXXX	2
1	BUSHING	BRONZE	BOSTON GEAR # XXXX	3
1	SHAFT SAE	1020	Ø 1.50 X 8.00	4
1	SHAFT SAE	1020	Ø.750 X 12.00	5
4	BOLT, HEX HD	ST.L	.375 NC X 1.25 LG	6
4	NUT, HEX	STL	.375 NC	7
4	WASHER	STL	.375 HELICAL SPRING	8
				9
				10

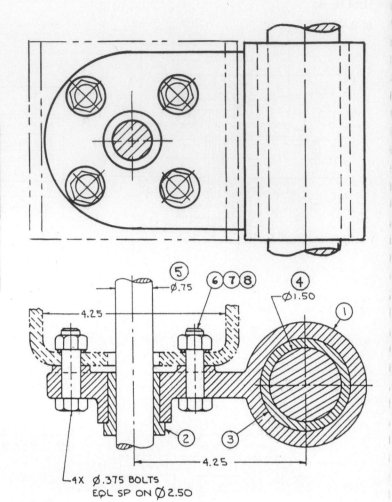

Unit 9–6

11. 9–6–A

9–6–B

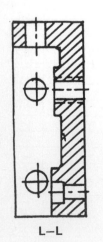 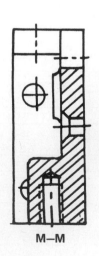 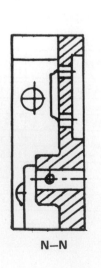 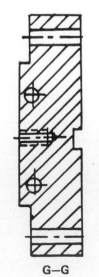 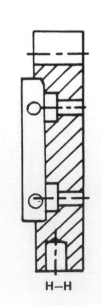 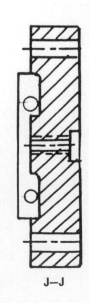

L–L   M–M   N–N   G–G   H–H   J–J

Unit 9–7

12. 9–7–A

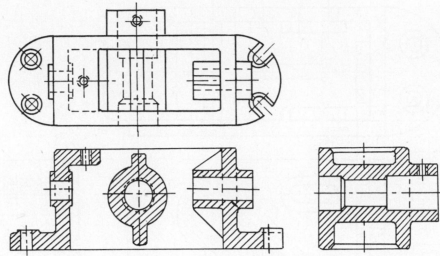

9–7–B

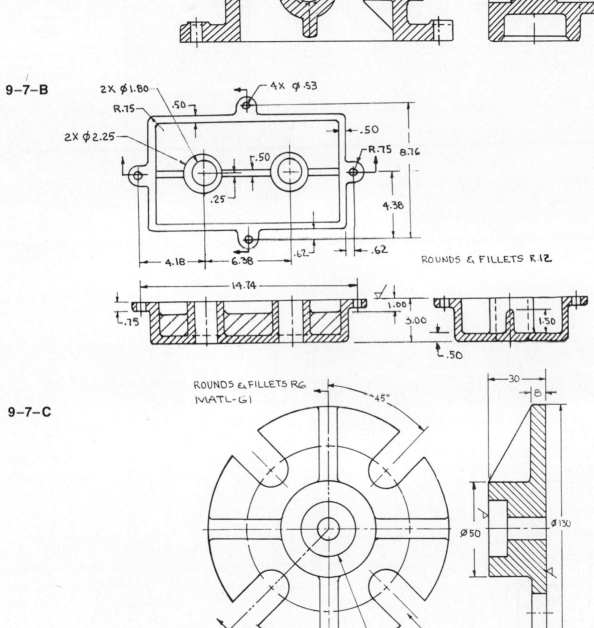

Unit 9–7

12. CONTD

9–7–D

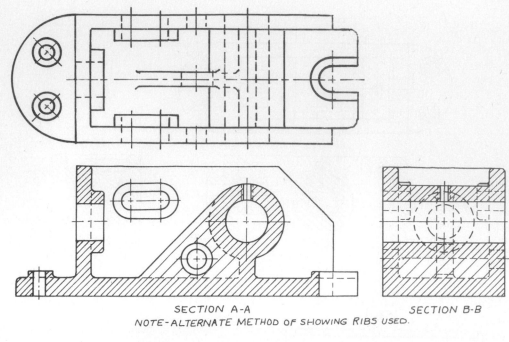

SECTION A-A
NOTE-ALTERNATE METHOD OF SHOWING RIBS USED.

SECTION B-B

9–7–E

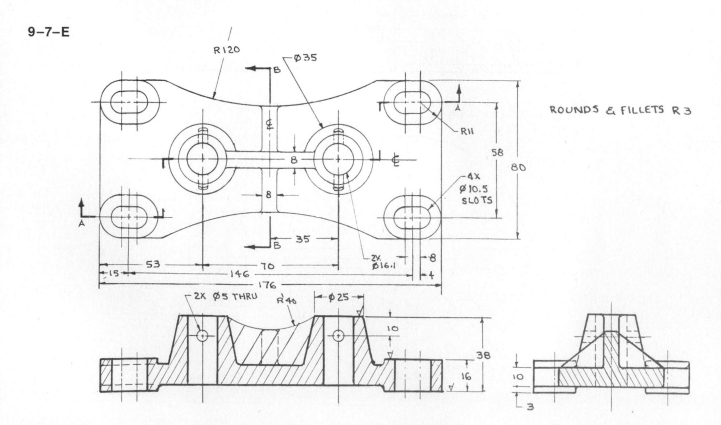

ROUNDS & FILLETS R 3

Unit 9–8

13. 9–8–A

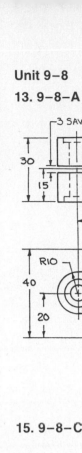
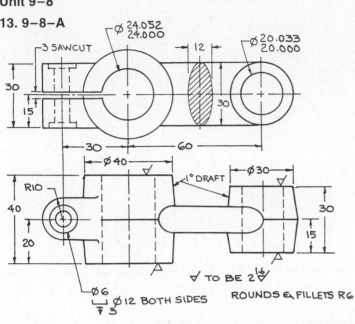

Ø 24.052 / 24.000

3 SAWCUT

12

Ø 20.033 / 20.000

30

15

30

30

60

Ø 40

Ø 30

R10

1° DRAFT

40

20

30

15

∀ TO BE 2 ∀

Ø 6
∓ 3   Ø 12 BOTH SIDES

ROUNDS & FILLETS R6

14. 9–8–B

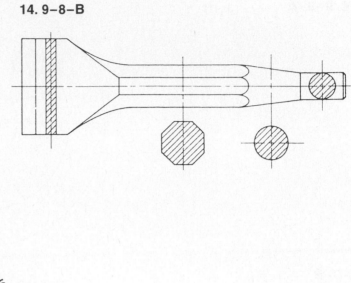

15. 9–8–C

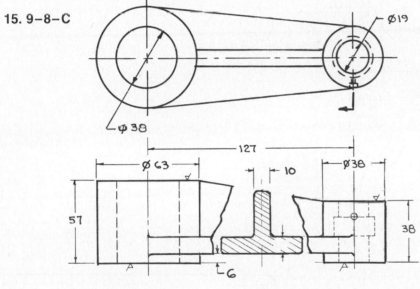

Ø 19

Ø 38

127

Ø 63

10

Ø 38

57

38

6

A   A

ROUNDS & FILLETS R3

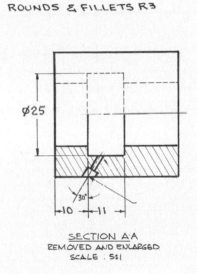

Ø 25

30°

10   11

SECTION A·A
REMOVED AND ENLARGED
SCALE : 5:1

9–8–D

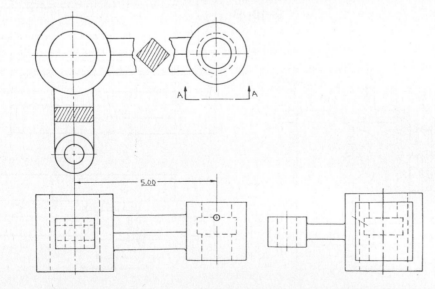

A   A

5.00

.40
30°

VIEW A-A
DOUBLE SIZE

Unit 9−9

16. 9−9−A

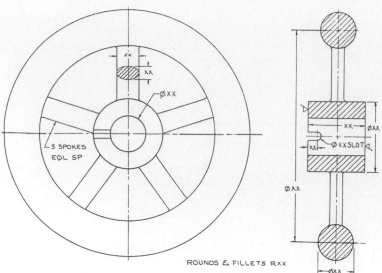

ROUNDS & FILLETS Rxx

5 SPOKES EQL SP

ØXX

9−9−B

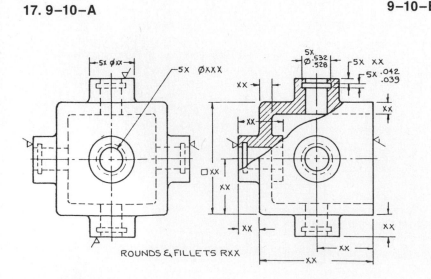

FILLETS Rxx

Unit 9−10

17. 9−10−A

5X ØXX

5X ØXXX

5X Ø.532 / .528

5X XX

5X .042 / .039

□ XX

ROUNDS & FILLETS Rxx

9−10−B

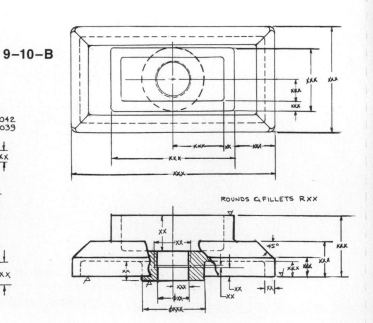

ROUNDS & FILLETS Rxx

45°

Unit 9–11

18. 9–11–A

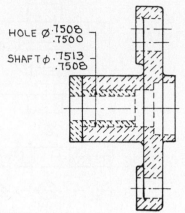

-3x Ø $\frac{24.021}{24.000}$

9–11–B

HOLE Ø $\frac{.7508}{.7500}$

SHAFT Ø $\frac{.7513}{.7508}$

9–11–C

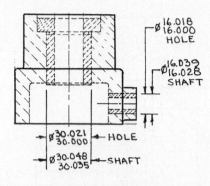

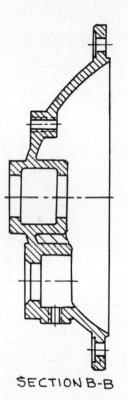

Ø $\frac{16.018}{16.000}$ HOLE

Ø $\frac{16.039}{16.028}$ SHAFT

Ø $\frac{30.021}{30.000}$ ← HOLE

Ø $\frac{30.048}{30.035}$ ← SHAFT

Unit 9–12

19. 9–12–A

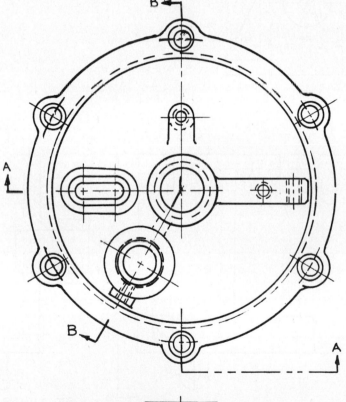

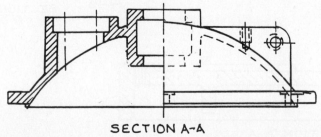

SECTION A-A

SECTION B-B

Unit 9–12
19. CONTD 9–12–B

9–12–C

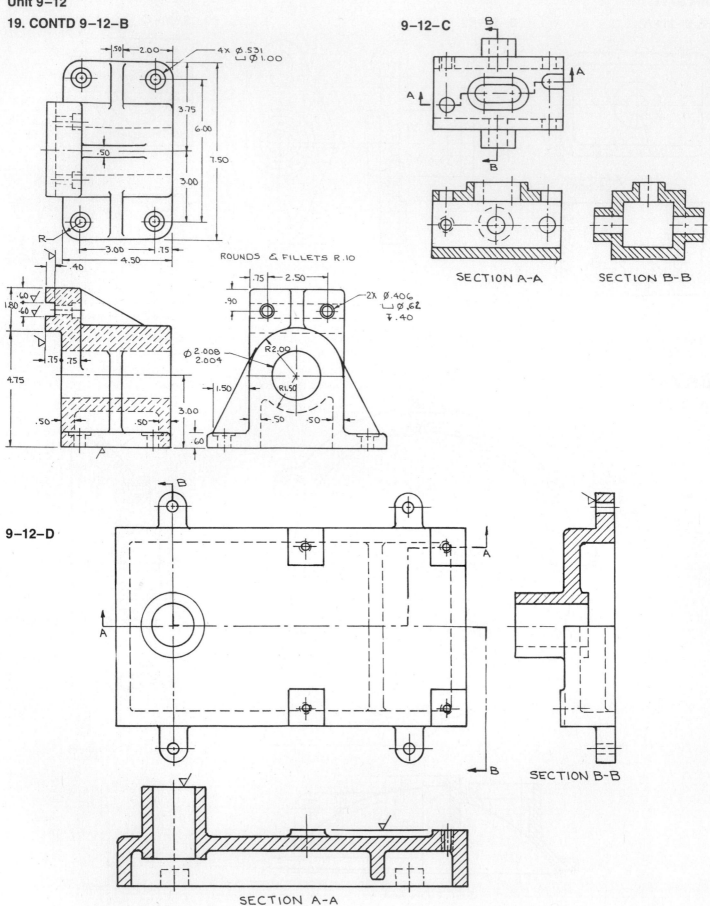

ROUNDS & FILLETS R.10

SECTION A-A   SECTION B-B

9–12–D

SECTION B-B

SECTION A-A

Unit 9–12

19. CONTD 9–12–E

9–12–F

9–12–G

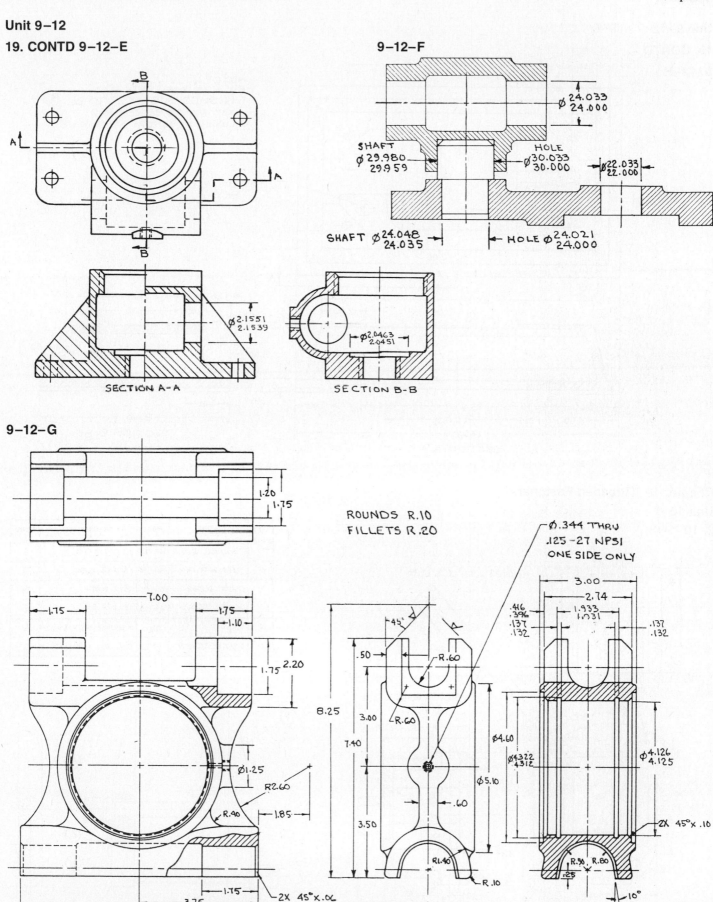

SECTION A-A

SECTION B-B

SHAFT Ø29.980 / 29.959

HOLE Ø30.033 / 30.000

Ø24.033 / 24.000

Ø22.033 / 22.000

SHAFT Ø24.048 / 24.035

HOLE Ø24.021 / 24.000

Ø2.1551 / 2.1539

Ø2.0463 / 2.0451

ROUNDS R.10
FILLETS R.20

Ø.344 THRU
.125 –27 NPSI
ONE SIDE ONLY

Unit 9–12
19. CONTD
9–12–H

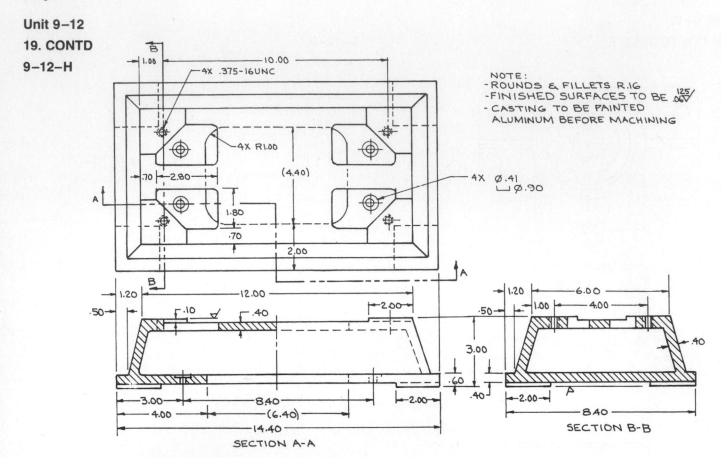

NOTE:
- ROUNDS & FILLETS R.16
- FINISHED SURFACES TO BE $\frac{125}{.06}\sqrt{}$
- CASTING TO BE PAINTED ALUMINUM BEFORE MACHINING

SECTION A-A

SECTION B-B

Chapter 10 Threaded Fasteners
Unit 10–1
1. 10–1–A

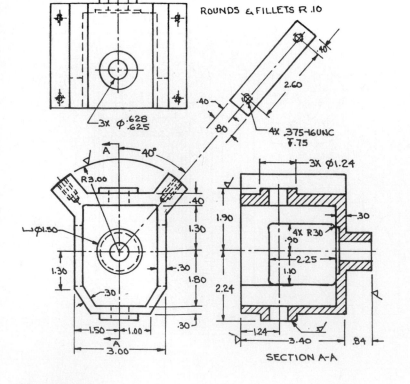

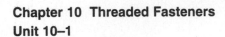

ROUNDS & FILLETS R.10

SECTION A-A

2. 10–1–B

QTY	ITEM	MATL	DESCRIPTION	PT NO.
1	MOVABLE JAW	SAE 1020	12 × 12 × 80	1
1	STATIONARY JAW	SAE 1020	12 × 12 × 80	2
1	OUTER SCREW	SAE 1112	⌀12 × 80	3
1	INNER SCREW	SAE 1112	⌀12 × 90	4
1	CLIP	STL	#6 USS (1.52) × 12 × 27	5
1	MACH. SCREW	STL	RH M6 × 8 LG	6

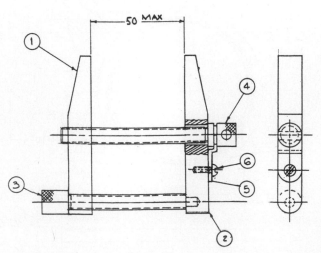

Unit 10–1

3. 10–1–B

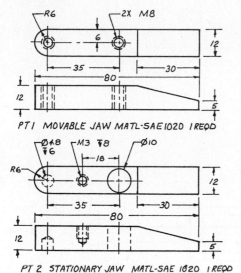

PT 1 MOVABLE JAW MATL-SAE 1020 1 REQD

PT 2 STATIONARY JAW MATL-SAE 1020 1 REQD

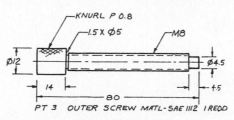

PT 3 OUTER SCREW MATL-SAE 1112 1 REQD

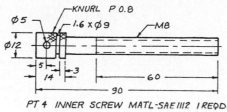

PT 4 INNER SCREW MATL-SAE 1112 1 REQD

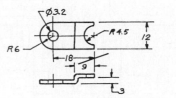

PT 5 CLIP MATL-1.52 (16 USS) STL 1 REQD

4. 10–1–C

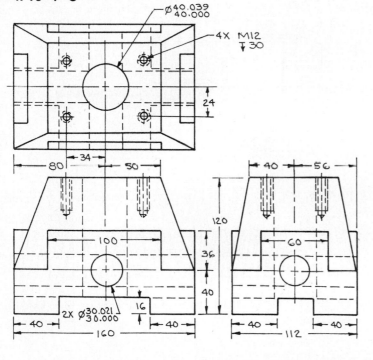

5. 10–1–D

HOLE	HOLE SIZE	LOCATION		
		X	Y	Z
A1	.190-24		0	3.34
A2	.190-24		0	1.78
B1	.500-13 ⊤ .62	-1.00		3.56
B2	.500-13 ⊤ .62	1.00		3.56
B3	.500-13 ⊤ .62	-1.00		1.56
B4	.500-13 ⊤ .62	1.00		1.56
B5	.500-13 ⊤ .62	-1.00		3.56
B6	.500-13 ⊤ .62	1.00		3.56
B7	.500-13 ⊤ .62	-1.00		1.56
B8	.500-13 ⊤ .62	1.00		1.56
C1	Ø .531	-3.12		3.62
C2	Ø .531	-3.12		1.50

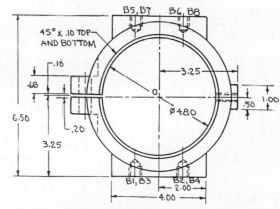

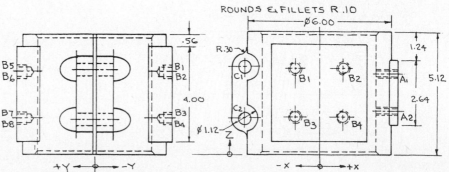

ROUNDS & FILLETS R .10

Unit 10–1

6. 10–1–E

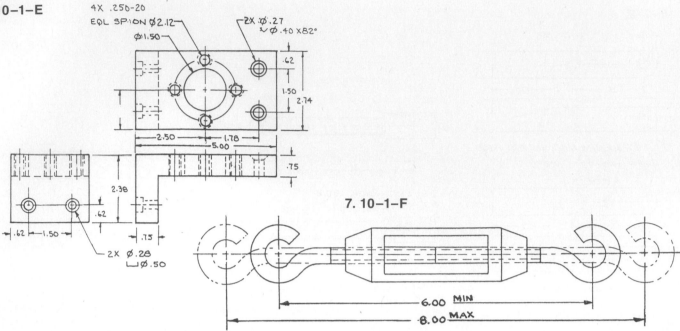

7. 10–1–F

8. 10–1–F

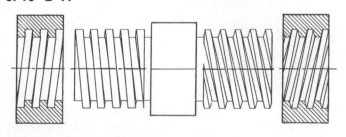

PT 1 SCREW – AS SHOWN   MATL – SAE 1112   1 REQD

PT 2 SCREW AS SHOWN EXCEPT
      WITH LEFT-HAND THREADS   1 REQD

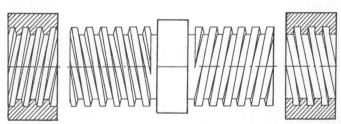

PT 3 TURNBUCKLE   MATL– M1   1 REQD

Unit 10–2

9. 10–2–A

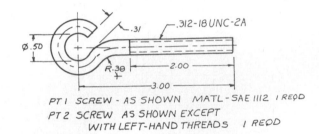

10–2–B

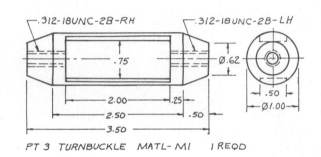

10. 10–2–C

SHARP V TRIPLE THREAD 2.5 PITCH

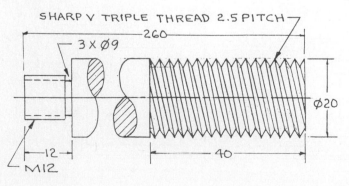

10–2–D

45° x .06

LEFT HAND BUTTRESS THREAD P=.25

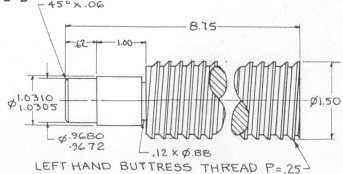

Unit 10–2

11. 10–2–E

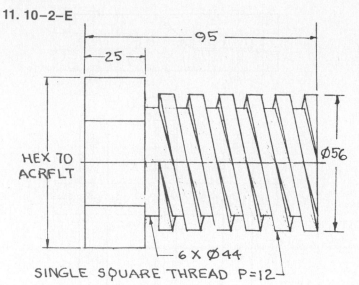

10–2–F

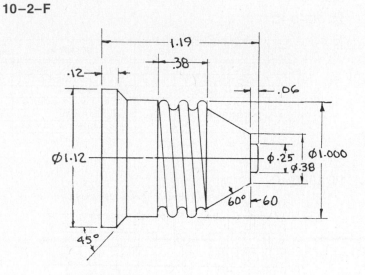

Unit 10–3

12. 10–3–A

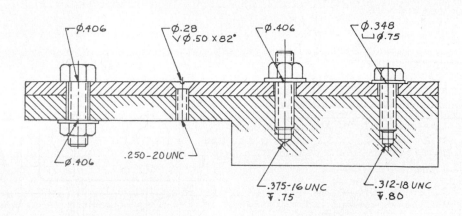

13. 10–3–B

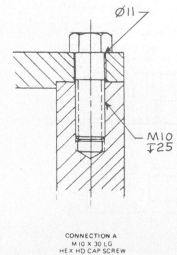

CONNECTION A
M 10 X 30 LG
HEX HD CAP SCREW

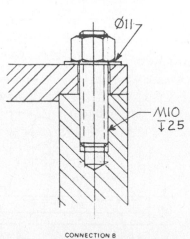

CONNECTION B
M 10 X 40 LG STUD
THREAD EACH END 20 LG
HEX NUT STYLE I AND
SPRING LOCKWASHER

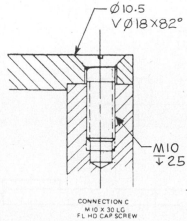

CONNECTION C
M 10 X 30 LG
FL HD CAP SCREW

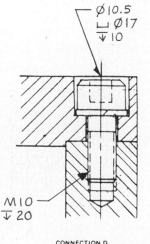

CONNECTION D
M 10 X 1.25 X 25 LG
SOCKET HEAD CAP SCREW
AND SPRING LOCKWASHER

Unit 10–3

14. 10–3–C

QTY	ITEM	MATL	DESCRIPTION	PT NO.
1	SPACER	MST	1.00 X 1.50 X 7.00	1
2	HOOK	MST	.38 X 2.00 X 7.50	2
1	PULLER SCREW	STL	SQ HD CAP SCREW	3
			.625–11 UNC X 6.00 LG	
2	CAP SCREW	STL	SOCKET HD	4
			.312–18 UNC X 1.62 LG	

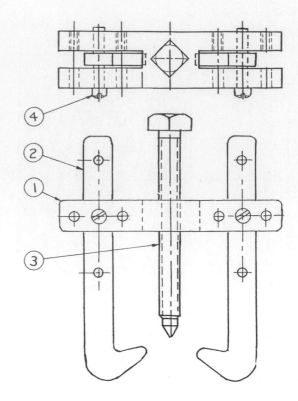

15. 10–3–C

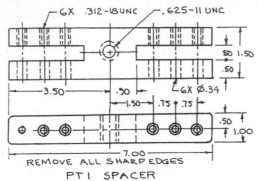

REMOVE ALL SHARP EDGES
PT 1   SPACER

REMOVE ALL SHARP EDGES

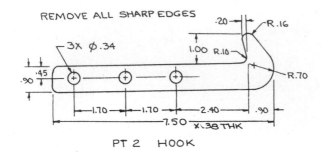

PT 2   HOOK

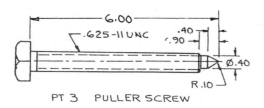

PT 3   PULLER SCREW

SURFACES MARKED ∇ TO BE .04 ∇   250/
ROUNDS & FILLETS R.10

16. 10–3–D

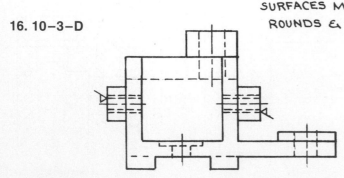

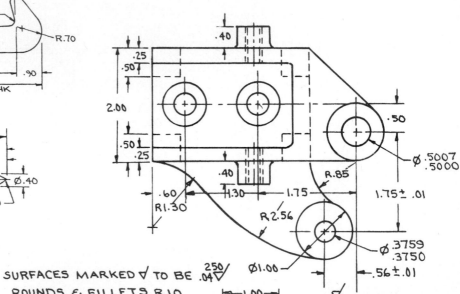

Unit 10–3

17. 10–3–E

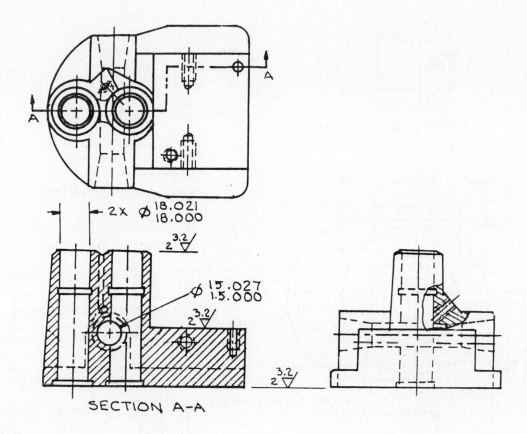

SECTION A-A

Unit 10–4

18. 10–4–A

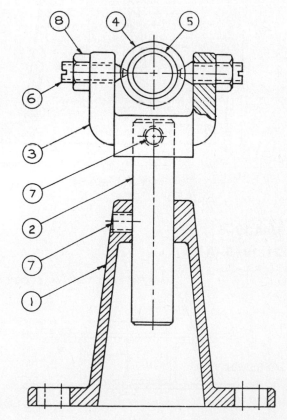

19. 10–4–B

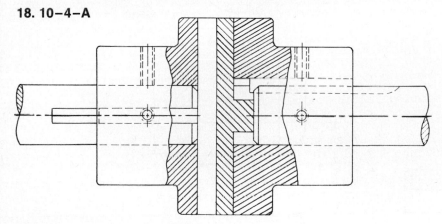

QTY	ITEM	MATL	DESCRIPTION	PT
1	BASE	G1	PATTERN NO XXX	1
1	VERTICAL SHAFT	STL	Ø 20 X 100 LG	2
1	YOKE	G1	PATTERN NO XXX	3
1	BEARING HOUSING	STL	25 ID X 32 OD X 50 LG	4
2	BEARING	BRONZE	20 ID X 25 OD X 20 LG	5
2	SET SCREW	STL	M10 X 30 CON PT SL HD	6
2	SET SCREW	STL	M10 X 10, DOG PT, HEX SOCK.	7
2	JAM NUT	STL	HEX HD, M10	8
				9
				10

Unit 10-4

20. 10-4-B

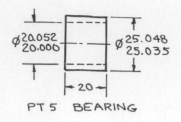

PT 5   BEARING

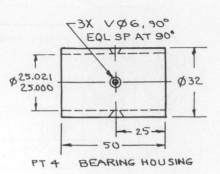

3X ∨Ø6, 90°
EQL SP AT 90°

Ø25.021
25.000

Ø32

25

50

PT 4   BEARING HOUSING

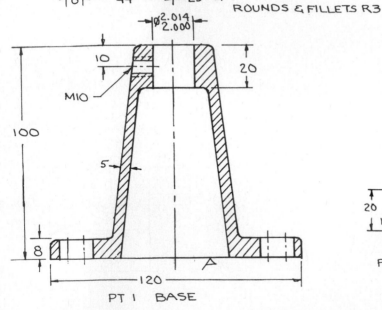

Ø60

Ø38

Ø8 SLOTS

R35

8   44   25

ROUNDS & FILLETS R3

Ø2.014
2.000

10

20

M10

100

5

8

120

PT 1   BASE

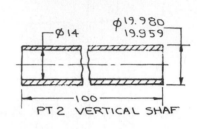

Ø14   Ø19.980
19.959

100

PT 2   VERTICAL SHAF

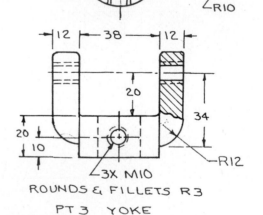

Ø20.014
20.000

R10

12   38   12

20

34

20

10

R12

3X M10

ROUNDS & FILLETS R3

PT 3   YOKE

Unit 10-5

21. 10-5-A

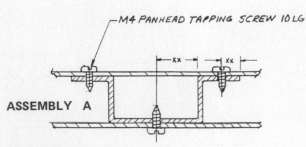

M4 PANHEAD TAPPING SCREW 10 LG

xx   xx

ASSEMBLY A

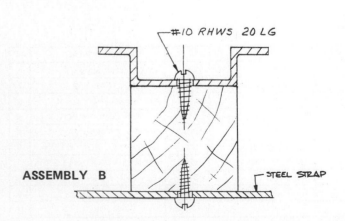

#10 RHWS 20 LG

STEEL STRAP

ASSEMBLY B

Unit 10-5

22. 10-5-B

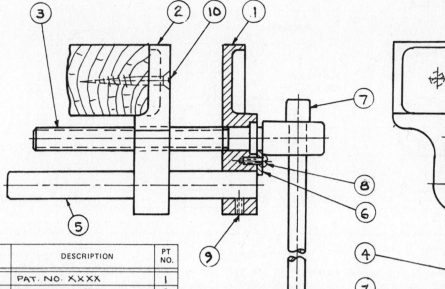

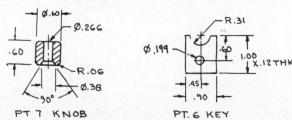

QTY	ITEM	MATL	DESCRIPTION	PT NO.
1	FRONT JAW	C1	PAT. NO. XXXX	1
2	REAR JAW	C1	PAT. NO. XXXX	2
1	SCREW	MST	Ø1.00 X 8.00	3
1	HANDLE	MST	Ø.44 X 7.00	4
1	BEAM	MST	Ø.750 X 7.00	5
1	KEY	MST	.12 X 1.00 X .90	6
2	KNOB	MST	Ø.625 X .60	7
1	SCREW-RHMS	STL	10-24 UNC X .50	8
1	SET SCREW	STL	.250-20 UNC X .50 SOC.HD	9
2	WOOD SCREW	STL	#12 X 2.00 FL HD	10

23. 10-5-B

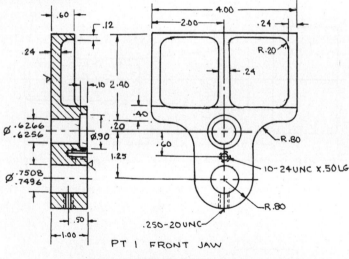

PT 1 FRONT JAW

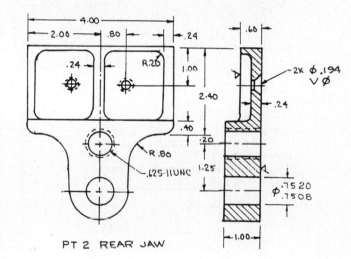

PT 2 REAR JAW

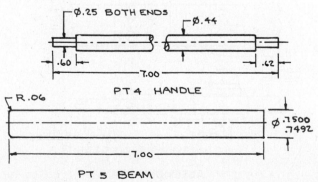

PT 4 HANDLE

PT 5 BEAM

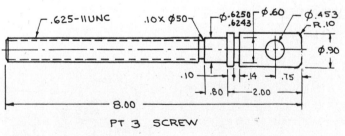

PT 3 SCREW

PT 7 KNOB

PT. 6 KEY

Chapter 11 Miscellaneous Types of Fasteners
Unit 11–1,   1. 11–1–A

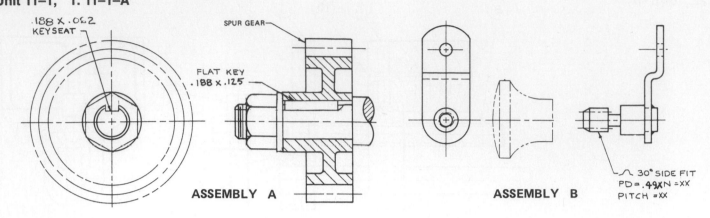

.188 X .062
KEYSEAT

SPUR GEAR

FLAT KEY
.188 X .125

ASSEMBLY A

ASSEMBLY B

30° SIDE FIT
PD = .49 N = XX
PITCH = XX

11–1–B

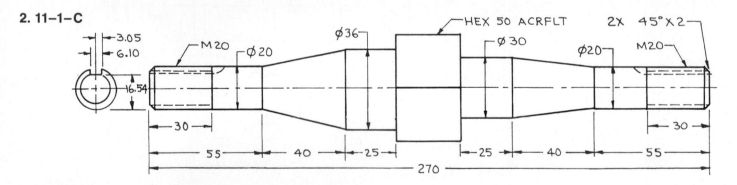

5 X 2.5
KEYSEAT

Ø 150
V-BELT PULLEY

SQUARE KEY
5 X 5 X 40

5 X 2.5

Ø 16
STEPPED
SHAFT

Ø 50

ASSEMBLY A

GEAR

808 WOODRUFF

Ø 36 FLAT WASHER

Ø 26 TAPERED SHAFT

30 A/F HEX NUT

808 WOODRUFF KEYSEAT

M 20

ASSEMBLY B

2. 11–1–C

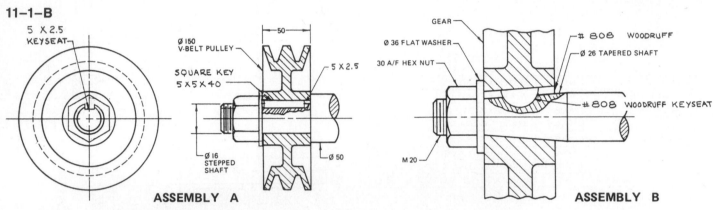

3.05
6.10
16.54

M 20

Ø 20

Ø 36

HEX 50 ACRFLT

Ø 30

2X   45° X 2

Ø 20

M 20

30

55

40

25

25

40

55

30

270

Unit 11–2

3. 11–2–A

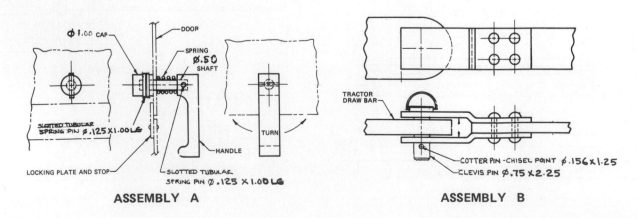

Ø 1.00 CAP

DOOR

SPRING

Ø .50
SHAFT

SLOTTED TUBULAR
SPRING PIN Ø .125 X 1.00 LG

LOCKING PLATE AND STOP

SLOTTED TUBULAR
SPRING PIN Ø .125 X 1.00 LG

HANDLE

TURN

ASSEMBLY A

TRACTOR
DRAW BAR

COTTER PIN - CHISEL POINT Ø .156 X 1.25
CLEVIS PIN Ø .75 X 2.25

ASSEMBLY B

Unit 11–2

3. CONTD 11–2–B

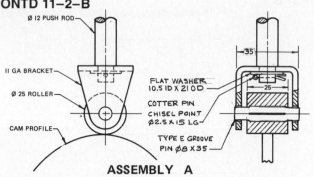

ASSEMBLY A

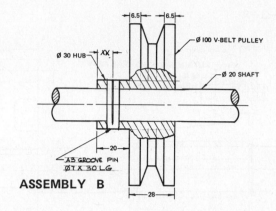

ASSEMBLY B

4. 11–2–C

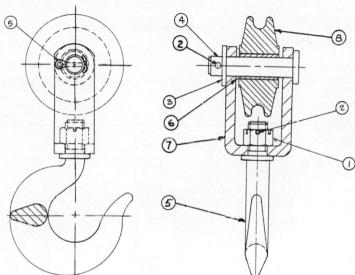

QTY	ITEM	MATL	DESCRIPTION	PT
1	LOCKNUT-SLOTTED	STL	M12 STYLE 1	1
2	COTTER PIN	STL	Ø3 MITRE END 25 LG	2
1	WASHER – FLAT	STL	M12	3
1	CLEVIS PIN	STL	Ø12 X 50 LG	4
1	HOOK	STL	FORGING # XXXX	5
1	BEARING	TEFLON	16 ID X 22 OD X 24 LG	6
1	FRAME	STL	DWG NO. XXXX	7
1	PULLEY	GI	DWG NO XXXX	8

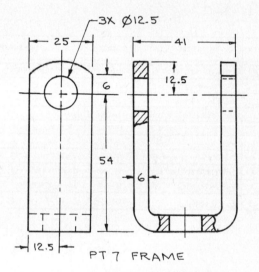

PT 7 FRAME

5. 11–2–C

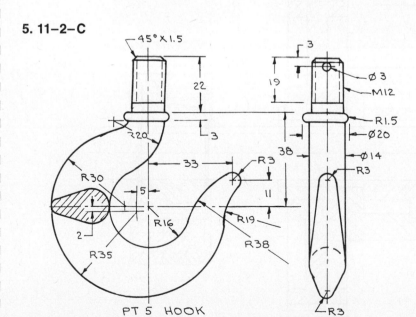

PT 5 HOOK

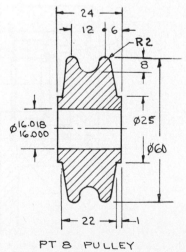

PT 8 PULLEY

Unit 11–3

6. 11–3–A

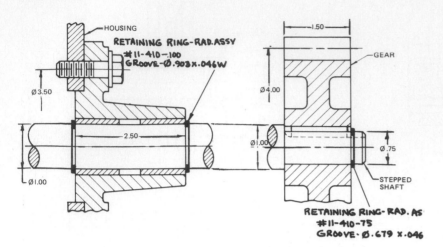

ASSEMBLY A   ASSEMBLY B

11–3–B

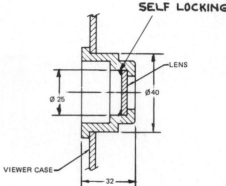

ASSEMBLY A

ASSEMBLY B

7. 11–3–C

QTY	ITEM	MATL	DESCRIPTION	PT NO.
2	RETAINING RING	WALDES	EX. 5100-100	1
2	RETAINING RING	WALDES	RADIAL 11-410-100	2
2	RETAINING RING	WALDES	INT. N5000-200	3
1	RETAINING RING	WALDES	RADIAL 11-410-200	4
1	KEY SQ	STL	.25 X .25 X 1.25	5
1	KEY SQ	STL	.25 X .25 X 1.75	6
1	KEY SQ	STL	.375 X .375 X 1.00	7
1	SETSCREW	STL	SLOTTED Ø.25 X .25	8
1	BEARING	SKF	#6005	9
1	O-RING	RUBBER	Ø.125 X 1.811 ID	10

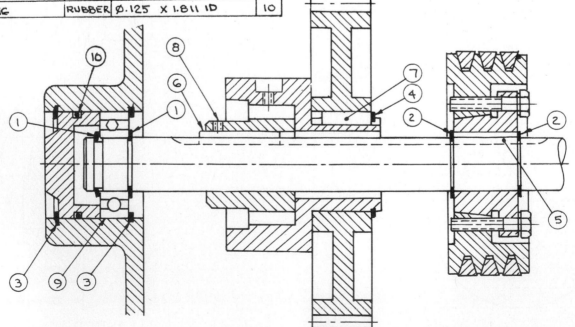

Unit 11-3

8. 11-3-C

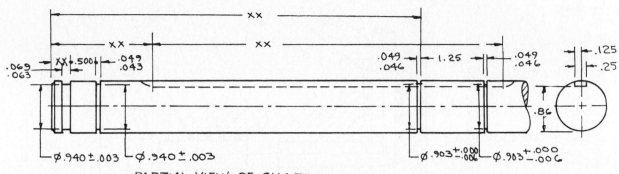

PARTIAL VIEW OF SHAFT

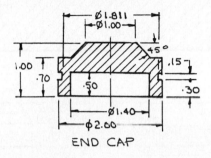

END CAP

Unit 11-4

9. 11-4-A

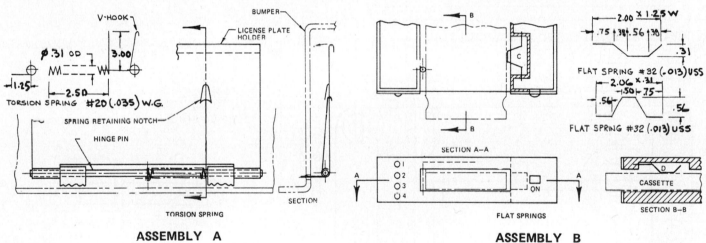

ASSEMBLY A

ASSEMBLY B

11-4-B

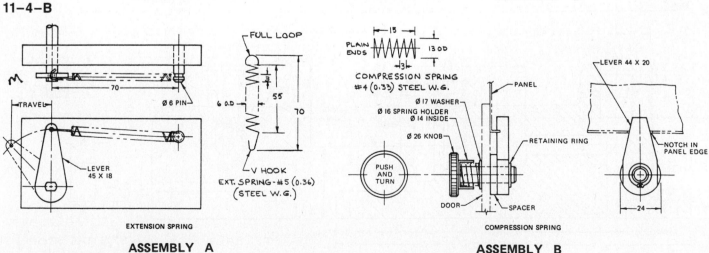

ASSEMBLY A

ASSEMBLY B

Unit 11–4

10. 11–4–C

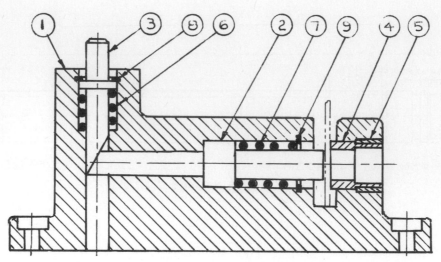

QTY	ITEM	MATL	DESCRIPTION	PT NO.
1	BASE	G1	PATTERN XXXX	1
1	PUNCH	STL	Ø.75 X 3.60	2
1	PLUNGER	STL	Ø.625 X 2.20	3
1	DIE	STL	Ø.783 X .26	4
1	DIE STOP	STL	Ø.875 X .50	5
1	SPRING	STL	SEE DWG	6
1	SPRING	STL	SEE DWG	7
1	RETAINING RING	WALDES	INT. RAD N5000-62	8
1	RETAINING RING	WALDES	INT. RAD N5000-75	9

11. 11–4–C

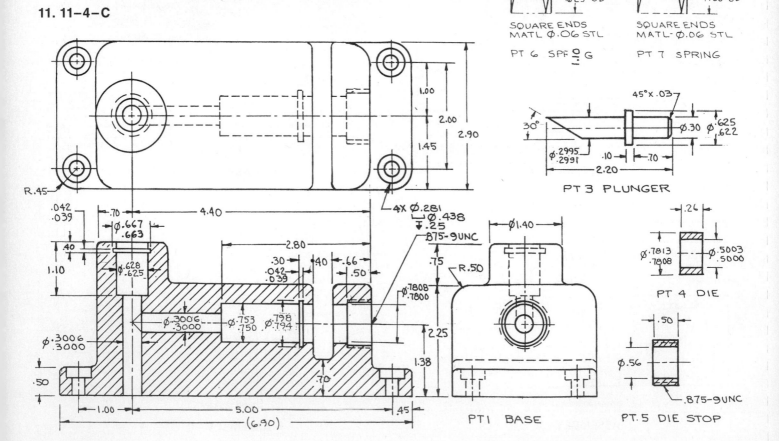

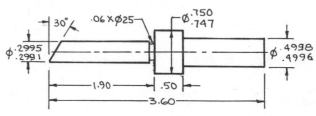

Unit 11–5

12. 11–5–A

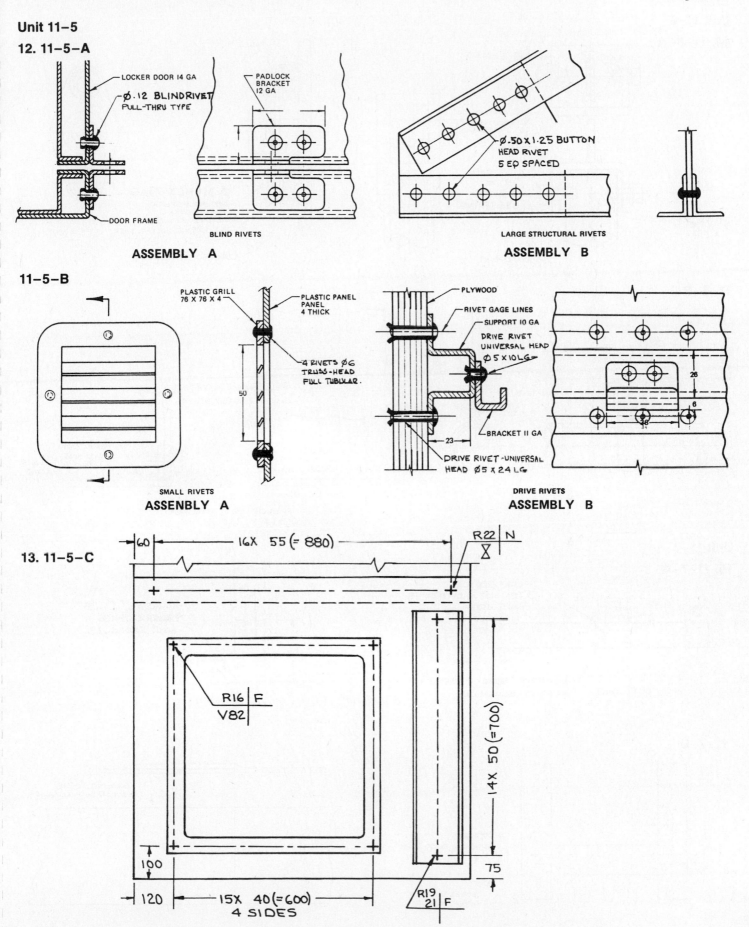

LOCKER DOOR 14 GA

Ø.12 BLINDRIVET
PULL-THRU TYPE

DOOR FRAME

PADLOCK
BRACKET
12 GA

BLIND RIVETS

ASSEMBLY A

Ø.50 X 1.25 BUTTON
HEAD RIVET
5 EQ SPACED

LARGE STRUCTURAL RIVETS

ASSEMBLY B

11–5–B

PLASTIC GRILL
76 X 76 X 4

PLASTIC PANEL
PANEL
4 THICK

4 RIVETS Ø6
TRUSS-HEAD
FULL TUBULAR

50

SMALL RIVETS

ASSENBLY A

PLYWOOD

RIVET GAGE LINES

SUPPORT 10 GA

DRIVE RIVET
UNIVERSAL HEAD
Ø 5 X 10 LG

BRACKET 11 GA

DRIVE RIVET –UNIVERSAL
HEAD Ø5 X 24 LG

23

26

6

DRIVE RIVETS

ASSEMBLY B

13. 11–5–C

60

16X 55 (= 880)

R22 N

R16 F
V82

14X 50 (=700)

100

120

15X 40 (=600)
4 SIDES

75

R19
21 F

Unit 11–6

14. 11–6–A

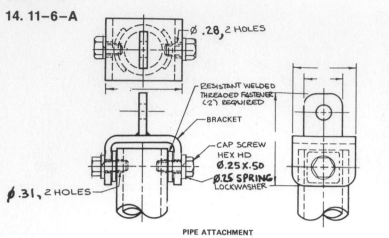

Ø.28, 2 HOLES

RESISTANT WELDED
THREADED FASTENER
(2) REQUIRED

BRACKET

CAP SCREW
HEX HD
Ø.25 X .50

Ø.25 SPRING
LOCKWASHER

Ø.31, 2 HOLES

PIPE ATTACHMENT

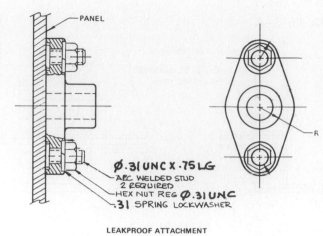

PANEL

Ø.31 UNC X .75 LG
ARC WELDED STUD
2 REQUIRED
HEX NUT REG Ø.31 UNC
.31 SPRING LOCKWASHER

R

LEAKPROOF ATTACHMENT

11–6–B

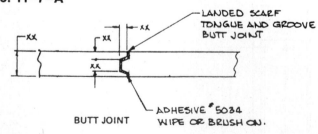

PANEL

CLAMP 16 GA

22 OD

SPOTWELD
NUT
'B' TYPE

TAB ATTACHMENT

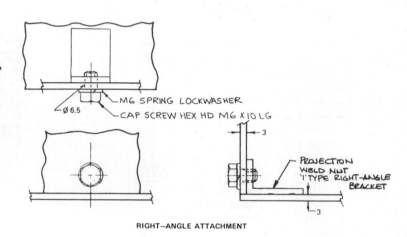

Ø 6.5

M6 SPRING LOCKWASHER

CAP SCREW HEX HD M6 X 10 LG

3

PROJECTION
WELD NUT
'I' TYPE RIGHT-ANGLE
BRACKET

3

RIGHT–ANGLE ATTACHMENT

Unit 11–7

15. 11–7–A

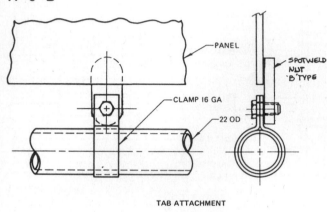

LANDED SCARF
TONGUE AND GROOVE
BUTT JOINT

XX

XX

XX

BUTT JOINT

ADHESIVE #5034
WIPE OR BRUSH ON.

XX

ADHESIVE # 2210
BRUSH OR WIPE
BOTH SURFACES.

XX

XX

SLIP JOINT

11–7–B

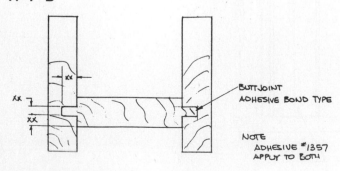

XX

XX

XX

BUTT JOINT
ADHESIVE BOND TYPE

NOTE
ADHESIVE #1357
APPLY TO BOTH

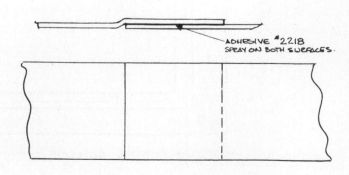

ADHESIVE #2218
SPRAY ON BOTH SURFACES.

Unit 11–8

16. 11–8–A

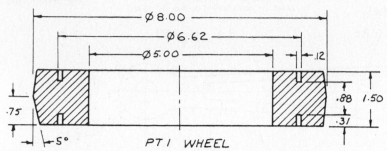

PT 1   WHEEL

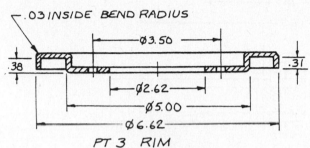

PT 3   RIM

FILLETS R.12

6X Ø.39
EQL SP ON Ø3.50

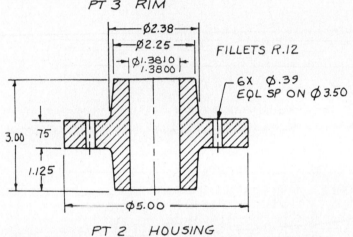

PT 2   HOUSING

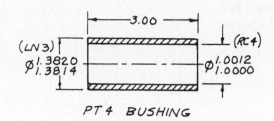

PT 4   BUSHING

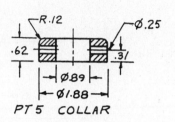

PT 5   COLLAR

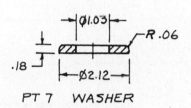

PT 7   WASHER

QTY	ITEM	MATL	DESCRIPTION	PT
1	WHEEL	ABS	Ø8.00 X 1.50	1
1	HOUSING	G1.	PATTERN XXXX	2
2	RIM	STL	STAMPING XXXX	3
1	BUSHING	B05-EF-1	1.00 ID X 1.38 OD X 3.00 LG	4
1	COLLAR	STL	Ø1.88 X .62	5
1	SPRING PIN	STL	Ø.25 X 2.00 LG	6
2	WASHER	STL	1.0 ID X 2.12 OD X .18	7
6	RIVET · BUTTON HD	ALUM.	Ø.31 X 1.00 LG	8

17. 11–8–B

QTY	ITEM	MATL	DESCRIPTION	PT NO.
2	FORK	G1	PATTERN XXX	1
1	RING	STL	.62 X 1.00 X 1.00	2
2	SPRING PIN	STL	Ø.25 X 1.00	3
4	SCREW-FH	STL	SLOTTED	4
			.250-20 X .62	

Chapter 12 Manufacturing Materials
Unit 12–1
1. 12–1–A

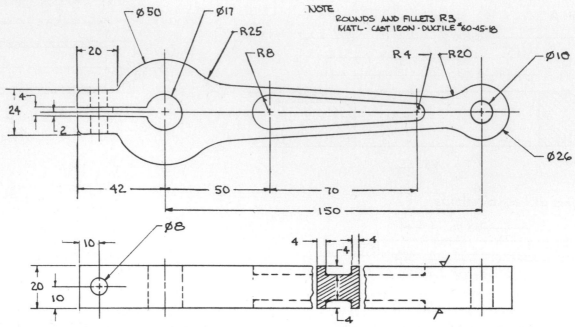

12–1–B

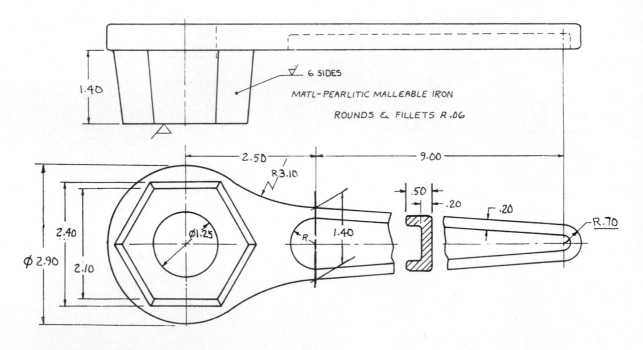

Unit 12–2
2. 12–2–A

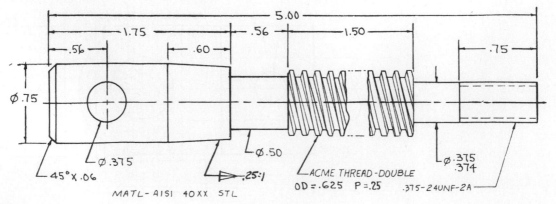

Unit 12-2
2. CONTD 12-2-B

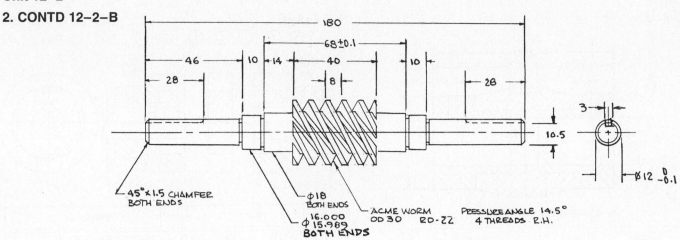

Unit 12-3
3. 12-3-B

4. 12-3-C

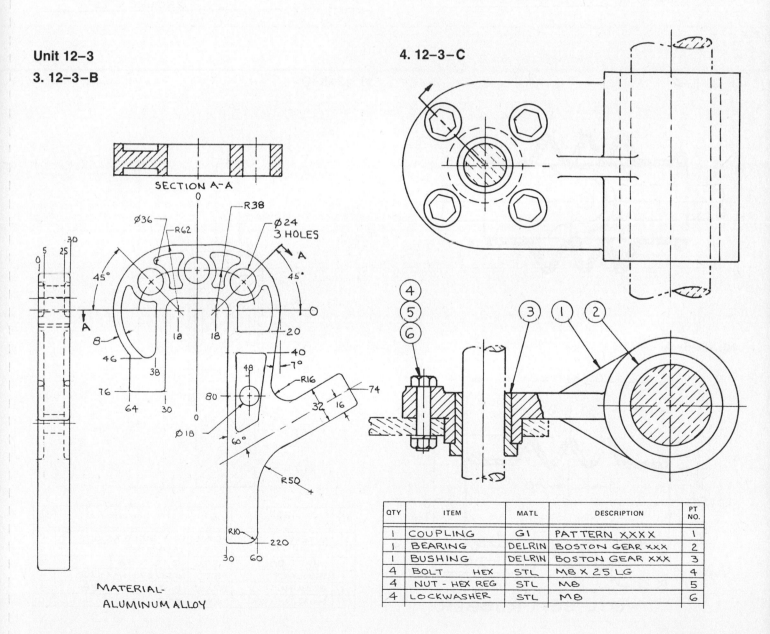

MATERIAL-
ALUMINUM ALLOY

QTY	ITEM	MATL	DESCRIPTION	PT NO.
1	COUPLING	GI	PATTERN XXXX	1
1	BEARING	DELRIN	BOSTON GEAR XXX	2
1	BUSHING	DELRIN	BOSTON GEAR XXX	3
4	BOLT HEX	STL	M8 X 25 LG	4
4	NUT - HEX REG	STL	M8	5
4	LOCKWASHER	STL	M8	6

Unit 12–4

7. 12–4–A

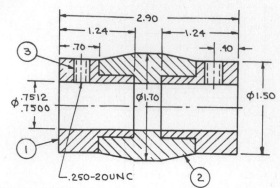

QTY	ITEM	MATL	DESCRIPTION	PT NO.
2	HUB	STL	Ø1.50 X 1.24	1
1	BOOT	MECH. RUBBER	MOLD XXX	2
2	SETSCREW-HDLESS STL		.250 X .38 LG HEX SOC	3

8. 12–4–B

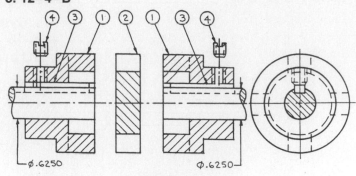

QTY	ITEM	MATL	DESCRIPTION	PT NO.
2	HUB CAST	STL	PATTERN XXXX	1
1	INSERT	RUBBER	MOLD XXXX	2
2	KEY	STL	□ .188 X 1.25	3
2	SETSCREW	STL	SLOTTED HEADLESS	4
			.250 UNC X .25	

9. 12–5–A

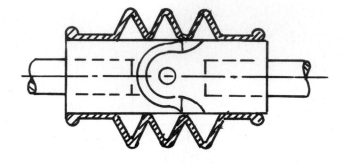

10. 12–5–A

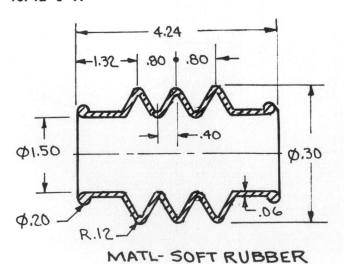

MATL- SOFT RUBBER

11. 12–5–B

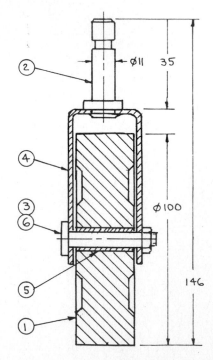

QTY	ITEM	MATL	DESCRIPTION	PT NO.
1	WHEEL	HD RUBBER	MOLD XXXX	1
1	POST	STL	Ø 19 X 48	2
1	SHAFT BOLT	STL	Ø 16 X 38	3
1	BRACKET	STL	#13 (2.29) X 44 X 168	4
1	BUSHING	BRASS	8 ID X 11 OD X 28	5
1	NUT HEX	STL	M6	6

Chapter 13 Forming Processes
Unit 13–1
1. 13–1–A

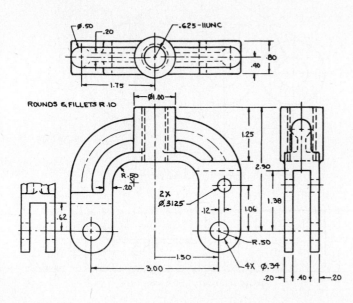

2. 13–1–B

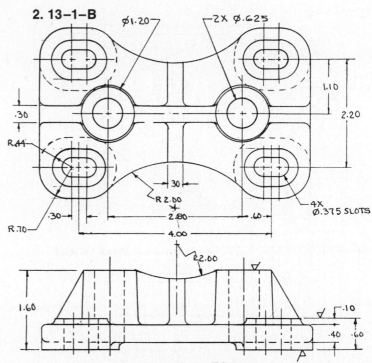

NOTE: - ROUNDS & FILLETS R.16
- DRAFT ANGLES 1° FOR EXTERNAL SURFACES
AND 2° FOR INTERNAL SURFACES

3. 13–1–C

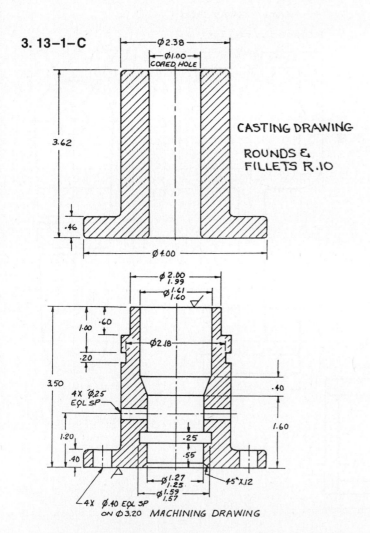

CASTING DRAWING

ROUNDS &
FILLETS R.10

MACHINING DRAWING

4. 13–1–D

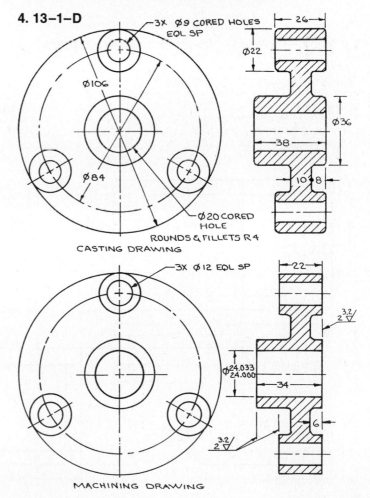

ROUNDS & FILLETS R 4
CASTING DRAWING

MACHINING DRAWING

Unit 13–1

5. 13–1–E

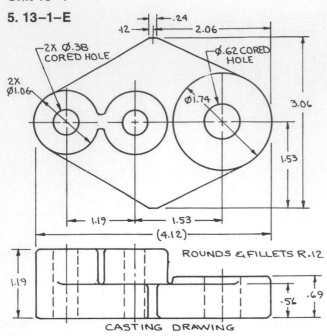

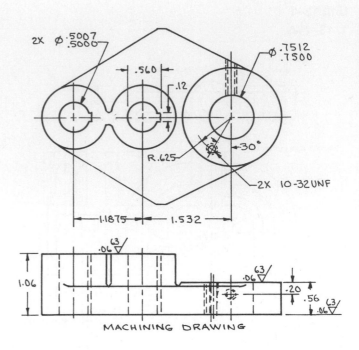

MATL – GRAY IRON

ROUNDS & FILLETS R5

6. 13–1–F

13–1–G

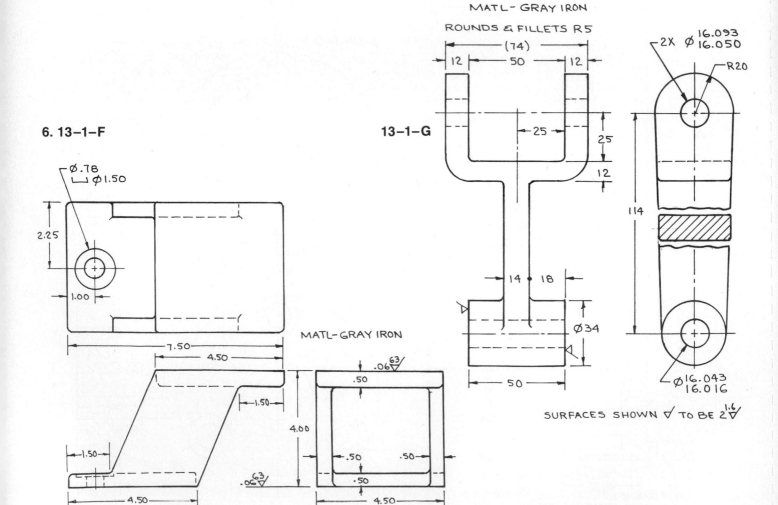

MATL– GRAY IRON

SURFACES SHOWN ∇ TO BE 2∇

Unit 13–1

6. CONTD 13–1–H

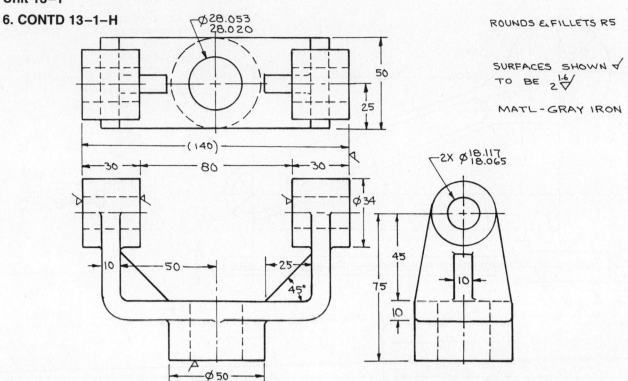

ROUNDS & FILLETS R5

SURFACES SHOWN ∀
TO BE 2 $\frac{1.6}{\nabla}$

MATL – GRAY IRON

Unit 13–2

7. 13–2–A

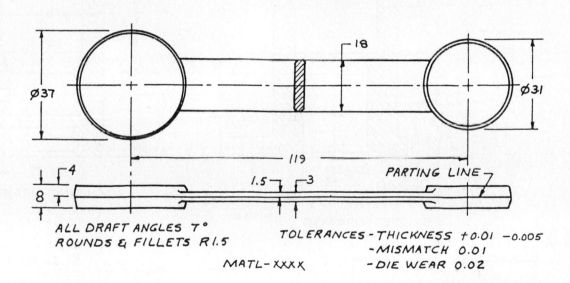

ALL DRAFT ANGLES 7°
ROUNDS & FILLETS R1.5

MATL– XXXX

TOLERANCES – THICKNESS +0.01 –0.005
– MISMATCH 0.01
– DIE WEAR 0.02

13–2–B

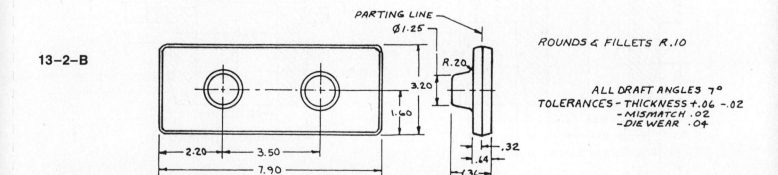

ROUNDS & FILLETS R.10

ALL DRAFT ANGLES 7°
TOLERANCES – THICKNESS +.06 –.02
– MISMATCH .02
– DIE WEAR .04

Unit 13–2

8. 13–2–C

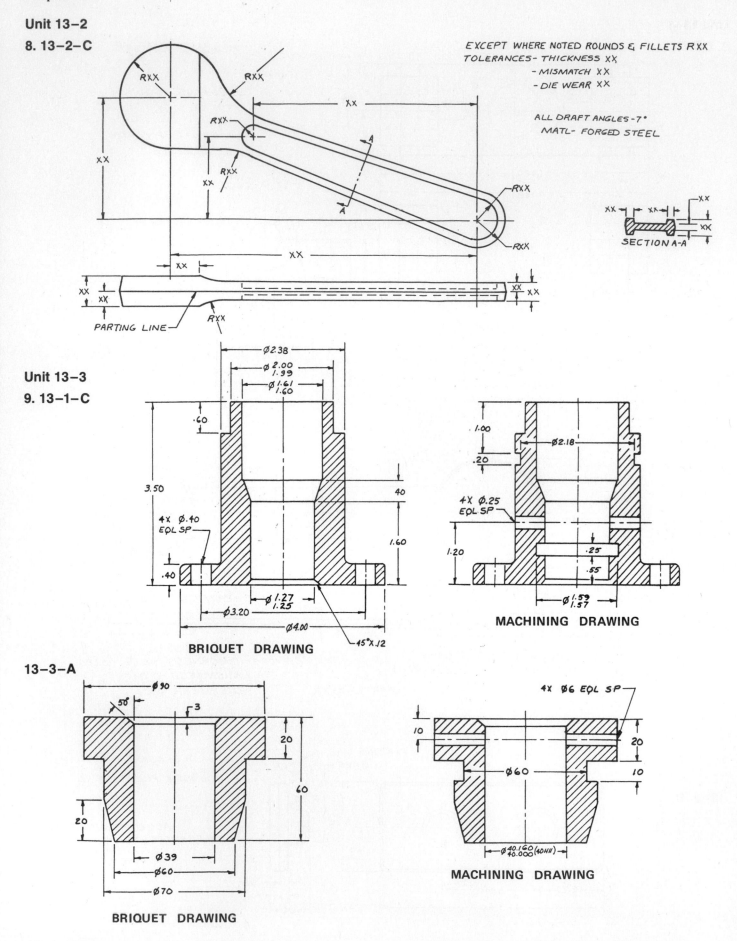

EXCEPT WHERE NOTED ROUNDS & FILLETS R XX
TOLERANCES- THICKNESS XX
– MISMATCH XX
– DIE WEAR XX

ALL DRAFT ANGLES-7°
MATL- FORGED STEEL

SECTION A-A

PARTING LINE

Unit 13–3

9. 13–1–C

BRIQUET DRAWING

MACHINING DRAWING

13–3–A

BRIQUET DRAWING

MACHINING DRAWING

Unit 13-3

9. CONTD

13-3-B

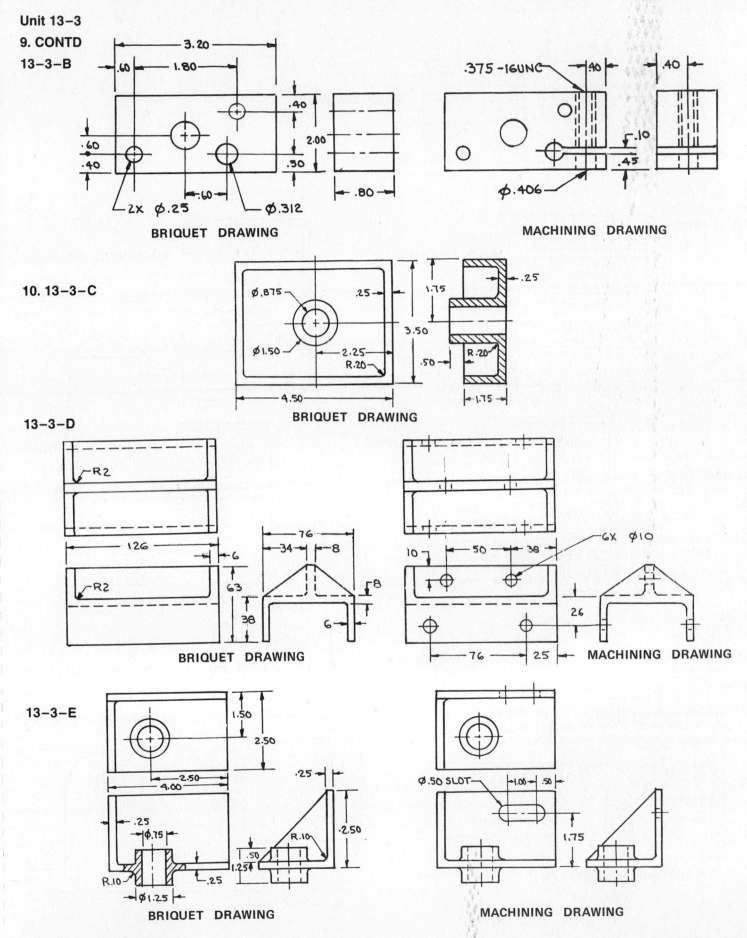

BRIQUET DRAWING

MACHINING DRAWING

10. 13-3-C

BRIQUET DRAWING

13-3-D

BRIQUET DRAWING

MACHINING DRAWING

13-3-E

BRIQUET DRAWING

MACHINING DRAWING

Unit 13−4

11. 13−3−C

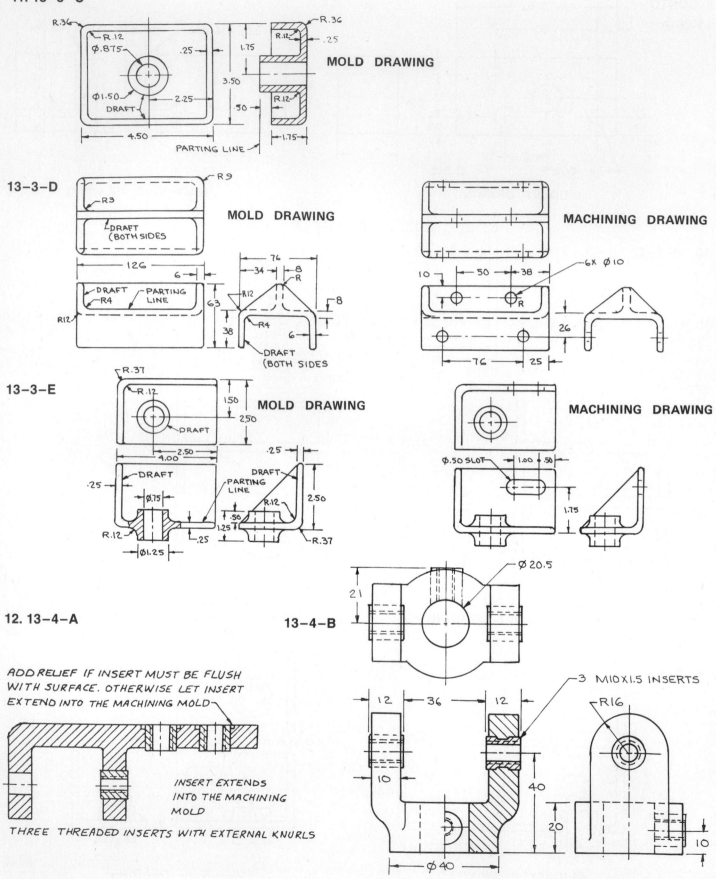

MOLD DRAWING

PARTING LINE

13−3−D

MOLD DRAWING

MACHINING DRAWING

13−3−E

MOLD DRAWING

MACHINING DRAWING

12. 13−4−A

ADD RELIEF IF INSERT MUST BE FLUSH
WITH SURFACE. OTHERWISE LET INSERT
EXTEND INTO THE MACHINING MOLD

INSERT EXTENDS
INTO THE MACHINING
MOLD

THREE THREADED INSERTS WITH EXTERNAL KNURLS

13−4−B

3 M10X1.5 INSERTS

Unit 13–4

12. CONTD 13–4–C

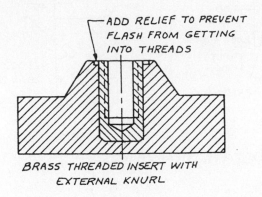

ADD RELIEF TO PREVENT FLASH FROM GETTING INTO THREADS

BRASS THREADED INSERT WITH EXTERNAL KNURL

13–4–D

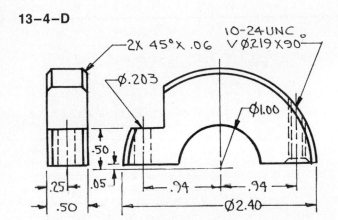

2X 45°X .06

Ø.203

10-24UNC
∇ Ø219 X 90°

Ø1.00

.50

.25 .05

.50

.94   .94

Ø2.40

13. 13–4–E

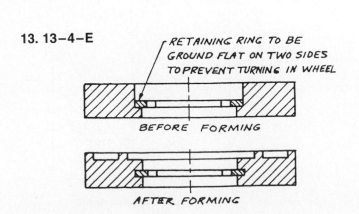

RETAINING RING TO BE GROUND FLAT ON TWO SIDES TO PREVENT TURNING IN WHEEL

BEFORE FORMING

AFTER FORMING

Chapter 14 Detail and Assembly Drawings
Unit 14–2,   1. 14–2–A

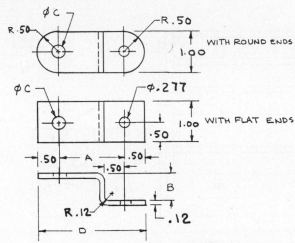

ØC   R.50   R.50

WITH ROUND ENDS

1.00

ØC   Ø.277

WITH FLAT ENDS

1.00

.50

.50   A   .50

.50

R.12

B

.12

D

PT NO	A	B	C	D	ROUND ENDS	SQUARE ENDS
1	2.50	.50	.277	3.50	✓	
2	1.25	.75	.562	2.25		✓
3	1.10	.50	.386	2.10	✓	
4	1.00	.62	.50	2.00		✓

2. 14–2–B

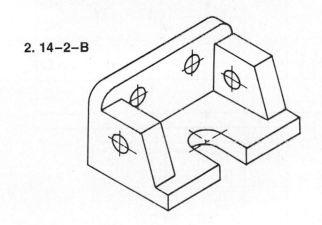

3. 14–2–C

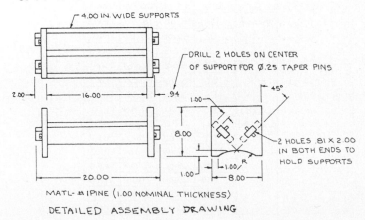

4.00 IN. WIDE SUPPORTS

DRILL 2 HOLES ON CENTER OF SUPPORT FOR Ø.25 TAPER PINS

2.00   16.00   .94

1.00

8.00

45°

2 HOLES .81 X 2.00 IN BOTH ENDS TO HOLD SUPPORTS

1.00 R

20.00

1.00   8.00

MATL- #1 PINE (1.00 NOMINAL THICKNESS)

DETAILED ASSEMBLY DRAWING

Unit 14–2

3. 14–2–C CONTD

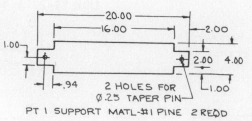

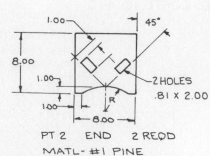

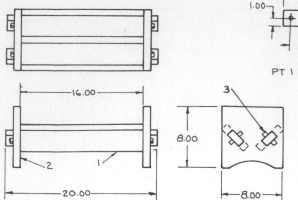

PT 1 SUPPORT MATL–#1 PINE 2 REQD

2 HOLES FOR
Ø.25 TAPER PIN

PT 2   END   2 REQD
MATL– #1 PINE

ASSEMBLY AND DETAIL DRAWINGS

4. 14–2–D

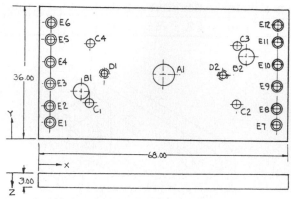

SIZE	SYMBOL	FROM X	FROM Y	FROM Z
Ø 4.015 / 4.000	A1	34.00	18.00	THRU
Ø3.00	B1	11.50	13.00	THRU
	B2	56.50	23.00	THRU
Ø.625	C1	14.00	10.00	THRU
	C2	54.00	10.00	THRU
	C3	54.00	26.00	THRU
	C4	14.00	26.00	THRU
.750–10 UNC	D1	18.00	18.00	THRU
	D2	50.00	18.00	THRU
	E1	3.00	4.50	THRU
	E2	3.00	9.00	THRU
	E3	3.00	15.00	THRU
	E4	3.00	21.00	THRU
Ø1.281	E5	3.00	27.00	THRU
⌴ Ø2.50	E6	3.00	31.50	THRU
↧ 1.20	E7	65.00	4.50	THRU
	E8	65.00	9.00	THRU
	E9	65.00	15.00	THRU
	E10	65.00	21.00	THRU
	E11	65.00	27.00	THRU
	E12	65.00	31.50	THRU

5. 14–2–E

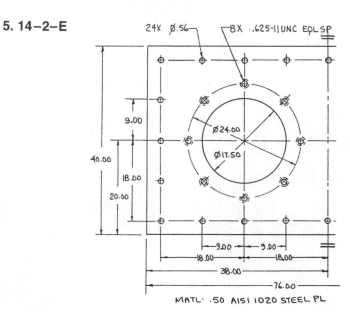

MATL– .50 AISI 1020 STEEL PL

14–2–F

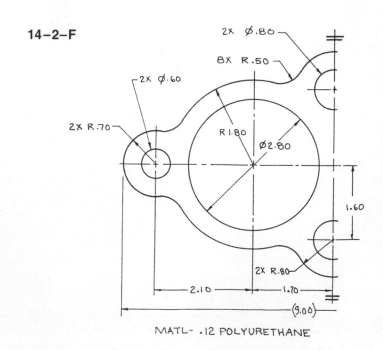

MATL– .12 POLYURETHANE

Unit 14-2

6. 14-2-G

14-2-H

7. 14-2-J

9. 14-2-L

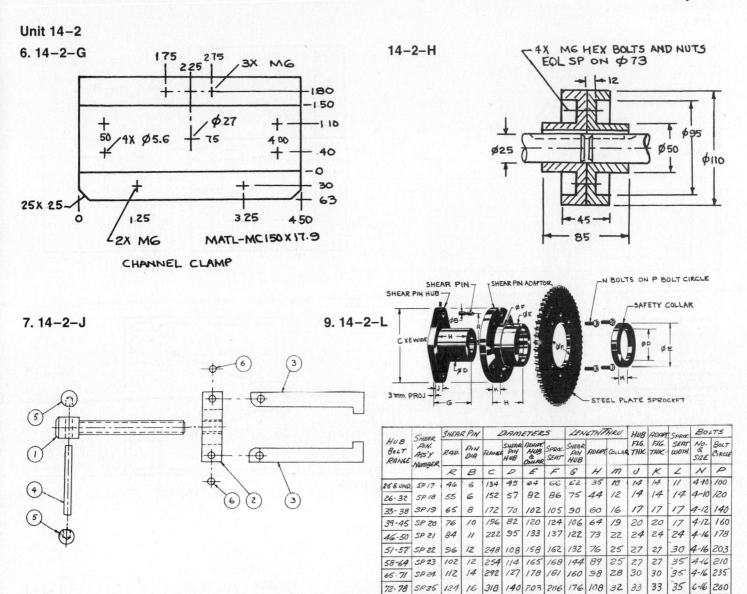

HUB BOLT RANGE	SHEAR PIN ASS'Y NUMBER	SHEAR PIN		DIAMETERS				LENGTH THRU			HUB FLG. THK.	ADAPT FLG. THK.	SPROC. SEAT WDTH.	BOLTS	
		RAD.	PIN DIA.	FLANGE	SHEAR PIN HUB	ADAPT. HUB & COLLAR	SPROC. SEAT	SHEAR PIN HUB	ADAPT.	COLLAR				No. & SIZE	BOLT CIRCLE
		R	B	C	D	E	F	G	H	M	J	K	L	N	P
25 & UND.	SP 17	46	6	134	95	84	66	62	35	10	14	14	11	4-10	100
26-32	SP 18	55	6	152	57	82	86	75	44	12	14	14	14	4-10	120
33-38	SP 19	65	8	172	70	102	105	90	60	16	17	17	17	4-12	140
39-45	SP 20	76	10	196	82	120	124	106	64	19	20	20	17	4-12	160
46-50	SP 21	84	11	222	95	133	137	122	73	22	24	24	24	4-16	178
51-57	SP 22	96	12	248	108	158	162	132	76	25	27	27	30	4-16	203
58-64	SP 23	102	12	254	114	165	168	144	89	25	27	27	35	4-16	210
65-71	SP 24	112	14	292	127	178	181	160	98	28	30	30	35	4-16	235
72-78	SP 25	124	16	318	140	203	206	176	108	32	33	33	35	6-16	260

Unit 14-3

10. 14-3-A

14-3-B

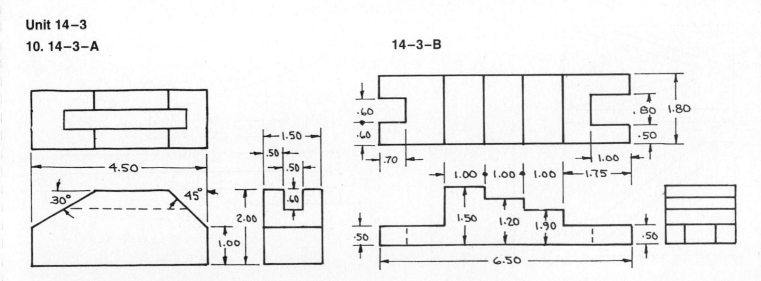

Unit 14-3
10. CONTD

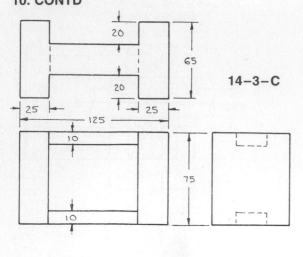

14-3-C

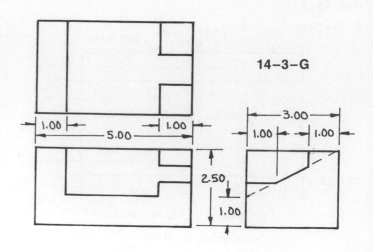

14-3-G

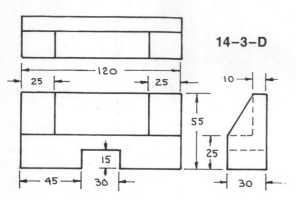

14-3-D

11.

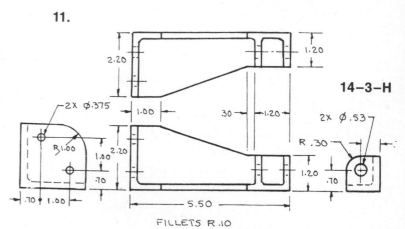

14-3-H

FILLETS R.10
WALL THICKNESS .25

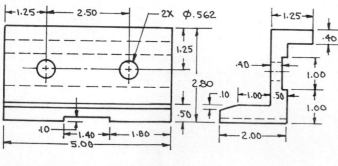

14-3-E

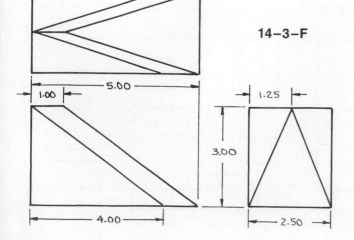

14-3-F

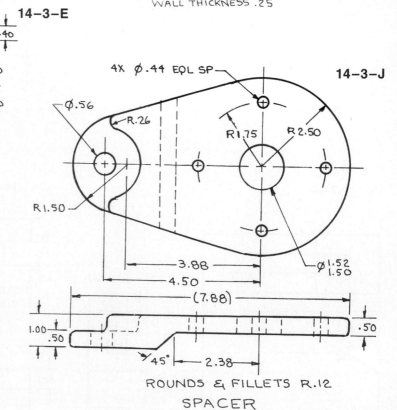

14-3-J

ROUNDS & FILLETS R.12
SPACER

Unit 14-3

11. CONTD 14-3-K

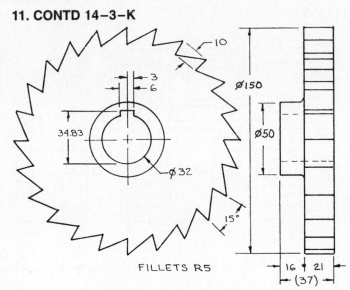

FILLETS R5

14-3-N

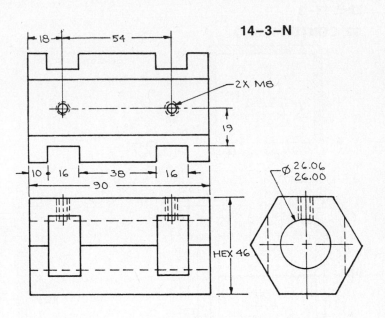

HEX 46

14-3-L

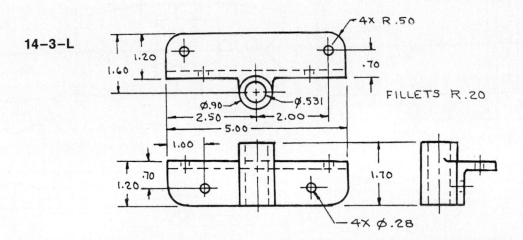

FILLETS R.20

14-3-M

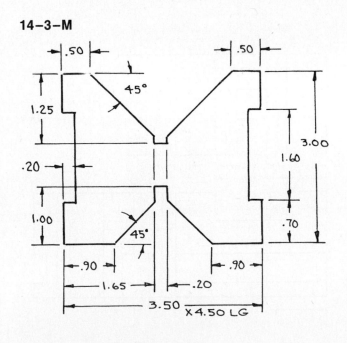

X 4.50 LG

12. 14-3-P

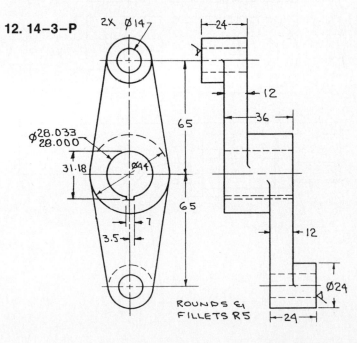

ROUNDS &
FILLETS R5

Unit 14–3

12. CONTD 14–3–Q

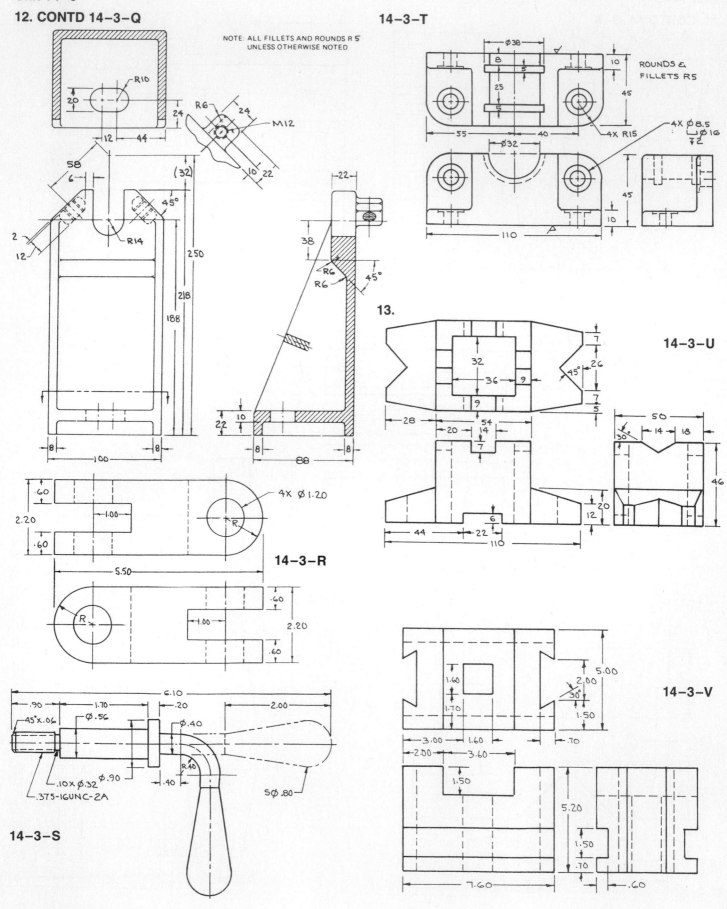

NOTE: ALL FILLETS AND ROUNDS R 5
UNLESS OTHERWISE NOTED

14–3–T

14–3–R

13.

14–3–U

14–3–S

14–3–V

Unit 14-3

13. CONTD

14-3-W

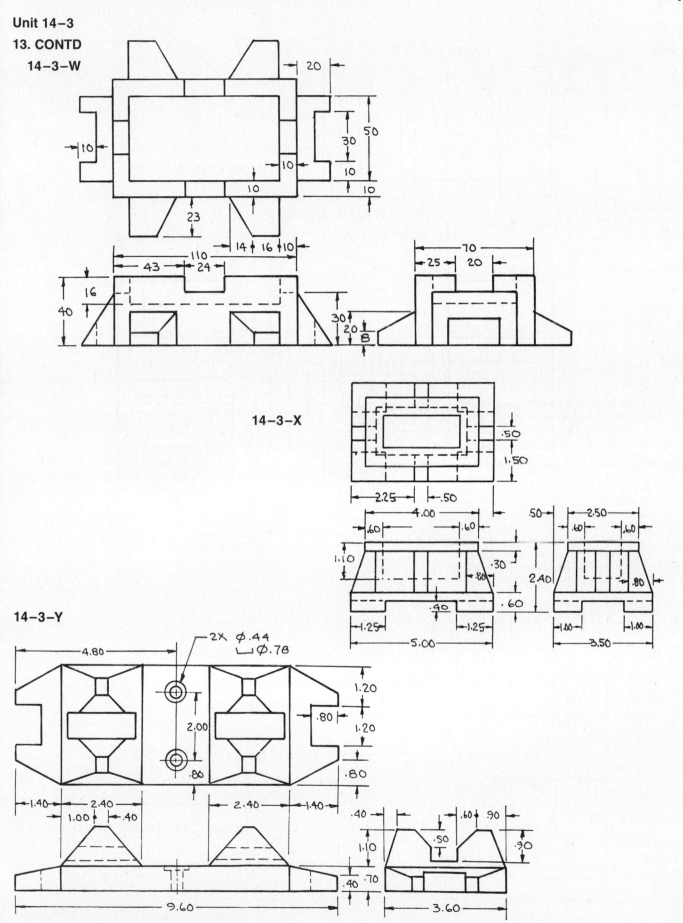

14-3-X

14-3-Y

Unit 14–3

14. 14–3–Z

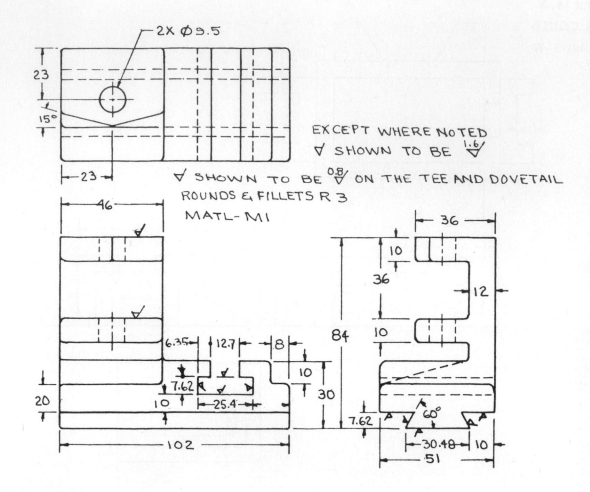

EXCEPT WHERE NOTED
∀ SHOWN TO BE 1.6∀

∀ SHOWN TO BE 0.8∀ ON THE TEE AND DOVETAIL
ROUNDS & FILLETS R 3
MATL– MI

14–3–AA

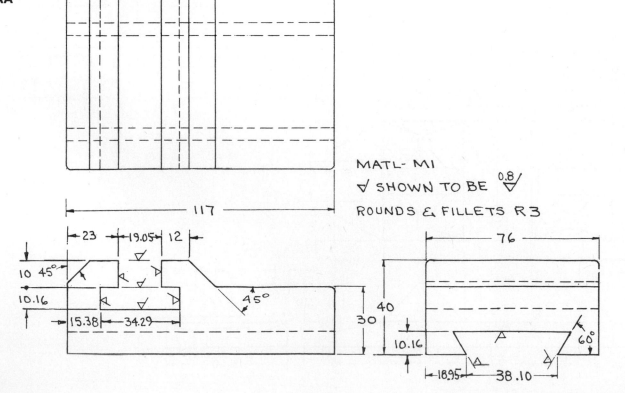

MATL– MI
∀ SHOWN TO BE 0.8∀
ROUNDS & FILLETS R3

Unit 14–3

15. 14–3–BB

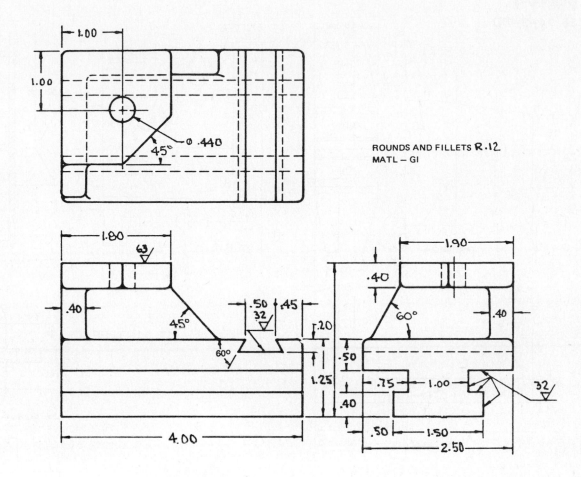

ROUNDS AND FILLETS R·12
MATL – GI

14–3–CC

MATL – MALLEABLE IRON
ROUNDS AND FILLETS R 5

2X Ø13
⌴ Ø24

Unit 14-3

16. 14-3-DD

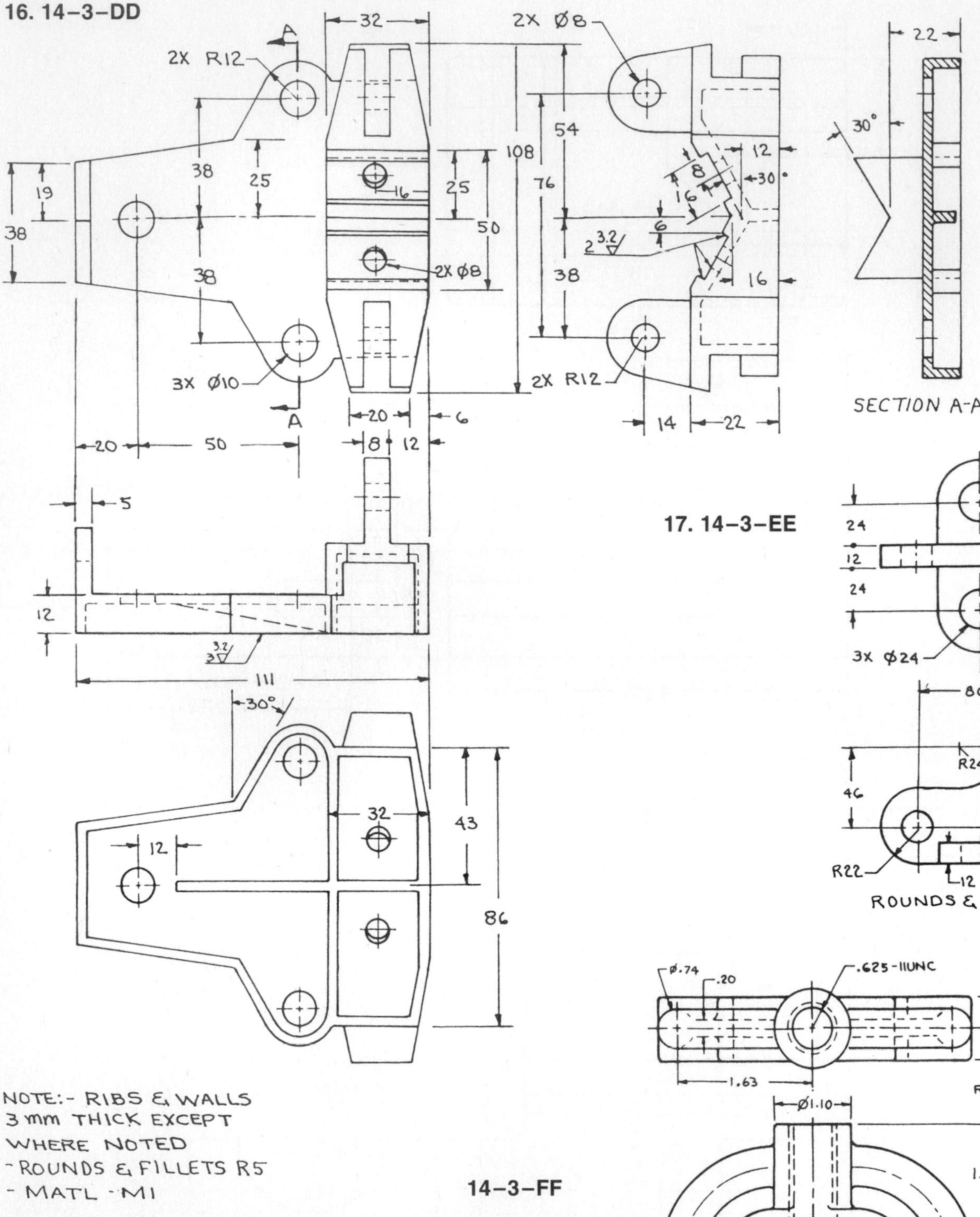

SECTION A-A

17. 14-3-EE

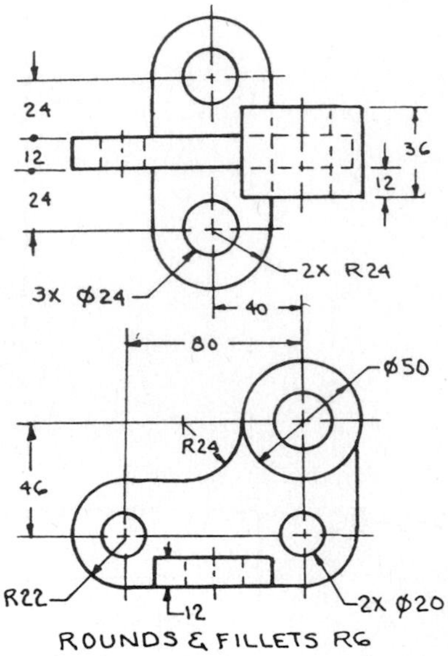

ROUNDS & FILLETS R6

NOTE:- RIBS & WALLS
3 mm THICK EXCEPT
WHERE NOTED
- ROUNDS & FILLETS R5
- MATL - M1

14-3-FF

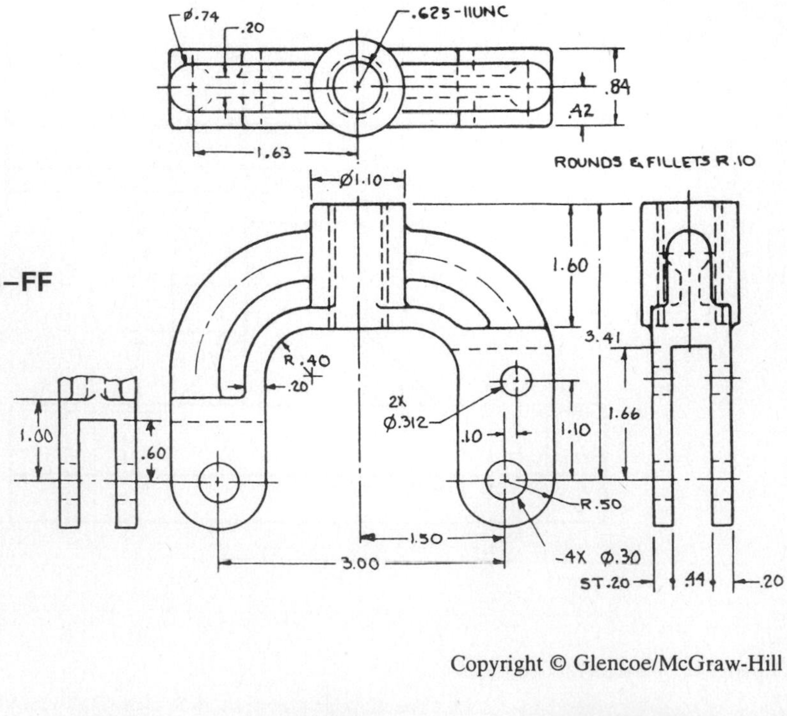

ROUNDS & FILLETS R.10

Unit 14–3

18. 14–3–GG

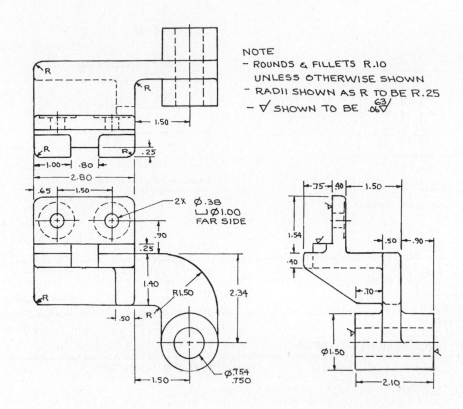

NOTE
- ROUNDS & FILLETS R.10
  UNLESS OTHERWISE SHOWN
- RADII SHOWN AS R TO BE R.25
- ∇ SHOWN TO BE .06∇

19. 14–3–HH

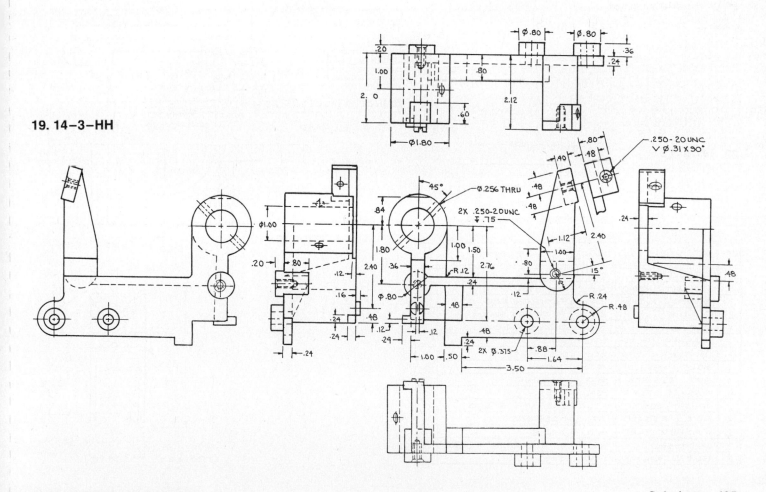

Unit 14–4

21. 14–4–A

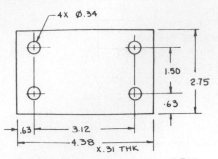

BASE   MATL- SAE 1025   1 REQD

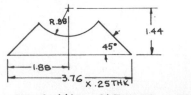

SHAFT SUPPORT   MATL-SAE 1025   1 REQD

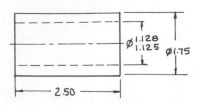

CRADLE   MATL-SAE 1025   1 REQD

14–4–B

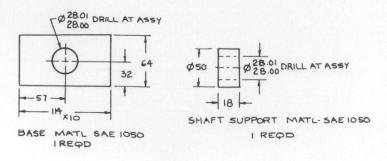

BASE MATL SAE 1050
1 REQD

SHAFT SUPPORT MATL- SAE 1050
1 REQD

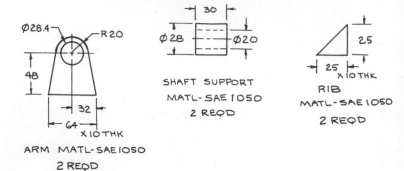

ARM MATL-SAE 1050
2 REQD

SHAFT SUPPORT
MATL-SAE 1050
2 REQD

RIB
MATL-SAE 1050
2 REQD

RIB   MATL SAE 1020 .25 X .56 X 1.13   2 REQD

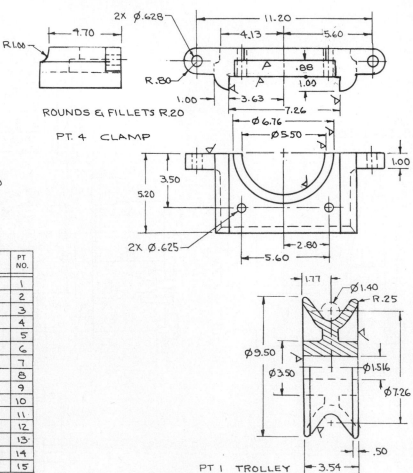

ROUNDS & FILLETS R.20

PT. 4 CLAMP

PT 1 TROLLEY

22. 14–4–C

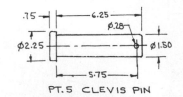

PT. 5 CLEVIS PIN

QTY	ITEM	MATL	DESCRIPTION	PT NO.
1	TROLLEY	CI	PATTERN XXXX	1
1	TROLLEY SUPPORT	CI	PATTERN XXXX	2
1	ADJUSTABLE SUPPORT	CI	PATTERN XXXX	3
2	CLAMP	CI	PATTERN XXXX	4
1	CLEVIS PIN	STL	Ø2.25 X 7.00	5
1	WASHER , FLAT	STL	Ø1.50	6
1	COTTER PIN - STD	STL	Ø.25 X 2.50	7
1	BOLT - HEX HD	STL	.750 UNC X 9.00	8
1	NUT REG HEX	STL	.750 UNC	9
1	WASHER , FLAT	STL	.750	10
2	BOLT - HEX HD	STL	.625 UNC X 3.00	11
2	NUT REG HEX	STL	.625 UNC	12
2	WASHER , FLAT	STL	.625	13
2	SET SCREW	STL	SO HD .500 UNC X 2.50	14
2	NUT REG HEX	STL	.500 UNC	15

Unit 14–4

22. 14–4–C CONTD

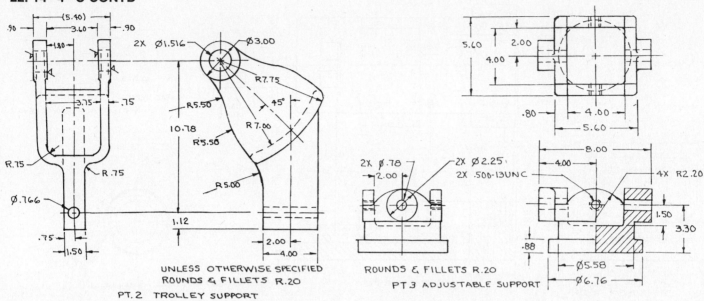

UNLESS OTHERWISE SPECIFIED
ROUNDS & FILLETS R.20

PT. 2   TROLLEY SUPPORT

ROUNDS & FILLETS R.20

PT.3 ADJUSTABLE SUPPORT

14–4–D

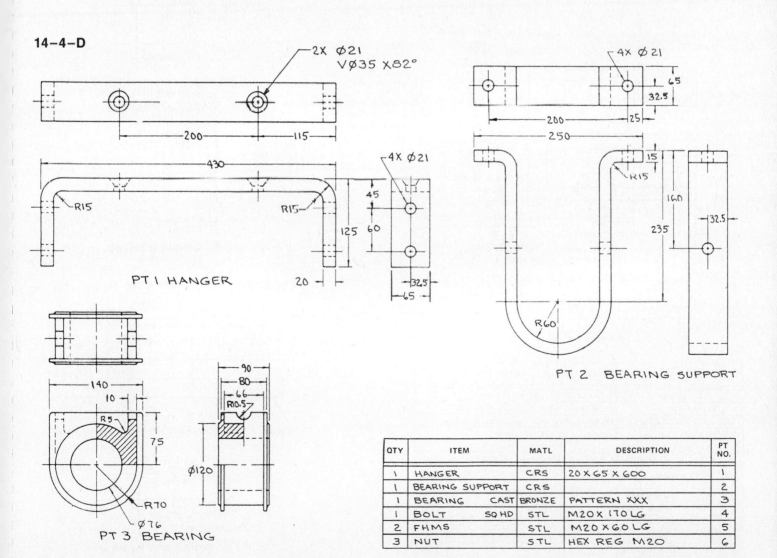

PT I HANGER

PT 2   BEARING SUPPORT

PT 3 BEARING

QTY	ITEM		MATL	DESCRIPTION	PT NO.
1	HANGER		CRS	20 X 65 X 600	1
1	BEARING SUPPORT		CRS		2
1	BEARING	CAST	BRONZE	PATTERN XXX	3
1	BOLT	SQ HD	STL	M20 X 170 LG	4
2	FHMS		STL	M20 X 60 LG	5
3	NUT		STL	HEX REG M20	6

Unit 14-4

23. 14-4-E

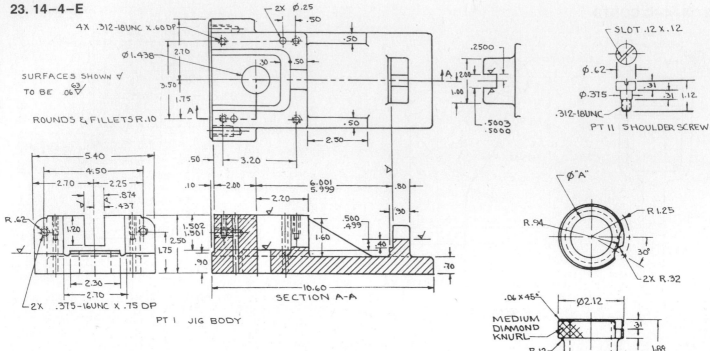

4X .312-18UNC X .60DP
2X Ø.25
.50
.50
Ø1.438
2.70
.50
.30
.50
SURFACES SHOWN ∀
TO BE .06 63
3.50
1.75
.2500
A 2.00
1.00
ROUNDS & FILLETS R.10

.5003
.5000

SLOT .12 X .12
Ø.62
Ø.375
.31
.31 1.12
.312-18UNC
PT 11 SHOULDER SCREW

5.40
.50
3.20
4.50
2.70
2.25
.874
.437
.10
2.00
6.001
5.999
.80
2.20
.500
.499
.90
R.62
1.20
1.502
1.501
2.50
1.60
.40
1.75
.90
.70
2.30
2.70
SECTION A-A
10.60
2X .375-16UNC X .75 DP
PT 1 JIG BODY

Ø"A"
R1.25
R.94
30°
2X R.32

.06 X 45°
Ø2.12
MEDIUM
DIAMOND
KNURL
.31
R.12
1.88
1.25
Ø1.7496
1.7490
PT 9 SLIP BUSHING A= Ø1.344
PT 10 SLIP BUSHING A= Ø1.375

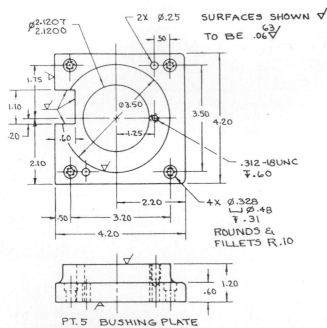

Ø2.1207
2.1200
2X Ø.25
SURFACES SHOWN ∀
TO BE .06 63
.50
1.75
1.10
Ø3.50
3.50
4.20
.20
1.25
.60
2.10
.312-18UNC
∓.60
2.20
4X Ø.328
⊔Ø.48
∓.31
.50
3.20
4.20
ROUNDS &
FILLETS R.10

Ø2.1218
2.1213
1.20
Ø1.7510
1.7500
PT. 8 LINER BUSHING

1.20
.60
PT.5 BUSHING PLATE

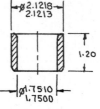

REMOVE ALL SHARP EDGES

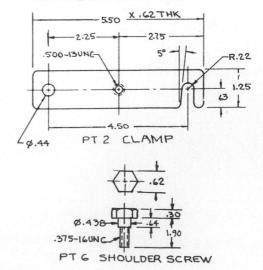

5.50 X .62 THK
2.25
2.75
.500-13UNC
5°
R.22
1.25
.63
4.50
Ø.44
PT 2 CLAMP

QTY	ITEM	MATL	DESCRIPTION	PT NO.
1	JIG BODY	CI	PATTERN XXXX	1
1	CLAMP	CRS	.62 X 1.25 X 5.50	2
4	CAP SCREW	STL	FIL HD .312UNC X 1.00	3
2	SPRING PIN	STL	Ø.25 X 1.00	4
1	BUSHING PLATE	CI	PATTERN XXXX	5
2	SHOULDER SCREW	STL	HEX .62 X 2.20	6
1	SETSCREW	STL	SQ HD .500 UNC X 2.00 FL POINT	7
1	LINER BUSHING	CAST BRONZE	1.75 ID X 2.120 D X 1.20	8
1	SLIP BUSHING		#XXXX (AM. DRILL BUSH.CO.)	9
1	SLIP BUSHING		#XXXX (AM. DRILL BUSH.CO)	10
1	SHOULDER SCREW	STL	Ø.62 X 1.12	11

.62
Ø.438
.30
.44
1.90
.375-16UNC
PT 6 SHOULDER SCREW

Unit 14-4

24. 14-4-F

GEAR CUTTING DATA	
NO. OF TEETH	50
PITCH DIAMETER	6.250
DIAMETRAL PITCH	8
PRESSURE ANGLE	20°
WHOLE DEPTH	.270
CHORDAL ADDENDUM	.149
CHORDAL THICKNESS	.196

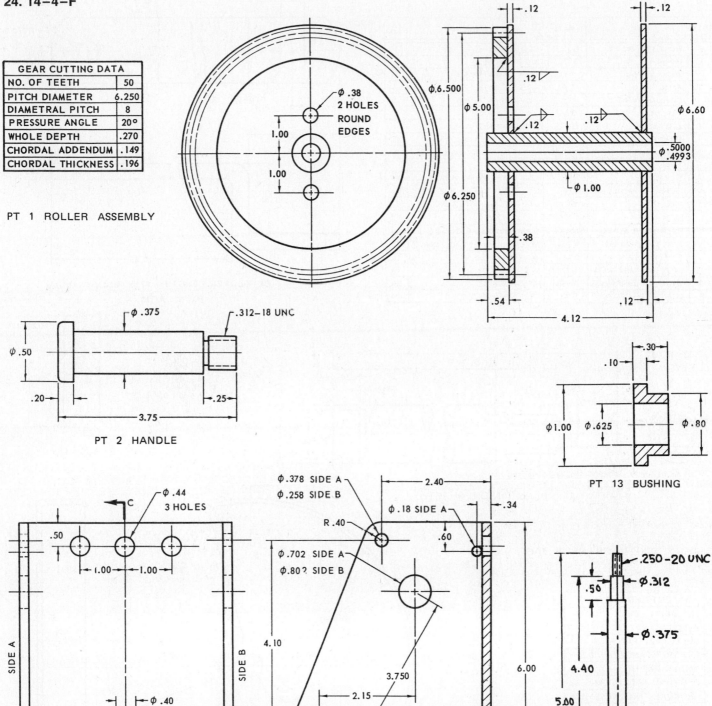

PT 1 ROLLER ASSEMBLY

PT 2 HANDLE

PT 13 BUSHING

PT 3 BRACKET

SECTION C-C

PT 15 BOLT

DETAILS CONTD ON NEXT PAGE

Unit 14–4

24. 14–4–F CONTD

QTY	ITEM	MATL	DESCRIPTION	PT NO.
1	ROLLER ASSY	STL	SEE DWG	1
1	HANDLE	STL	Ø.50 X 3.75	2
1	BRACKET	STL	.188 X 6.00 X 12.65	3
1	ARM	STL	.25 X 1.00 X 12.55	4
1	PINION SHAFT	STL	Ø.625 X 5.40	5
1	PINION	STL	Ø1.50 X .38	6
1	GUARD	STL	STAMPING XXX	7
1	BUSHING	STL	Ø1.00 X .40	8
1	BUSHING	BRONZE	.38 ID X .5006–.5016 OD X 4.20 LG	9
1	SLEEVE	STL	.38 ID X .50 OD X 3.25 LG	10
1	SPRING HOLDER	STL	#20 (.0359) X 2.00 X 2.28	11
1	PAWL	STL	.25 X 2.16 X 2.25	12
1	SPRING	W. BARNES	# XXX	13
1	RETAINING RING	WALDES	RAD. ASSY 11–410–62	14
1	BOLT HEX HD	STL	Ø.375 X 5.00 (SEE DWG)	15
1	BOLT HEX HD	STL	Ø.375 X 5.00	16
2	NUT HEX REG	STL	.375 UNC	17
1	NUT HEX REG	STL	.500 UNC	18
2	WASHER – FLAT	STL	TYPE A Ø.312	19

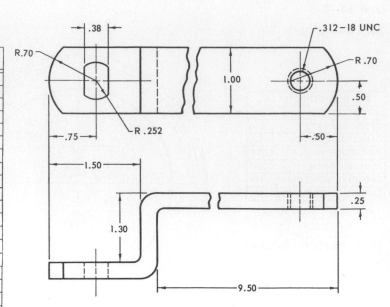

DEVELOPED LENGTH – 12.55 (MEAN TRAVEL)
PT 4 ARM

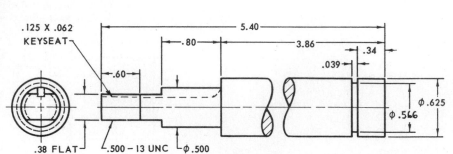

PT 5 PINION SHAFT

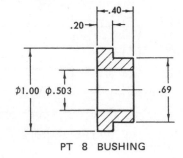

PT 8 BUSHING

DETAILS CONTD ON NEXT PAGE

CUTTING DATA	
NO. OF TEETH	10
PITCH DIAMETER	1.250
DIAMETRAL PITCH	8
PRESSURE ANGLE	20°
WHOLE DEPTH	.270
CHORDAL ADDENDUM	.133
CHORDAL THICKNESS	.196

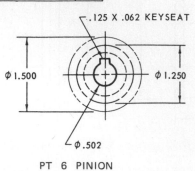

PT 6 PINION

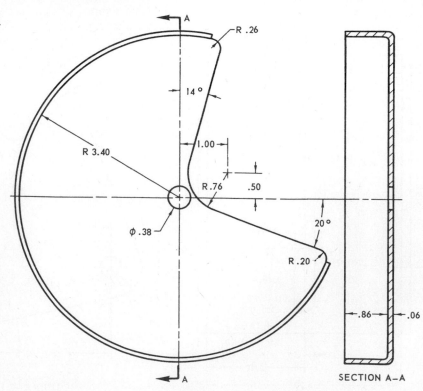

PT 7 GUARD

SECTION A–A

Unit 14–4

24. 14–4–F CONTD

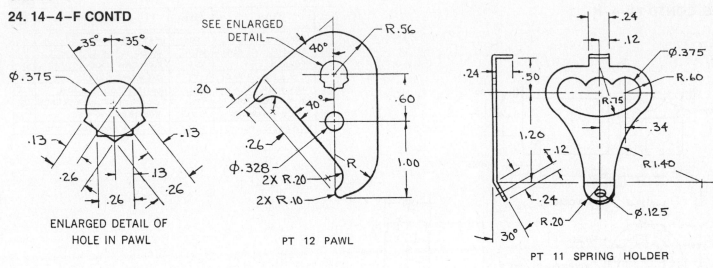

ENLARGED DETAIL OF
HOLE IN PAWL

PT 12 PAWL

PT 11 SPRING HOLDER

25. 14–4–G

QTY	ITEM	MATL	DESCRIPTION	PT NO.
1	BASE	GI	PATTERN XXX	1
1	PULLEY	GI	PATTERN XXX	2
1	SHAFT	STL	⌀ 35 X 540	3
1	JOURNAL BEARING	BABBITT	35 ID X 45 OD X 90	4
1	NUT HEX	STL	M24 PREVAILING TORQUE	5
1	KEY WOODRUFF	STL	#808	6

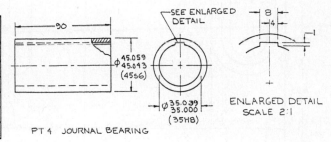

PT 4   JOURNAL BEARING

ENLARGED DETAIL
SCALE 2:1

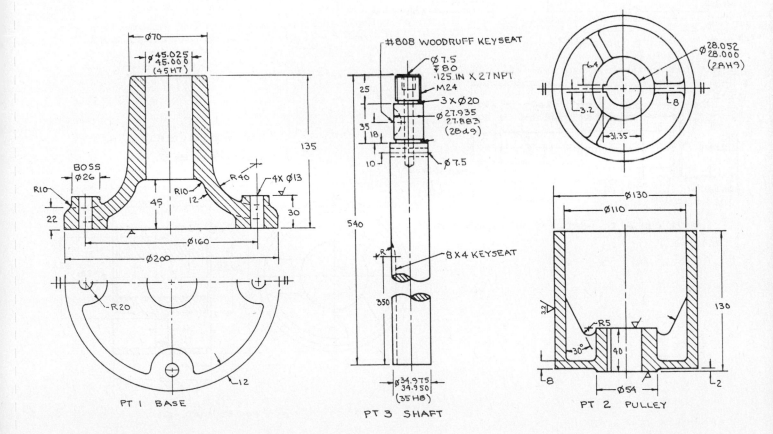

PT 1 BASE

PT 3 SHAFT

PT 2 PULLEY

Unit 14–4
25. CONTD 14–4–H

QTY	ITEM		MATL	DESCRIPTION	PT NO.
1	FRAME		GI	PATTERN XXXX	1
1	SHAFT	AISI	1117	Ø1.80 X 6.82	2
1	PULLEY		GI	PATTERN XXXX	3
1	COLLAR	AISI	1117	Ø1.80 X 1.20	4
1	WASHER, FLAT		STL	1.06 ID X 2.00 OD X .134	5
1	NUT – HEX REG		STL	1.000–12 UNF WASHER FACE	6
1	BOLT – HEX HD		STL	.500–13UNC X 4.00	7
1	WASHER, FLAT		STL	.531 ID X 1.062 OD X .095	8

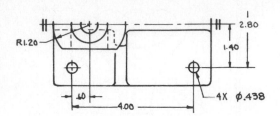

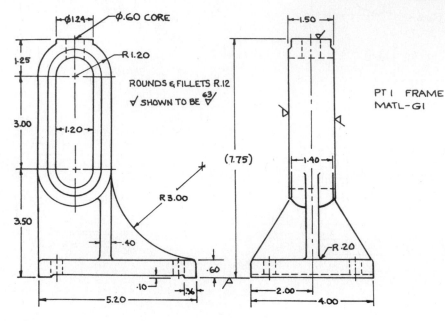

PT 1 FRAME
MATL– GI

ROUNDS & FILLETS R.12
√ SHOWN TO BE ⁶³∇

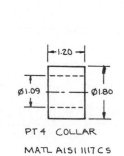

PT 4 COLLAR

MATL AISI 1117 C S

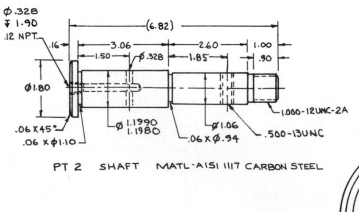

PT 2   SHAFT   MATL · AISI 1117 CARBON STEEL

ROUNDS & FILLETS R.12
√ SHOWN TO BE ⁶³∇

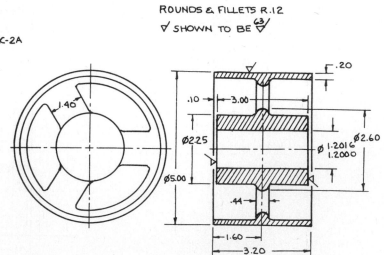

PT 3 PULLEY MATL–GI

Unit 14-4

26. 14-4-J

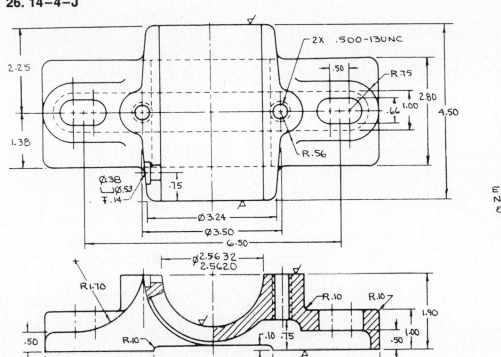

EXCEPT WHERE
NOTED ROUNDS
& FILLETS R.20

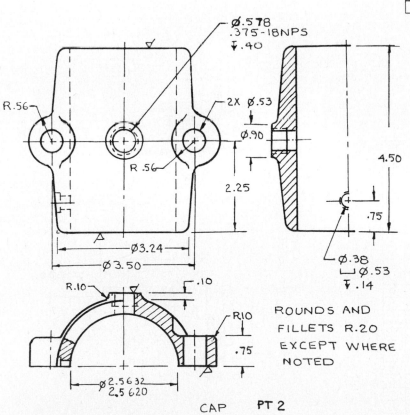

BASE   PT 1

CAP   PT 2

ROUNDS AND
FILLETS R.20
EXCEPT WHERE
NOTED

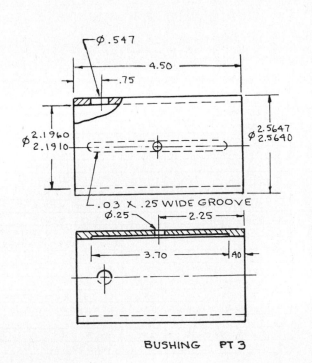

BUSHING   PT 3

QTY	ITEM	MATL	DESCRIPTION	PT NO.
1	BASE	CI	PATTERN XXXX	1
1	CAP	CI	PATTERN XXXX	2
1	BUSHING - CAST	BRONZE	PATTERN XXXX	3
1	LOCATOR	STL	Ø.50 X .56	4
2	CAP SCREW - HEX	STL	.500 UNC X 1.75	5
2	LOCKWASHER	STL	.500	6

Unit 14-4
26. CONTD 14-4-K

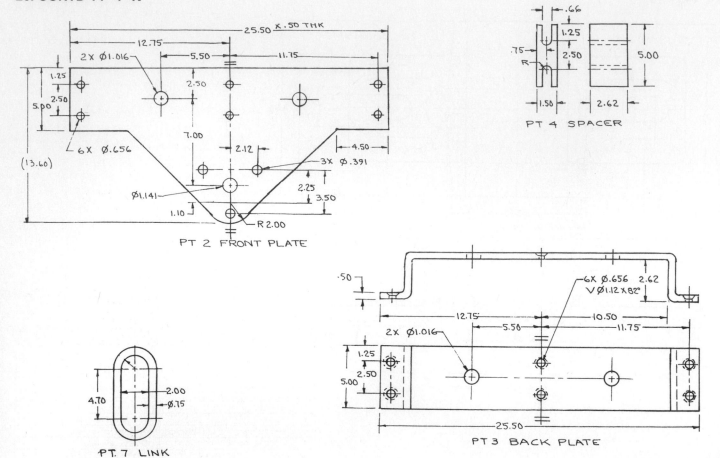

PT 2 FRONT PLATE

PT 4 SPACER

PT. 7 LINK

PT 3 BACK PLATE

PT 13 BEVELED WASHER

QTY	ITEM	MATL	DESCRIPTION	PT NO.
4	WHEEL	CI	CAT. NO. XXXX	1
2	FRONT PLATE	STL	.50 X 13.60 X 25.50	2
2	BACK PLATE	STL	.50 X 5.00 X 28.25	3
2	SPACER	STL	1.50 X 2.62 X 5.00	4
4	AXLE	STL	CAT. NO. XXXX	5
2	STUD	CI	PATTERN XXXX	6
1	LINK	STL	CAT. NO. XXXX	7
4	LUBRICATING CUP		CAT. NO. XXXX	8
4	ROLLER BEARING		CAT. NO. XXXX	9
1	STUD	STL	Ø1.125 X 11.00 THREAD BOTH ENDS 1.125 UNF X 2.00	10
2	NUT- STD HEX	STL	1.125 UNF	11
2	LOCKWASHER	STL	1.125	12
2	BEVELED WASHER	STL	Ø2.50 X 1.00	13
6	RIVET	STL	Ø.375 X 2.00 BUTTON HD	14
4	SCREW FL HD	STL	.625 UNC X 4.50	15
12	NUT STD HEX	STL	.625 UNC	16
12	LOCKWASHER	STL	.625	17
8	BOLT HEX HD	STL	.625 UNC X 1.75	18

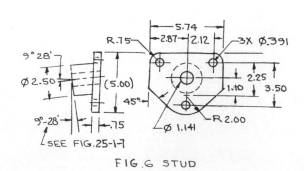

FIG. 6 STUD

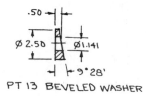

Unit 14–4

27. Details for 14–6–A

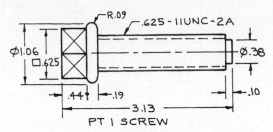

PT 1 SCREW

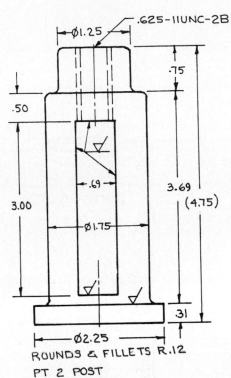

ROUNDS & FILLETS R.12
PT 2 POST

QTY	ITEM	MATL	DESCRIPTION	PT
1	SCREW	SAE 1112	Ø1.06 X 3.13	1
1	POST	CAST STL	PATTERN NO. XXX	2
1	WEDGE	SAE 1050	.50 X .62 X 4.00	3
1	RING	SAE 1020	Ø3.50 X .44	4
1	BLOCK	SAE 1020	.50 X 2.75 X 2.75	5

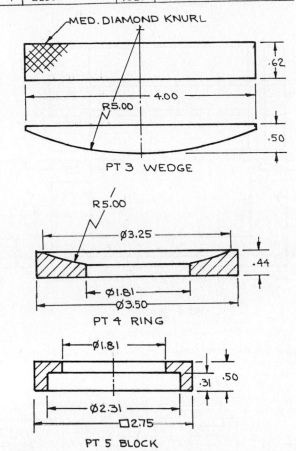

MED. DIAMOND KNURL

PT 3 WEDGE

PT 4 RING

PT 5 BLOCK

Details for 14–6–B

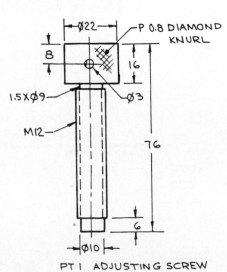

PT 1 ADJUSTING SCREW

QTY	ITEM	MATL	DESCRIPTION	PT
1	ADJUSTING SCREW	SAE 1112	Ø22 X 76	1
1	YOKE	CAST STL	PATTERN NO. XXX	2
1	BASE	SAE 1020	50 X 54 X 75	3

PT 3 BASE

PT 2 YOKE

Unit 14–4

27. CONTD Details for 14–6–C

QTY	ITEM	MATL	DESCRIPTION	PT NO.
1	BASE	C I	PATTERN XXXX	1
1	MOVABLE JAW	C I	PATTERN XXXX	2
1	SCREW	STL	Ø24 × 140	3
1	PLATE	STL	5 × 20 × 32	4
1	SCREW – FL HD	STL	M6 × 40 LG	5
1	HANDLE	STL	Ø8 ×100 – THREAD BOTH ENDS M8 × 10 LG	6
1	SCREW – FL HD	STL	M6 × 20 LG	7
2	NUT	STL	Ø12 × 10 LG	8

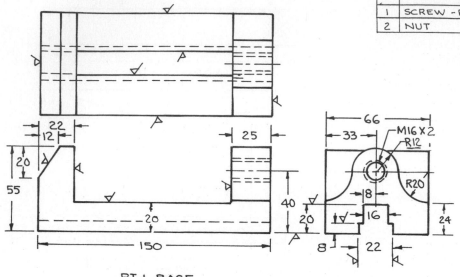

PT 1 BASE

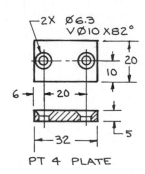

PT 5 SCREW

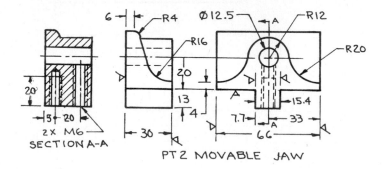

2 × M6
SECTION A–A

PT 2 MOVABLE JAW

PT 4 PLATE

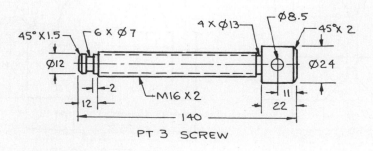

PT 3 SCREW

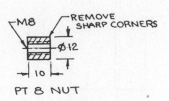

PT 8 NUT

Unit 14–4

27. CONTD Details for 14–6–D

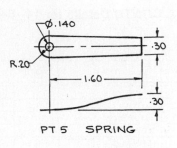

PT 5   SPRING

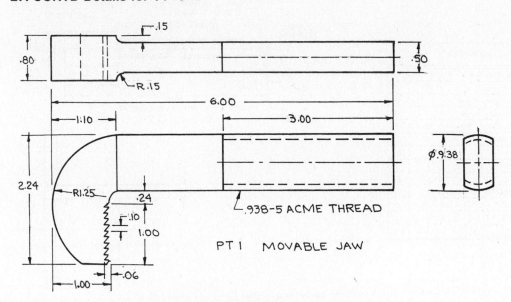

PT 1   MOVABLE JAW

.938–5 ACME THREAD

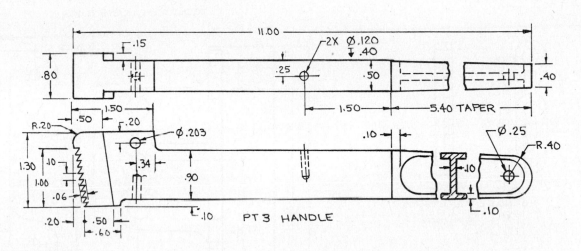

PT 3 HANDLE

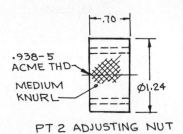

.938–5
ACME THD

MEDIUM
KNURL

Ø1.24

PT 2 ADJUSTING NUT

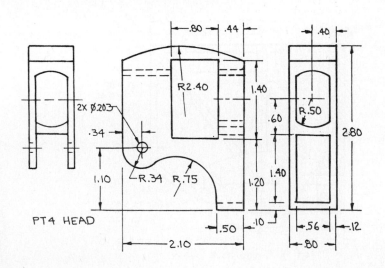

PT4 HEAD

QTY	ITEM	MATL	DESCRIPTION	PT NO.
1	MOVABLE JAW	STL	FORGING DIE XXXX	1
1	ADJUSTING NUT	STL	Ø1.25 X .70	2
1	HANDLE	STL	FORGING DIE XXXX	3
1	HEAD	CI	PATTERN XXXX	4
2	SPRING	STL	#20 (.032) X .40 X 1.80	5
1	BUTTON RIVET	STL	Ø.188 X 1.00	6
2	GROOVED STUD	STL	#6 X .375 DRIVE-LOK	7

Unit 14−4

27. CONTD Details for 14−6−E

QTY	ITEM	MATL	DESCRIPTION	PT NO.
1	BASE	GI	PATTERN XXX	1
2	GUIDE	CRS	Ø.75 X 10.20	2
1	MOVABLE JAW	GI	PATTERN XXX	3
1	SPACER	GI	PATTERN XXX	4
1	HANDLE	CRS	Ø.375X 6.00 THREAD BOTH ENDS WITH 375-16UNC X .40 LG	5
2	KNOB	STL	SØ.64	6
1	SCREW	CRS	Ø1.00 X 12.20	7
2	JAW PLYWOOD	FIR	.75 X 2.50 X 7.00	8
1	RETAINING RING	WALDES	EXTERNAL 5100 -50	9
2	WOOD SCREW	STL	#10 FLAT HEAD 1.00 LG	10
2	WOOD SCREW	STL	#10 FLAT HEAD 2.00LG	11
4	LAG BOLT-HEX HD	STL	Ø.312 X 1.50	12

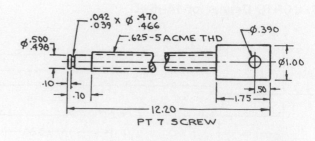

PT 7 SCREW

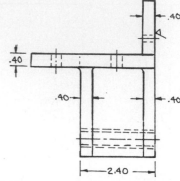

PT 1 BASE
ROUNDS & FILLETS R.10

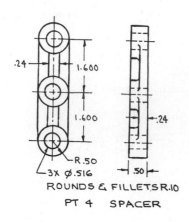

ROUNDS & FILLETS R.10
PT 4   SPACER

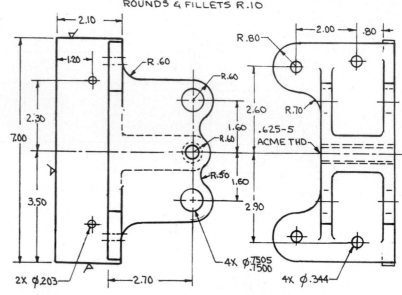

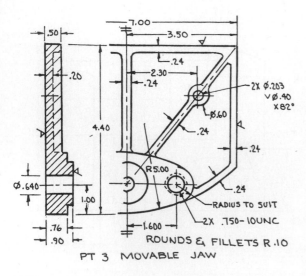

PT 3 MOVABLE JAW
ROUNDS & FILLETS R.10

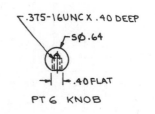

.375-16UNC X .40 DEEP
SØ.64
.40 FLAT
PT 6 KNOB

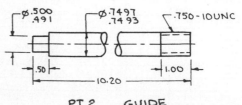

PT 2   GUIDE

Unit 14-4
27. CONTD Details for 14-6-F

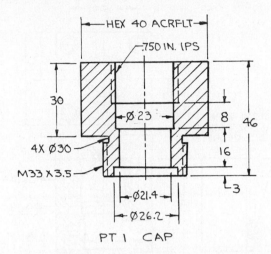

PT 1 CAP

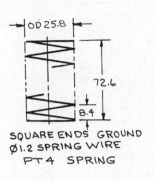

SQUARE ENDS GROUND
Ø1.2 SPRING WIRE
PT 4 SPRING

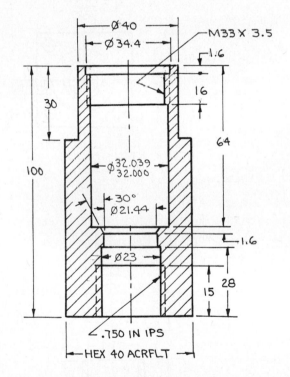

PT 2 BODY

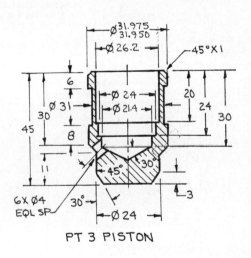

PT 3 PISTON

QTY	ITEM	MATL	DESCRIPTION	PT NO.
1	CAP	STL	HEX 40 ACR FLT X 46	1
1	BODY	STL	HEX 40 ACR FLT X 100	2
1	PISTON	STL	Ø 32 X 45	3
1	SPRING	STL	SEE DRAWING	4
1	O-RING		Ø 2 X 30 ID	5

Unit 14−4

27. CONTD Details for 14−6−G

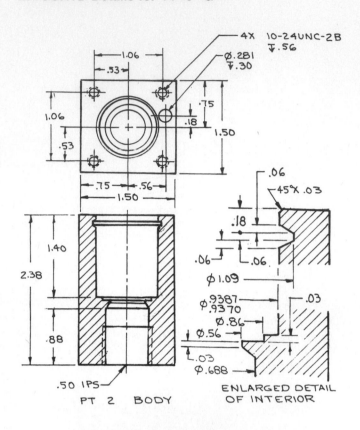

PT 2   BODY

ENLARGED DETAIL
OF INTERIOR

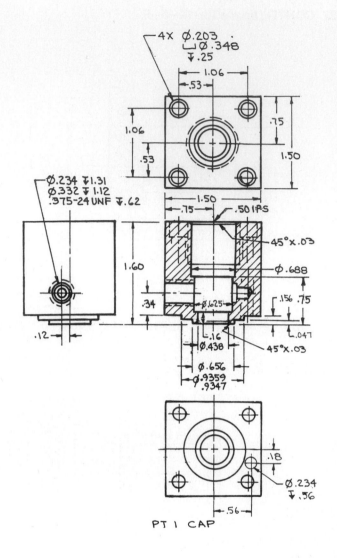

PT 1 CAP

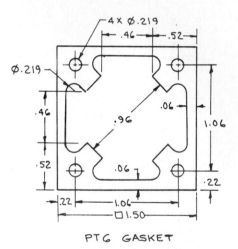

PT 6 GASKET

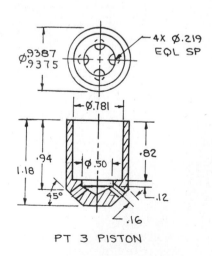

PT 3 PISTON

QTY	ITEM	MATL	DESCRIPTION	PT NO.
1	CAP	STL	1.50 X 1.50 X 1.60	1
1	BODY	STL	1.50 X 1.50 X 2.38	2
1	PISTON	ALUM	Ø1.00 X 1.20	3
1	SPRING	STL	SEE DWG	4
1	VALVE	STL	Ø.375 X 1.90	5
1	GASKET	NEOPRENE	.06 X 1.50 X 1.50	6
4	CAP SCREW	STL	FIL HD−HEX CAP	7
			10−24 UNC X 1.75 LG	
4	LOCKWASHER	STL	HEL SPRING #10	8

Unit 14-4
27. CONTD Details for 14-6-G CONTD

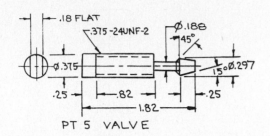

PT 5   VALVE

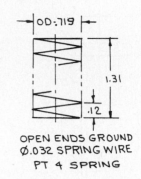

OPEN ENDS GROUND
Ø.032 SPRING WIRE
PT 4 SPRING

Unit 14-5
28. 14-5-A

SYMBOL	REVISION
1	92 WAS 88
2	14 WAS 12
3	Ø10 WAS Ø8
4	30 WAS 28

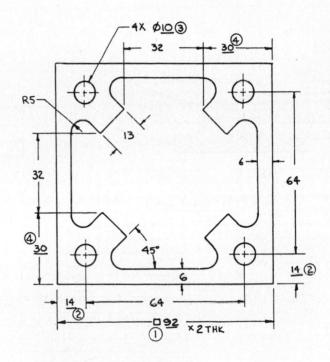

14-5-B

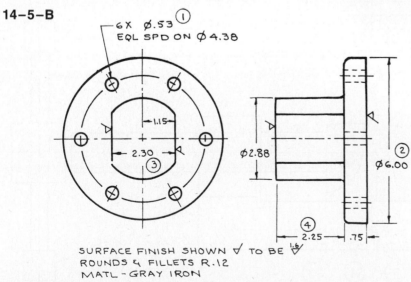

REVISIONS		
SYMBOL	DESCRIPTION	DATE & APPROV.
1	Ø.53 WAS Ø.50	21/11/93
2	Ø6.00 WAS Ø5.75	21/11/93
3	2.30 WAS 2.25	21/11/93
4	2.25 WAS 2.38	21/11/93

SURFACE FINISH SHOWN ∇ TO BE ∇1.6
ROUNDS & FILLETS R.12
MATL - GRAY IRON

Unit 14-6

29. 14-6-A

14-6-B

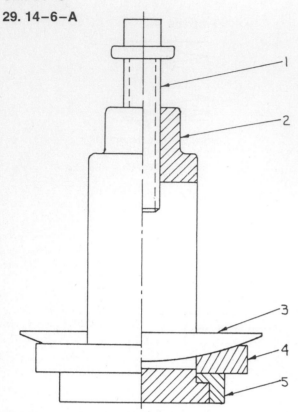

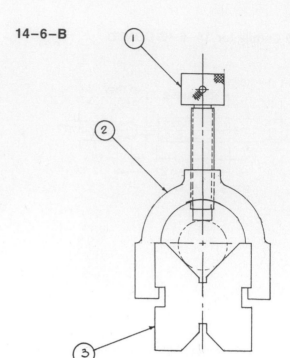

QTY	ITEM	MATL	DESCRIPTION	PT
1	SCREW	SAE 1112	Ø1.06 X 3.13	1
1	POST	CAST STL	PATTERN NO. XXX	2
1	WEDGE	SAE 1050	.50 X .62 X 4.00	3
1	RING	SAE 1020	Ø3.50 X .44	4
1	BLOCK	SAE 1020	.50 X 2.75 X 2.75	5

QTY	ITEM	MATL	DESCRIPTION	PT
1	ADJUSTING SCREW	SAE 1112	Ø22 X 76	1
1	YOKE	CAST STL	PATTERN NO. XXX	2
1	BASE	SAE 1020	50 X 54 X 75	3

30. 14-6-C

QTY	ITEM	MATL	DESCRIPTION	PT NO.
1	BASE	C I	PATTERN XXXX	1
1	MOVABLE JAW	C I	PATTERN XXXX	2
1	SCREW	STL	Ø24 X 140	3
1	PLATE	STL	5 X 20 X 32	4
1	SCREW - FL HD	STL	M6 X 40 LG	5
1	HANDLE	STL	Ø8 X100 - THREAD	6
			BOTH ENDS M8 X 10 LG	
1	SCREW - FL HD	STL	M6 X 20 LG	7
2	NUT	STL	Ø12 X 10 LG	8

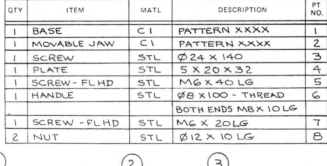

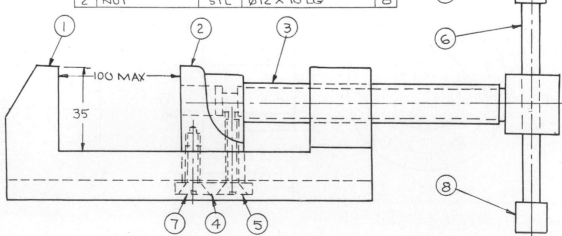

Unit 14–6

31. 14–6–D

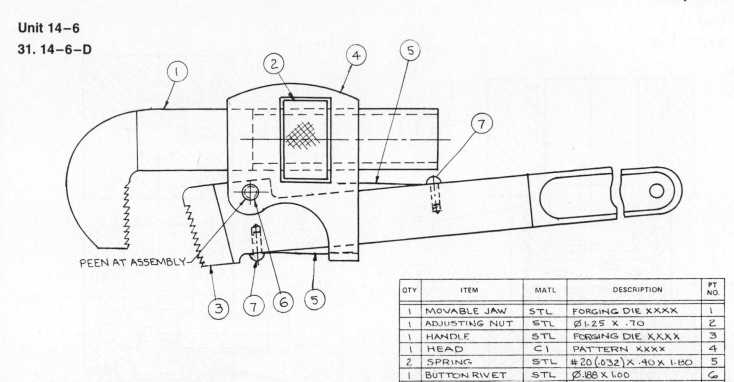

PEEN AT ASSEMBLY

QTY	ITEM	MATL	DESCRIPTION	PT NO.
1	MOVABLE JAW	STL	FORGING DIE XXXX	1
1	ADJUSTING NUT	STL	Ø1.25 X .70	2
1	HANDLE	STL	FORGING DIE XXXX	3
1	HEAD	C1	PATTERN XXXX	4
2	SPRING	STL	#20 (.032) X .40 X 1.80	5
1	BUTTON RIVET	STL	Ø.188 X 1.00	6
2	GROOVED STUD	STL	#6 X .31 DRIVE-LOK	7

32. 14–6–E

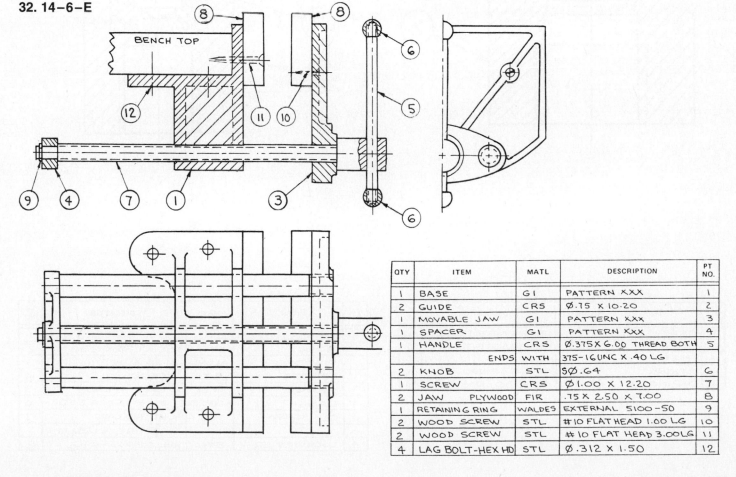

BENCH TOP

QTY	ITEM	MATL	DESCRIPTION	PT NO.
1	BASE	G1	PATTERN XXX	1
2	GUIDE	CRS	Ø.75 X 10.20	2
1	MOVABLE JAW	G1	PATTERN XXX	3
1	SPACER	G1	PATTERN XXX	4
1	HANDLE	CRS	Ø.375 X 6.00 THREAD BOTH	5
		ENDS	WITH 375-16UNC X .40 LG	
2	KNOB	STL	SØ.64	6
1	SCREW	CRS	Ø1.00 X 12.20	7
2	JAW PLYWOOD	FIR	.75 X 2.50 X 7.00	8
1	RETAINING RING	WALDES	EXTERNAL 5100-50	9
2	WOOD SCREW	STL	#10 FLAT HEAD 1.00 LG	10
2	WOOD SCREW	STL	#10 FLAT HEAD 3.00LG	11
4	LAG BOLT-HEX HD	STL	Ø.312 X 1.50	12

Unit 14–6

33. 14–6–F

34. 14–6–G

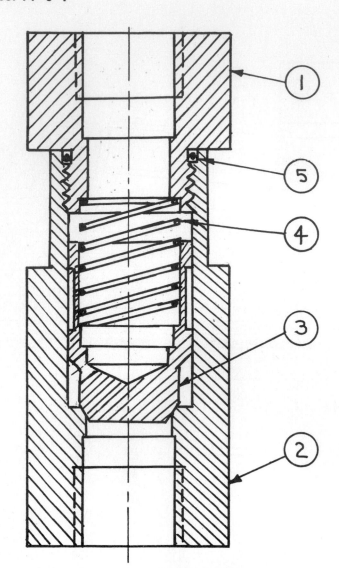

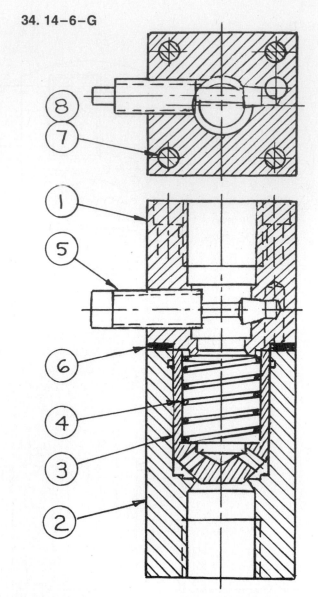

QTY	ITEM	MATL	DESCRIPTION	PT NO.
1	CAP	STL	HEX 40 ACR FLT X 46	1
1	BODY	STL	HEX 40 ACR FLT X 100	2
1	PISTON	STL	Ø 32 X 45	3
1	SPRING	STL	SEE DRAWING	4
1	O-RING		Ø 2 X 30 ID	5

QTY	ITEM	MATL	DESCRIPTION	PT NO.
1	CAP	STL	1.50 X 1.50 X 1.60	1
1	BODY	STL	1.50 X 1.50 X 2.38	2
1	PISTON	ALUM	Ø 1.00 X 1.20	3
1	SPRING	STL	SEE DWG	4
1	VALVE	STL	Ø .375 X 1.90	5
1	GASKET	NEOPRENE	.06 X 1.50 X 1.50	6
4	CAP SCREW	STL	FIL HD-HEX CAP	7
			10-24 UNC X 1.75 LG	
4	LOCKWASHER	STL	HEL SPRING #10	8

Unit 14–6

35. 14–6–H

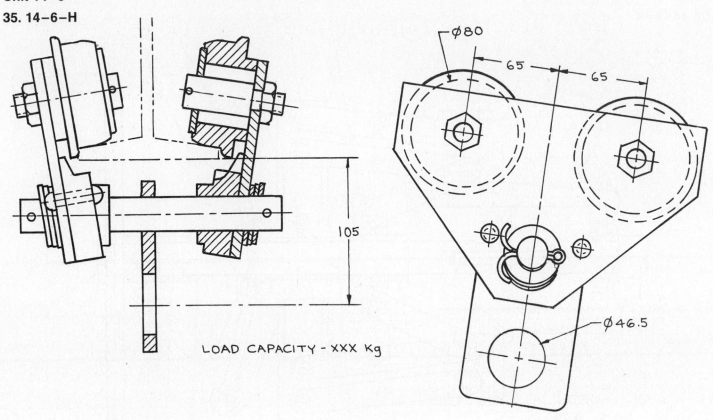

LOAD CAPACITY - XXX Kg

36. 14–6–J

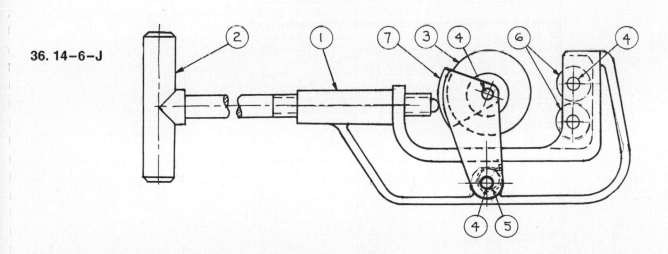

QTY	ITEM	MATL	DESCRIPTION	PT NO.
1	FRAME	C1	PATTERN XXXX	1
1	HANDLE	CRS	DWG XXXX	2
1	CUTTER	TOOL STL	⌀40 × 18 LG	3
4	SPRING PIN	STL	⌀10 × 45 LG	4
1	TORSION SPRING	STL		5
2	ROLLER	STL	⌀30 × 30 LG	6
1	CUTTER SUPPORT	C1	PATTERN XXXX	7

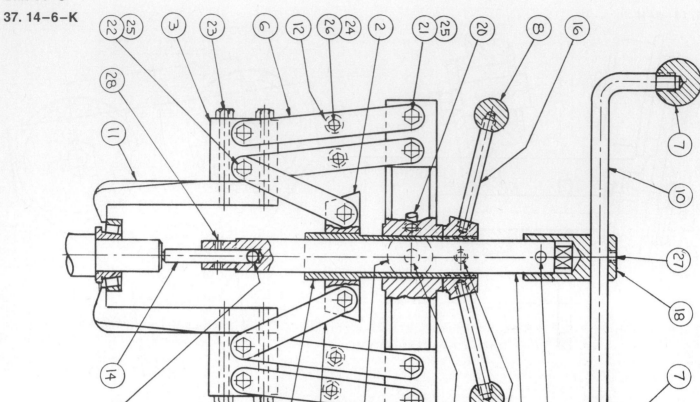

QTY	ITEM	MATL	DESCRIPTION	PT NO.
1	BODY	CI	PATTERN XXXX	1
1	SLIDE BAR	STL	1.00 X 2.00 X 3.80	2
2	JAW	STL	.60 X 2.00 X 1.90	3
1	ADJ. SCREW	STL	⌀.875 X 11.00	4
4	LINK	STL	.188 X .70 X 4.00	5
8	LINK	CI	PATTERN XXXX	6
3	KNOB	ALLYLICS	MOLD NO. XXXX	7
2	KNOB	ALLYLICS	MOLD NO. XXXX	8
1	COLLAR	STL	⌀2.50 X 1.00	9
1	HANDLE	STL	⌀.50 X 13.00	10
2	FINGER	CI	PATTERN XXXX	11
4	SPACER	STL	⌀.40 X .60	12
1	ADJUSTING SCREW	STL	⌀1.25 X 5.00	13
1	CENTER ROD	STL	⌀.375 X 2.60	14
1	NUT	STL	⌀.60 X .50	15
2	HANDLE	STL	⌀.375 X 3.50	16
1	HANDLE	STL	⌀.50 X 3.00	17
1	HANDLE SUPPORT	STL	⌀1.25 X 2.80	18
1	BALL BEARING	STL	5⌀.375	19
1	GREASE CUP	STL	#XXXX .250-28UNF	20
6	BOLT HEX HD	STL	.312UNF X 1.50	21
4	BOLT HEX HD	STL	.312UNF X 1.75	22
4	BOLT HEX HD	STL	.312 UNF X 2.50	23
4	MACH SCREW	STL	HEX HD 8-32 X 1.25	24
10	NUT HEX	STL	.312 UNF	25
4	NUT HEX	STL	8-32	26
2	SET SCREW-HEADLESS	STL	.375 UNF X .50 LG CUP POINT, HEX CAP	27
2	SET SCREW-HEADLESS	STL	8-32 X .625 LG, FULL DOG HEX CAP	28

Unit 14–6

38. 14–6–L

QTY	ITEM	MATL	DESCRIPTION	PT NO.
1	BODY	CI	PATTERN XXXX	1
1	BASE	CI	PATTERN XXXX	2
1	STANDARD TOP	CI	PATTERN XXXX	3
1	RATCHET	STL	Ø2.375 X .188	4
1	PAWL	CI	PATTERN XXXX	5
1	LIFTING SOCKET	CI	PATTERN XXXX	6
1	LOCKING SCREW	STL	HEX HD .500-13UNC X 1.00	7
1	PIN	STL	Ø.56 X 2.12	8
1	SPRING	STL	SEE DWG	9
1	PLUNGER	STL	Ø.625 X 1.20	10
1	JACK SCREW	STL	Ø1.50 X 8.75	11
1	STANDARD	STL	Ø2.50 X 7.10	12
1	BEARING	BRONZE	Ø1.18 ID X 1.50 OD X 1.90	13
1	LIFTING NUT	BRONZE	Ø 2.00 X 2.50	14
1	BALL PLATE	STL	Ø4.00 X .56	15
1	BEVEL PINION	STL	SEE DWG	16
1	BEARING	BRONZE	.968 ID X 1.25 OD X .50	17
1	BEVEL GEAR	STL	SEE DWG	18
1	WASHER	STL	1.19 ID X 2.25 OD X .188	19
1	PIN	STL	Ø.188 X 1.00	20
12	BALL BEARING	STL	SØ .625	21
1	KEY –WOODRUFF	STL	# 608	22
1	PIN	STL	Ø.25 X .40	23
1	COTTERPIN	STL	Ø.125 X 1.25	24
1	COTTERPIN	STL	Ø.094 X .75	25
1	HANDLE	STL	.875 ID X 1.00 OD X 18	26

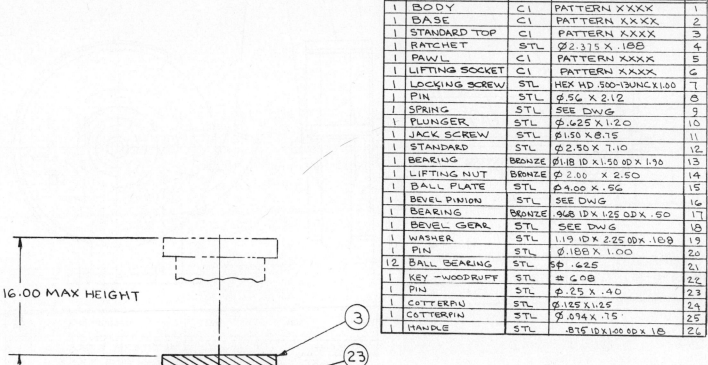

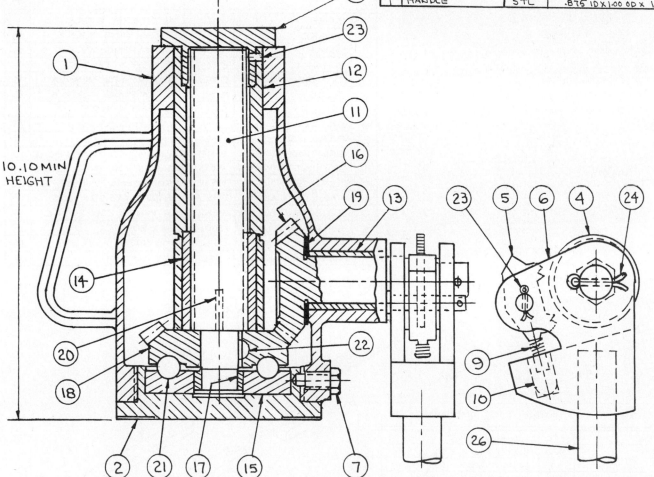

16.00 MAX HEIGHT

10.10 MIN HEIGHT

Unit 14-6

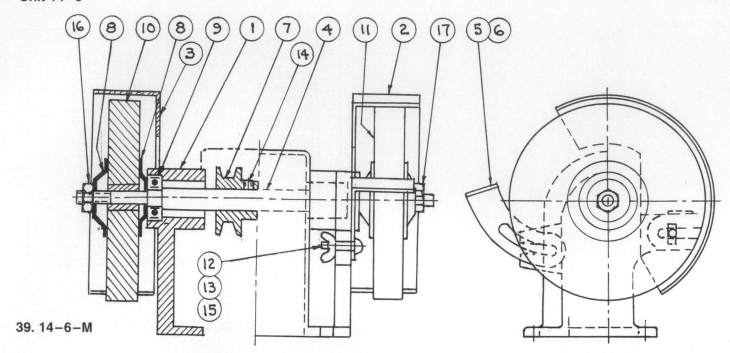

39. 14-6-M

QTY	ITEM	MATL	DESCRIPTION	PT NO.
1	BASE	CI	PATTERN XXXX	1
1	GUARD	STL	STAMPING XXXX	2
1	GUARD	STL	STAMPING XXXX	3
1	SHAFT	CRS	Ø.625 X 10.70	4
1	GUIDE	STL	STAMPING XXXX	5
1	GUIDE	STL	STAMPING XXXX	6
1	PULLEY	STL	Ø 2.00 X 1.25	7
4	SPACER	STL	STAMPING XXXX	8
2	BEARING		SKF 6002-2Z	9
1	GRINDING WHEEL	FINE	Ø6.00 X 1.00, Ø.501 BORE	10
1	GRINDING WHEEL	MED	Ø6.00 X 1.00, Ø.501 BORE	11
4	CARRIAGE BOLT	STL	.250-20UNC X 1.00	12
4	WASHER, PLAIN	TYPE A	.281 ID X .625 OD X .065	13
1	SET SCREW - SLOTTED HEAD,		CUP POINT, .250-20UNC	14
4	WING NUT	STL	.250-20 UNC	15
1	NUT - HEX REG	STL	.500 UNC	16
1	NUT - HEX REG	STL	.500 UNC - LH	17

Unit 14-7

41. 14-2-J

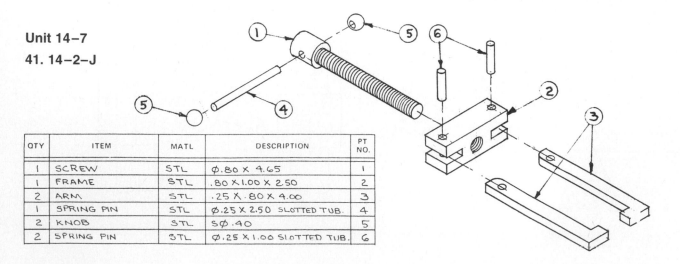

QTY	ITEM	MATL	DESCRIPTION	PT NO.
1	SCREW	STL	Ø.80 X 4.65	1
1	FRAME	STL	.80 X 1.00 X 2.50	2
2	ARM	STL	.25 X .80 X 4.00	3
1	SPRING PIN	STL	Ø.25 X 2.50 SLOTTED TUB.	4
2	KNOB	STL	SØ.40	5
2	SPRING PIN	STL	Ø.25 X 1.00 SLOTTED TUB.	6

Unit 14–7

41. CONTD 14–7–A

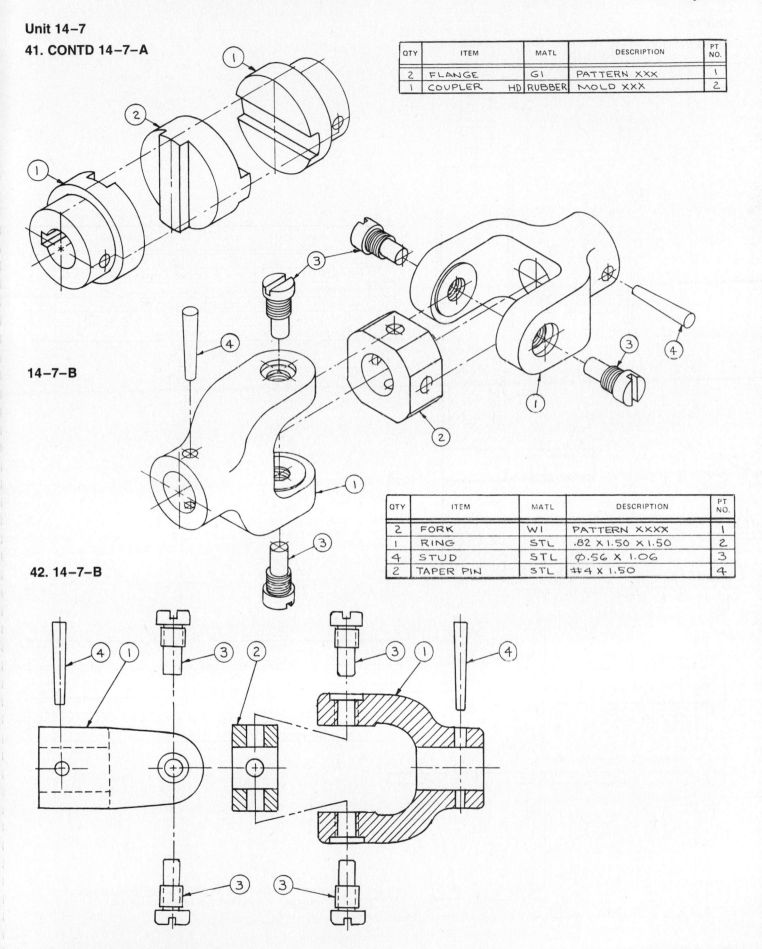

QTY	ITEM	MATL	DESCRIPTION	PT NO.
2	FLANGE	GI	PATTERN XXX	1
1	COUPLER	HD RUBBER	MOLD XXX	2

14–7–B

QTY	ITEM	MATL	DESCRIPTION	PT NO.
2	FORK	WI	PATTERN XXXX	1
1	RING	STL	.82 X 1.50 X 1.50	2
4	STUD	STL	⌀.56 X 1.06	3
2	TAPER PIN	STL	#4 X 1.50	4

42. 14–7–B

Unit 14–7

42. CONTD 14–7–C

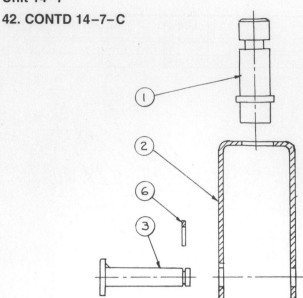

QTY	ITEM		MATL	DESCRIPTION	PT NO.
1	POST	SAE	1112	Ø18 X 43	1
1	BRACKET		STL	2.38 (13GA) X 44X165	2
1	SHAFT	SAE	1112	Ø16 X 40	3
1	WHEEL	HARD	RUBBER	MOLD XXXX	4
1	BUSHING		BRASS	8 ID X 14 OD X 28	5
1	RETAINING RING-EXT		STL	#5133	6

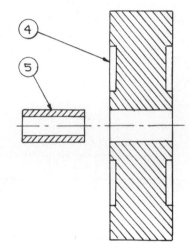

Unit 14–8

43. 14–8–A

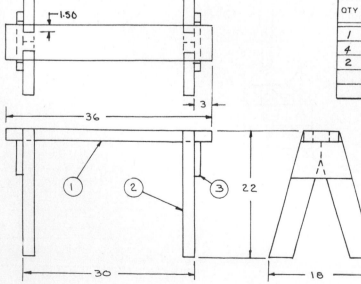

QTY	ITEM	MATL	DESCRIPTION	PT NO.
1	TOP	SPRUCE	2 X 6 X 36	1
4	LEG	SPRUCE	2 X 4 X 22.5	2
2	SUPPORT	SPRUCE	1 X 6 X 11	3
	NAILS, COMMON	STL	2.5 LG	4
	NAILS, COMMON	STL	3.5 LG	5

FASTEN 2X4'S & 2X6 WITH 3.5 COMMON NAILS
FASTEN 1 X6 WITH 2.5 COMMON NAILS

Unit 14–8
43. CONTD 14–8–B

QTY	ITEM	MATL	DESCRIPTION	PT NO.
2	RAFTER (UPPER CHORD	SPRUCE	2 × 4 × 13'-10	1
3	LOWER CHORD	SPRUCE	2 × 4 × 8'-0	2
2	TRUSS	SPRUCE	2 × 4 × 7'-9	3
2	TRUSS	SPRUCE	2 × 4 × 3'-7	4
2	TOP GUSSET	PLYWOOD	.50 × 12 × 18	5
4	END GUSSET	PLYWOOD	.50 × 12 × 18	6
4	TRUSS GUSSET	PLYWOOD	.50 × 12 × 12	7
4	TRUSS GUSSET	PLYWOOD	.50 × 12 × 18	8
10 LB	NAILS	STL	2 IN COMMON	9

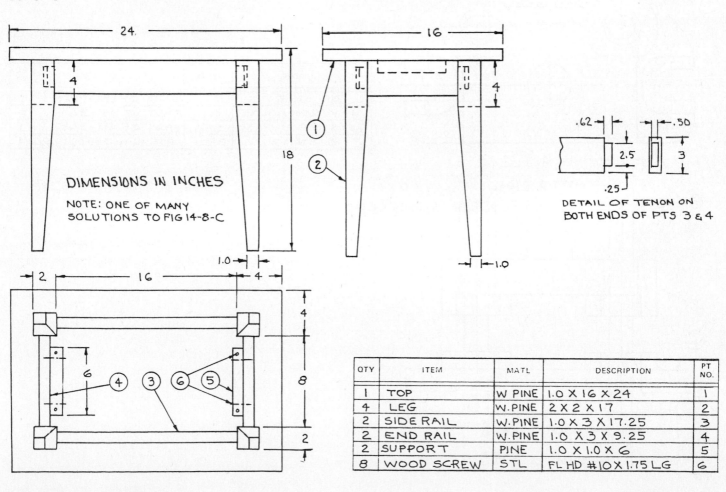

MAX. 1.50

NAILING ARRANGEMENT

30°

30° 60°

8'-0

24'-0

14–8–C

24.

16

18

4

4

DIMENSIONS IN INCHES

NOTE: ONE OF MANY
SOLUTIONS TO FIG 14-8-C

1.0

1.0

.62

.50

2.5

3

.25

DETAIL OF TENON ON
BOTH ENDS OF PTS 3 & 4

2   16   4

4

6

8

2

QTY	ITEM	MATL	DESCRIPTION	PT NO.
1	TOP	W PINE	1.0 × 16 × 24	1
4	LEG	W. PINE	2 × 2 × 17	2
2	SIDE RAIL	W. PINE	1.0 × 3 × 17.25	3
2	END RAIL	W. PINE	1.0 × 3 × 9.25	4
2	SUPPORT	PINE	1.0 × 1.0 × 6	5
8	WOOD SCREW	STL	FL HD #10 × 1.75 LG	6

Unit 14–8
43. CONTD 14–8–D

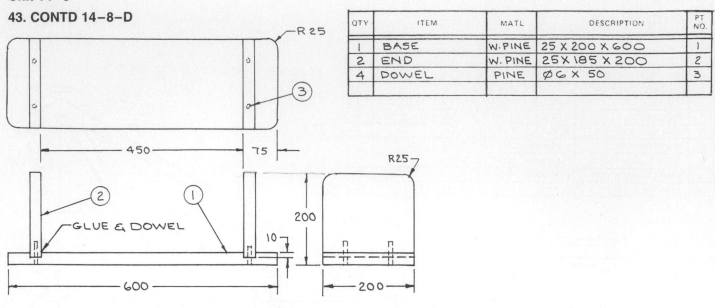

QTY	ITEM	MATL	DESCRIPTION	PT NO.
1	BASE	W. PINE	25 X 200 X 600	1
2	END	W. PINE	25 X 185 X 200	2
4	DOWEL	PINE	ϕ 6 X 50	3

Unit 14–9
44. 14–9–A

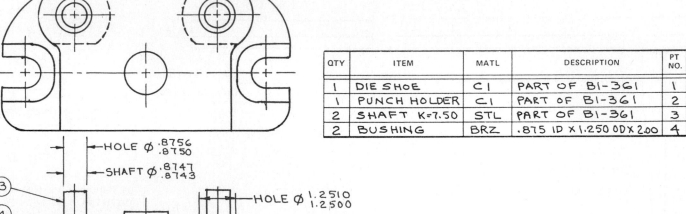

QTY	ITEM	MATL	DESCRIPTION	PT NO.
1	DIE SHOE	CI	PART OF BI-361	1
1	PUNCH HOLDER	CI	PART OF BI-361	2
2	SHAFT K=7.50	STL	PART OF BI-361	3
2	BUSHING	BRZ	.875 ID X 1.250 OD X 2.00	4

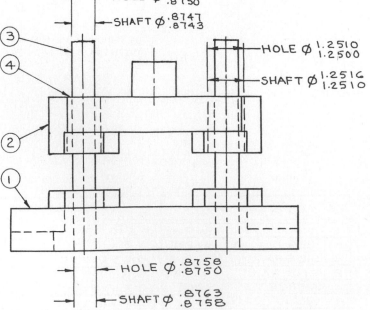

Unit 14−9

45. 14−9−B

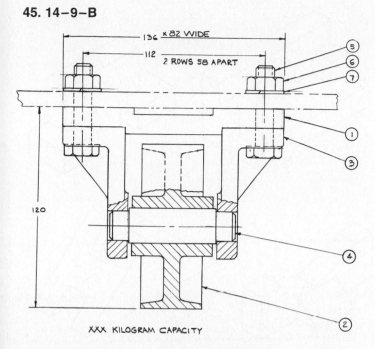

XXX KILOGRAM CAPACITY

QTY	ITEM	MATL	DESCRIPTION	PT
1	TOP PLATE	M1	PATTERN XXXX	1
1	WHEEL	M.I.	PATTERN XXXX	2
2	AXLE SUPPORT	M.I.	PATTERN XXXX	3
1	AXLE SAE 1112		Ø 22 X 78	4
4	BOLT HEX HD	STL	M10 X 50 LG	5
4	NUT HEX HD	STL	M10	6
4	WASHER	STL	M10 HELICAL SPRING	7

46. 14−9−C

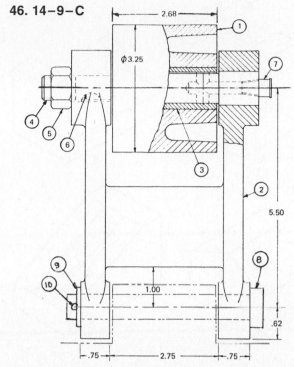

QTY	ITEM	MATL	DESCRIPTION	PT NO.
1	IDLER PULLEY	GI	A−5432	1
1	IDLER PULLEY FRAME	GI	A−1734	2
1	BUSHING	BRZ	Ø1.06 X 2.68	3
1	SHAFT	SAE 1120	Ø.875 X 5.25	4
1	NUT − REG HEX	STL	.625 UNC	5
1	WOODRUFF KEY	STL	#405	6
1	OILER	STD	.125 NPT	7
1	CLEVIS PIN	STD	Ø.625 X 4.75	8
1	WASHER, PLAIN	TYPE A	.656 ID X 1.312 OD X .095	9
1	COTTER PIN	STD	Ø.125 X 1.25 LG	10

47. 14−9−D

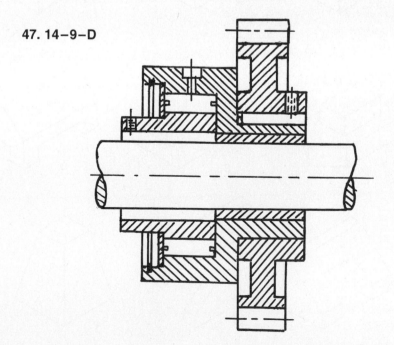

Chapter 15 Pictorial Drawings
Unit 15–1, 1. 15–1–A

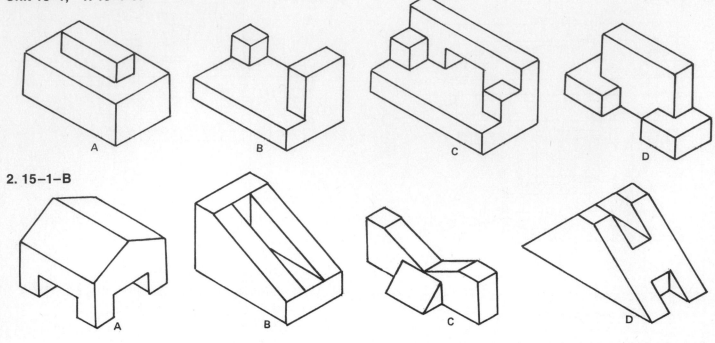

2. 15–1–B

3. 15–1–C

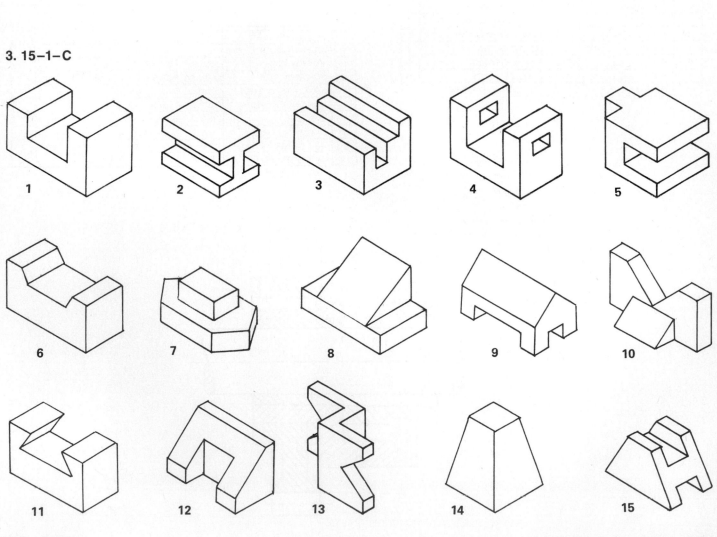

Unit 15-1

4. 15-1-D

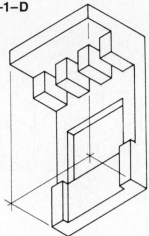

15-1-E

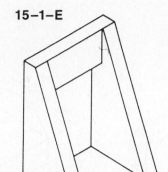

15-1-F

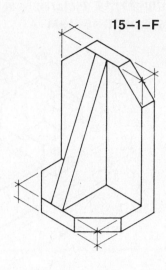

5. 15-1-G

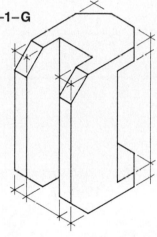

15-1-H

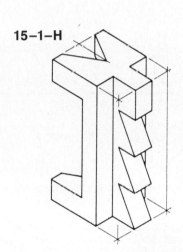

15-1-J

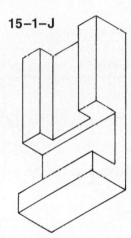

6. 15-1-K

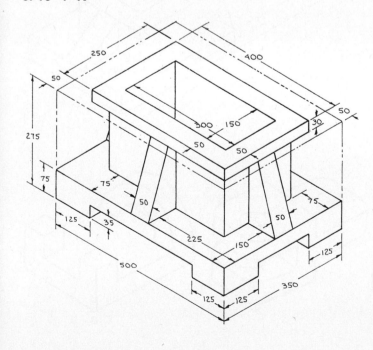

15-1-L

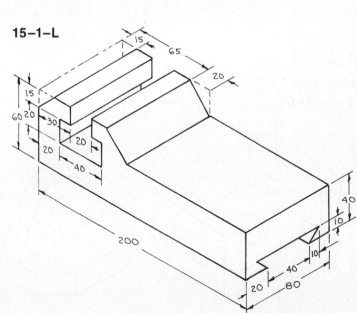

Unit 15–1
6. CONTD 15–1–M

15–1–N

15–1–P

15–1–Q

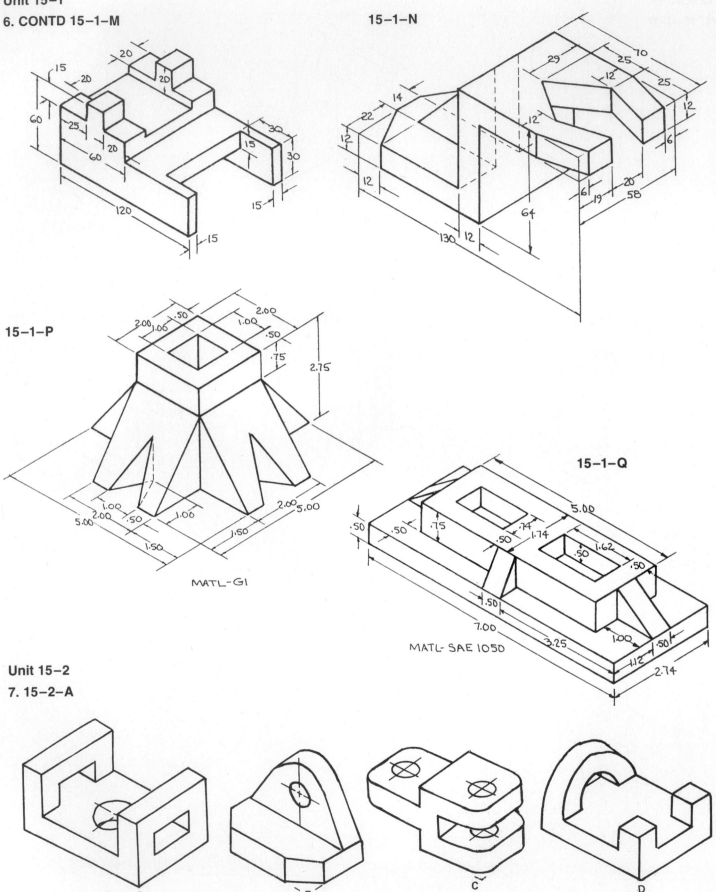

MATL-GI

MATL- SAE 1050

Unit 15–2
7. 15–2–A

A

B

C

D

Unit 15–2

8. 15–2–B

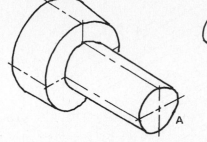 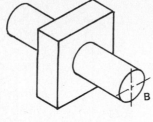 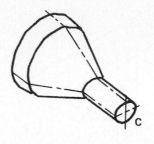 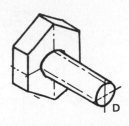

9. 15–2–C

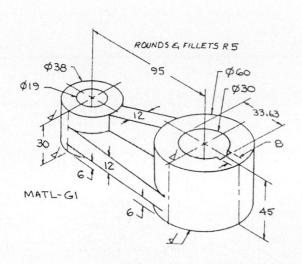

ROUNDS & FILLETS R 5

MATL–G1

15–2–D

Ø45
⌴Ø60
▽6

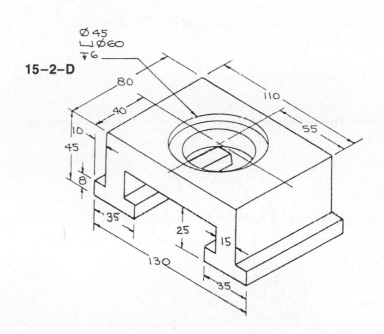

15–2–E

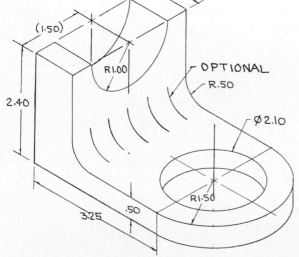

OPTIONAL
R.50

15–2–F

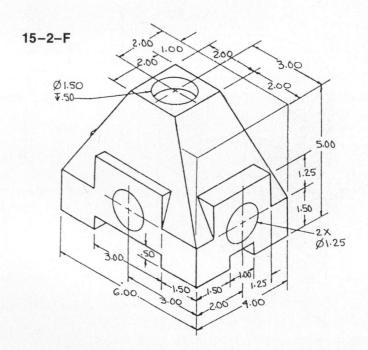

Unit 15–3

10. 15–3–A

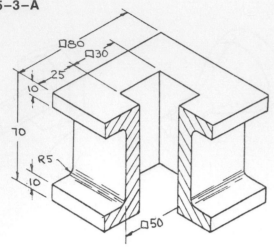

15–3–B

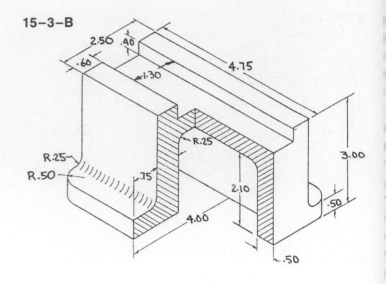

11. 15–3–C

15–3–D

15–3–E

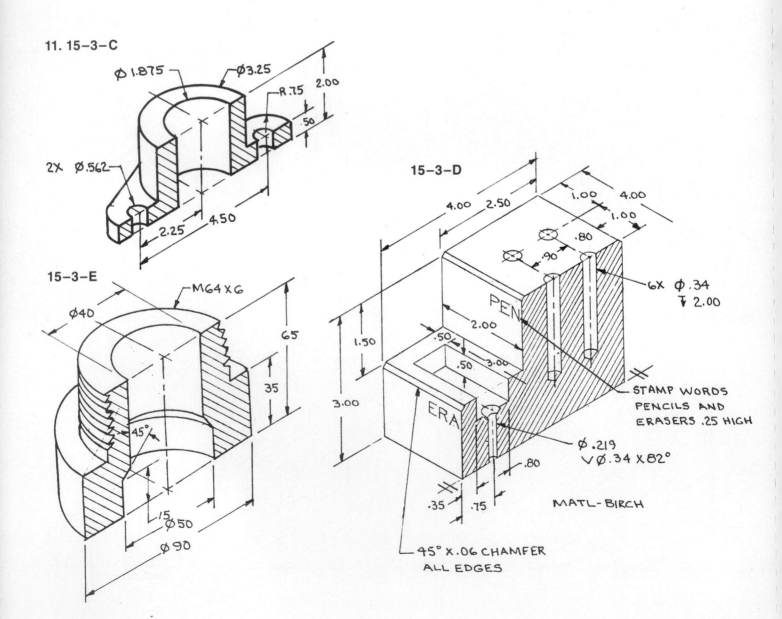

Unit 15–3

12. 15–3–F

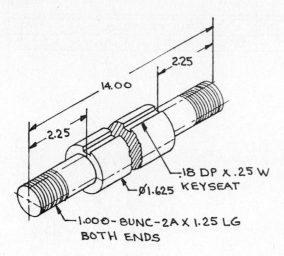

14.00

2.25

2.25

.18 DP X .25 W KEYSEAT

Ø1.625

1.000–8UNC–2A X 1.25 LG
BOTH ENDS

15–3–G

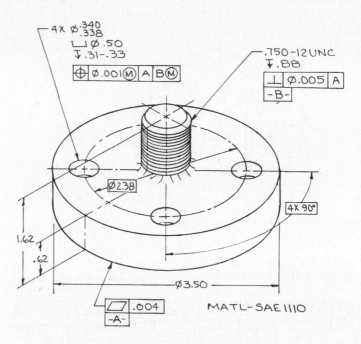

4X Ø.340
  .338
⌴ Ø.50
↧ .31–.33
⊕ Ø.001 Ⓜ A BⓂ

.750–12UNC
↧ .88
⊥ Ø.005 A
–B–

Ø2.38

4X 90°

1.62

.62

Ø3.50

▱ .004
–A–

MATL–SAE 1110

13. 15–3–H

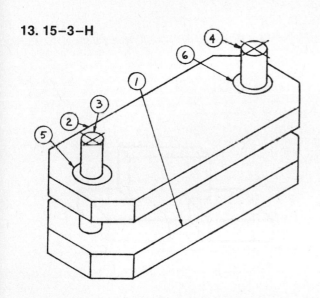

QTY	ITEM	MATL	DESCRIPTION	PT
1	BASE	GI	DWG XXX	1
1	TOP	GI	DWG XXX	2
1	POST	STL	Ø1.00 X 6.50 LG	3
1	POST	STL	Ø1.25 X 6.50 LG	4
1	BUSHING	STL	BOSTON GEAR–CAT XXX	5
1	BUSHING	STL	BOSTON GEAR–CAT XXX	6

14. 15–3–J

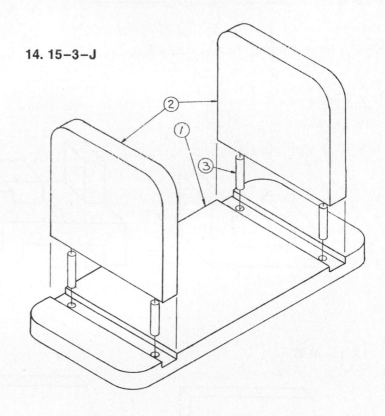

QTY	ITEM	MATL	DESCRIPTION	PT
1	BASE	PINE	20 X 150 X 320	1
2	END	PINE	20 X 150 X 134	2
4	DOWEL	PINE	Ø6 X 50 LG	3
				4

Unit 15–3

15. 15–3–K

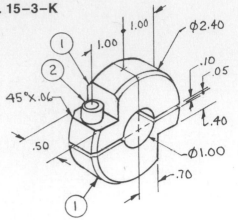

QTY	ITEM	MATL	DESCRIPTION	PT NO.
2	CLAMP	CAST	STL PATTERN XXX	1
2	CAP SCREW	STL	10-24 UNC X 1.00 SOC. HD	2

16. 15–3–L

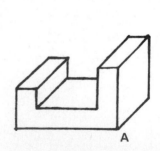

QTY	ITEM	MATL	DESCRIPTION	PT
2	FORK	GI	DWG XXXX	1
1	RING	STL	1.00 X 1.00 X .62	2
2	SPRING PIN	STL	Ø .25 X 1.00 LG	3
4	FHMS	STL	.250-20 X .62 LG	4

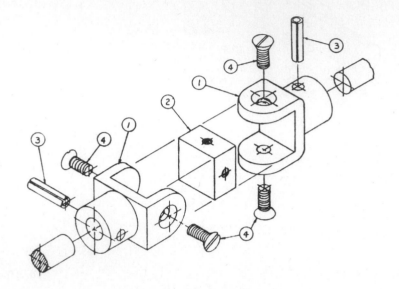

Unit 15–4

17. 15–4–A

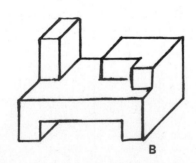
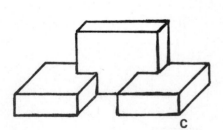

A   B   C

18. 15–4–B

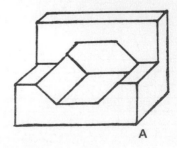
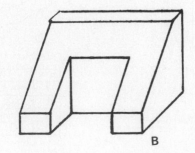
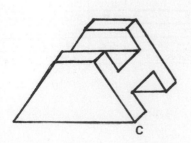

A   B   C

Unit 15–4

19. 15–4–C

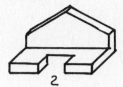
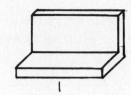
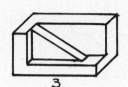
1   2   3   4
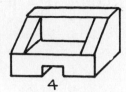

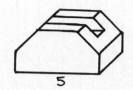
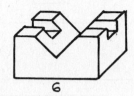
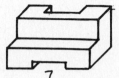
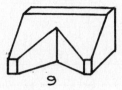
5   6   7   8   9

20. 15–4–D

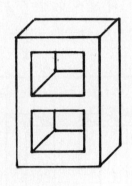

15–4–E

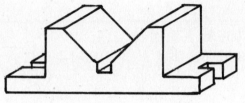

21. 15–4–F

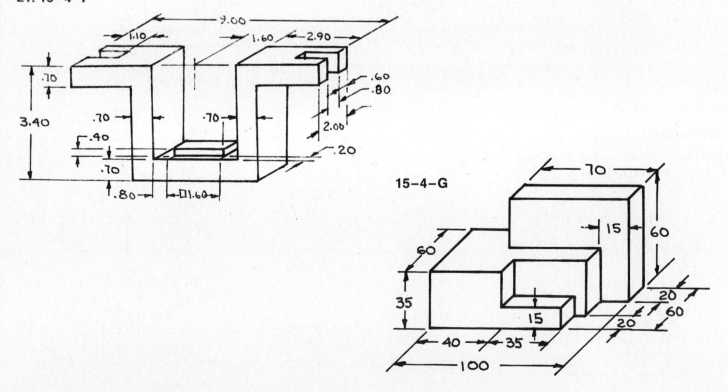

15–4–G

Unit 15–4

21. CONTD 15–4–H

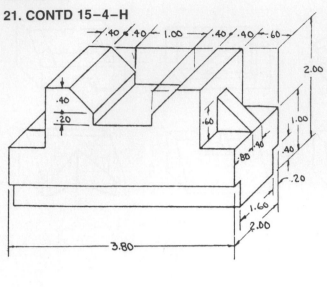

15–4–J

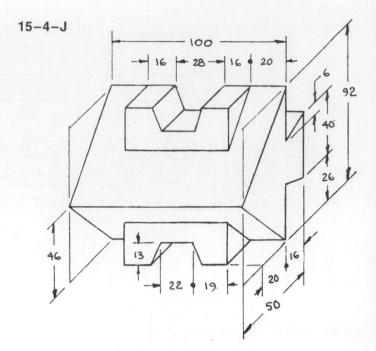

Unit 15–5

22. 15–5–A

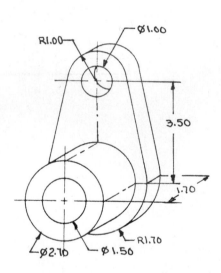

15–5–B

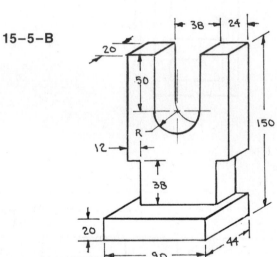

23. 15–5–C

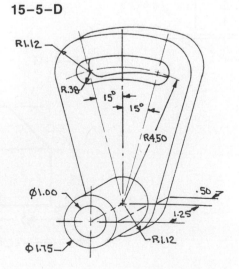

15–5–D

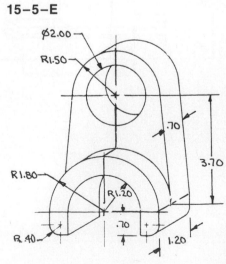

15–5–E

Unit 15-5

23. CONTD 15-5-F

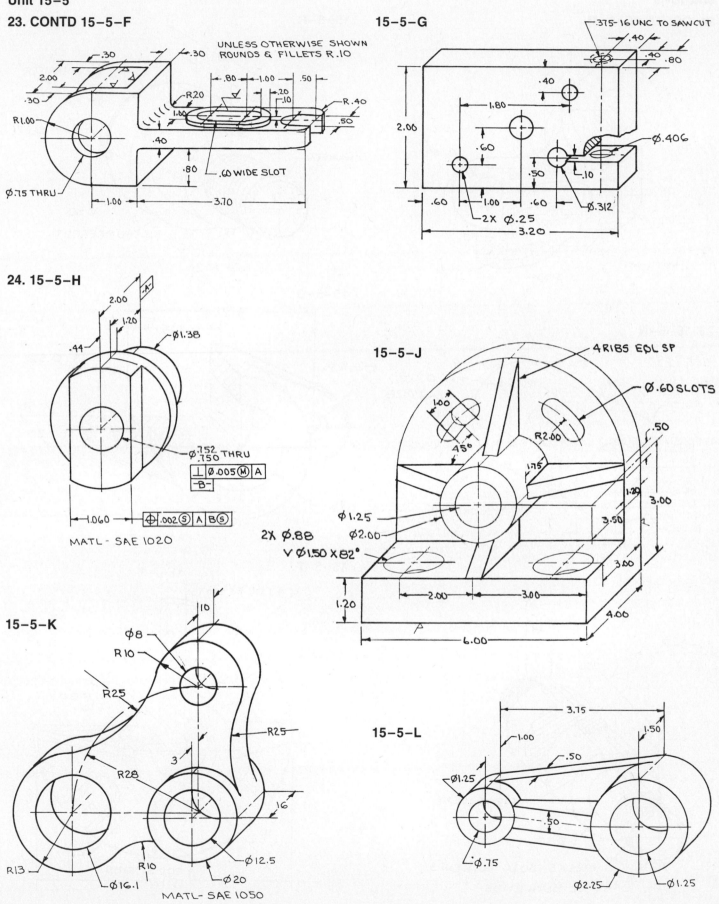

UNLESS OTHERWISE SHOWN
ROUNDS & FILLETS R.10

15-5-G

24. 15-5-H

MATL - SAE 1020

15-5-J

15-5-K

15-5-L

MATL- SAE 1050

Unit 15–5

24. CONTD 15–5–M

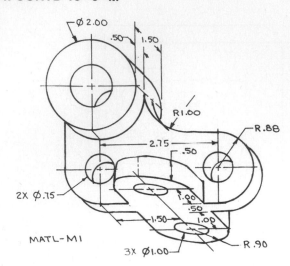

15–5–R

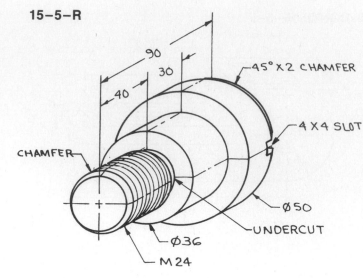

25. 15–5–N

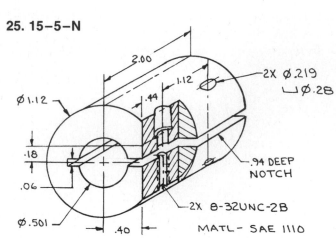

15–5–S

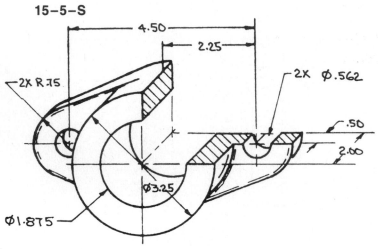

15–5–P

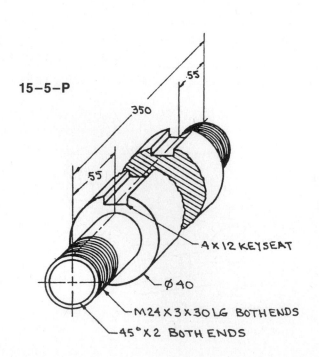

15–5–T

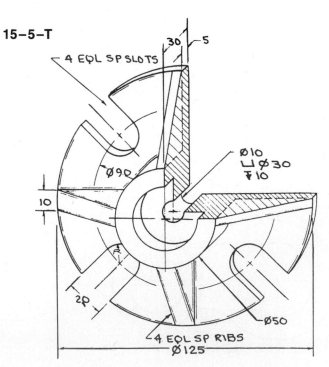

Unit 15–5

25. CONTD 15–5–U

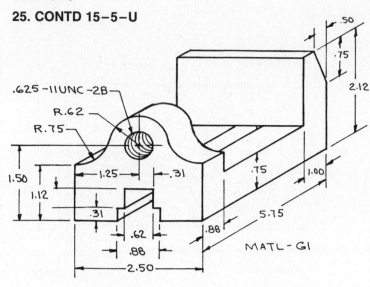

.625 –11UNC –2B

R.62

R.75

1.50
1.12
.31
.62
.88
2.50
1.25
.31
.75
.88
.50
.75
2.12
1.00
5.75

MATL – GI

Unit 15–6

26. 15–6–A

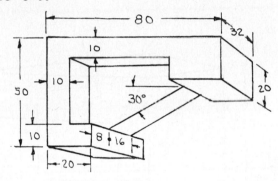

80
32
10
10
50
20
30°
10
8 16
20

15–6–B

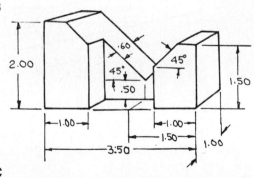

2.00
.60
45°
45°
.50
1.50
1.00
1.00
1.00
1.50
3.50

15–6–C

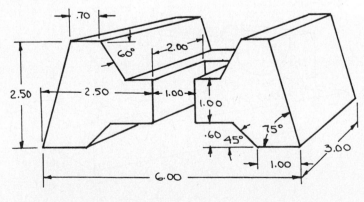

.70
60°
2.00
2.50
2.50
1.00
1.00
.60
45°
75°
1.00
3.00
6.00

27. 15–6–D

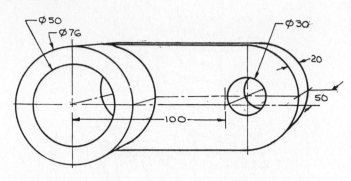

Ø50
Ø76
Ø30
20
50
100

15–6–E

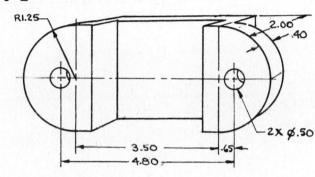

R1.25
2.00
.40
3.50
.65
4.80
2X ø.50

15–6–F

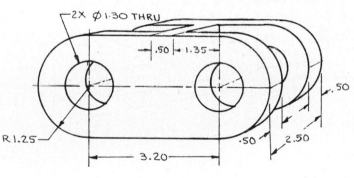

2X Ø1.30 THRU
.50 1.35
.50
R1.25
.50
2.50
3.20

28. 15–6–G

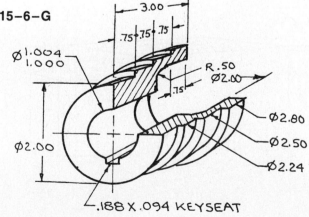

3.00
.75 .75 .75
Ø1.004
1.000
R.50
Ø2.00
.75
Ø2.80
Ø2.50
Ø2.24
Ø2.00
.188 X .094 KEYSEAT

Unit 15-6

28. CONTD 15-6-H

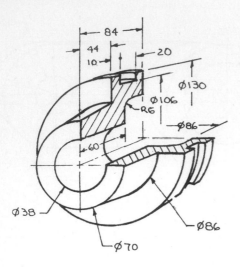

15-6-J

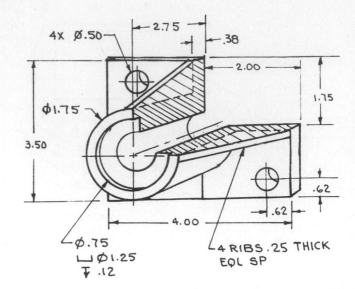

Unit 15-7

29. 15-7-A

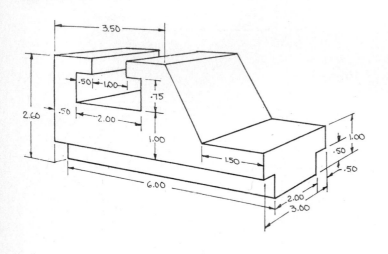

15-7-B

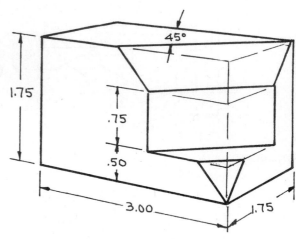

15-7-C

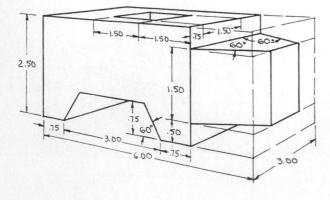

15-7-D

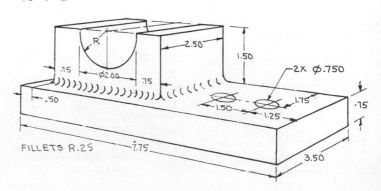

Unit 15–7
29. CONTD 15–7–E

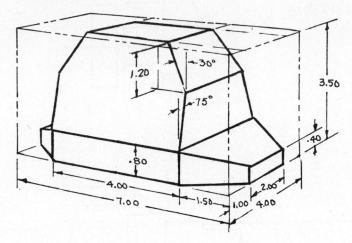

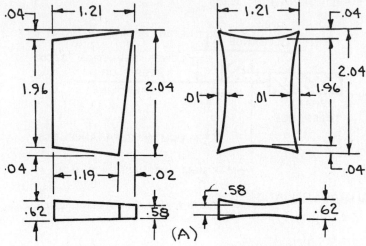

(A)

15–7–F

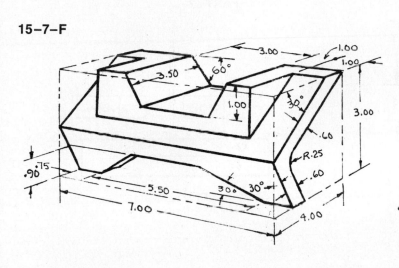

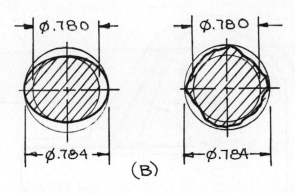

(B)

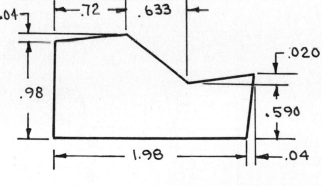

15–7–G

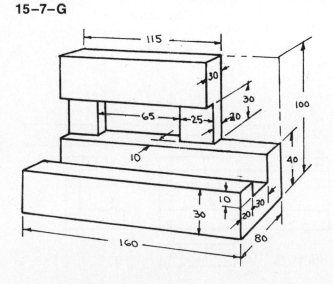

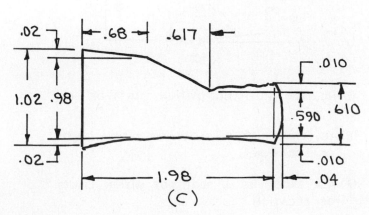

(C)

Unit 16–1

2. 16–1–B

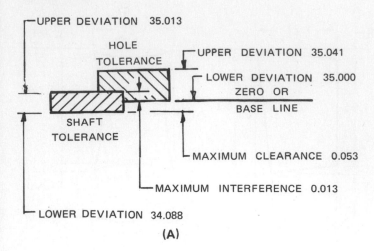

(A)

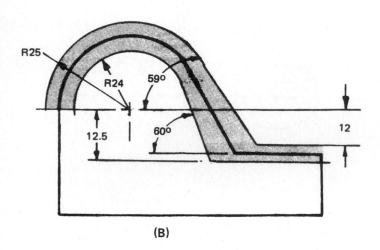

(B)

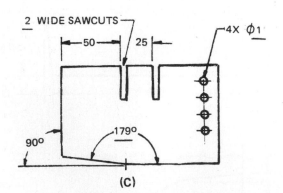

(C)

(D) NO. TOP WIDTH NOT WITHIN LIMITS OF SIZE.

(E) ALL EXCEPT PT 2. SLOT NOT WITHIN LIMITS OF LOCATION.

(F) ALL EXCEPT PT 4. SLOT NOT WITHIN LIMITS OF LOCATION.

Unit 16–2

3. 16–2–A

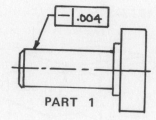

PART 1

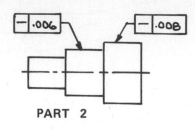

PART 2

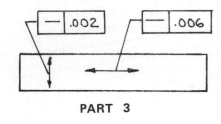

PART 3

PART 4 (A) .002

(B) .002

(C) .002

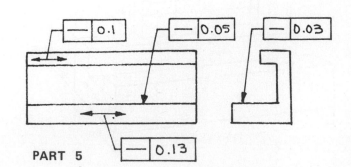

PART 5

Unit 16–3

4. 16–3–A

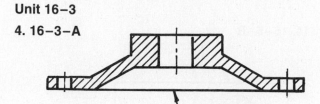

5. 16–3–B

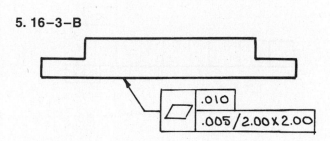

6. 16–3–C

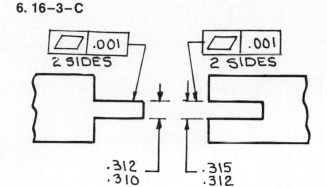

7. 16–3–D

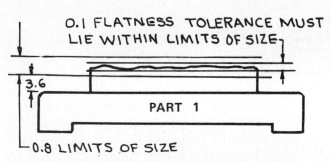

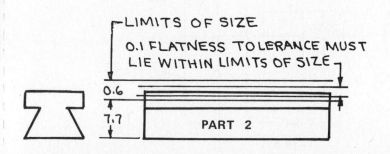

Unit 16–4

8. 16–4–A

A. ⌀1.511   B. ⌀1.127

C. ⌀1.752   D. ⌀1.880

9. 16–4–B   .003 AT LMC

10. 16–4–C

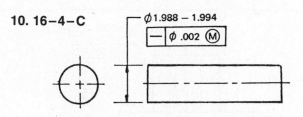

⌀1.988 – 1.994

— ⌀ .002 Ⓜ

FEATURE SIZE (DIA)	PERMISSIBLE STRAIGHTNESS TOLERANCE (DIA)
1.994	.002
1.993	.003
1.992	.004
1.991	.005
1.990	.006
1.989	.007
1.988	.008

16–4–D

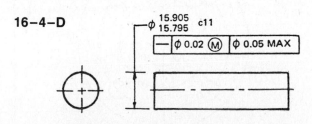

⌀ 15.905 / 15.795 c11

— ⌀ 0.02 Ⓜ ⌀ 0.05 MAX

FEATURE SIZE (DIA)	PERMISSIBLE STRAIGHTNESS TOLERANCE (DIA)
15.905	0.02
15.895	0.03
15.885	0.04
15.875	0.05
15.865	0.05
15.845	0.05
15.795	0.05

11. 16–4–E

PART A NOT ACCEPTABLE. EXCEEDS STRAIGHTNESS TOLERANCE

PART B NOT ACCEPTABLE. NOT WITHIN LIMITS OF SIZE.

PART C NOT ACCEPTABLE. NOT WITHIN LIMITS OF SIZE.

PART D ACCEPTABLE.   PART E ACCEPTABLE.

12. 16–4–F   (A) .001   (B) .001   (C) .001

Unit 16–5

13. 16–5–A

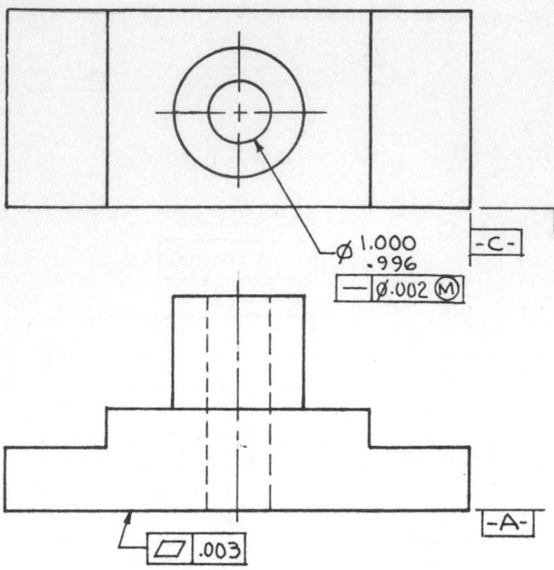
-B-
-C-
Ø 1.000
.996
— Ø.002 Ⓜ
-A-
.003

14. 16–5–B

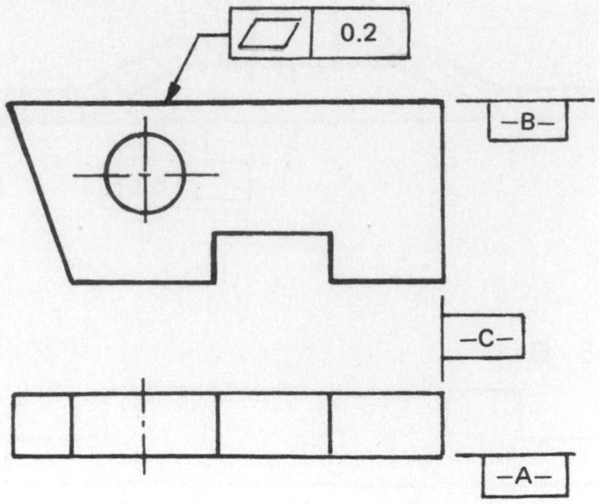
0.2
–B–
–C–
–A–

15. 16–5–C

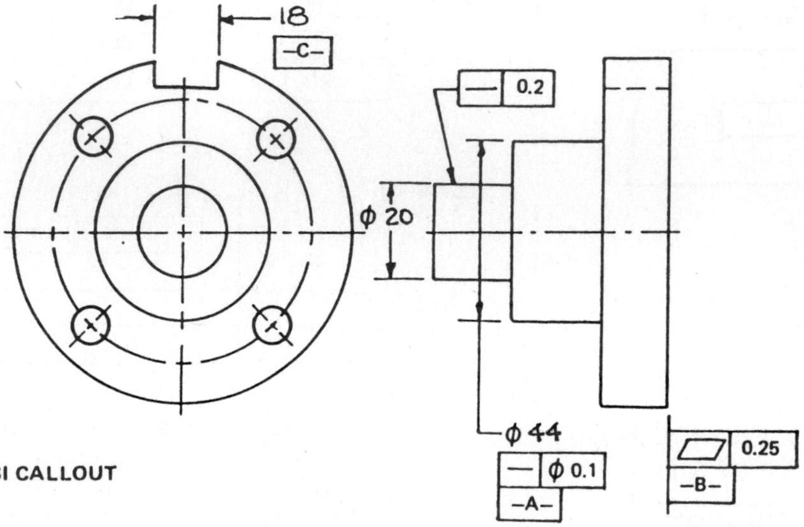
18
–C–
0.2
ϕ 20
ϕ 44
ϕ 0.1
–A–
0.25
–B–

ANSI CALLOUT

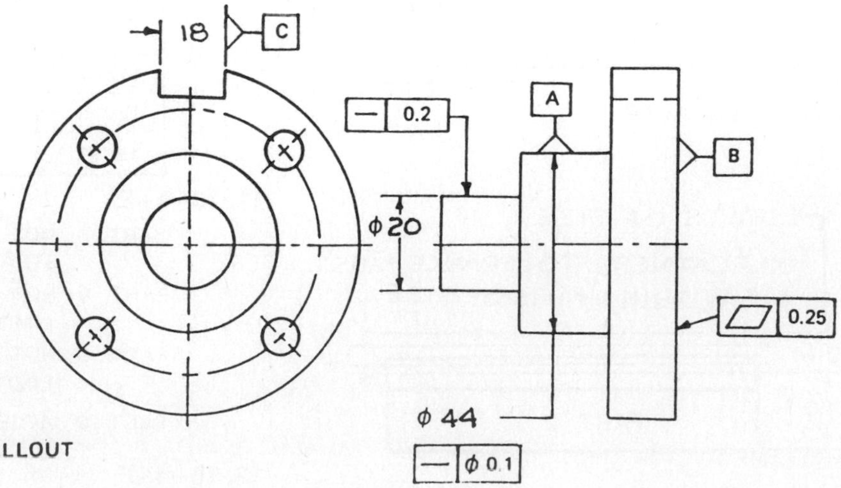
18
C
0.2
A
B
ϕ20
0.25
ϕ 44
ϕ 0.1

ISO CALLOUT

Unit 16-5

16. 16-5-D

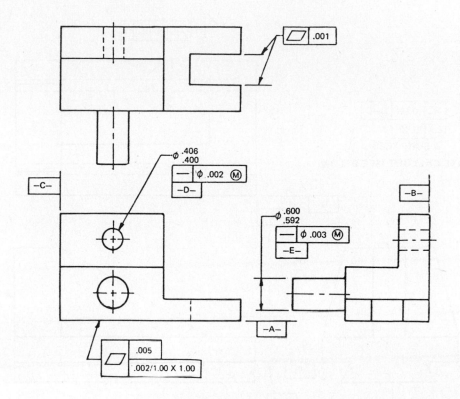

Unit 16-6

17. 16-6-A

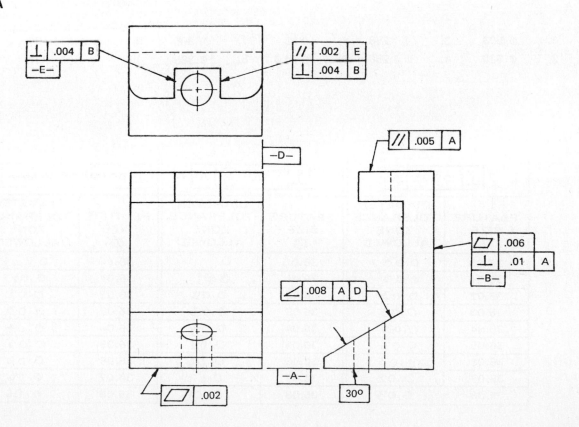

Unit 16-6
18. 16-6-B

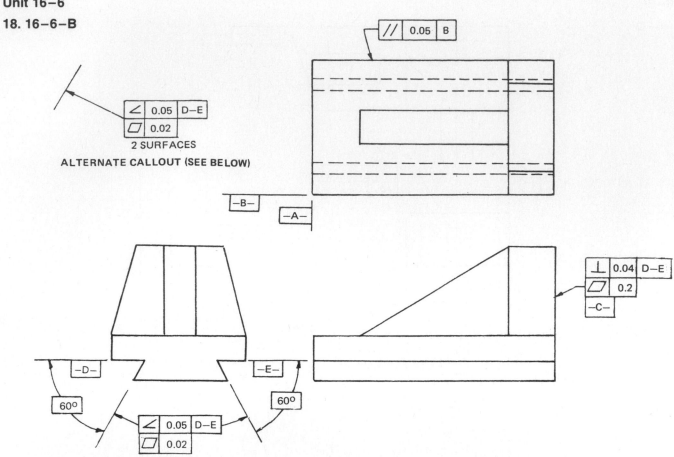

Unit 16-7
19. 16-7-A

1.	φ .503	3.	φ 1.258	5.	φ 25	7.	φ .396	9.	φ .866	11.	φ 34.35
2.	φ .500	4.	φ 1.258	6.	φ 24.9	8.	φ .388	10.	φ .866	12.	φ 34.74

Unit 16-8
20. 16-8-A

⊥ ⌀0.03 A		⊥ ⌀0 Ⓜ A		⊥ ⌀0 Ⓜ ⌀0.06 MAX A	
FEATURE SIZE Ø	DIAMETER TOLERANCE ZONE ALLOWED	FEATURE SIZE Ø	DIAMETER TOLERANCE ZONE ALLOWED	FEATURE SIZE Ø	DIAMETER TOLERANCE ZONE ALLOWED
36.00	0.03	36.00	0	36.00	0
36.01	0.03	36.01	0.01	36.01	0.01
36.02	0.03	36.02	0.02	36.02	0.02
36.03	0.03	36.03	0.03	36.03	0.03
36.04	0.03	36.04	0.04	36.04	0.04
36.05	0.03	36.05	0.05	36.05	0.05
36.06	0.03	36.06	0.06	36.06	0.06
36.07	0.03	36.07	0.07	36.07	0.06
36.08	0.03	36.08	0.08	36.08	0.06

Unit 16−8

21. 16−8−B

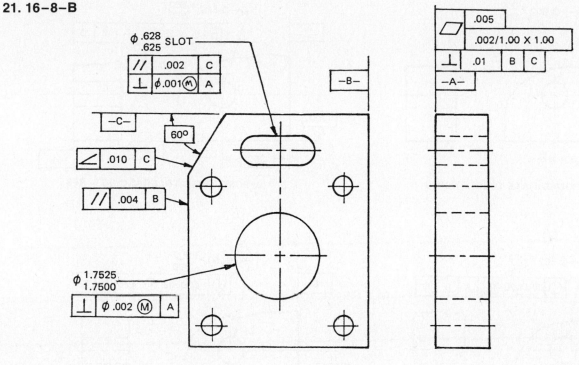

22. 16−8−C

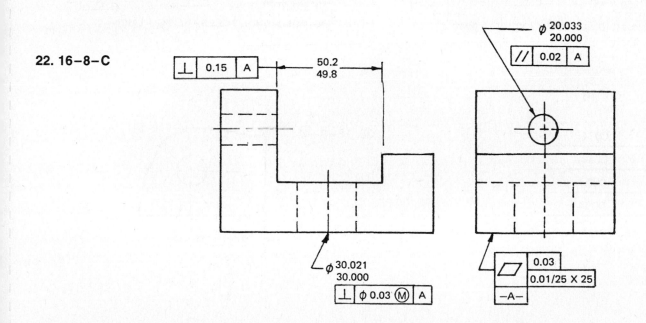

Unit 16−9

23. 16−9−A

(A) .02 SQUARE TOLERANCE ZONE   (B) .04 X .06 RECTANGULAR TOLERANCE ZONE

.028 DISTANCE BETWEEN   .072 DISTANCE BETWEEN

EXTREME VARIATIONS   EXTREME VARIATIONS

Unit 16−9

24. 16−9−B

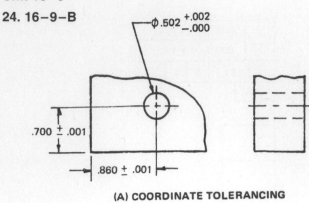

(A) COORDINATE TOLERANCING

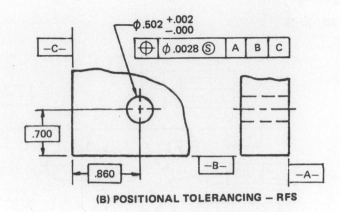

(B) POSITIONAL TOLERANCING − RFS

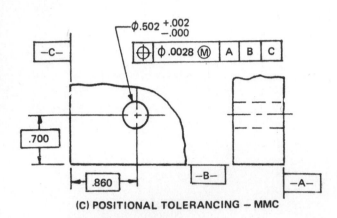

(C) POSITIONAL TOLERANCING − MMC

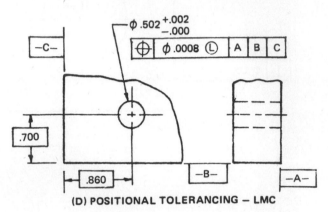

(D) POSITIONAL TOLERANCING − LMC

25.

A	.0014
B	.0014
C	.0024
D	.0004

26. 16−9−C

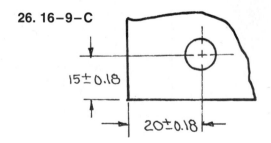

27. 16−9−D

(A)		(B)		(C)	(D)	
PT. 1	ACCEPTABLE	PT.1	0.11	19.7	PT.1	0.30
PT. 2	REJECT	PT. 2	0.125		PT. 2	0.36
PT. 3	REJECT	PT. 3	0.13		PT.3	0.26
PT. 4	ACCEPTABLE	PT. 4	0.12		PT. 4	0.24
PT. 5	ACCEPTABLE	PT. 5	0.05		PT. 5	0.30
PT. 6	ACCEPTABLE	PT. 6	0.11		PT. 6	0.35

Unit 16–10

28. 16–10–A

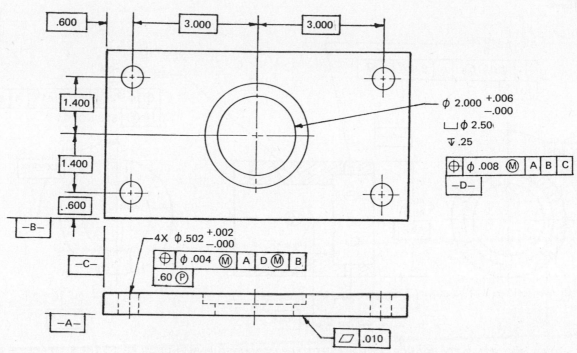

NOTE: DIRECTING THE PROJECTED TOLERANCE NOTE TO THE TOP VIEW MAY BE AMBIGUOUS AS TO WHICH DIRECTION THE NOTE APPLIES. IN THIS PARTICULAR CASE WE RECOMMEND DIRECTING THE NOTE TO THE FRONT VIEW.

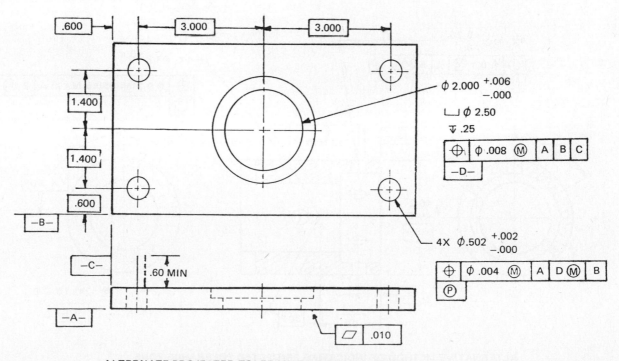

ALTERNATE PROJECTED TOLERANCE ZONE CALLOUT

Unit 16–10

29. 16–10–B

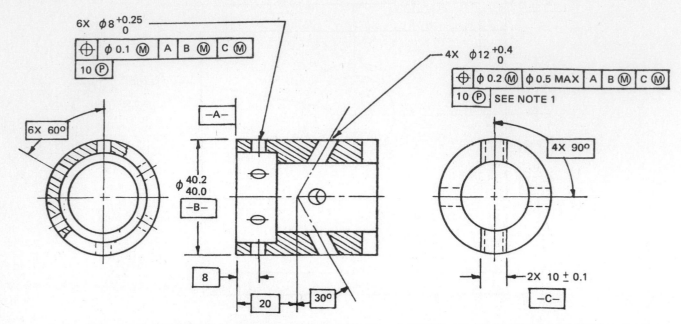

NOTE 1 PROJECTED TOLERANCE ZONE HEIGHT TO BE MEASURED PERPENDICULAR TO THE SURFACE OF DATUM B

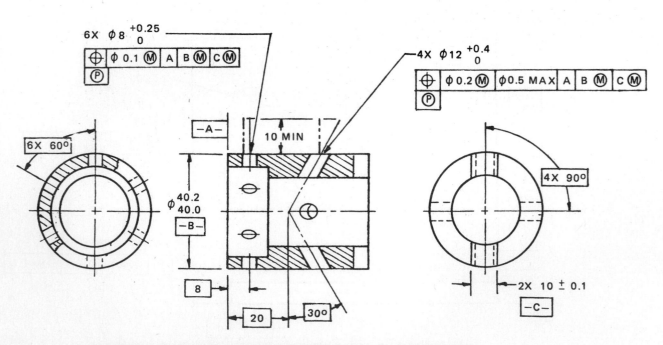

ALTERNATIVE METHOD OF INDICATING PROJECTED TOLERANCE ZONE

Unit 16-11

30. 16-11-A

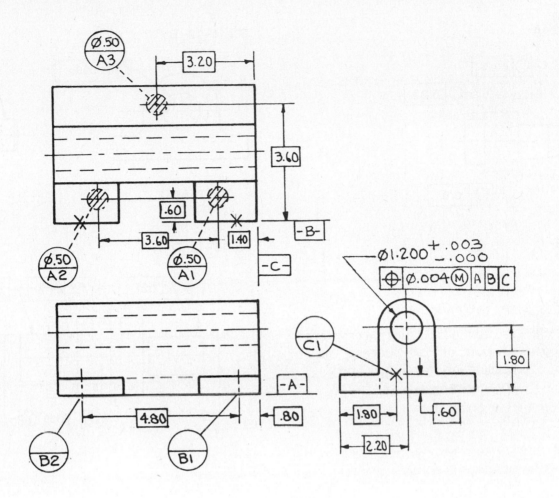

31. 16-11-B

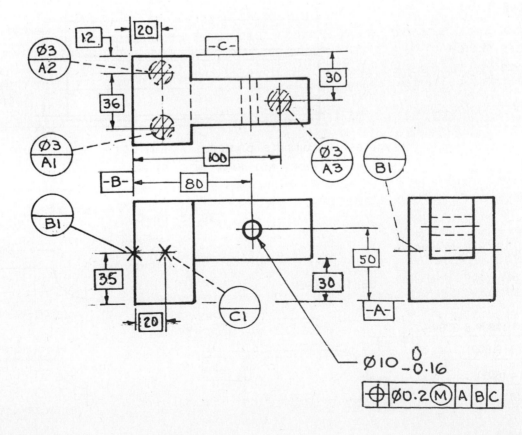

Unit 16-12

32. 16-12-A

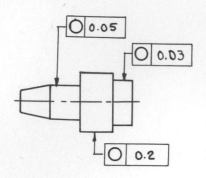

33. 16-12-B

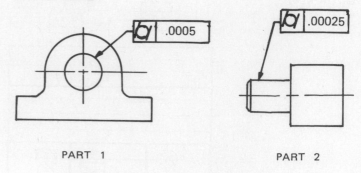

PART 1   PART 2

34. 16-12-C

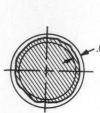

.001 TOLERANCE ZONE

SECTION B–B IS NOT ACCEPTABLE
AS PART OF THE SURFACE LIES
OUTSIDE OF THE CIRCULARITY
TOLERANCE

35. 16-12-D

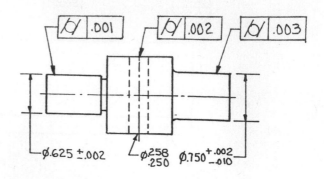

36. 16-12-E

(A) NO

(B) SEE SKETCH

(C) YES

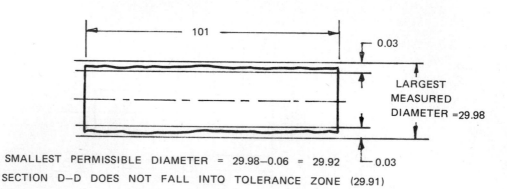

SMALLEST PERMISSIBLE DIAMETER = 29.98–0.06 = 29.92
SECTION D–D DOES NOT FALL INTO TOLERANCE ZONE (29.91)

39. 16-13-C

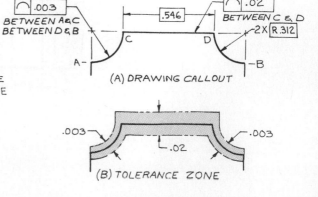

Unit 16-13

37. 16-13-A

.A CAM 1 (180° & 300°)

CAM 3 (180°)

CAM 4 (60°)

.B CAM 3

38. 16-13-B

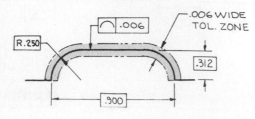

Unit 16–13

40. 16–13–D

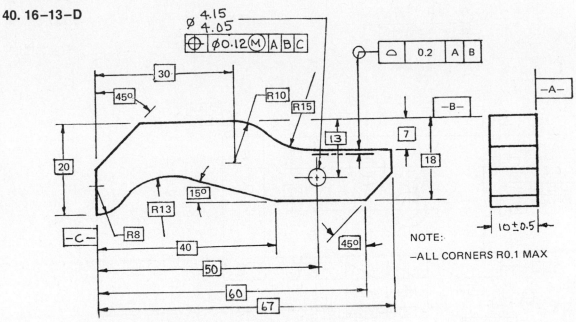

NOTE:
–ALL CORNERS R0.1 MAX

41. 16–13–E

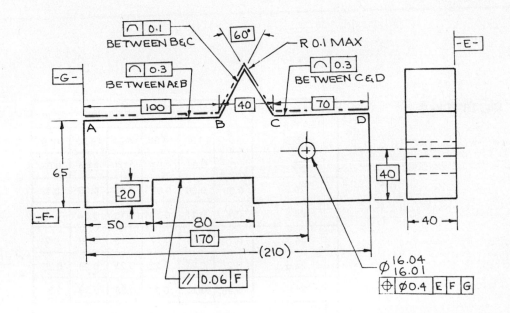

Unit 16–14

42. 16–14–A

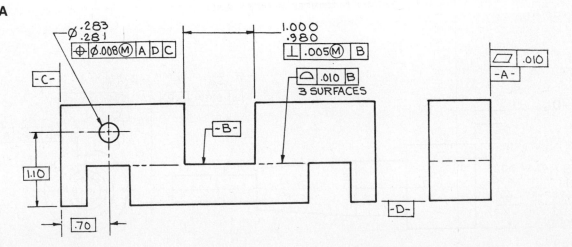

Unit 16–14

43. 16–14–B

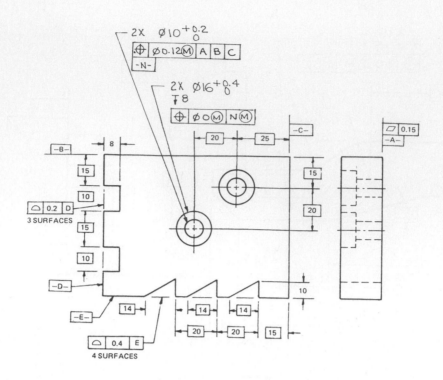

44. 16–14–C

CONSIDERED FEATURE SIZES	DATUM FEATURE SIZES					
	15	14.98	14.96	14.94	14.92	14.9
25	0	0.01	0.02	0.03	0.04	0.05
24.9	0.05	0.06	0.07	0.08	0.09	0.1
24.8	0.1	0.11	0.12	0.13	0.14	0.15
24.7	0.15	0.16	0.17	0.18	0.19	0.2
24.6	0.2	0.21	0.22	0.23	0.24	0.25
24.5	0.25	0.26	0.27	0.28	0.29	0.3

ALLOWABLE DISTANCES BETWEEN AXES

45. 16–14–D

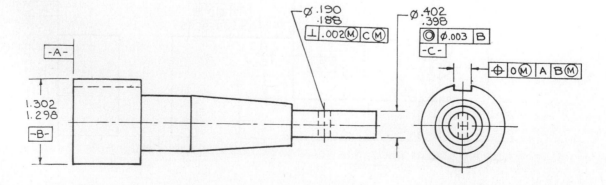

Unit 16-14

46. 16-14-E

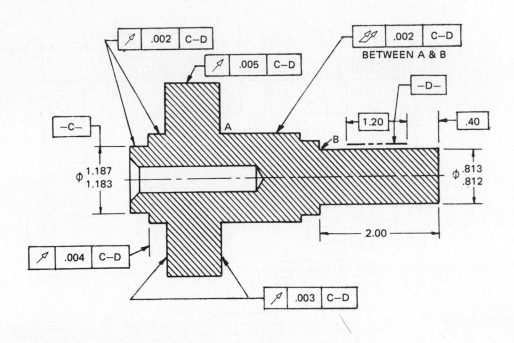

47. 16-14-F

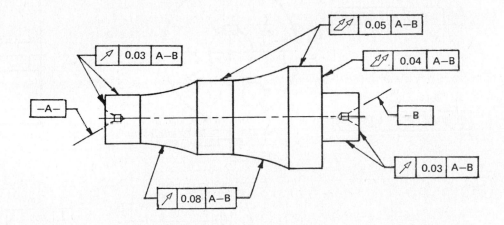

48. 16-14-G

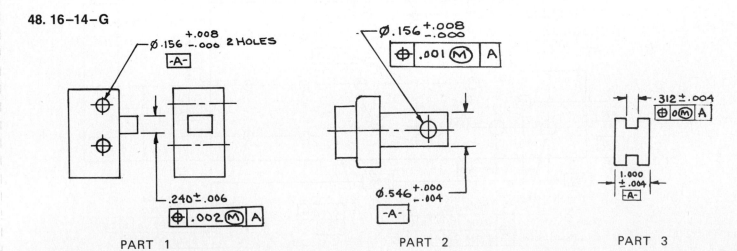

PART 1   PART 2   PART 3

Unit 16–15

49. 16–15–A

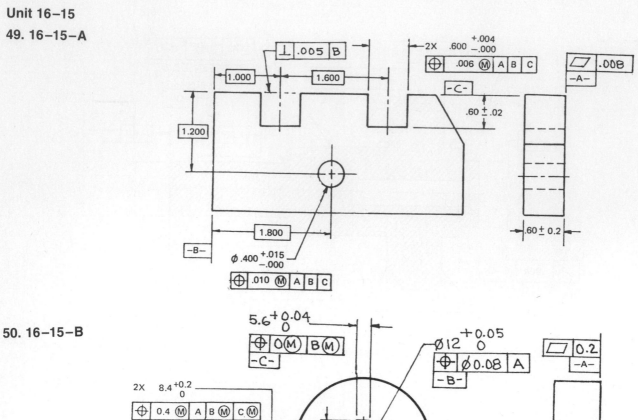

50. 16–15–B

Unit 16–16

51. 16–16–A

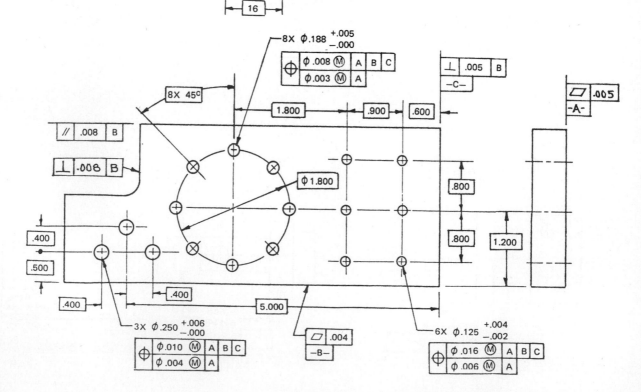

Unit 16–16

52. 16–16–B

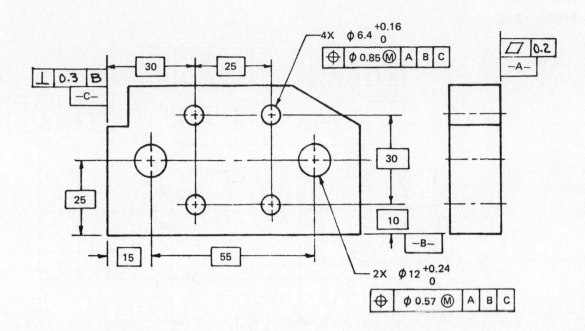

53. 16–16–C

HOLE	MAXIMUM DIAMETER TOLERANCE ZONE PERMITTED
A	0.67
B	0.79
C	0.90
D	0.99
E	0.85
F	1.01

54. 16–16–D

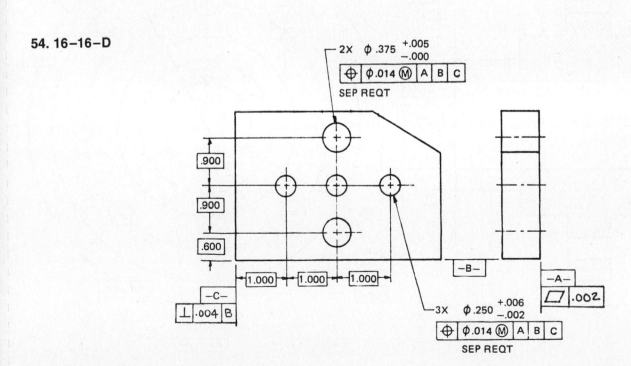

Unit 16–16

55. 16–16–E

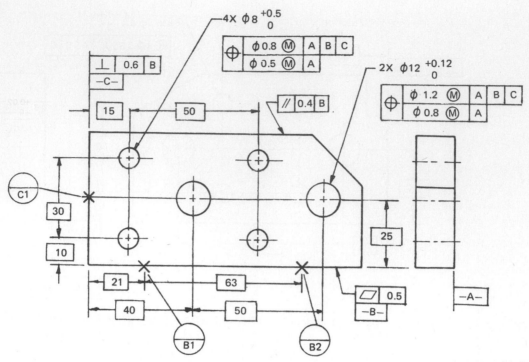

Unit 16–17

56. 16–17–A .060 IN   **57. 16–17–A** .080 IN   **58. 16–17–A** .030 IN

59. 16–17–B 5.12 MM   **60. 16–17–C** ⌀ 0.066

Unit 16–18

61. 16–18–A

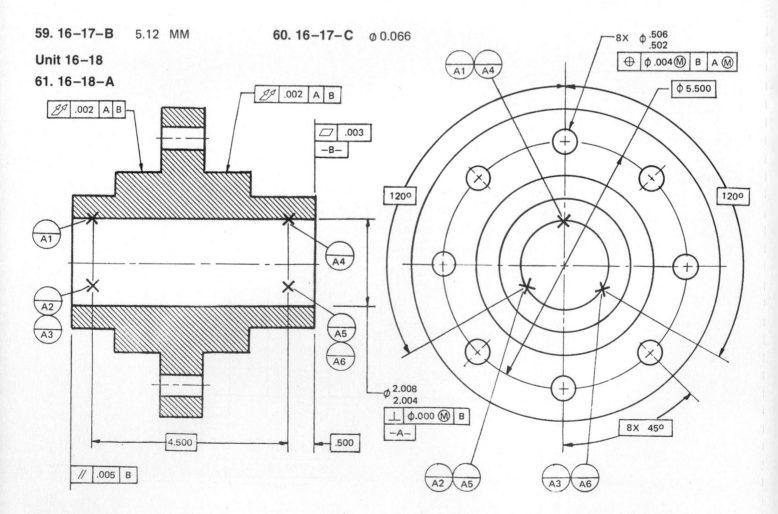

Unit 16–18

62. 16–18–B

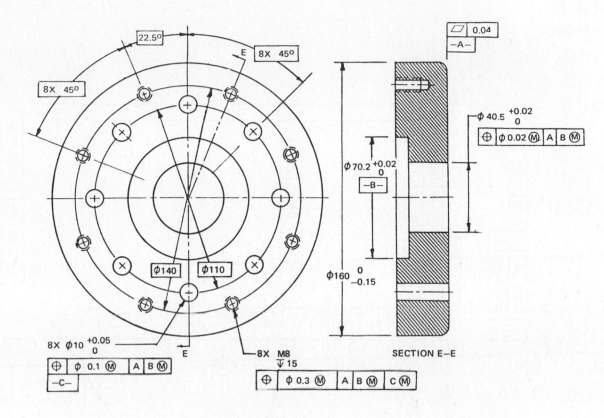

63. 16–18–C

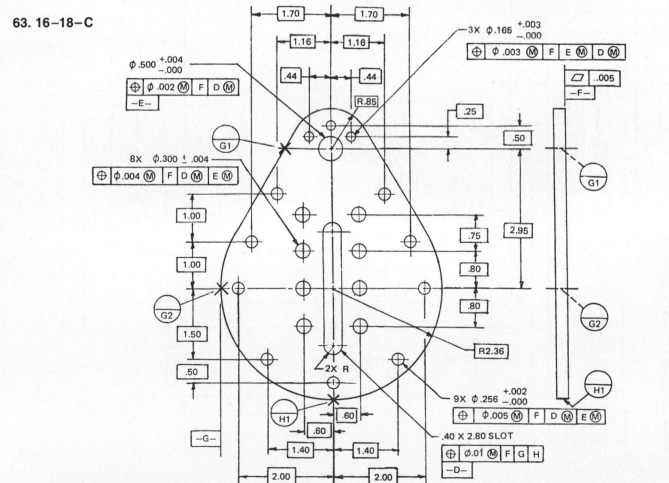

Unit 16–18

64. 16–18–D

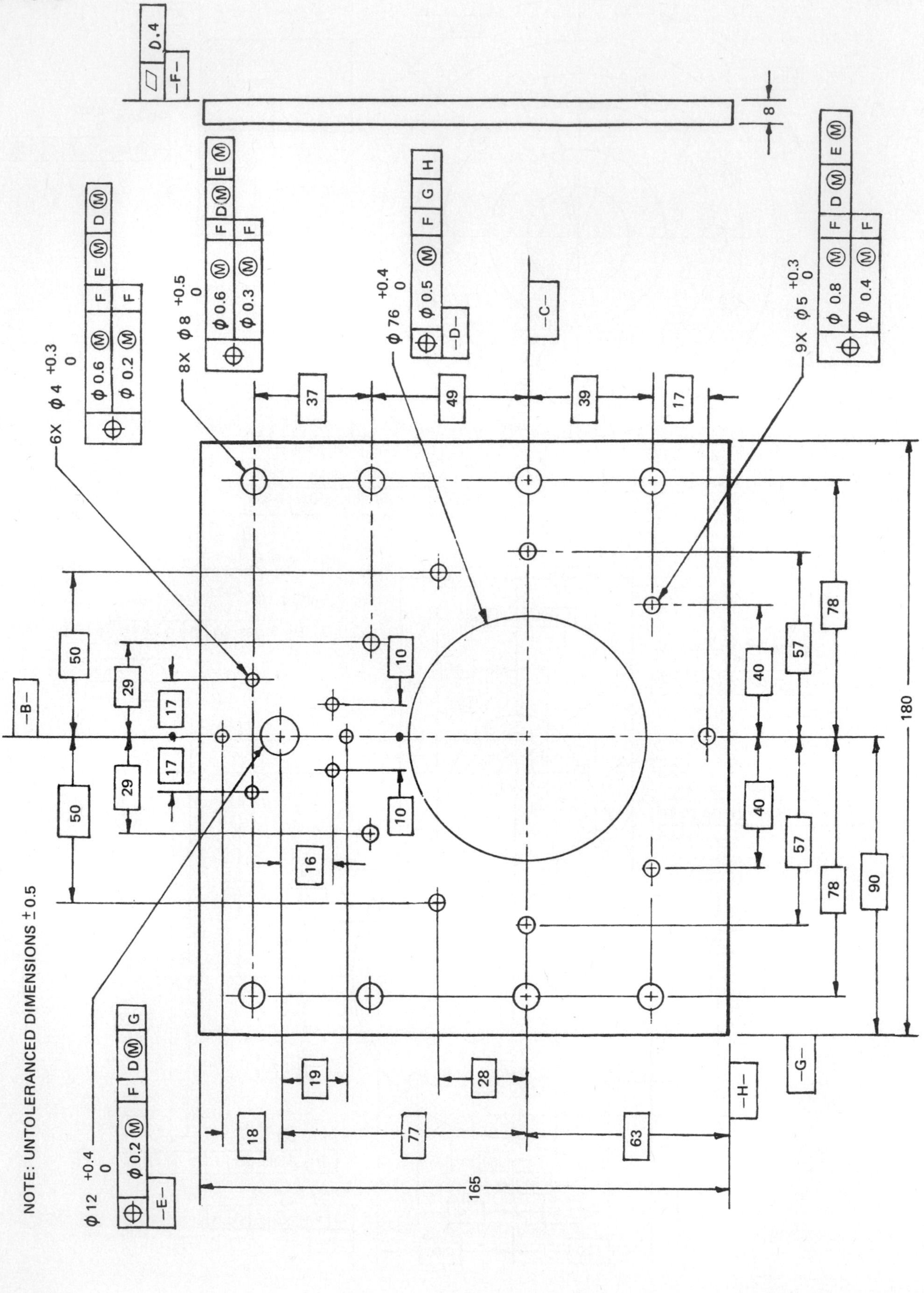

Unit 16–18

65. 16–18–E

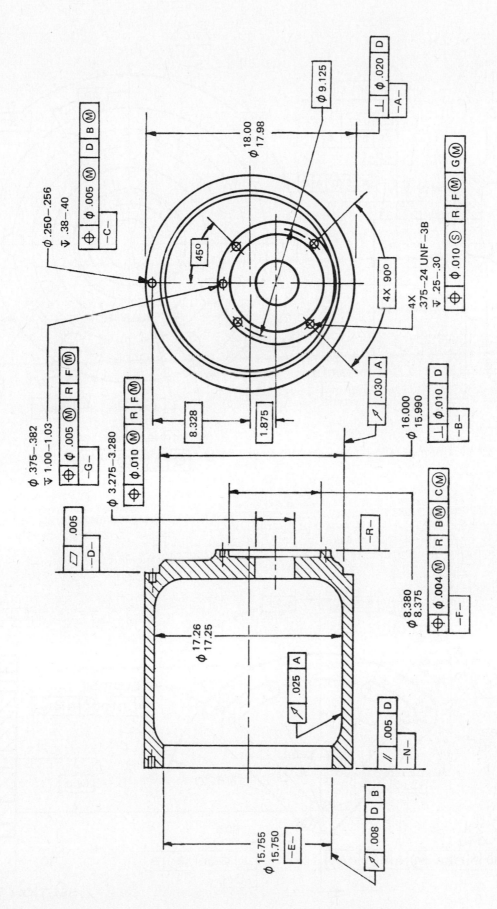

Unit 16−18

66. 16−18−F

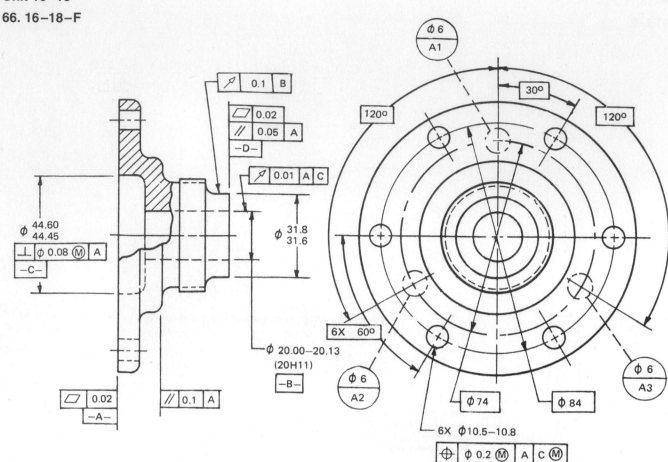

67. 16−18−G

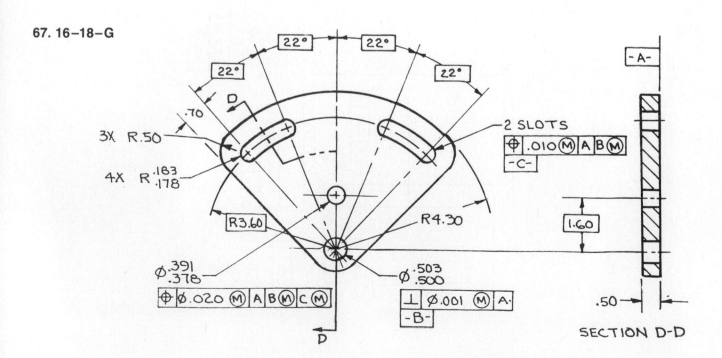

Unit 16–18

68. 16–18–H

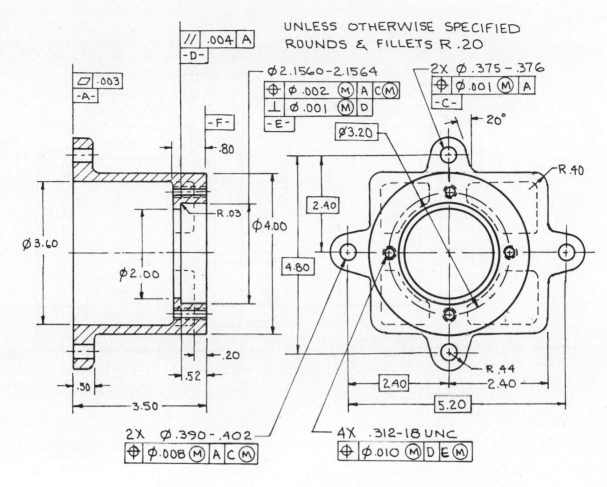

69. 16–18–J

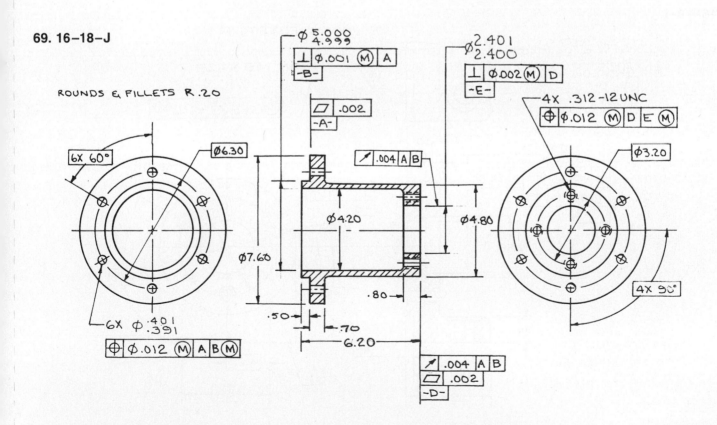

Unit 16–18

70. 16–18–K

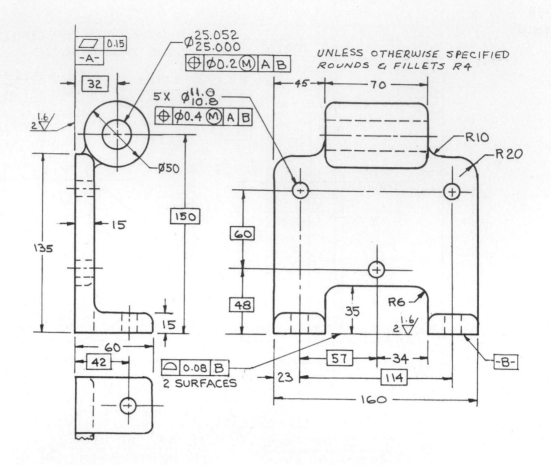

71. 16–18–L

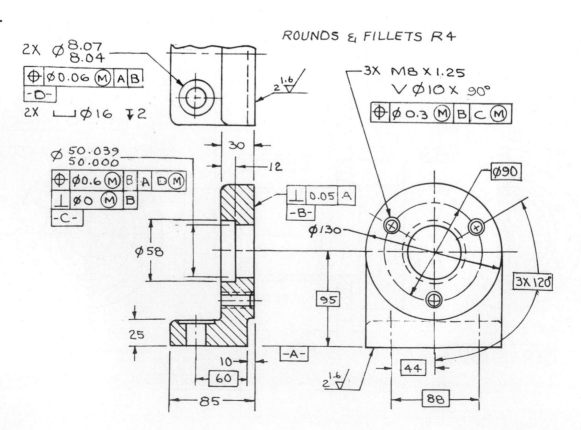

Unit 16–18

72. 16–18–M

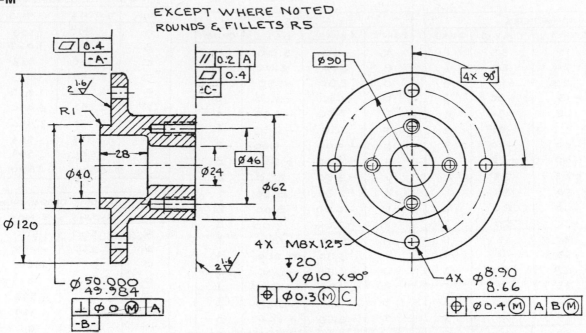

73. 16–18–N

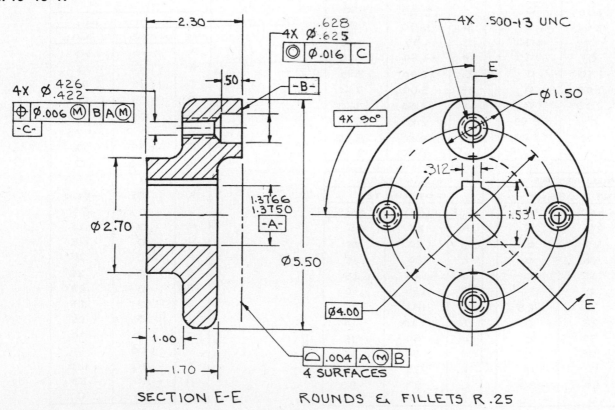

SECTION E-E   ROUNDS & FILLETS R.25

Chapter 17 Drawings for Numerical Control
Unit 17–1, 1. 17–1–A

2. 17–1–B

MILLIMETERS

POINT	X AXIS	Y AXIS	QUAD
A	15	50	1
B	20	20	1
C	28	-15	4
D	-15	-30	3
E	38	-34	4
F	17	-83	4
G	-40	-90	3
H	-60	-25	3
J	-25	35	2
K	58	85	1
L	-60	80	2
M	-70	20	2
N	80	15	1
P	58	-78	4
Q	-68	-63	3
R	33	0	1
S	0	-53	3
T	-35	-2	3
U	80	-18	4
V	0	65	1

INCHES

POINT	X-AXIS	Y-AXIS	QUAD
A	1.50	5.00	1
B	2.00	2.00	1
C	2.80	-1.50	4
D	-1.50	-3.00	3
E	3.80	-3.40	4
F	1.70	-8.30	4
G	-4.00	-9.00	3
H	-6.00	-2.50	3
J	-2.50	3.50	2
K	5.80	8.50	1
L	-6.00	8.00	2
M	-7.00	2.00	2
N	8.00	1.50	1
P	5.80	-7.80	4
Q	-6.80	-6.30	3
R	3.30	0	1
S	0	-5.30	3
T	-3.50	-.20	3
U	8.00	-1.80	4
V	0	6.50	1

COORDINATE DIMENSIONING

HOLE	X-AXIS	Y-AXIS
A	.500	1.500
B	.750	1.067
C	.750	1.933
D	4.250	1.933
E	4.250	1.067
F	2.241	2.496
G	2.759	2.496
H	2.759	.504
J	2.241	.504
K	4.500	1.500

POINT-TO-POINT DIMENSIONING

HOLE	X-AXIS	Y-AXIS
A	.500	1.500
B	.250	-.433
C	0	.866
D	3.500	0
E	0	-.866
F	-2.009	1.399
G	.518	0
H	0	-1.992
J	-.518	0
K	2.259	.996

3. 17–1–C

INCHES — POINT TO POINT

HOLE	X AXIS	Y AXIS
A	2.50	2.50
B	2.50	0
C	7.00	0
D	2.50	0
E	0	2.50
F	0	2.50
G	-2.50	0
H	-7.00	0
J	-2.50	0
K	0	-2.50
L	6.00	3.00
M	0	-6.00

INCHES — COORDINATE

HOLE	X AXIS	Y AXIS
A	2.50	2.50
B	5.00	2.50
C	12.00	2.50
D	14.50	2.50
E	14.50	5.00
F	14.50	7.50
G	12.00	7.50
H	5.00	7.50
J	2.50	7.50
K	2.50	5.00
L	8.50	8.00
M	8.50	2.00

4. 17–1–D

INCHES — POINT TO POINT

HOLE	X AXIS	Y AXIS
A	2.00	1.50
B	-1.75	1.25
C	1.75	0
D	1.75	0
E	0	-1.25
F	0	-1.25
G	-1.75	0
H	-1.75	0
J	0	1.25
K	.75	-1.00
L	0	.25
M	0	.25
N	0	.25
P	0	.25
Q	0	.25
R	0	.25
S	0	.25
T	0	.25
U	1.50	-.30
V	1.00	0
W	-.50	-1.10
X	-.40	-.50
Y	.80	0

MILLIMETERS — POINT TO POINT

HOLE	X AXIS	Y AXIS
A	25	25
B	25	0
C	70	0
D	25	0
E	0	25
F	0	25
G	-25	0
H	-70	0
J	-25	0
K	0	-25
L	60	30
M	0	-60

MILLIMETERS — COORDINATE

HOLE	X AXIS	Y AXIS
A	25	25
B	50	25
C	120	25
D	145	25
E	145	50
F	145	75
G	120	75
H	50	75
J	25	75
K	25	50
L	85	80
M	85	20

Unit 17-2

5. 17-2-A

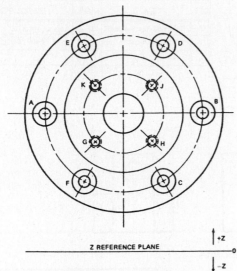

POINT TO POINT			
HOLE	X	Y	Z
A	-1.50	0	-2.01
B	3.00	0	-2.01
C	-.75	-1.299	-2.01
D	0	2.598	-2.01
E	-1.50	0	-2.01
F		-2.598	-2.01
G	.220	.766	-1.41
H	1.060	0	-1.41
J	0	1.060	-1.41
K	-1.060	0	-.41

6. 17-2-B

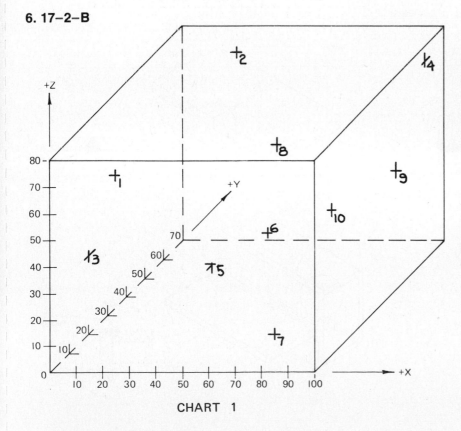

CHART 1

NOTE: ALL COORDINATES ARE +

POINT	X AXIS	Y AXIS	Z AXIS
P1	20	30	40
P2	10	0	55
P3	10	60	70
P4	30	10	20
P5	70	30	.80
P6	7C	20	0
P7	90	70	40
P8	65	70	60
P9	65	30	25
P10	95	25	55

CHART 2

Chapter 18 Welding Drawings
Unit 18–1, 1. 18–1–A

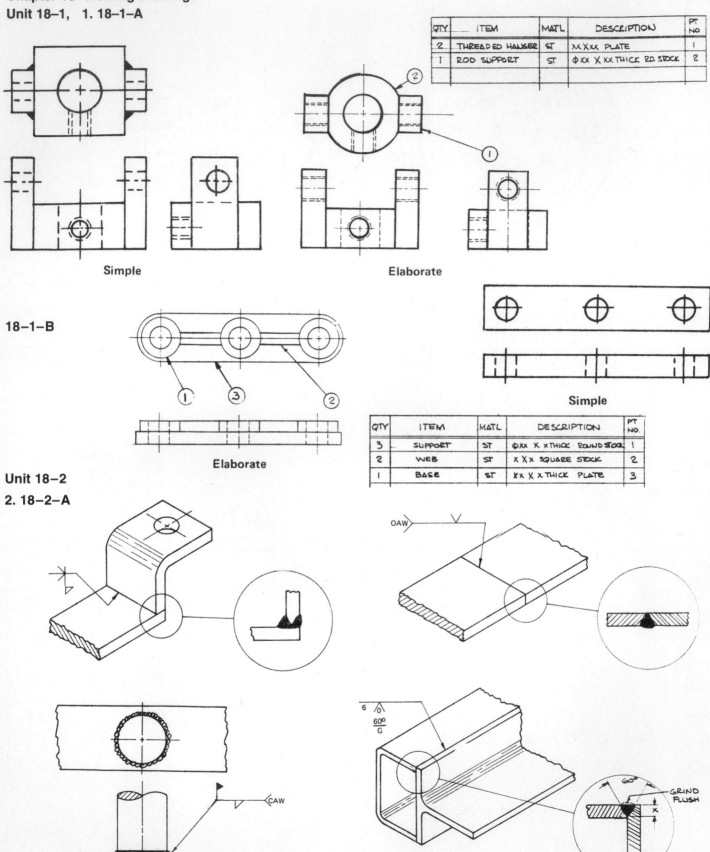

QTY	ITEM	MATL	DESCRIPTION	PT NO.
2	THREADED HANGER	ST	XX XXX PLATE	1
1	ROD SUPPORT	ST	ΦXX X XX THICK RD. STOCK	2

Simple

Elaborate

18–1–B

Elaborate

Simple

QTY	ITEM	MATL	DESCRIPTION	PT NO.
3	SUPPORT	ST	ΦXX X X THICK ROUND STOCK	1
2	WEB	ST	X XX SQUARE STOCK	2
1	BASE	ST	XX X X THICK PLATE	3

Unit 18–2
2. 18–2–A

Unit 18–2
3. 18–2–B

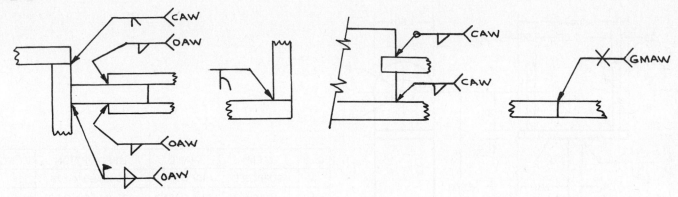

Unit 18–3
4. 18–3–A

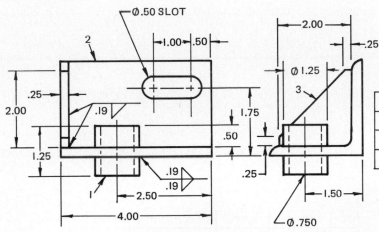

QTY	ITEM	MATL	DESCRIPTION	PT NO.
1	SHAFT	STL	Ø1.25 X 1.25	1
1	BASE	STL	L2.50 X 2.50 X .25	2
1	RIB	STL	.25 X 2.00 X 2.00	3

QTY	ITEM	MATL	DESCRIPTION	PT NO.
1	CHANNEL	STL	C75 X 7.5 X 125	1
1	SUPPORT	STL	8 X 25 X 125	2
4	RIB	STL	6 X 25 X 34*	3

*MAKES TWO

18–3–B

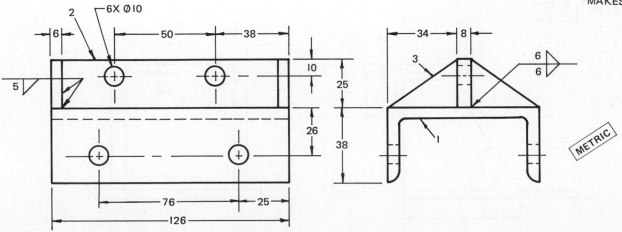

Unit 18–3
4. CONTD 18–3–C

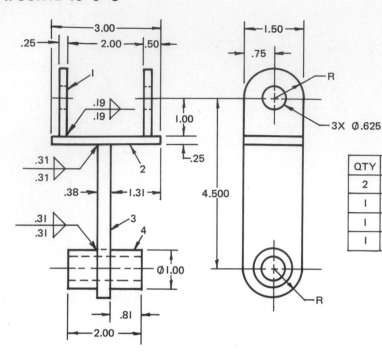

QTY	ITEM	MATL	DESCRIPTION	PT NO.
2	SUPPORT	AISI C-1040	.25 X 1.50 X 1.75	1
1	BASE	AISI C-1040	.25 X 1.50 X 3.00	2
1	STEM	AISI C-1040	.38 X 1.50 X 4.00	3
1	SHAFT	AISI C-1040	Ø1.00 X 2.00	4

18–3–D

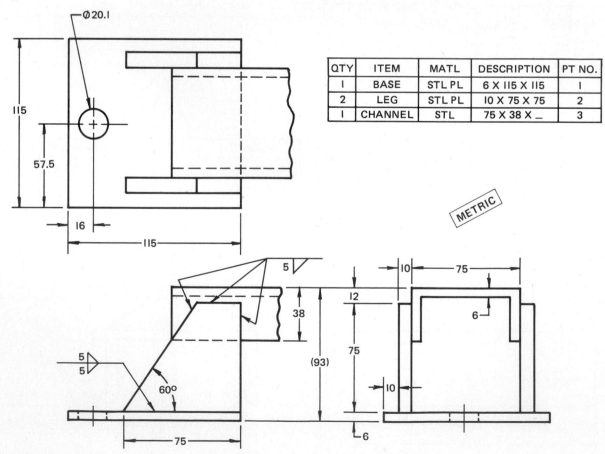

QTY	ITEM	MATL	DESCRIPTION	PT NO.
1	BASE	STL PL	6 X 115 X 115	1
2	LEG	STL PL	10 X 75 X 75	2
1	CHANNEL	STL	75 X 38 X __	3

METRIC

Unit 18–3

5. 18–3–E

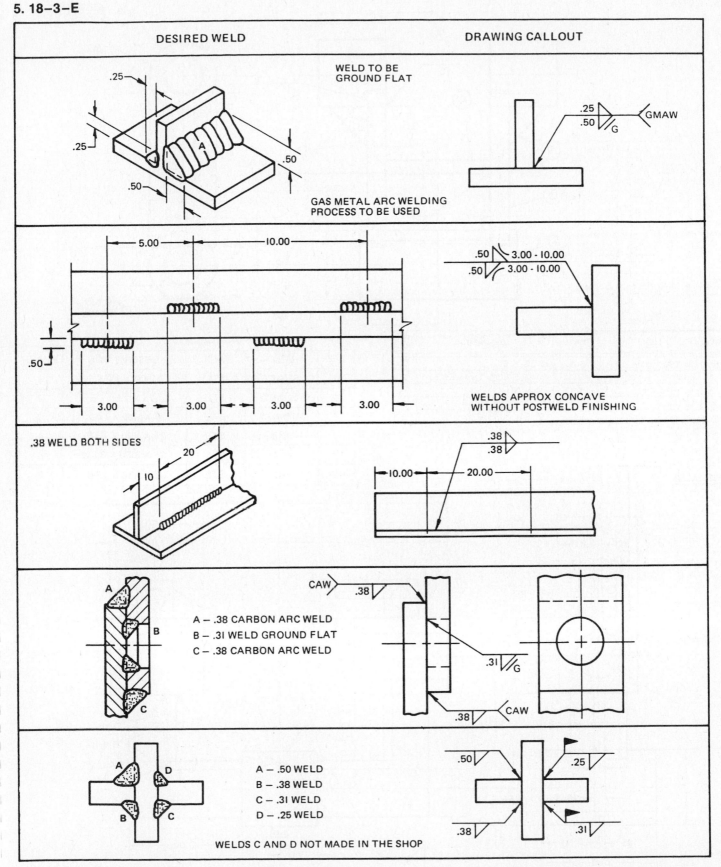

DESIRED WELD	DRAWING CALLOUT

WELD TO BE GROUND FLAT

GAS METAL ARC WELDING PROCESS TO BE USED

WELDS APPROX CONCAVE WITHOUT POSTWELD FINISHING

.38 WELD BOTH SIDES

A – .38 CARBON ARC WELD
B – .31 WELD GROUND FLAT
C – .38 CARBON ARC WELD

A – .50 WELD
B – .38 WELD
C – .31 WELD
D – .25 WELD

WELDS C AND D NOT MADE IN THE SHOP

Unit 18−4

6. 18−4−A

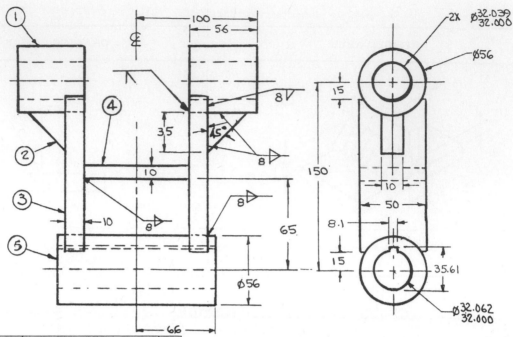

QTY	ITEM	MATL	DESCRIPTION	PT NO.
2	UPPER YOKE	STL	Ø56 X 56	1
2	RIB	STL	10 X 35 X 35 (MAKES 2)	2
2	SUPPORT	STL	10 X 50 X 120	3
1	BRACE	STL	10 X 50 X 88	4
1	LOWER YOKE	STL	Ø56 X 132	5

QTY	ITEM	MATL	DESCRIPTION	PT NO.
1	SHAFT	STL	☐ 2.00 X 3.00	1
2	SHAFT	STL	☐ .75 X 2.50	2
4	SUPPORT	STL	.25 X 1.00 X 1.00	3
2	RIB	STL	.25 X 1150 X 1.875	4
2	RIB	STL	.25 X 3.00 X 3.375	5
2	RIB	STL	.25 X 1.50 X 2.00	6

18−4−B

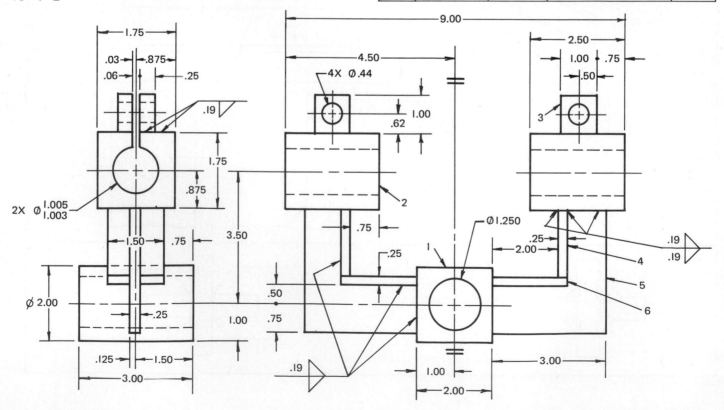

Unit 18–4
6. CONTD 18–4–C

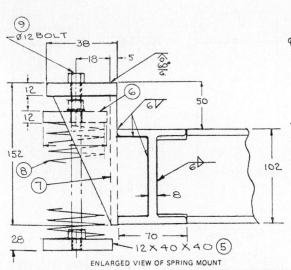

ENLARGED VIEW OF SPRING MOUNT

QTY	ITEM	MATL	DESCRIPTION	PT NO.
2	FRAME	STL	S100 X 14.1 X 687	1
3	FRAME	STL	S100 X 14.1 X 220	2
4	MOUNTING PLATE	STL	X 38 X 64	3
8	ARM	STL	8 X 65 X 152	4
4	FOOT	STL	12 X 40 X 40	5
4	ADJ. PLATE	STL	12 X 40 X 40	6
4	BACKING PLATE	STL	5 X 48 X 152	7
4	SPRING	STL	DWG XXX	8
1	BOLT HEX HD	STL	Ø12 UNC X 180	9
8	NUT HEX HD	STL	Ø12 – UNC	10

18–4–D

QTY	ITEM	MATL	DESCRIPTION	PT
1	COVER	STL	.38 X 6.76 X 12.24	1
2	SIDE	"	.38 X 1.62 X 11.36	2
2	END	"	.38 X 1.62 X 5.50	3
1	RIB	"	.38 X 1.00 X 5.50	4
1	SHAFT SUPPORT	"	Ø2.18 X 2.50	5
4	TAB	"	.50 X 1.00 X 1.50	6

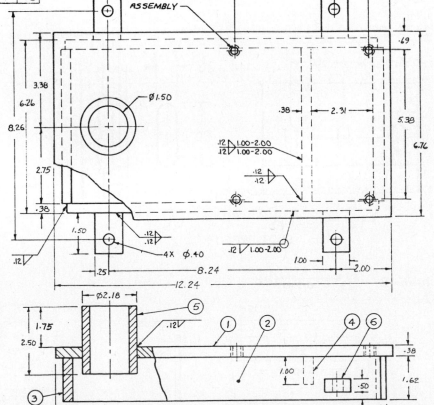

Unit 18–4

7. 18–4–E

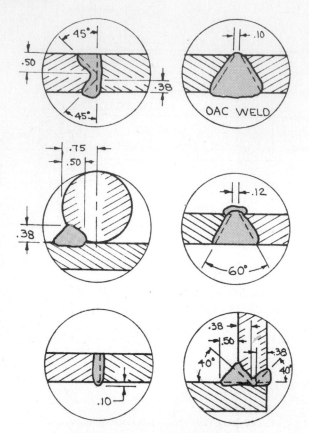

OAC WELD

Unit 18–5

8. 18–5–A

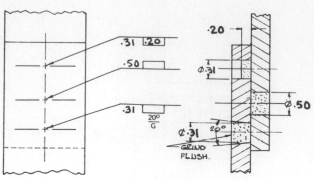

PLUG WELDS

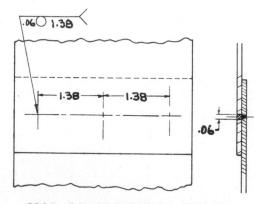

SPOT OR PROJECTION WELDS

18–5–B

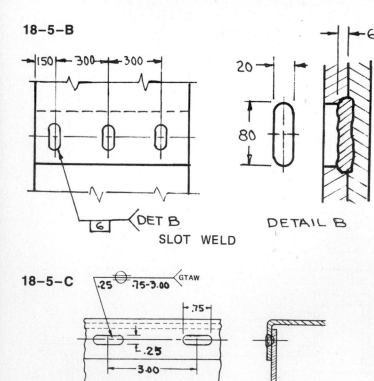

DET B

SLOT WELD

DETAIL B

FLANGED WELD

18–5–C

SEAM WELDS

NOTE – WORN SHAFT TO BE BUILT UP AND TURNED TO ORIGINAL SIZE SHOWN.

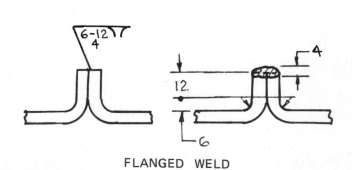

SURFACE WELDS

Unit 18-5

9. 18-5-D

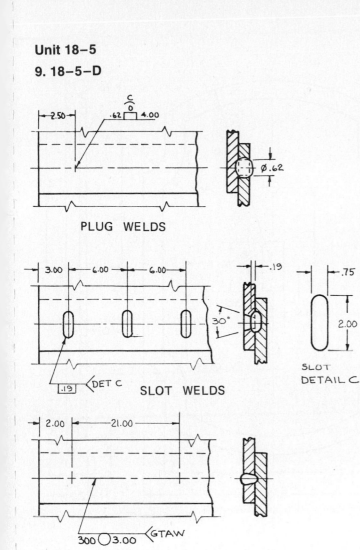

PLUG WELDS

SLOT WELDS

SLOT
DETAIL C

DET C

SPOT WELDS

10. 18-5-E

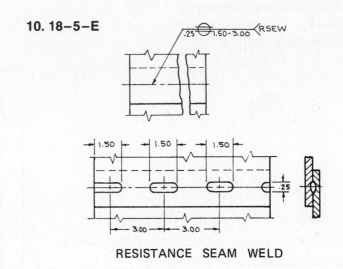

RESISTANCE SEAM WELD

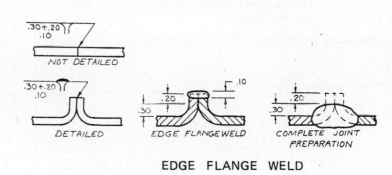

EDGE FLANGE WELD

NOT DETAILED

DETAILED

EDGE FLANGE WELD

COMPLETE JOINT PREPARATION

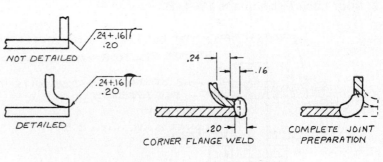

NOT DETAILED

DETAILED

CORNER FLANGE WELD

COMPLETE JOINT PREPARATION

CORNER FLANGE WELD

11. 18-5-F

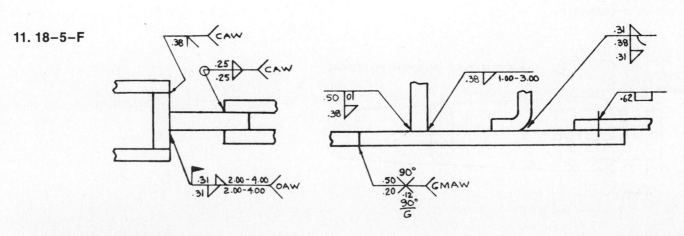

Chapter 19 Design Concepts
Unit 19–1
1. Switch Plate, 19–1–A

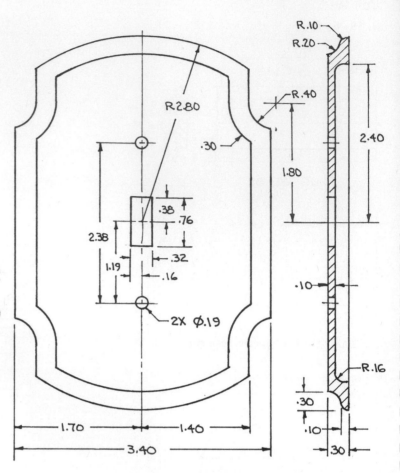

ONE OF MANY SOLUTIONS

PRODUCTION AND SPECIFICATION DATA

MATL– CELLULOSE PROPIONATE
        EXCELLENT FINISH AND A WIDE RANGE
        OF COLORS

METHOD OF PRODUCTION – COMPRESSION MOLD

2. Door Lock Mechanism, 19–1–B

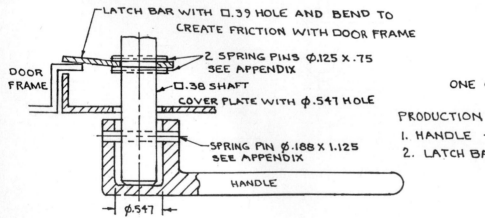

ONE OF MANY SOLUTIONS

PRODUCTION AND SPECIFICATION DATA
1. HANDLE – DIE CAST ALUMINUM
2. LATCH BAR – .06 X 1.75 MTL STAMPING

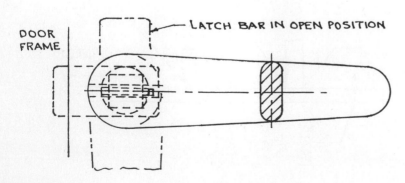

Unit 19–1

3. Microphone Holder, 19–1–C

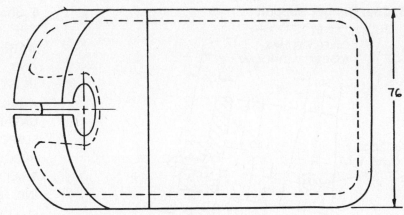

ONE OF MANY SOLUTIONS
PRODUCTION AND SPECIFICATION DATA
MATERIAL – CELLULOSE PROPIANATE
EXCELLENT FINISH, TOUGH, WIDE
RANGE OF COLORS

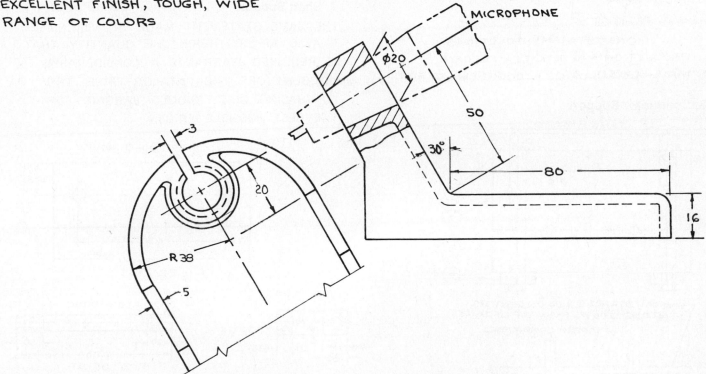

ONE OF MANY SOLUTIONS
PRODUCTION AND SPECIFICATION DATA
MATERIAL – MOLDED URETHANE DIPPED IN
HIGH-GLOSS ENAMEL PAINT. – PERCH FINISH

4. Key Holder

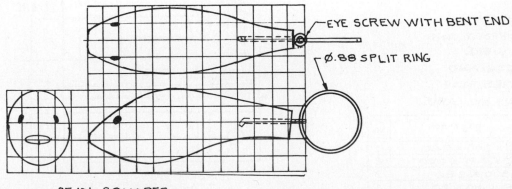

.25 IN. SQUARES

Unit 19–2

5. Coat Hanger Support

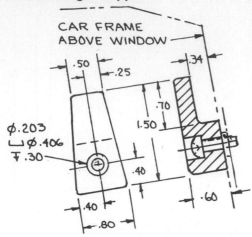

CAR FRAME ABOVE WINDOW

ONE OF MANY SOLUTIONS
PRODUCTION AND SPECIFICATION DATA
MATL- CELLULOSIC – TOUGHNESS, CLARITY

8. Conductor Support

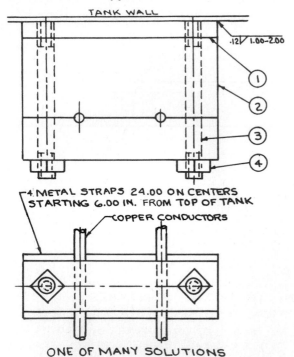

TANK WALL

.12 / 1.00-2.00

4 METAL STRAPS 24.00 ON CENTERS
STARTING 6.00 IN. FROM TOP OF TANK

COPPER CONDUCTORS

ONE OF MANY SOLUTIONS

PRODUCT AND SPECIFICATION DATA
– HARD MAPLE FOR THE SUPPORT AND
FIBER FOR THE FASTENER WERE
CHOSEN FOR THEIR STRENGTH AND
NONCONDUCTIBLE PROPERTIES. ALSO
THESE MATERIALS ARE READILY AVAILABLE.

QTY	ITEM	MATL	DESCRIPTION	PT NO.
4	BRACKET	CRS	.50 X 2.00 X 6.00	1
4	SUPPORT	HD MAPLE	1.50 X 3.75 X 6.00	2
8	STUD	HD FIBER	Ø.500 X 2.80	3
8	NUT	HD FIBER	.38 X .80 X .80	4

6. Shaft Support

THE QUANTITY OF BRACKETS DOES NOT WARRANT THE HIGH COST OF DESIGNING AND CONSTRUCTING THE BEARING BRACKETS IF STANDARD BEARING BRACKETS ARE AVAILABLE. ON CHECKING MANUFACTURER'S CATALOGS WE FOUND SUITABLE BEARING SUPPORTS TO DO THE JOB REQUIRED. WE RECOMMENDED TO THE ENGINEERING DEPT. THAT THEY PURCHASE FROM BOSTON GEAR INC. EIGHT LIGHT DUTY SPLIT CAST IRON PILLOW BLOCKS, CAT NO. PPB12–3/4.

7. Shaft Support

THE SAME STATEMENT MADE IN ASSIGNMENT 6 ALSO APPLIES HERE. THE QUANTITY THAT IS REQUIRED WARRANTS A CONSIDERABLE DISCOUNT OFF THE CATALOG PRICE. THE PURCHASING DEPT. SHOULD BARGAIN FOR THE BEST POSSIBLE PRICE.

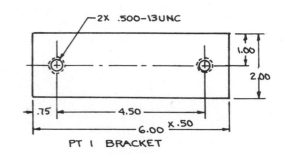

PT 1 BRACKET

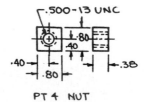

PT 4 NUT

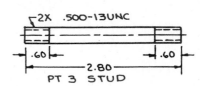

PT 3 STUD

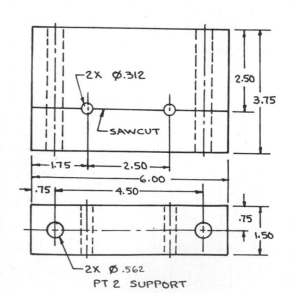

SAWCUT

PT 2 SUPPORT

Unit 19–2

9. Cassette Holder

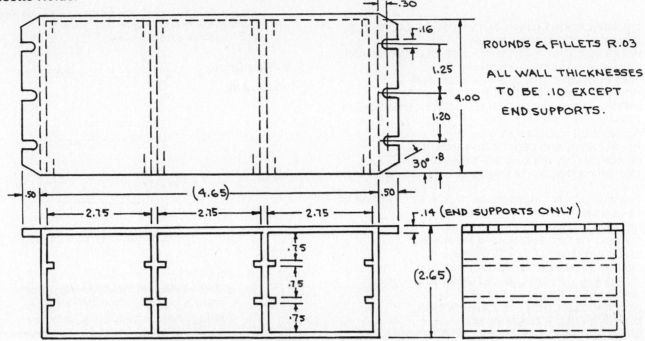

ROUNDS & FILLETS R.03

ALL WALL THICKNESSES TO BE .10 EXCEPT END SUPPORTS.

ONE OF MANY SOLUTIONS

PRODUCT AND SPECIFICATION DATA

- OPEN CONCEPT DESIGN WAS CHOSEN FOR TWO REASONS – COST AND EASE OF HANDLING THE TAPE CONTAINERS.
- MATERIAL – POLYSTYRENE WAS CHOSEN BECAUSE OF EASE OF FORMING, LOW COST, STRENGTH AND FINISH

10. Napkin Holder

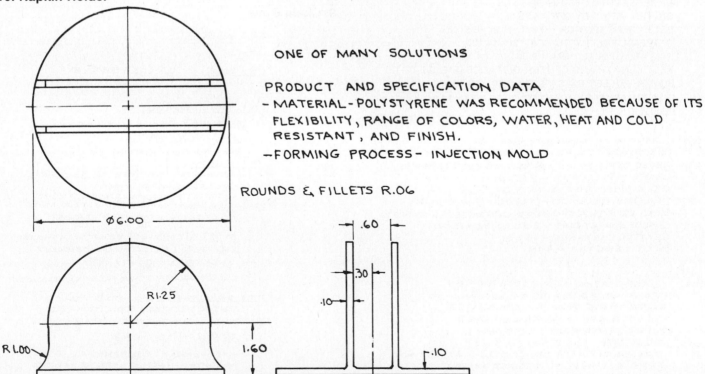

ONE OF MANY SOLUTIONS

PRODUCT AND SPECIFICATION DATA

- MATERIAL – POLYSTYRENE WAS RECOMMENDED BECAUSE OF ITS FLEXIBILITY, RANGE OF COLORS, WATER, HEAT AND COLD RESISTANT, AND FINISH.
- FORMING PROCESS – INJECTION MOLD

ROUNDS & FILLETS R.06

Chapter 20 Belts, Chains, and Gears
Unit 20–1,   1. V–Belt Drives

(A) CLASSIFICATION – LIGHT DUTY, SERVICE FACTOR 1.20
DESIGN HP = .33 X 1.20 = .396 HP
REFER TO FIG. 20—1—11. RPM = 1750, CLOSEST HP = .38
DRIVER V-PULLEY DIA = 2.50 in.
BELT CROSS SECTION = .50 X .31 in. (A SIZE)
— DRIVEN V-PULLEY DIA (FIG. 20—1—12)
750 RPM IS CLOSEST TO DESIRED BLOWER SPEED OF 765 RPM
OD OF DRIVEN PULLEY = 5.50 in.
— BELT LENGTH AND CENTER DISTANCE (FIG. 20—1—13)
SUM OF PULLEY DIAS. = 2.50 + 5.50 = 8.00
CENTER DISTANCE – 13.6 in. BELT LENGTH = 40 in.

(B) CLASSIFICATION – NORMAL DUTY, SERVICE FACTOR 1.0
DESIGN kW = 0.37 kW
REFER TO FIG. 20—1—11. r/min = 1160
CLOSEST kW TO DESIGN kW = 0.34 OR 0.40
OD OF DRIVER PULLEY = 76 OR 83 mm
(AS SPINDLE SPEED IS CRITICAL, THE DIA OF THE
 DRIVER PULLEY WILL BE DETERMINED BY DIA OF
 DRIVEN PULLEY IN NEXT STEP.)
BELT CROSS SECTION = 12 X 8 mm (A SIZE)
— DRIVEN V-PULLEY DIA (FIG. 20—1—12)
516 r/min UNDER 83 mm OD PULLEY IS THE ONLY
r/min SUITABLE UNDER 76 AND 83 HEADINGS
∴ OD OF DRIVER PULLEY = 83 mm
   OD OF DRIVEN PULLEY = 178 mm
— BELT LENGTH AND CENTER DISTANCE (FIG. 20—1—13)
SUM OF PULLEY DIAS = 83 + 178 = 261 (USE COLUMN 265)
CENTER DISTANCE = 550 mm (USE 551)
BELT LENGTH = 1520 mm

(C) CLASSIFICATION – HEAVY DUTY, SERVICE FACTOR .85
DESIGN HP = 1.5 X .85 = 1.28 HP
REFER TO FIG. 20—1—11. RPM = 1750
HP CLOSEST TO DESIGN HP = 1.25
DRIVER V-PULLEY DIA = 4.25 in.
BELT CROSS SECTION = .66 X .41 in. (B SIZE)
— DRIVEN V-PULLEY DIA (FIG. 20—1—12)
RPM = 800 PULLEY DIA = 9.00 in.
— BELT LENGTH AND CENTER DISTANCE (FIG. 20—1—13)
SUM OF PULLEY DIAS = 4.25 + 9.00 = 13.25 in.
(USE COLUMN 13.5) CENTER DISTANCE = 14 in.
BELT LENGTH = 50 in.

(D) CLASSIFICATION – HEAVY DUTY S.F = .85
DESIGN HP = .50 X .85 = 42.5 HP
REFER TO FIG. 20-1-11. RPM = 1750 CLOSEST HP = .38
—DRIVER V-PULLEY DIA = 2.50 IN.
BELT SIZE - .50 X .31 (A SIZE)
DRIVEN V-PULLEY DIA - (FIG. 20-1-12) = 5.50 IN.
— BELT LENGTH AND CENTER DISTANCE (FIG 20-1-13)
SUM OF PULLEY DIAS = 2.50 + 5.50 = 8.00
CENTER DISTANCE - 15.50
BELT LENGTH = 44 IN.

(E) CLASSIFICATION – HEAVY DUTY, S.F = .85
DESIGN KW = 0.75 X .85 = 0.64 KW
REFER TO FIG. 20-1-11 r/min = 1750
CLOSEST KW TO DESIGN KW = 0.63
—OD OF DRIVER PULLEY = 89 mm
BELT SIZE - 12 X 8 mm (A SIZE)
DRIVEN V- PULLEY DIA - (FIG. 20-1-12) = 254 mm
(585 IS CLOSEST TO DESIRED SPEED)
— BELT LENGTH AND CENTER DISTANCE (FIG. 20-1-13)

(E) CONTD

SUM OF PULLEY DIAS = 89 + 254 = 343 (USE 345)
CENTER DISTANCE = 430 (USE 434)
BELT LENGTH = 1920 mm

2. V–Belt Drive, 20–1–A/B

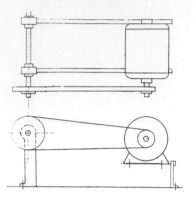

SOLUTION
STEP 1. – SMALLEST O.D. OF SMALL PULLEY
FOR 0.2 KW RATING (FIG. 20-1-11) = ⌀64
STEP 2. r/min OF LARGER PULLEY (815–
835) (FIG 20-1-12)   PULLEY ⌀127 PRODUCES
AN 830 r/min
STEP 3 (FIG 20-1-13)
    SUM = 64 + 127 = 191   USE COLUMN 190
CENTER TO CENTER DISTANCE = 435 mm
BELT LENGTH (USE 432 DISTANCE SHOWN
IN CHART) = 1170 mm
BELT SIZE - 12 X 8 mm

Unit 20–2

3. Chain Drive

STEP 1. SERVICE FACTOR   1.5

STEP 2. DESIGN kW   3.7 X 1.5 = 5.55 kW

STEP 3. TENTATIVE CHAIN SELECTION
    (FIG 20-2-15) - USE SINGLE CHAIN AND r/min
    OF 100 OF SMALL SPROCKET
    CHAIN NO. 80 (25 PITCH)
STEP 4. SMALL SPROCKET
    (FIG 20-2-13   100 r/min   NO. 80 CHAIN
    KW = 5.55   SHAFT DIA = 44
    NO. OF TEETH ON SMALL SPROCKET = 23 MIN
STEP 5. LARGE SPROCKET (FIG 20-2-17) SHAFT BORE ACCEPT.
    NO. OF TEETH = $\frac{100}{40}$ X 23 = 57.5 = 58
    AS 58 TOOTH SPROCKETS ARE NOT STOCK SPROCKETS
    CHECK WITH A 24 TOOTH SMALL SPROCKET
    NO. OF TEETH = $\frac{100}{40}$ X 24 = 60
    USE 24 AND 60 AS THEY ARE STOCK SPROCKETS
    AND GIVE DESIRED r/min.
STEP 6. CHAIN LENGTH IN PITCHES (FIG 20-2-8)
    CENTER DISTANCE = 900   PITCH = 25
    CHAIN LENGTH IN PITCHES = $2C + \frac{N}{2} + \frac{S}{C}$
    = $\left(2 \times \frac{900}{25}\right) + \left(\frac{24+60}{2}\right) + \frac{32.83}{36}$
    = 72 + 42 + 0.92 = 114.92 PITCHES
    AS CENTER DISTANCE IS APPROX. USE 114 PITCHES
STEP 7. CHAIN LENGTH IN MILLIMETRES = 25 X 114 = 2850

Unit 20–2

4. Chain Drive

STEP 1. SERVICE FACTOR 1.5

STEP 2. DESIGN kW (FIG 20-2-9) M.S.F. = 1.7

3.7 × 1.5 ÷ 1.7 = 3.27 kW

STEP 3. TENTATIVE CHAIN SELECTION

(FIG 20-2-15) USE DOUBLE CHAIN AND r/min

OF 100 OF SMALL SPROCKET

CHAIN NO. 60 (20 PITCH)

STEP 4. SMALL SPROCKET

(FIG 20-2-13) 100 r/min NO. 60 CHAIN kW = 3.27

MIN. NO OF TEETH ON SMALL SPROCKET = 32

STEP 5. LARGE SPROCKET

NO. OF TEETH = $\frac{100}{40}$ × 32 = 80

(FIG 20-2-17) BOTH 32 AND 80 TOOTH SPROCKETS

ARE STOCK SPROCKETS AND GIVE DESIRED r/min.

STEP 6. CHAIN LENGTH IN PITCHES (FIG. 20-2-8)

CENTER DISTANCE = 900   PITCH = 20

CHAIN LENGTH IN PITCHES = $2C + \frac{M}{2} + \frac{S}{C}$

$= \left(2 \times \frac{900}{20}\right) + \left(\frac{32+80}{2}\right) + \frac{58.36}{45}$

= 90 + 56 + 1.3 = 147.3   PITCHES

AS CENTER DISTANCE IS APPROX. USE 148 PITCHES

STEP 7. CHAIN LENGTH IN MILLIMETRES = 20 × 148 = 2960

5. Chain Drive

STEP 1. SERVICE FACTOR 1.3

STEP 2. DESIGN kW (FIG. 20-2-9) USE TRIPLE CHAIN

5.6 × 1.3 ÷ 2.5 = 2.9 kW

STEP 3. TENTATIVE CHAIN SELECTION

(FIG. 20-2-15) USE TRIPLE CHAIN AND r/min OF

OF 100 OF SMALL SPROCKET

CHAIN NO. 50 (16 PITCH)

STEP 4. SMALL SPROCKET

(FIG 20-2-13) 100 r/min NO. 50 CHAIN kW = 2.9

MIN. NO. OF TEETH ON SMALL SPROCKET = 45

STEP 5. LARGE SPROCKET

NO OF TEETH = $\frac{100}{66}$ × 45 = 68 (NOT STOCK SIZE-TRY

WITH A SMALL SPROCKET WITH 46 TEETH)

NO OF TEETH = $\frac{100}{66}$ × 46 = 70 TEETH

BOTH 46 AND 70 ARE STOCK SPROCKETS AND GIVE

DESIRED r/min.

STEP 6. CHAIN LENGTH IN PITCHES (FIG. 20-2-8)

CENTER DISTANCE = 1055   PITCH = 16

CHAIN LENGTH IN PITCHES = $2C + \frac{M}{2} + \frac{S}{C}$

$= \left(2 \times \frac{1055}{16}\right) + \left(\frac{46+70}{2}\right) + \frac{14.59}{66}$

= 132 + 58 + 0.22 = 190.22 PITCHES

USE 190 PITCHES

STEP 7. CHAIN LENGTH IN MILLIMETRES = 16 × 190 = 3040

6. Chain Drive

STEP 1. SERVICE FACTOR 1.5

STEP 2. DESIGN HP = 3 × 1.5 = 4.5

STEP 3. TENTATIVE CHAIN SELECTION (FIG 20-2-14)

1000 RPM FOR SMALL SPROCKET

SINGLE CHAIN

CHAIN NO. 35 (.38 PITCH)

STEP 4. SMALL SPROCKET (FIG. 20-2-11)

1000 RPM NO. 35 CHAIN   4.5 HP

(USE 900 COLUMN)

MIN. NO. OF TEETH - 45

STEP 5. LARGE SPROCKET

860 RPM Ø1.25 SHAFT

NO. OF TEETH = $\frac{1000 \times 45}{860}$ = 52.3

USE 54 TEETH WHICH IS A STOCK SPROCKET

STEP 6. CHAIN LENGTH IN PITCHES (FIG. 20-2-8)

CENTER DISTANCE = 10 IN. MIN

PITCH = .38

CHAIN LENGTH IN PITCHES =

$\frac{2 \times 10}{.38} + \frac{45+54}{2} + \frac{9 \times .38}{10}$ = 97.4 (USE 98)

STEP 7. CHAIN LENGTH IN INCHES - 98 × .38 = 37.24

7. Chain Drive

STEP 1. SERVICE FACTOR 1.0

STEP 2. DESIGN HP = 10

STEP 3. TENTATIVE CHAIN SELECTION (FIG. 20-2-14)

480 RPM FOR SMALL SPROCKET

CHAIN NO. 60 (.75 PITCH)

STEP 4. SMALL SPROCKET (FIG. 20-2-11)

480 RPM, NO. 60 CHAIN, 10 HP

USE 500 RPM   17 OR 18 TEETH Ø1.69

STEP 5. LARGE SPROCKET 160 RPM

NO. OF TEETH = $\frac{480}{160}$ × 17 = 51 OR

$\frac{480}{160}$ × 18 = 54. USE 17 WITH 51 OR

18 WITH 54 (ALL ARE STOCK SPROCKETS)

STEP 6. CHAIN LENGTH IN PITCHES (FIG. 20-2-8)

CENTER DISTANCE = 48   PITCH = .75

$\frac{2 \times 48}{.75} + \frac{17+51}{2} + \frac{29.28 \times .75}{48}$ = 162.46 (USE 162)

STEP 7. CHAIN LENGTH IN INCHES = 162 × .75 = 121.5

Unit 20-2

8. Chain Drive

STEP 1. SERVICE FACTOR 1.0

STEP 2. DESIGN HP = 10

STEP 3. TENTATIVE CHAIN SELECTION (FIG. 20-2-14)
2800 RPM FOR SMALL SPROCKET
CHAIN NO. 35 (.38 PITCH)

STEP 4. SMALL SPROCKET (FIG. 20-2-11)
2800 RPM, 10 HP
23 TEETH   Ø1.25 SHAFT

STEP 5. LARGE SPROCKET 1800 RPM

NO. OF TEETH $= \dfrac{2800}{1800} \times 23 = 36$ TEETH

BOTH 23 AND 36 TEETH ARE
STOCK SPROCKETS

STEP 6. CHAIN LENGTH IN PITCHES (FIG. 20-2-8)
CENTER DISTANCE = 20 PITCH = .38

$\dfrac{2 \times 20}{.38} + \dfrac{23+36}{2} + \dfrac{4.28 \times .38}{20} = 134.9$

(USE 136)

STEP 7. CHAIN LENGTH IN INCHES = 136 × .38 = 103

Unit 20-3

9. 20-3-A

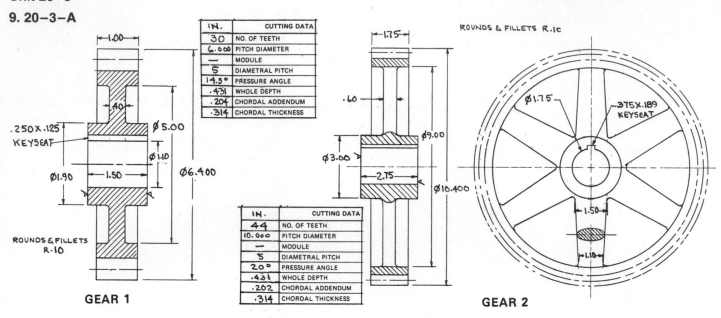

IN.	CUTTING DATA
30	NO. OF TEETH
6.000	PITCH DIAMETER
—	MODULE
5	DIAMETRAL PITCH
14.3°	PRESSURE ANGLE
.431	WHOLE DEPTH
.204	CHORDAL ADDENDUM
.314	CHORDAL THICKNESS

GEAR 1

IN.	CUTTING DATA
44	NO. OF TEETH
10.000	PITCH DIAMETER
—	MODULE
5	DIAMETRAL PITCH
20°	PRESSURE ANGLE
.431	WHOLE DEPTH
.202	CHORDAL ADDENDUM
.314	CHORDAL THICKNESS

GEAR 2

20-3-B

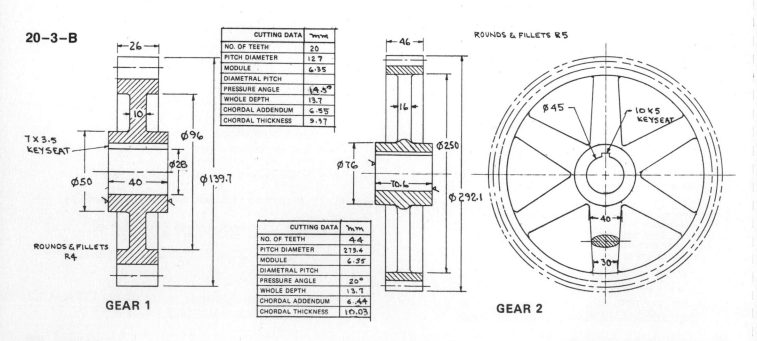

CUTTING DATA	mm
NO. OF TEETH	20
PITCH DIAMETER	127
MODULE	6.35
DIAMETRAL PITCH	
PRESSURE ANGLE	14.5°
WHOLE DEPTH	13.7
CHORDAL ADDENDUM	6.55
CHORDAL THICKNESS	9.97

GEAR 1

CUTTING DATA	mm
NO. OF TEETH	44
PITCH DIAMETER	279.4
MODULE	6.35
DIAMETRAL PITCH	
PRESSURE ANGLE	20°
WHOLE DEPTH	13.7
CHORDAL ADDENDUM	6.44
CHORDAL THICKNESS	10.03

GEAR 2

Unit 20–3

10. 20–3–C,D

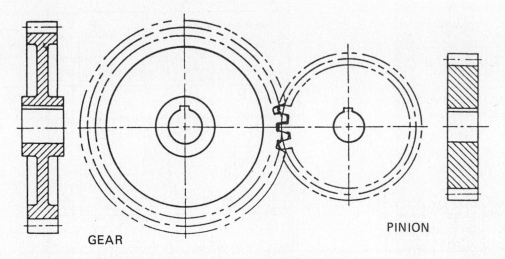

GEAR

PINION

20–3–C	GEAR	PINION
NO. OF TEETH	36	24
PITCH DIA.	7.200	4.800
DIAM. PITCH	5	5
PRESSURE ANGLE	14.5	14.5
WHOLE DEPTH	.431	.431
CHORD. ADD	.203	.205
CHORD. THICK.	.314	.314

20–3–D	GEAR	PINION
NO. OF TEETH	24	16
PITCH DIA.	30.48	20.32
MODULE	1.27	1.27
PRESS. ANGLE	14.5	14.5
WHOLE DEPTH	2.74	2.74
CHORDAL ADD.	1.30	1.32
CHORDAL THICKNESS	1.99	1.98

11. 20–3–E

GEAR	PD	N	DP	DIRECTION	R/MIN	CENTER DISTANCE
A	7.00	28	4	C'WISE	300	5.00
B	3.00	12	4	A'WISE	700	
C	6.00	18	3			5.00
D	4.00	12	3	C'WISE	1050	

20–3–F

GEAR	PD	N	MDL	DIRECTION	R/MIN	CENTER DISTANCE
A	324	54	6	C'WISE	160	216
B	108	18	6	A'WISE	480	
C	400	80	5			240
D	80	16	5	C'WISE	2400	

GEAR	PD	N	DP	DIRECTION	R/MIN	CENTER DISTANCE
A	7.50	30	4	A'WISE	240	6.00
B	4.50	18	4	C'WISE	400	
C	10.00	50	5			6.60
D	3.20	16	5	A'WISE	1250	
E	8.00	48	6			7.335
F	6.67	40	6	C'WISE	1500	

GEAR	PD	N	MDL	DIRECTION	R/MIN	CENTER DISTANCE
A	182.88	36	5.08	A'WISE	240	137.16
B	91.44	18	5.08	C'WISE	480	
C	203.52	64	3.18			152.64
D	101.76	32	3.18	A'WISE	960	
E	203.04	48	4.23			186.12
F	169.2	40	4.23	C'WISE	1152	

Unit 20-3

12. 20-3-G

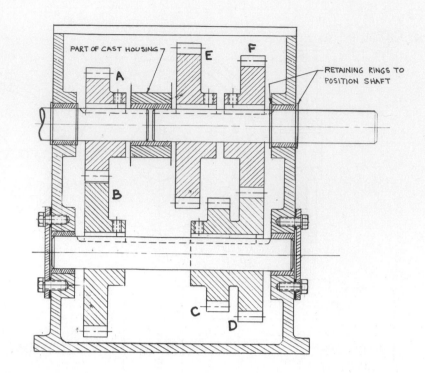

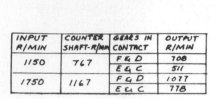

INPUT R/MIN	COUNTER SHAFT-R/MIN	GEARS IN CONTACT	OUTPUT R/MIN
1150	767	F & D	708
		E & C	511
1750	1167	F & D	1077
		E & C	778

13. 20-3-H

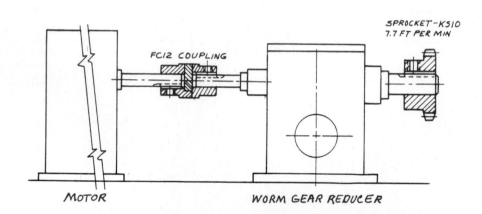

Unit 20-4

14. CONTD Spur Gear Transmission

(A)

CLASS OF SERVICE - III
SERVICE FACTOR 1.3
DESIGN HP = 8 X 1.3 = 10.4 HP
MOTOR - 1200 RPM
MACHINE - 1200 ÷ 4 = 300 RPM
PINION SELECTION (FIG. 20-4-2) 10.4 HP, 1200 RPM
DP = 6   N MIN. = 20
GEAR SELECTION (TABLE 77 IN APPENDIX)
RATIO 4:1   N SELECTION   21 AND 84
GEAR - N = 84   PD = 14.000
PINION - N = 21   PD = 3.500

(B)

CLASS OF SERVICE - III
SERVICE FACTOR 1.3
DESIGN HP = 22 X 1.3 = 28.6 HP
MOTOR - 1200 RPM   PRESS - 900 RPM
PINION SELECTION (FIG. 20-4-2) 28.6 HP,
1200 RPM, DP = 4   N MIN. = 16
RATIO 4:3
GEAR SELECTION (SEE TABLE 79 IN APPENDIX)
SELECTION = 18 AND 24
PINION ON MOTOR   N = 18, PD = 4.500
GEAR ON PRESS   N = 24, PD = 6.000

Unit 20-4

14. Spur Gear Transmission

(C)

CLASS OF SERVICE II.
SERVICE FACTOR 1.2
DESIGN KW = 7.5 × 1.2 = 9 KW
MOTOR-1200 Y/min MACHINE-400 Y/min
PINION SELECTION (FIG. 20-4-2) 9 KW, 1200 Y/min
MDL - 4.23 N=16 PD= 67.68
GEAR - N= 16 × 1200 ÷ 400 = 48, PD= 203.04
SEE TABLE 77 IN APPENDIX

15. Spur Gear Transmission

(A)

CLASS OF SERVICE - V
SERVICE FACTOR (FIG. 20-4-1) = 0.5
DESIGN HP= 7 × 0.5 = 3.5 HP
MACHINE RPM = 800
MOTORS AVAILABLE - 7 HP, 1200 RPM
                 - 5 HP, 750 RPM
- RECOMMEND THE SMALLER OF THE TWO
  MOTORS AS THE LARGER MOTOR MAY BE
  REQUIRED FOR ANOTHER MACHINE AT A
  LATER DATE.
- RECOMMEND 20° GEARS AS THEY HAVE
  REPLACED THE 14.5° GEARS AND ARE
  READILY AVAILABLE.
PINION SELECTION (FIG. 20-4-2) 3.5 HP,
800 RPM, 10 DP, N=16 MIN
    GEAR RATIO 800:750 = 16:15
GEAR SELECTION N = 800 × $\frac{16}{15}$ = 17

  (17 TOOTH GEAR NOT STOCK ITEM)
  TRY PINION WITH 20 TEETH
GEAR SELECTION N = 20 × $\frac{16}{15}$ = 21.33 (21)

BOTH 20 AND 21 ARE STOCK GEARS
AND ARE RECOMMENDED.

Unit 20-5

16. 20-5-A Inch

(D)

CLASS OF SERVICE III.
SERVICE FACTOR 1.3
DESIGN KW = 2 × 1.3 = 2.6 KW
MOTOR- 1800 R/MIN MACHINE 450 R/MIN
PINION SELECTION (FIG. 20-4-2) 2.6 KW,
1800 R/MIN MDL-2.54 N= 16 MIN.
GEAR N=16 × 1800 ÷ 450 = 64
ALTHOUGH GEARS WITH MDL OF 2.54 ARE
NOT SHOWN IN THE APPENDIX, GEARS OF
16 AND 64 TEETH ARE STANDARD IN ALL
GEAR SIZES AND ARE RECOMMENDED.

(B)

CLASS OF SERVICE IV
SERVICE FACTOR (FIG. 20-4-1) = 0.7
DESIGN HP = 0.7 × 7.5 = 5.25 HP
MACHINE - 600 RPM
MOTOR - 900 RPM
PINION SELECTION (FIG. 20-4-2) 5.25 HP,
  900 RPM, 8 DP, N. MIN = 16
GEAR RATIO = 900:600 = 3:2
GEAR SELECTION (TABLE 7L, APPENDIX)
PINION· N=16
GEAR - N=16 × 3 ÷ 2 = 24
BOTH THE GEAR AND PINION ARE
STOCK ITEMS AND ARE RECOMMENDED.

GEAR	RACK	CUTTING DATA
36	—	NO. OF TEETH
7.200	—	PITCH DIAMETER
—	—	MODULE
5	5	DIAMETRAL PITCH
14.5°	14.5°	PRESSURE ANGLE
.431	.431	WHOLE DEPTH
.219	—	CHORDAL ADDENDUM
.314	—	CHORDAL THICKNESS
.628	—	CIRCULAR PITCH
	.628	LINEAR PITCH

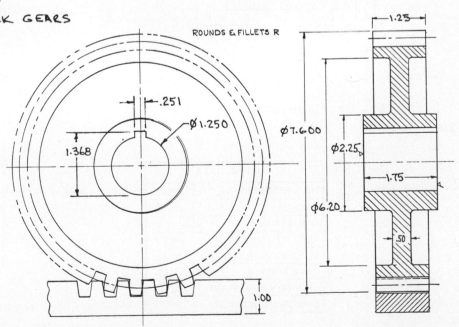

ROUNDS & FILLETS R

Unit 20–5

16. 20–5–A CONTD Metric

CUTTING DATA	GEAR	RACK
NO. OF TEETH	30	
PITCH DIAMETER	152.4	
MODULE	5.08	5.08
DIAMETRAL PITCH	5	5
PRESSURE ANGLE	14.5°	14.5°
WHOLE DEPTH	10.96	10.96
CHORDAL ADDENDUM	5.18	
CHORDAL THICKNESS	7.96	
CIRCULAR PITCH	15.96	
LINEAR PITCH		15.96

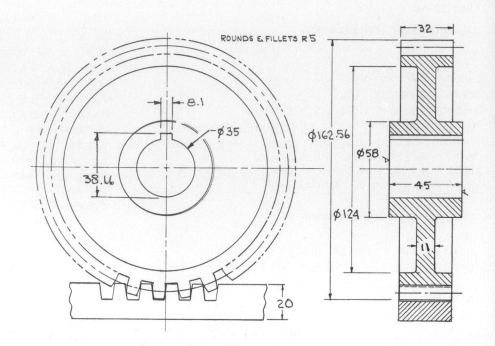

Unit 20–6

17. 20–6–A Inch

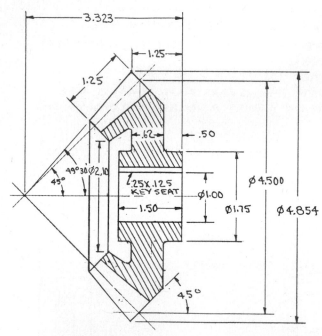

CUTTING DATA	
18	NO. OF TEETH
—	MODULE
4	DIAMETRAL PITCH
14.5°	TOOTH FORM
39°-49'	CUTTING ANGLE
.539	WHOLE DEPTH
.259	CHORDAL ADDENDUM
.392	CHORDAL THICKNESS

20–6–A Metric

CUTTING DATA	
NO. OF TEETH	18
MODULE	6.35
DIAMETRAL PITCH	
TOOTH FORM	14.5°
CUTTING ANGLE	39°-48'
WHOLE DEPTH	13.66
CHORDAL ADDENDUM	6.5
CHORDAL THICKNESS	9.97

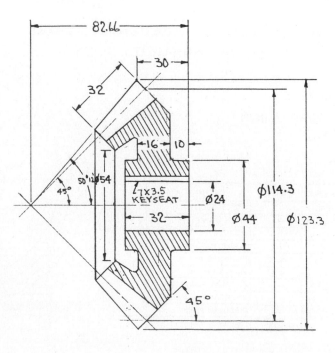

Unit 20−6

18. 20−6−B
Inch

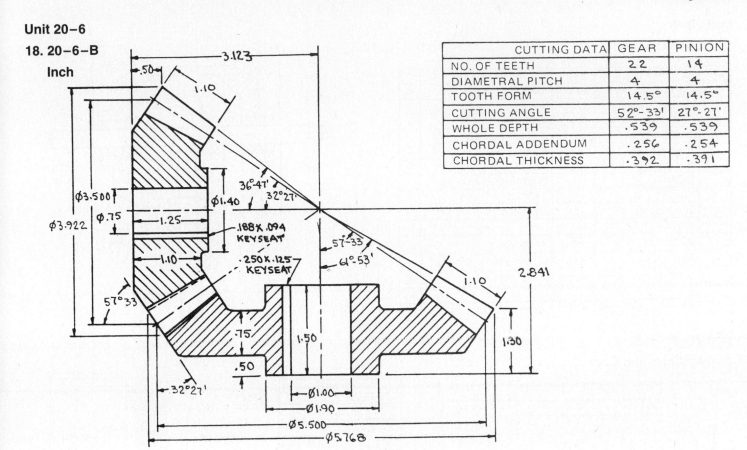

CUTTING DATA	GEAR	PINION
NO. OF TEETH	22	14
DIAMETRAL PITCH	4	4
TOOTH FORM	14.5°	14.5°
CUTTING ANGLE	52°-33'	27°-27'
WHOLE DEPTH	.539	.539
CHORDAL ADDENDUM	.256	.254
CHORDAL THICKNESS	.392	.391

20−6−B Metric

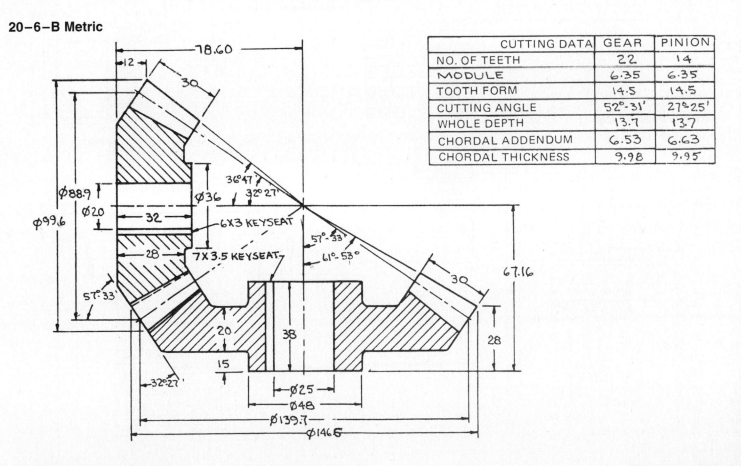

CUTTING DATA	GEAR	PINION
NO. OF TEETH	22	14
MODULE	6.35	6.35
TOOTH FORM	14.5	14.5
CUTTING ANGLE	52°-31'	27°-25'
WHOLE DEPTH	13.7	13.7
CHORDAL ADDENDUM	6.53	6.63
CHORDAL THICKNESS	9.98	9.95

Unit 20−6

19. 20−6−C

QTY	ITEM	MATL	DESCRIPTION	PT
1	HOUSING	W1	PATTERN XXX	1
2	MITER GEAR	STL	BOSTON # L110Y	2
1	SHAFT	STL	Ø.375 X 3.50	3
1	SHAFT	STL	Ø.375 X 5.60	4
2	SET SCREW	STL	SLOTTED, CUP PT. #10 X .19	5
1	BEARING	BR	BOSTON B612-10	6
2	BEARING	BR	BOSTON B612-4	7
1	COVER	W1	PATTERN XXX	8
3	OIL SEAL	—	GARLOCK #63 X 13	9
				10
1	GASKET	CORK	DWG XXX	11
2	RET. RING	STL	W. KOH-I-NOOR 5100-37	12
1	PIPE PLUG	W1	.125 IPS	13
				14
2	PIN - SL. TUBULAR	STL	Ø.125 X .375 LG	15

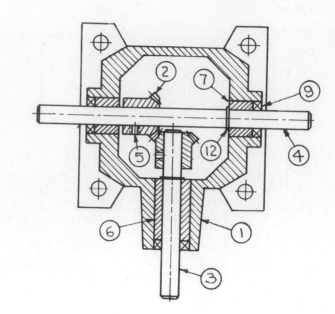

Unit 20−7

20. (A) Worm and Gear

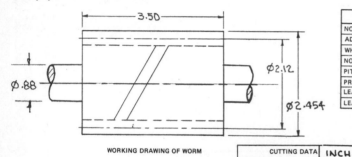

WORKING DRAWING OF WORM

CUTTING DATA	INCH
NO. OF TEETH	30
ADDENDUM	.167
WHOLE DEPTH	.359
NO. OF THREADS	1
PITCH (AXIAL)	.5236
PRESSURE ANGLE	14.5
LEAD ANGLE	4°
LEAD – R H	

ROUNDS & FILLETS R .10

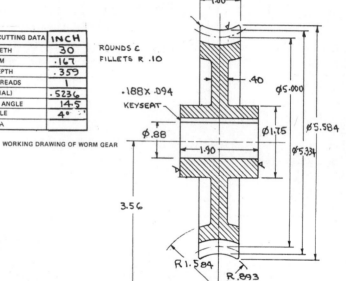

WORKING DRAWING OF WORM GEAR

CUTTING DATA	INCH
NO. OF THREADS	1
PITCH	.5236
PRESSURE ANGLE	14.5°
LEAD ANGLE	4°
LEAD – R H	.5236
WHOLE DEPTH	.359
ADDENDUM	.167

(B) Worm and Gear

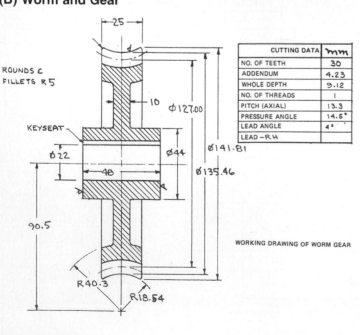

ROUNDS & FILLETS R 5

KEYSEAT

WORKING DRAWING OF WORM GEAR

CUTTING DATA	mm
NO. OF TEETH	30
ADDENDUM	4.23
WHOLE DEPTH	9.12
NO. OF THREADS	1
PITCH (AXIAL)	13.3
PRESSURE ANGLE	14.5°
LEAD ANGLE	4°
LEAD – R H	

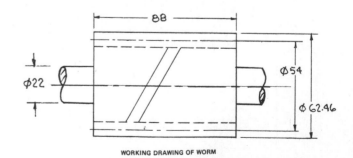

WORKING DRAWING OF WORM

CUTTING DATA	mm
NO. OF THREADS	1
PITCH	13.3
PRESSURE ANGLE	14.5°
LEAD ANGLE	4°
LEAD – R H	13.3
WHOLE DEPTH	9.12
ADDENDUM	4.23

Unit 20–7

21. (A) Worm and Gear

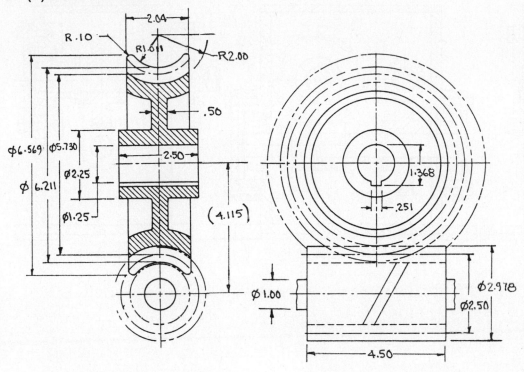

WORM

CUTTING DATA	
NO. OF THREADS	2
PITCH	.75
PRESSURE ANGLE	20°
LEAD ANGLE	10°49'
LEAD – LH	
WHOLE DEPTH	.515
ADDENDUM	.239

GEAR

CUTTING DATA	
NO. OF TEETH	24
ADDENDUM	.239
WHOLE DEPTH	.515
NO. OF THREADS	2
PITCH (AXIAL)	.75
PRESSURE ANGLE	20°
LEAD ANGLE	10°49'
LEAD – LH	

(B) Worm and Gear

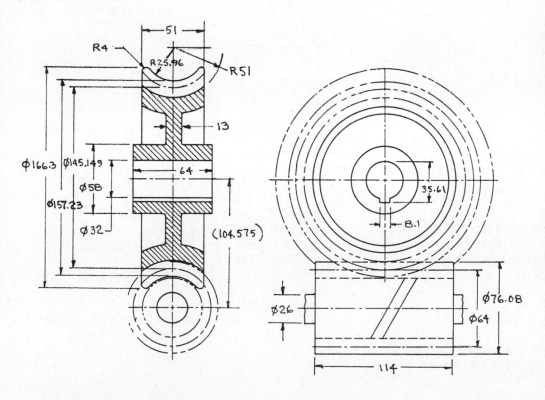

WORM

CUTTING DATA	
NO. OF THREADS	2
PITCH	19
PRESSURE ANGLE	20°
LEAD ANGLE	7°28'
LEAD – LH	
WHOLE DEPTH	13.03
ADDENDUM	6.04

GEAR

CUTTING DATA	
NO. OF TEETH	24
ADDENDUM	6.04
WHOLE DEPTH	13.03
NO. OF THREADS	2
PITCH (AXIAL)	19
PRESSURE ANGLE	20°
LEAD ANGLE	7°28'
LEAD – LH	

Unit 20-7

22. 20-7-A/B

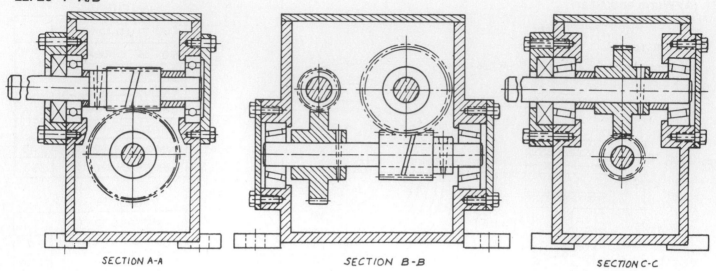

SECTION A-A   SECTION B-B   SECTION C-C

Unit 20-8

23. 20-8-A

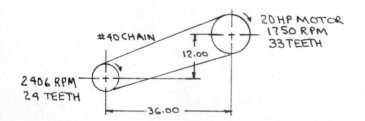

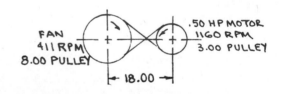

REPORT

BELT DRIVE - LIGHT DUTY BELTS NOT
   OF SUFFICIENT STRENGTH FOR HP
   REQUIRED. DESIGN HP = 20 X 1.2 = 24HP.
GEAR DRIVE - CENTER DISTANCE TOO
   GREAT FOR GEARS
CHAIN DRIVE - DESIGN HP = 20 X 1.5 = 30 HP
  - SMALL SPROCKET = 2400 RPM.
  - USING A TRIPLE STRAND DESIGN AND
   2400 RPM A 40 CHAIN (.50 P) IS
   SELECTED.
  - MIN. NO. OF TEETH ON SMALL
   SPROCKET = (2400 RPM & 24 ÷ 2.5 =
   9.6 HP) = 24 TEETH
  -USING SPROCKETS OF 24 AND 33 TEETH
   WILL PRODUCE 2406 RPM FOR COAL
   BREAKER

REPORT

GEAR DRIVE - CENTER DISTANCE NOT
   SUITABLE FOR GEAR DRIVE
CHAIN DRIVE - CANNOT REVERSE DIRECTION
   OF SHAFT WITH CHAIN DRIVE
BELT DRIVE - RECOMMENDED.
   QUIET, EASY TO INSTALL,
   INEXPENSIVE, REVERSE DIRECTION
   OF SHAFT BY USING CROSSOVER.
REQUIREMENTS
   DESIGN HP (LIGHT DUTY) .85 X .50 = .425 HP
   Ø 3.00 PULLEY
   Ø 8.00 PULLEY
   TYPE A INDUSTRIAL V-BELT, SECTION
   .50 X .31, BELT LENGTH = 54.00

Unit 20−8

23. CONTD 20−8−B

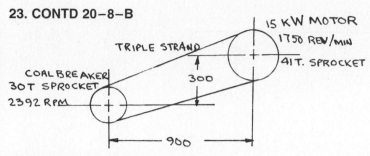

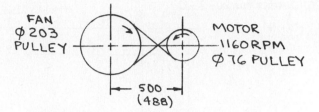

REPORT

BELT DRIVE − LIGHT DUTY BELTS NOT
   OF SUFFICIENT STRENGTH FOR KW
   REQUIRED. DESIGN KW = 15 X 1.2
   = 18 KW

CHAIN DRIVE − DESIGN KW = 15 X 1.5
   = 22.5 KW.
 − SMALL SPROCKET = 2400 r/min
 − USING A TRIPLE STRAND DESIGN
   AND 2400 r/min A 40 CHAIN (13 P)
   IS SELECTED.
 − MIN. NO. OF TEETH ON SMALL
   SPROCKET = 28
 − USING SPROCKETS HAVING 30 &
   41 TEETH WILL PRODUCE 2392
   r/min FOR COAL BREAKER.
GEAR DRIVE − CENTER DISTANCE
   TOO GREAT FOR GEARS
RECOMMEND CHAIN DRIVE

REPORT
 − RECOMMEND BELT DRIVE
 − CENTER DISTANCE NOT SUITABLE
   FOR GEAR DRIVE
 − CANNOT REVERSE DIRECTION OF
   SHAFT WITH CHAIN DRIVE
 − BELT DRIVE IS QUIET, EASY
   TO INSTALL, INEXPENSIVE,
   REVERSE DIRECTION BY USING
   CROSSOVER
REQUIREMENTS
   Ø 76 PULLEY
   Ø 203 PULLEY
   TYPE A INDUSTRIAL V-BELT
      SECTION 12 X 8, BELT
      LENGTH 14 0 mm

24. 20−8−C

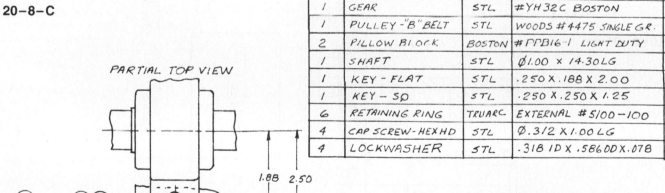

QTY	ITEM	MATL	DESCRIPTION	PT
1	GEAR	STL	#YH32C BOSTON	1
1	PULLEY -"B" BELT	STL	WOODS #4475 SINGLE GR.	2
2	PILLOW BLOCK	BOSTON	#PPB16-1 LIGHT DUTY	3
1	SHAFT	STL	Ø1.00 X 14.30 LG	4
1	KEY - FLAT	STL	.250 X .188 X 2.00	5
1	KEY - SQ	STL	.250 X .250 X 1.25	6
6	RETAINING RING	TRUARC	EXTERNAL #5100-100	7
4	CAP SCREW-HEX HD	STL	Ø.312 X 1.00 LG	8
4	LOCKWASHER	STL	.318 ID X .586 OD X .078	9

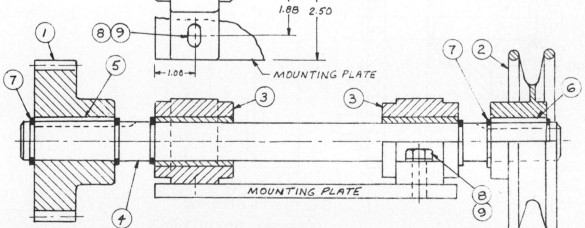

Unit 20−8

25. Details for 20−8−C

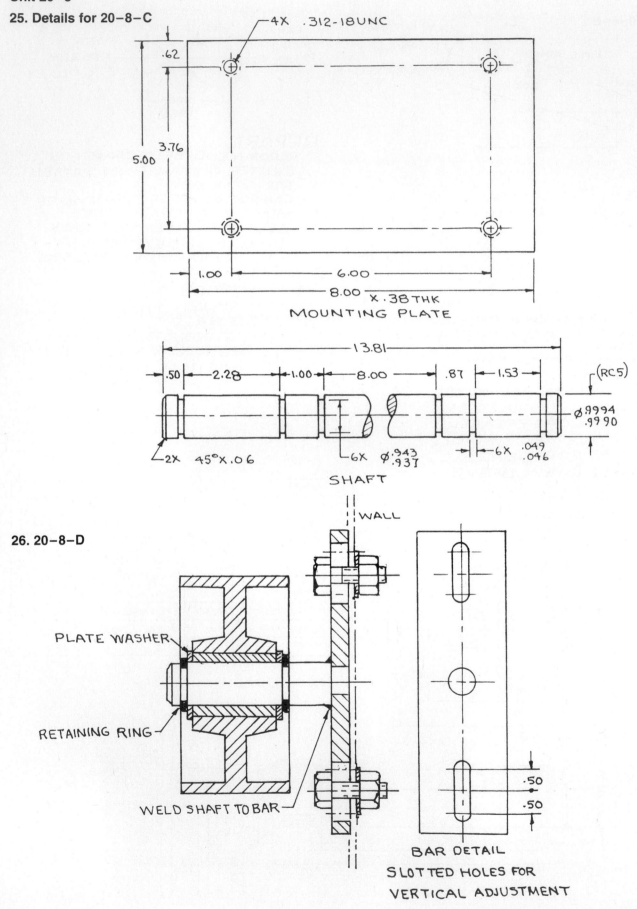

4X .312-18UNC

.62

3.76

5.00

1.00

6.00

8.00 X .38THK

MOUNTING PLATE

13.81

.50 2.28 1.00 8.00 .87 1.53

(RC5)

Ø.9994
.9990

2X 45°X.06

6X Ø.943
.937

6X .049
.046

SHAFT

26. 20−8−D

WALL

PLATE WASHER

RETAINING RING

WELD SHAFT TO BAR

BAR DETAIL

SLOTTED HOLES FOR

VERTICAL ADJUSTMENT

.50

.50

Chapter 21 Couplings, Bearings, and Seals

Unit 21–1

1. 21–1–A

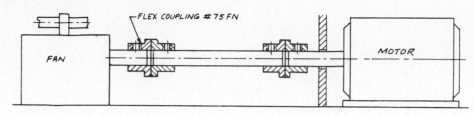

2. 21–1–B

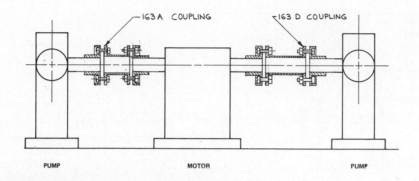

Unit 21–2

3. 21–2–A,B

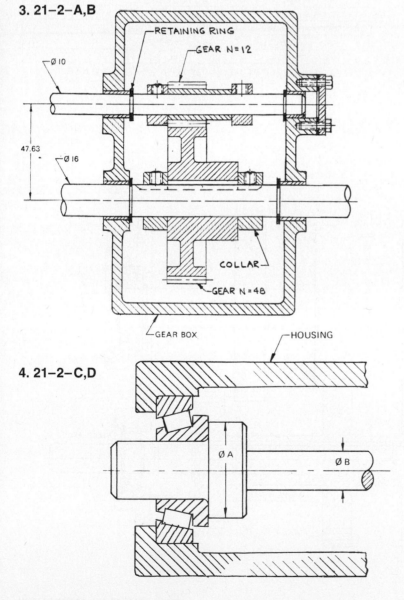

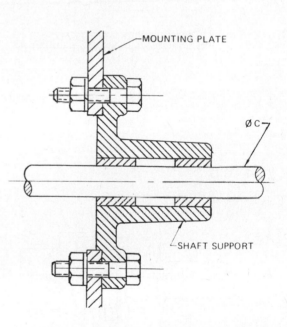

4. 21–2–C,D

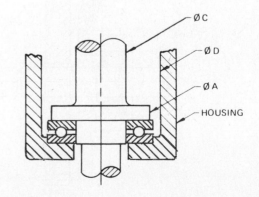

Unit 21–3

5. 21–3–A,B

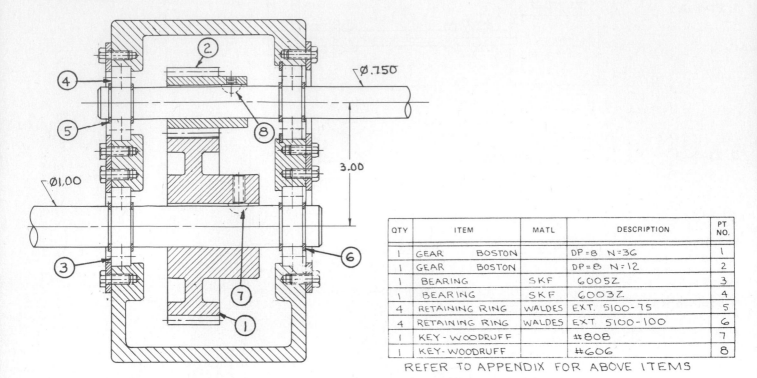

QTY	ITEM		MATL	DESCRIPTION	PT NO.
1	GEAR	BOSTON		DP=8 N=36	1
1	GEAR	BOSTON		DP=8 N=12	2
1	BEARING		SKF	6005Z	3
1	BEARING		SKF	6003Z	4
4	RETAINING RING		WALDES	EXT. 5100-75	5
4	RETAINING RING		WALDES	EXT. 5100-100	6
1	KEY-WOODRUFF			#808	7
1	KEY-WOODRUFF			#606	8

REFER TO APPENDIX FOR ABOVE ITEMS

6. 21–3–C,D

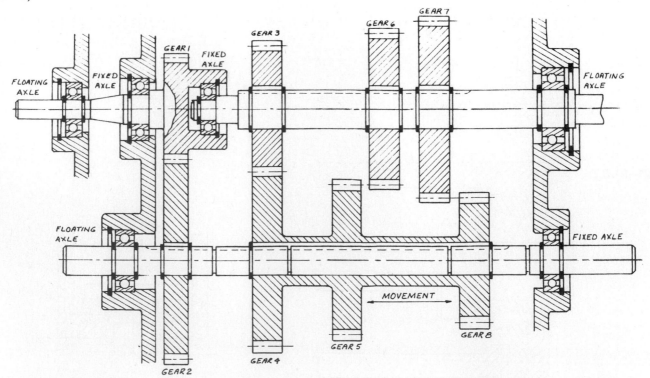

Unit 21–3

7. 21–3–E

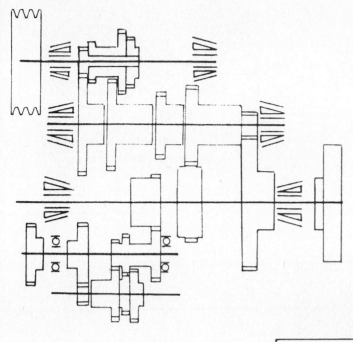

21–3–F

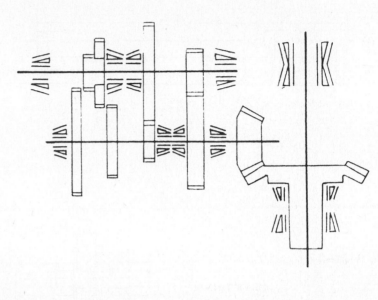

8. 21–3–G

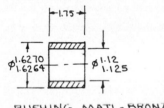

BUSHING MATL-BRONZE

CUTTING DATA	
NUMBER OF TEETH	27
PITCH DIAMETER	4.500
DIAMETRAL PITCH	6
PRESSURE ANGLE	20°
WHOLE DEPTH	.360
CHORDAL ADDENDUM	.182
CHORDAL THICKNESS	.261
CIRCULAR THICKNESS	.262
WORKING DEPTH	.333

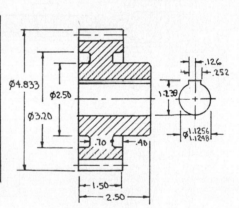

GEAR MATL-GI

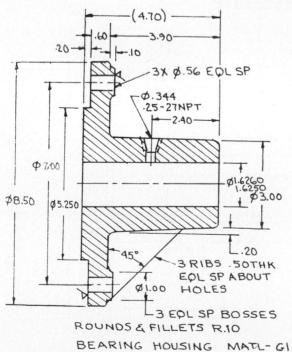

ROUNDS & FILLETS R.10

BEARING HOUSING MATL-GI

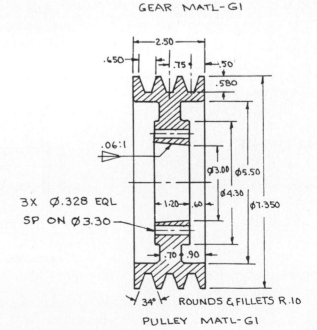

ROUNDS & FILLETS R.10

PULLEY MATL-GI

Unit 21–3
8. 21–3–G CONTD

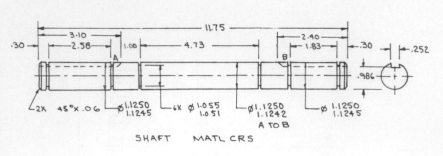

SHAFT   MATL CRS

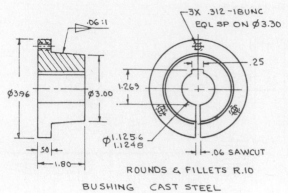

ROUNDS & FILLETS R.10

BUSHING   CAST STEEL

9. 21–3–H

10. 21–3–H

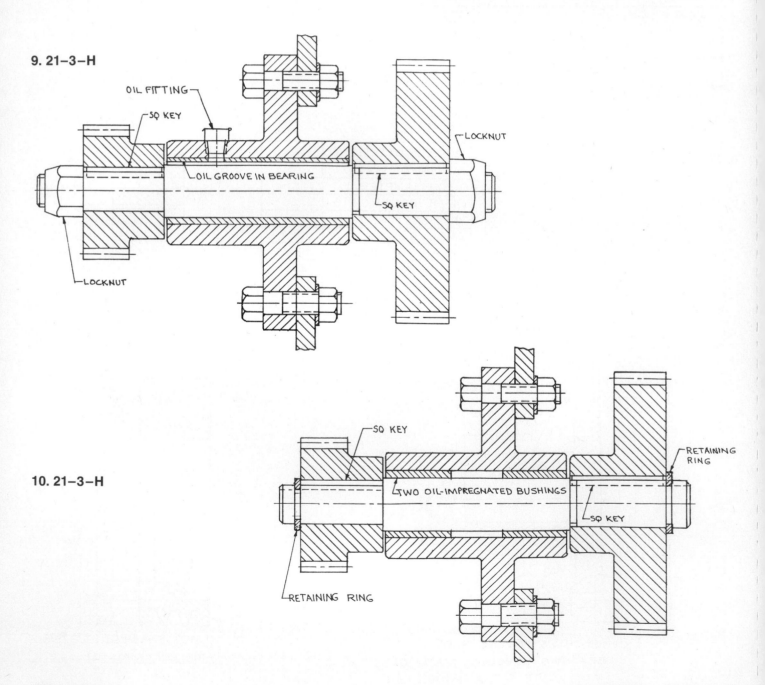

Unit 21–3

11. 21–3–J

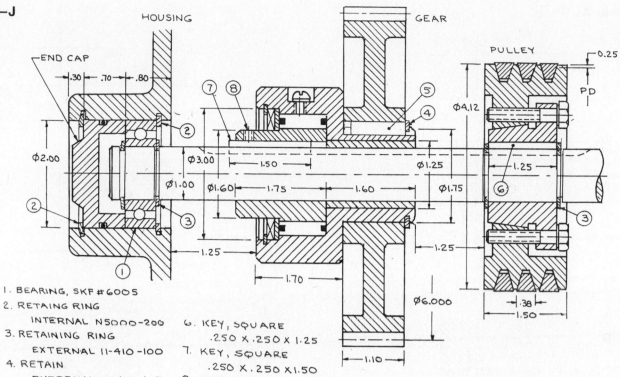

HOUSING   GEAR   PULLEY

1. BEARING, SKF #6005
2. RETAING RING
   INTERNAL N5000-200
3. RETAINING RING
   EXTERNAL 11-410-100
4. RETAIN
   EXTERNAL 11-410-1 5
5. KEY, SQUARE
   .250 X .250 X 1.00

6. KEY, SQUARE
   .250 X .250 X 1.25
7. KEY, SQUARE
   .250 X .250 X 1.50
8 SETSCREW-HEX SOCKET
   HD, CUP POINT
   .250-20UNC X .25

Unit 21–4

12. 21–4–A

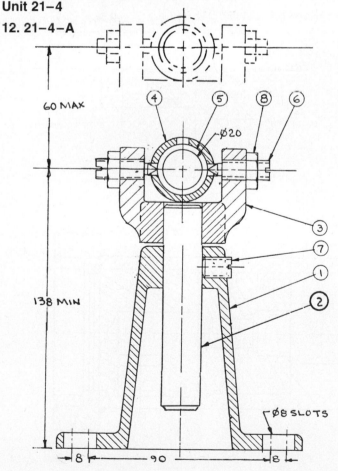

60 MAX

Ø20

138 MIN

Ø8 SLOTS

8   90   8

13. 21–4–B

QTY	ITEM	MATL	DESCRIPTION	PT NO.
1	BASE	WI	PATTERN NO. XXXX	1
1	BEARING HOUSING	WI	PATTERN NO XXXX	2
2	SHAFT	CRS	Ø.623 X 7.00	3
1	BEARING	SKF	X4012	4
2	BASE PIN	STL	#18 DRILL ROD X 1.25 LG	5
1	OIL CAP		XGF ·D5	6
2	BOLT, HEX HD REG	STL	.250-20UNC X 1.62 LG	7
2	NUT - REG HEX	STL	.250-20UNC	8

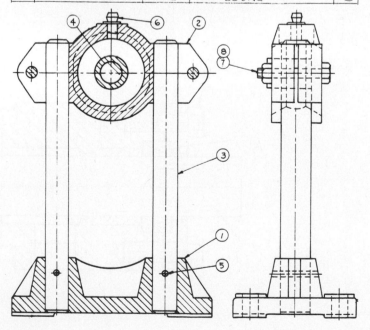

Unit 21–5

14. 21–5–A

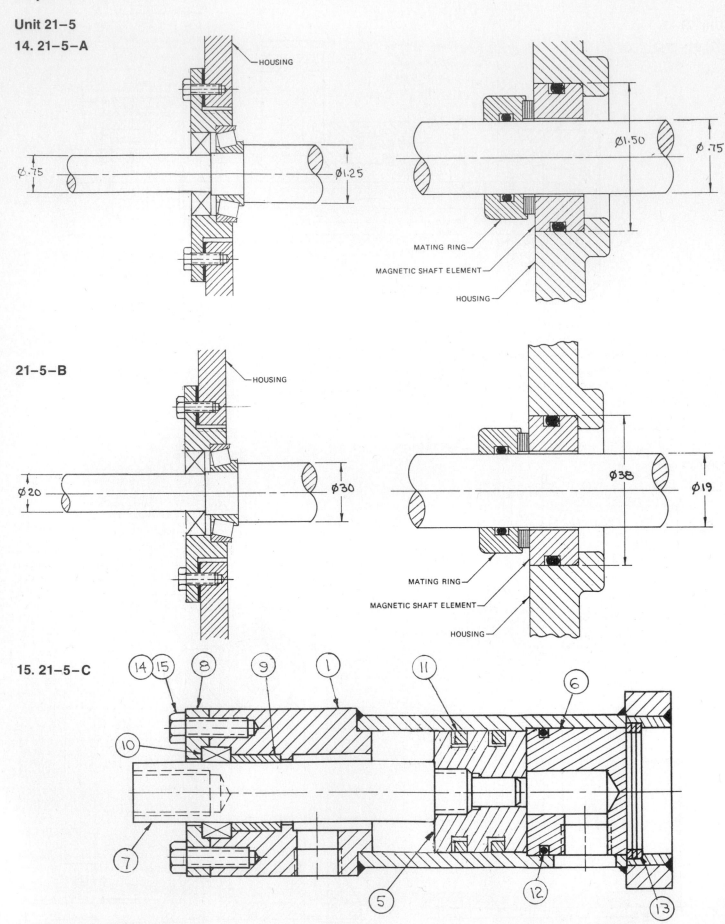

HOUSING

Ø.75   Ø1.25

Ø1.50   Ø.75

MATING RING

MAGNETIC SHAFT ELEMENT

HOUSING

21–5–B

HOUSING

Ø20   Ø30

Ø38   Ø19

MATING RING

MAGNETIC SHAFT ELEMENT

HOUSING

15. 21–5–C

Unit 21-5

15. 21-5-C CONTD

QTY	ITEM	MATL	DESCRIPTION	PT NO.
1	HOUSING ASSY			1
1	END PLATE	STL	Ø 2.75 X 2.65	2
1	CYLINDER	STL	1.875 ID X 2.50 OD X 4.35	3
1	FLANGE	STL	.75 X 3.20 X 4.00	4
1	PISTON	MEEHANITE	Ø 1.875 X 1.50	5
1	CYLINDER HEAD	STL	Ø 2.00 X 1.60	6
1	PISTON ROD SAE	1045	Ø 1.00 X 6.40	7
1	BUSHING RETAINER	STL	Ø 2.75 X .40	8
2	BUSHING	BRONZE	1.000 ID X 1.252 OD X .75	9
1	FELT RING PACKING	FELT	1.00 ID X 1.50 OD X .50	10
2	SPLIT RING SEAL		1.50 ID X 1.88 OD X .188	11
1	O-RING	RUBBER	1.81 ID X Ø .125	12
3	RETAINING RING	WALDES	INT. N 5000-200	13
4	CAP SCREW - HEX HD	STL	.312 UNC X 1.00	14
4	LOCKWASHER	STL	SPRING .312	15

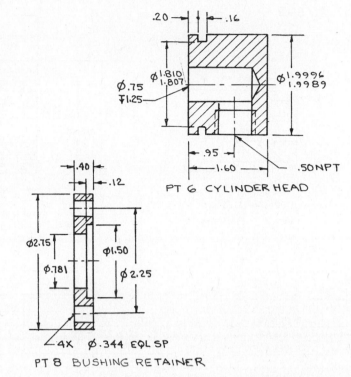

PT 6 CYLINDER HEAD

PT 8 BUSHING RETAINER

16. 21-5-C

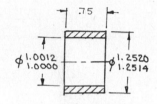

PT 9 BUSHING

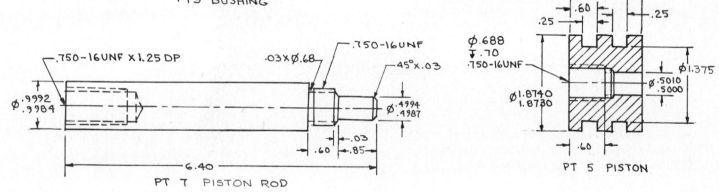

PT 7 PISTON ROD

PT 5 PISTON

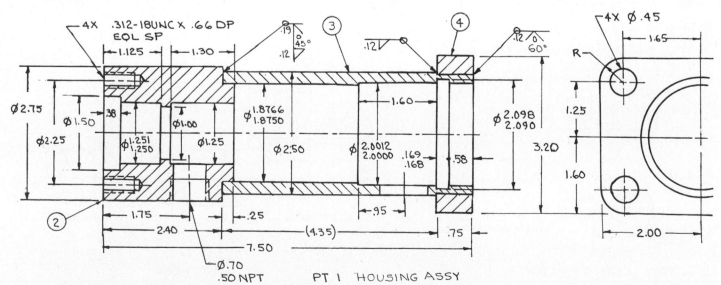

PT 1 HOUSING ASSY

Unit 21–5

17. 21–5–D

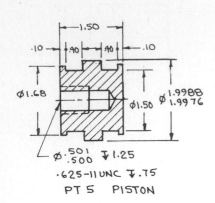

PT 5 PISTON

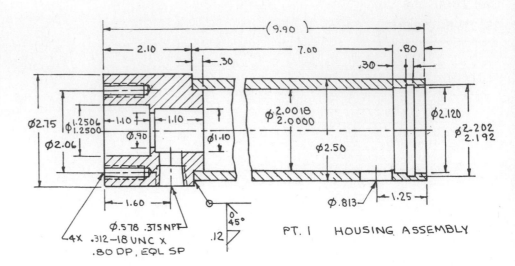

PT. 1   HOUSING ASSEMBLY

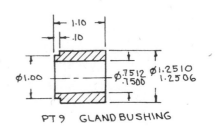

PT 9   GLAND BUSHING

QTY	ITEM	MATL	DESCRIPTION	PT NO.
1	HOUSING ASSY			1
1	END PLATE	STL	Ø2.75 X 2.40	2
1	CYLINDER	STL	2.00 ID X 2.50 OD X 7.80	3
1	CYLINDER HEAD	STL	Ø2.00 X 3.65	4
1	PISTON	MEEHANITE	Ø2.00 X 1.50	5
1	PISTON ROD	SAE 1045	Ø.75 X 8.87	6
1	CLEVIS ROD HEAD	STL	1.10 X 1.50 X 2.20	7
1	COVER PLATE	STL	Ø2.75 X .38	8
1	GLAND BUSHING	BRONZE	.75 ID X 1.25 OD X 1.10	9
1	FELT RING PACKING	FELT	.75 ID X 1.25 OD X .38	10
2	SPLIT RING SEAL		1.50 ID X 2.00 OD X .40	11
1	O-RING	RUBBER	1.75 ID X Ø.125	12
3	RETAINING RING	WALDES	INT. N5000-212	13
4	CAP SCREW-SOCKET	STL	HD .312 UNC X 1.25	14
4	LOCKWASHER	STL	SPRING .312	15

PT 8   COVER PLATE

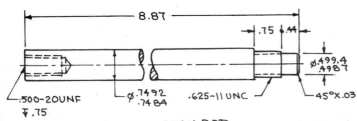

PT 6   PISTON ROD

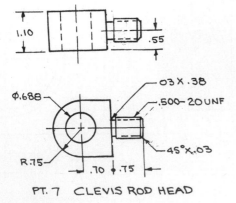

PT. 7   CLEVIS ROD HEAD

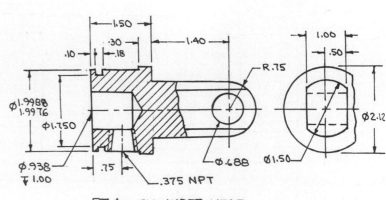

PT 4   CYLINDER HEAD

Unit 21–6

18. 21–6–A

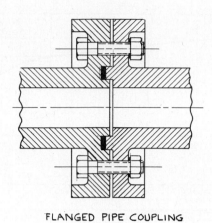 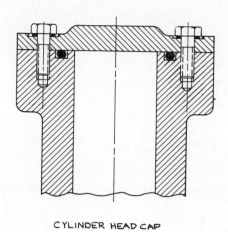

FLANGED PIPE COUPLING   CYLINDER HEAD CAP

19. 21–6–B

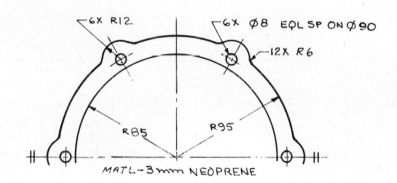

6X R12

6X Ø8 EQL SP ON Ø90

12X R6

R85   R95

MATL→3mm NEOPRENE

Chapter 22 Cams, Linkages, and Actuators
Unit 22–1

1. 22–1–A

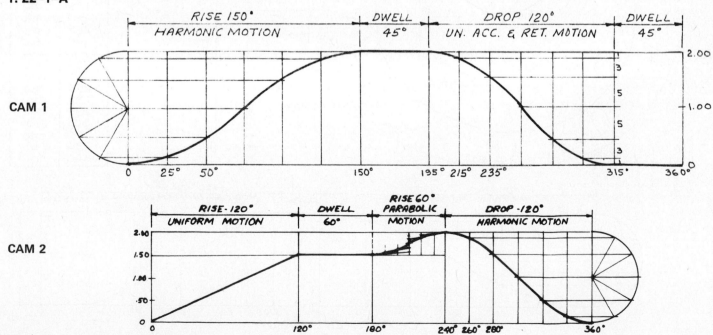

Unit 22–1

1. CONTD

22–1–B

CAM 1

CAM 2

Unit 22–2

2. Plate Cam

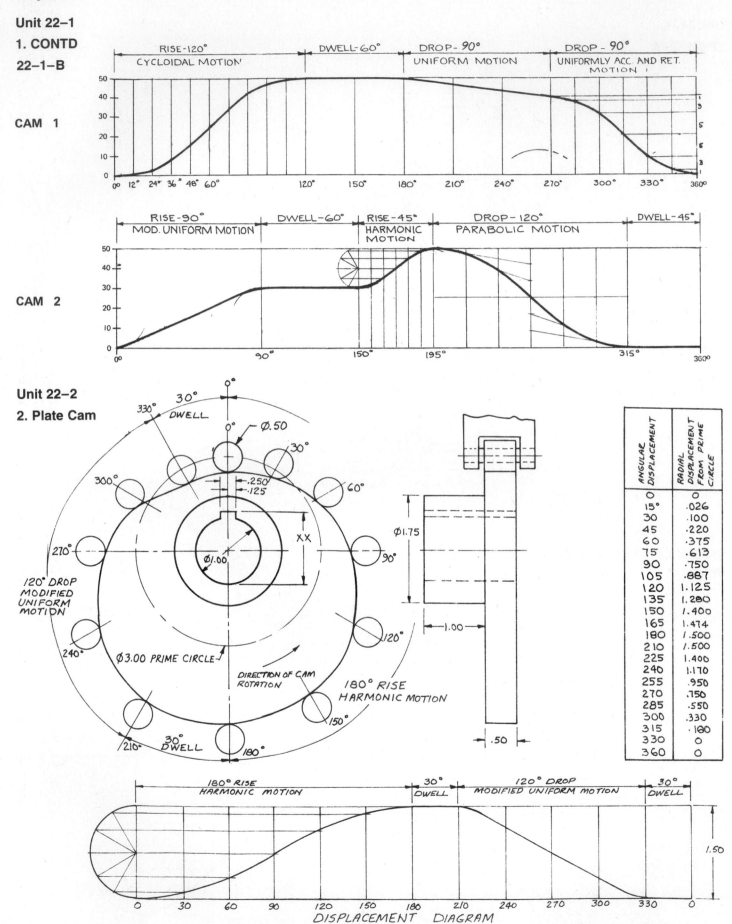

ANGULAR DISPLACEMENT	RADIAL DISPLACEMENT FROM PRIME CIRCLE
0	0
15°	.026
30	.100
45	.220
60	.375
75	.613
90	.750
105	.887
120	1.125
135	1.280
150	1.400
165	1.474
180	1.500
210	1.500
225	1.400
240	1.170
255	.950
270	.750
285	.550
300	.330
315	.180
330	0
360	0

DISPLACEMENT DIAGRAM

Unit 22–2

3. Plate Cam

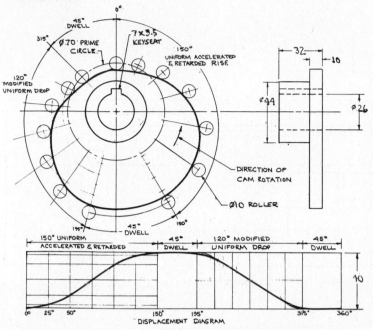

ANGULAR DISPLACEMENT	RADIAL DISPLACEMENT FROM PRIME CIRCLE
0°	0
15°	1.5
30°	3.5
45°	7.5
60°	13.5
75°	20
90°	26.5
105°	32.5
120°	36.5
135°	38.5
150°	40
195°	40
210°	37.5
225°	33
240°	27
255°	20
270°	13
285°	7
300°	2.5
315°	0
330°	0
345°	0
360°	0

4. 22–2–A

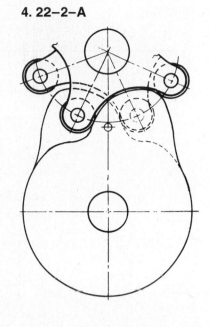

5. 22–2–B

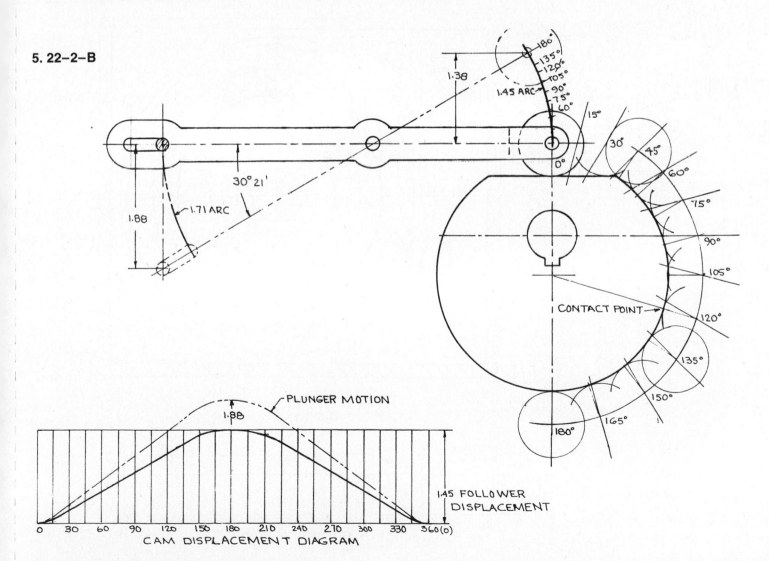

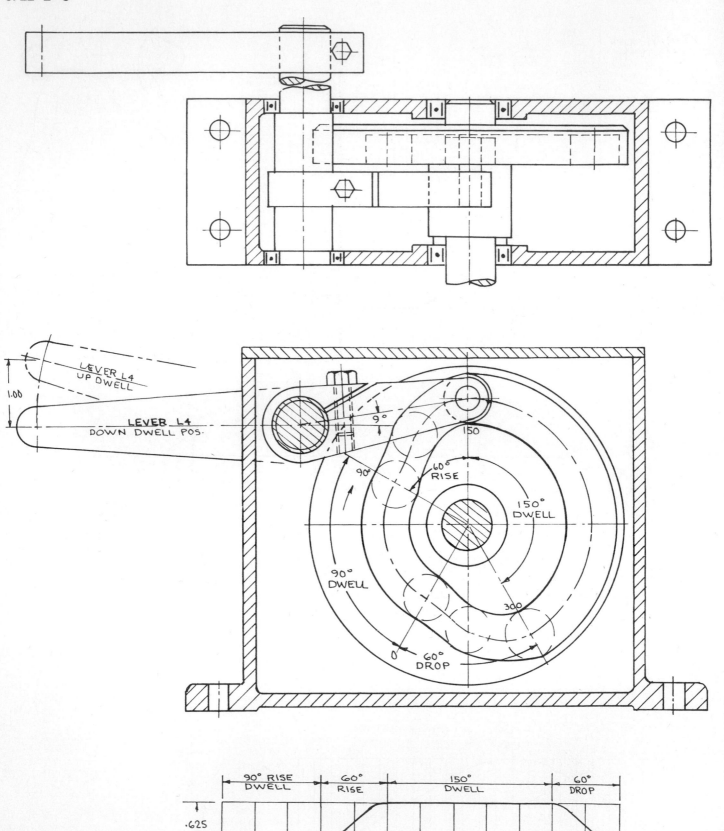

Unit 22–3

7. Face Cam

ANGLE	DISTANCE FROM PRIME CIRCLE
0	0
15	.030
30	.115
45	.247
60	.415
75	.600
90	.785
105	.953
120	1.085
135	1.170
150	1.200
180	1.200
195	1.162
210	1.050
225	.862
240	.600
255	.338
270	.150
285	.038
300	0
360	0

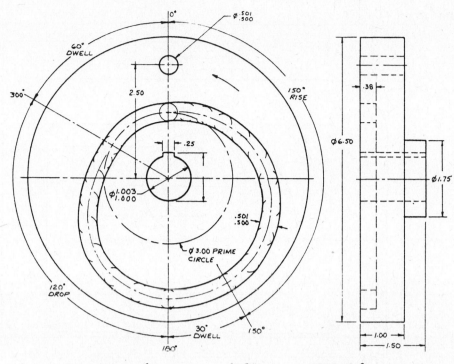

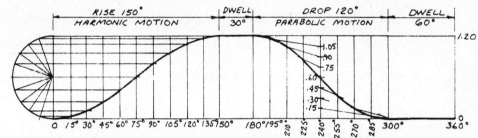

8. Face Cam

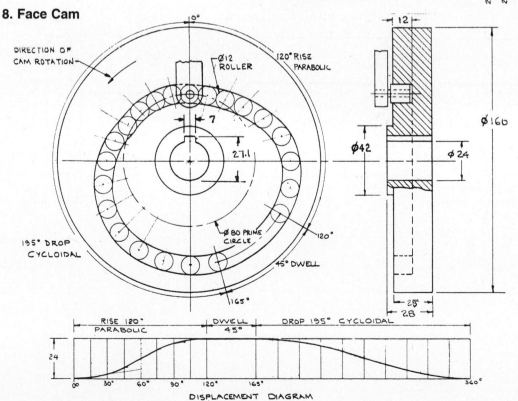

ANGULAR DISPLACEMENT	RADIAL DISPLACEMENT FROM PRIME CIRCLE
0°	0
15°	1
30°	3
45°	6.5
60°	12
75°	17.5
90°	21
105°	23
120°	24
165°	24
180°	23.5
195°	23
210°	22
225°	20.5
240°	17.5
255°	13.5
270°	10.5
285°	6.5
300°	3.5
315°	2
330°	1
345°	0.5
360°	0

DISPLACEMENT DIAGRAM

Unit 22–3
9. Yoke Cam

10. Yoke Cam

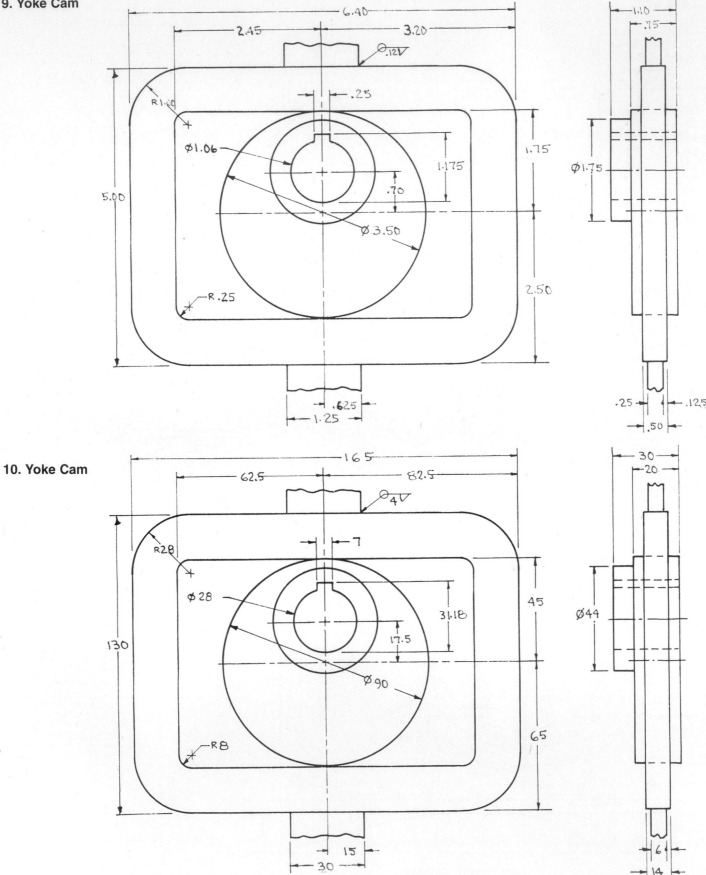

Unit 22–4

11. Drum Cam

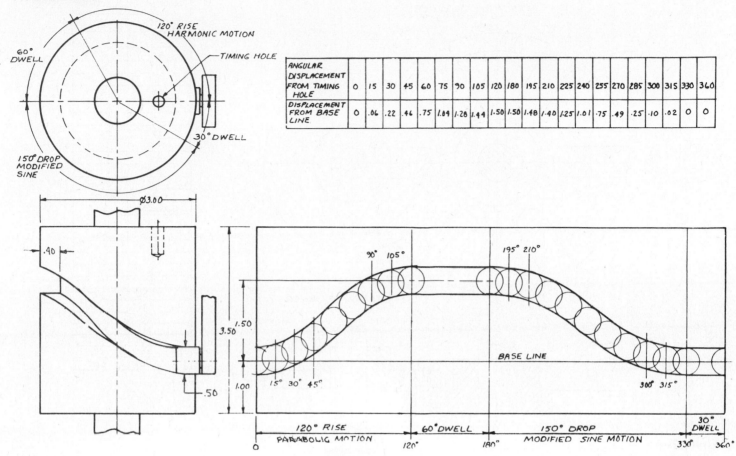

ANGULAR DISPLACEMENT FROM TIMING HOLE	0	15	30	45	60	75	90	105	120	180	195	210	225	240	255	270	285	300	315	330	360
DISPLACEMENT FROM BASE LINE	0	.06	.22	.46	.75	1.04	1.28	1.44	1.50	1.50	1.48	1.40	1.25	1.01	.75	.49	.25	.10	.02	0	0

12. Drum Cam

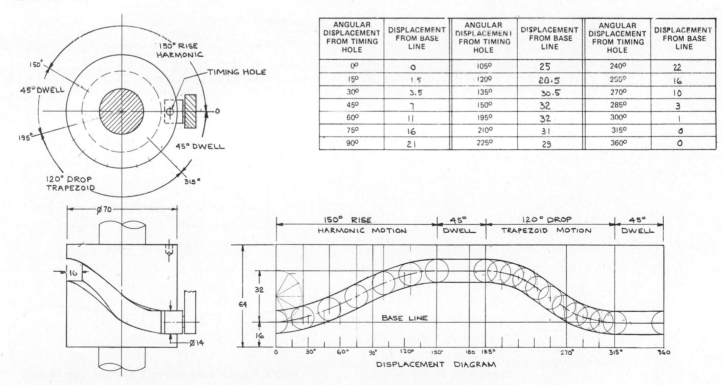

ANGULAR DISPLACEMENT FROM TIMING HOLE	DISPLACEMENT FROM BASE LINE	ANGULAR DISPLACEMENT FROM TIMING HOLE	DISPLACEMENT FROM BASE LINE	ANGULAR DISPLACEMENT FROM TIMING HOLE	DISPLACEMENT FROM BASE LINE
0°	0	105°	25	240°	22
15°	1.5	120°	28.5	255°	16
30°	3.5	135°	30.5	270°	10
45°	7	150°	32	285°	3
60°	11	195°	32	300°	1
75°	16	210°	31	315°	0
90°	21	225°	29	360°	0

Unit 22–5

13. 22–5–2

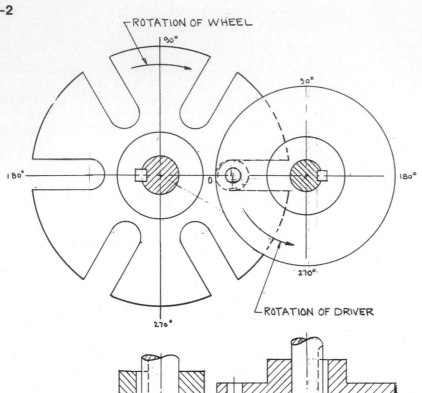

ANGULAR DISPLACEMENT	
DRIVER POSITION	WHEEL POSITION
0°	0
5°	5°
10°	9.8°
15°	14.5°
20°	18.5°
25°	21.5°
30°	24°
35°	26°
40°	27.5°
45°	28.7°
50°	29.5°
55°	29.8°
60°	30°
65 TO 295°	30°
300°	30°
305°	30.2°
310°	30.5°
315°	31.3°
320°	32.5°
325°	34°
330°	36°
335°	38.5°
340°	41.5°
345°	45.5°
350°	50.2°
355°	55°
360°	60°

Unit 22–6

14. 22–6–A

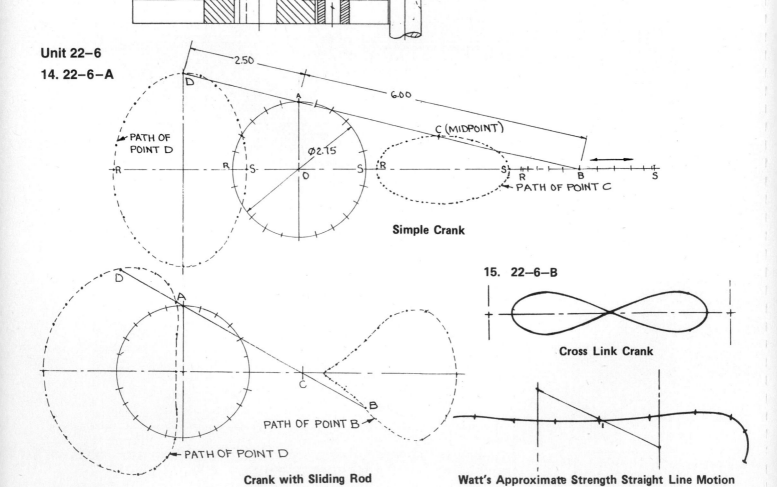

Simple Crank

15. 22–6–B

Cross Link Crank

Crank with Sliding Rod

Watt's Approximate Strength Straight Line Motion

Unit 22–6

16. 22–6–C

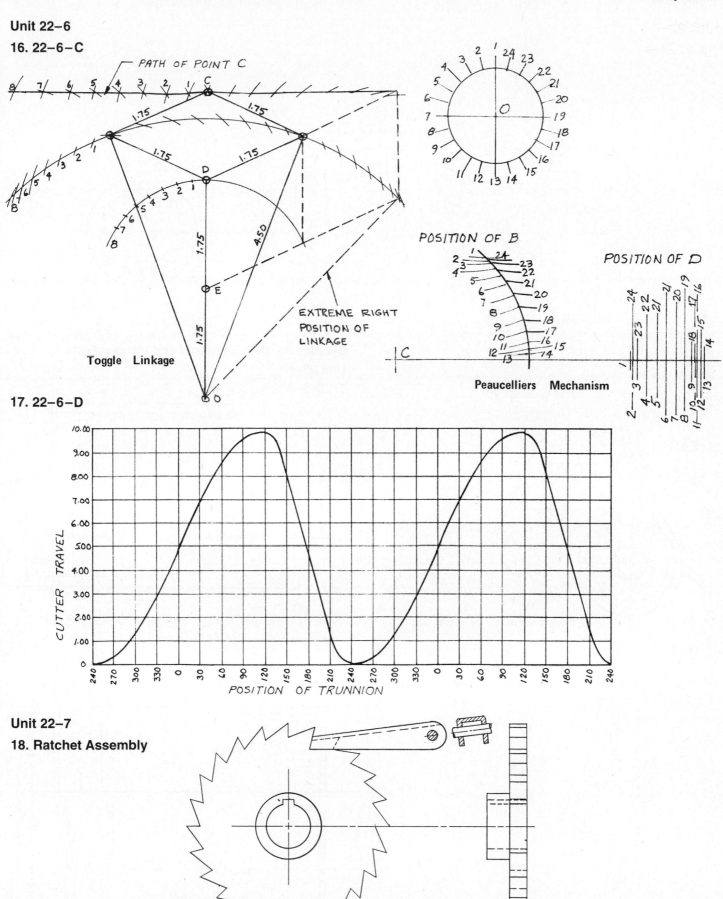

Toggle Linkage

PATH OF POINT C

EXTREME RIGHT POSITION OF LINKAGE

POSITION OF B

POSITION OF D

Peaucelliers Mechanism

17. 22–6–D

Unit 22–7

18. Ratchet Assembly

Unit 22–7

19. 22–7–A

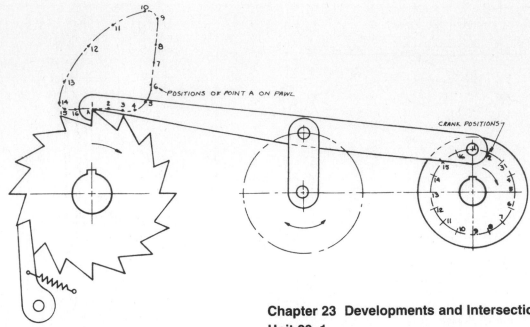

POSITIONS OF POINT A ON PAWL

CRANK POSITIONS

20. 14–3–F

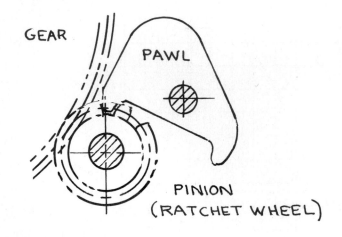

GEAR

PAWL

PINION
(RATCHET WHEEL)

Chapter 23 Developments and Intersections
Unit 23–1

1. 23–1–A

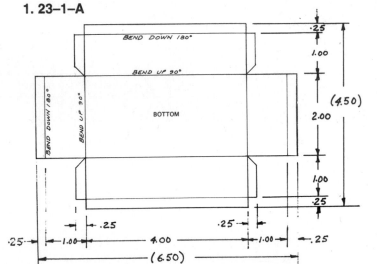

BEND DOWN 180°

BEND UP 90°

BEND DOWN 180°

BEND UP 90°

BOTTOM

.25

1.00

(4.50)

2.00

1.00

.25

.25

.25

.25

1.00

4.00

1.00

.25

(6.50)

23–1–B

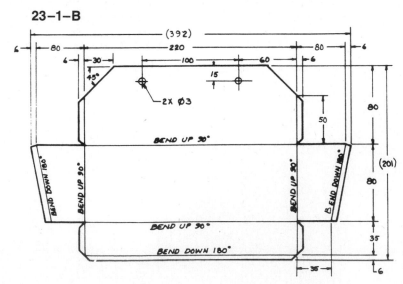

(392)

6

80

220

80

6

6

30

100

60

6

45°

15

2X Ø3

50

80

BEND UP 90°

BEND DOWN 180°

BEND UP 90°

BEND UP 90°

BEND DOWN 180°

(201)

80

BEND UP 90°

35

BEND DOWN 180°

35

6

Unit 23–1

1. CONTD 23–1–C

23–1–D

Unit 23–2

2. 23–2–A

23–2–B

3. 23–2–C

23–2–D

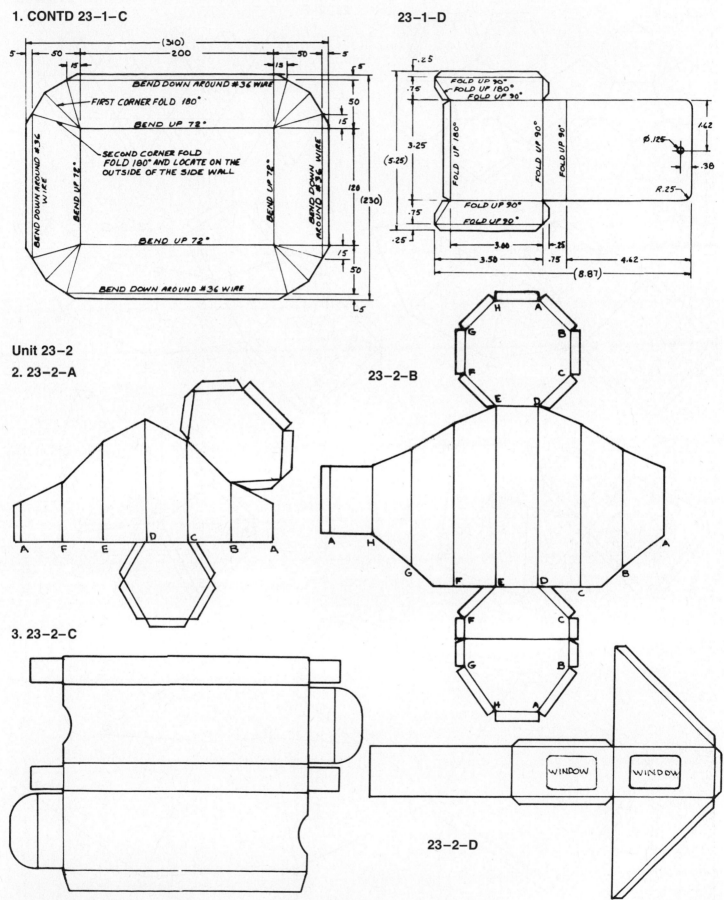

Unit 23–2

4. 23–2–E

23–2–F

Unit 23–3

5. 23–3–A

23–3–B

23–3–C

23–3–D

23–3–E

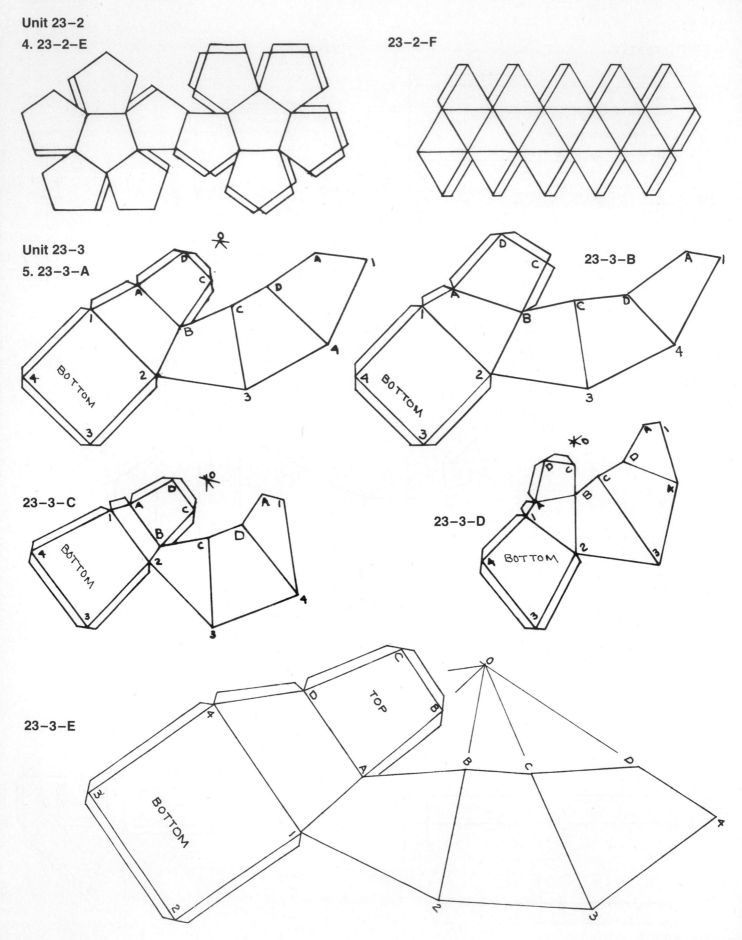

Unit 23–4

6. 23–4–A

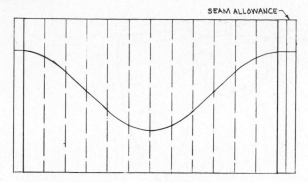

SEAM ALLOWANCE

23–4–B

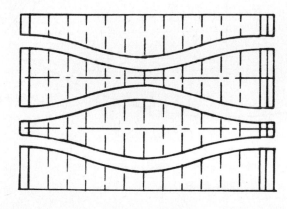

7. 23–4–C

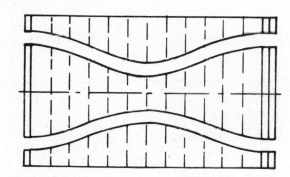

8. 23–4–D

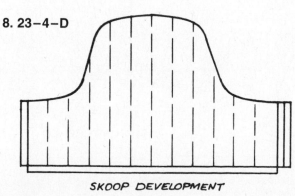

SKOOP DEVELOPMENT

8. CONTD 23–4–E

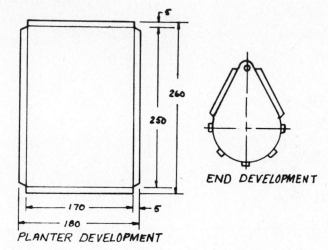

5

260

250

170

5

180

PLANTER DEVELOPMENT

END DEVELOPMENT

Unit 23–5

9. 23–5–A

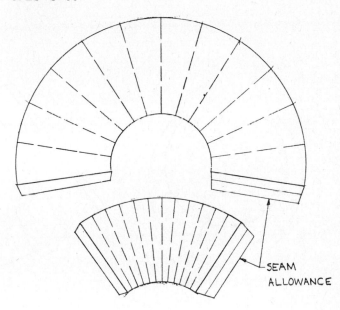

SEAM ALLOWANCE

23–5–B

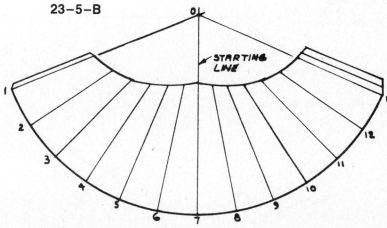

O

STARTING LINE

1 2 3 4 5 6 7 8 9 10 11 12 1

Unit 23–5
9. CONTD 23–5–C

23–5–D

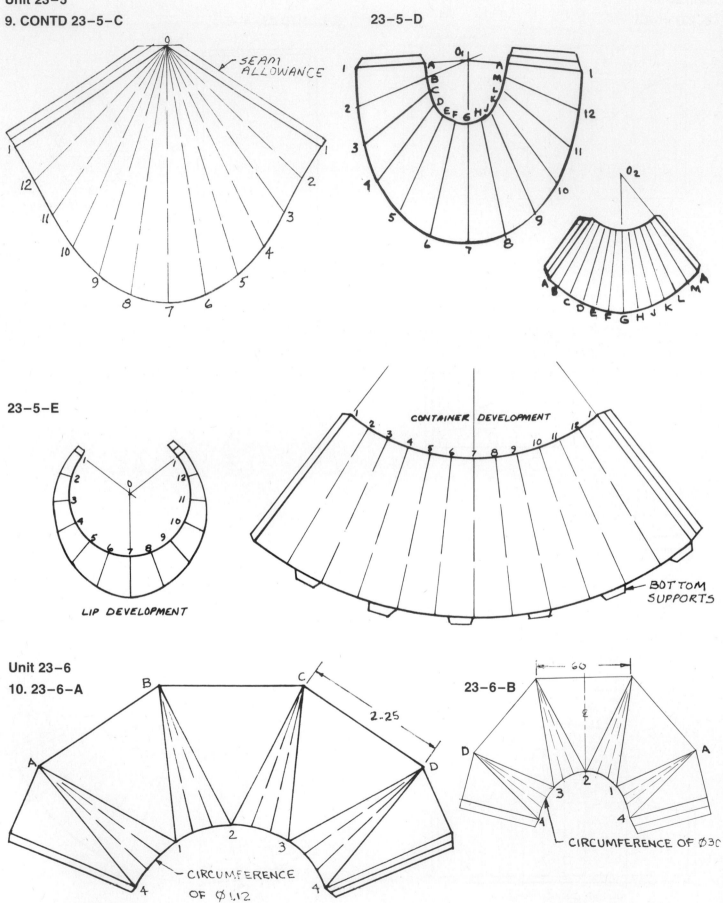

SEAM ALLOWANCE

CONTAINER DEVELOPMENT

23–5–E

LIP DEVELOPMENT

BOTTOM SUPPORTS

Unit 23–6
10. 23–6–A

2·25

CIRCUMFERENCE OF ⌀1·12

23–6–B

60

CIRCUMFERENCE OF ⌀30

Unit 23-6

10. CONTD 23-6-C

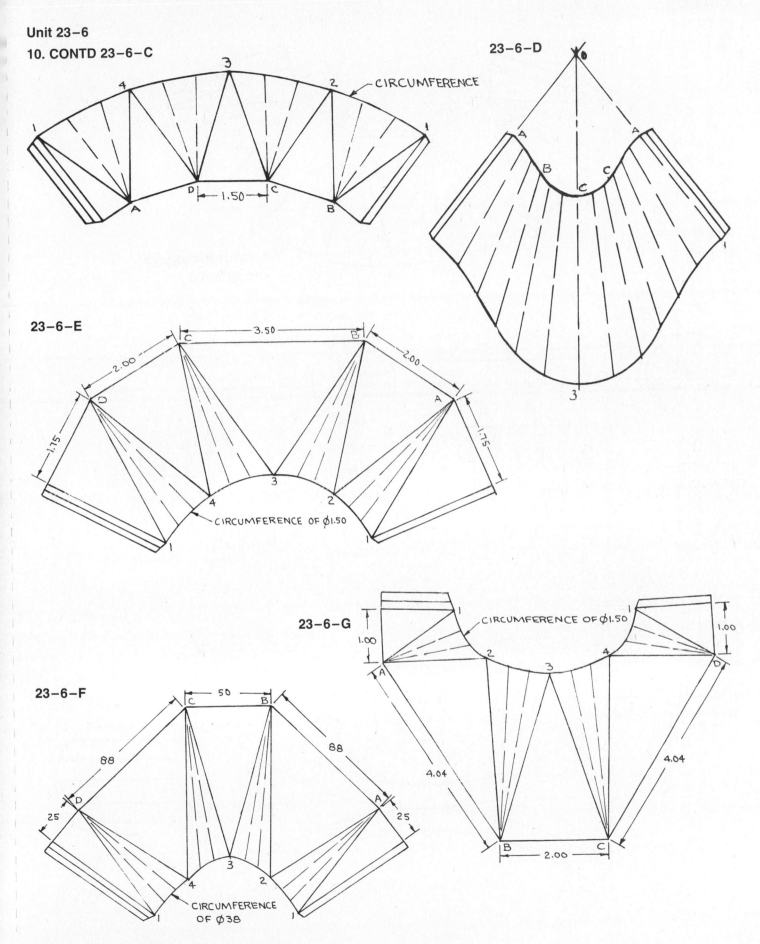

Unit 23−6
10. CONTD 23−6−H

Unit 23−7
11. Ball

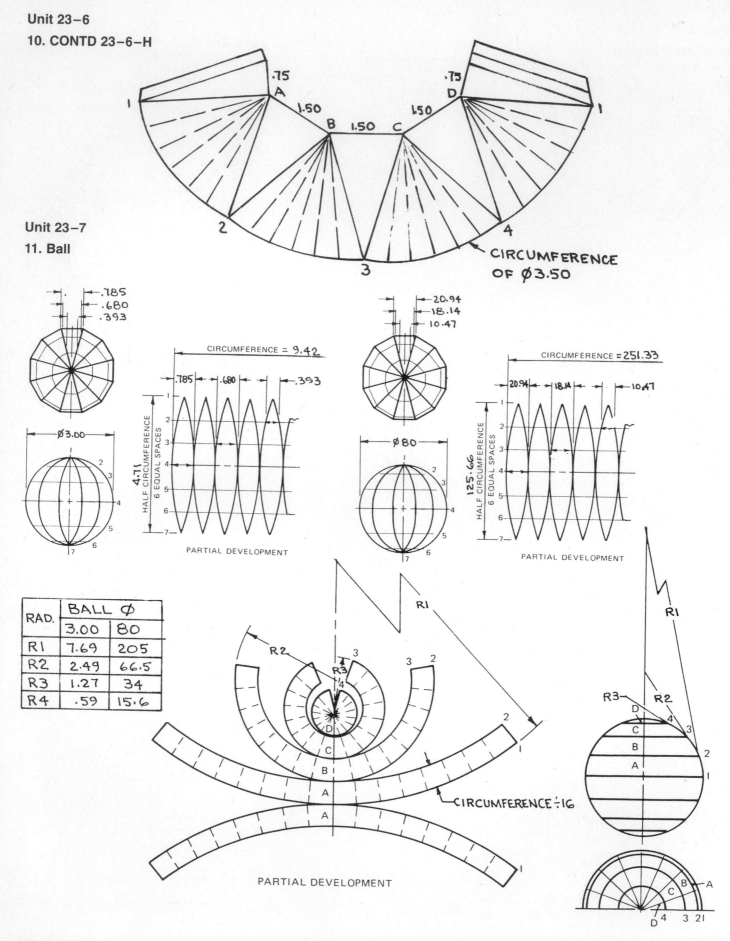

RAD.	BALL Ø	
	3.00	80
R1	7.69	205
R2	2.49	66.5
R3	1.27	34
R4	.59	15.6

Unit 23-7

12. 23-7-A

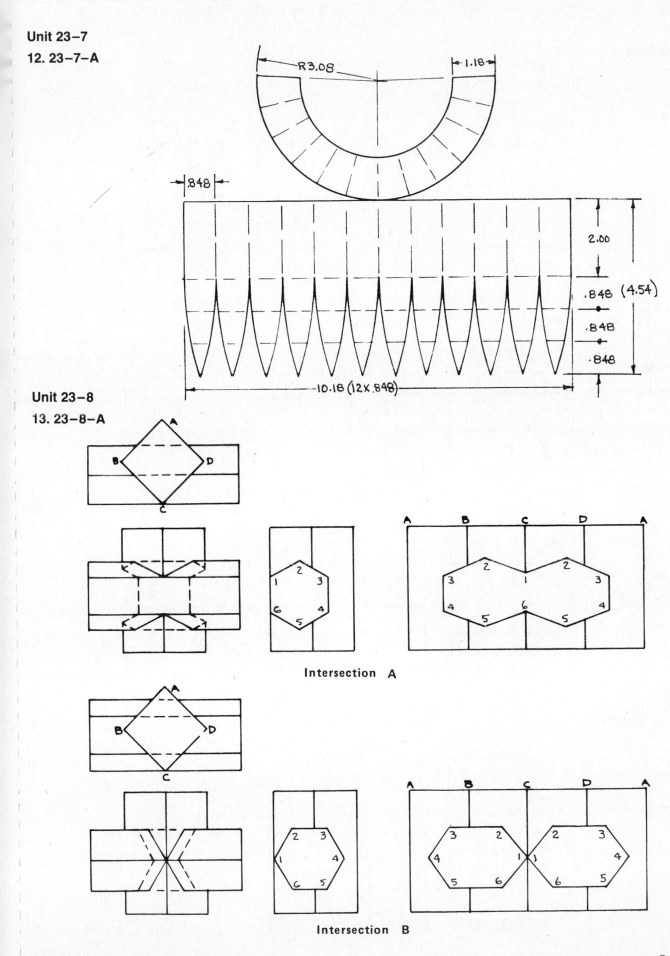

Unit 23-8

13. 23-8-A

Intersection A

Intersection B

Unit 23–8
13. 23–8–A CONTD

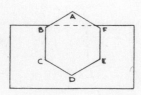

Intersection   C

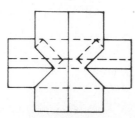

23–8–B

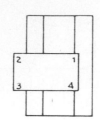

Intersection   A

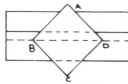

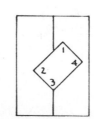

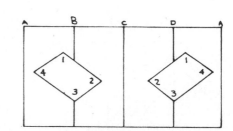

-38-B B

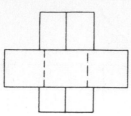

Intersection   B

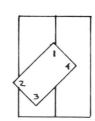

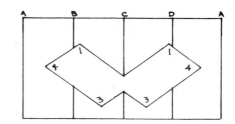

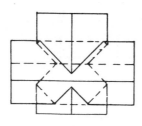

Intersection   C

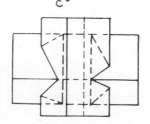

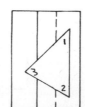

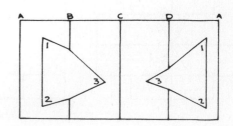

Unit 23–8
13. CONTD 23–8–C

Intersection A

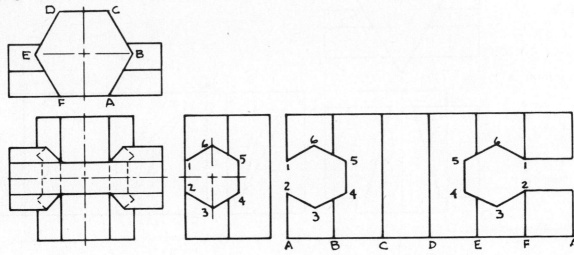

Intersection B

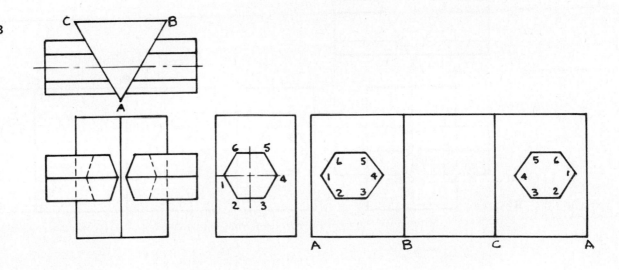

Intersection C

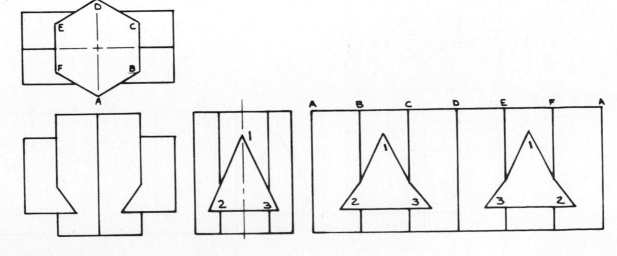

Unit 23−8
13. CONTD
23−8−D

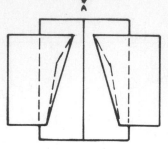
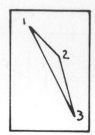
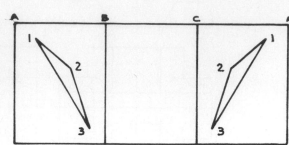

Intersection A

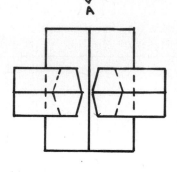

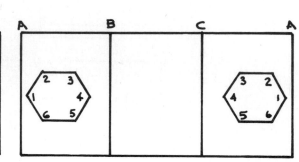

Intersection B

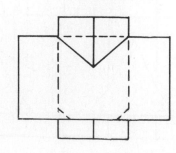
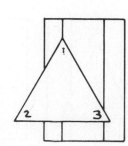
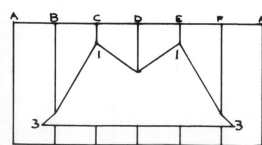

Intersection C

Unit 23-9

14. 23-9-A

23-9-B

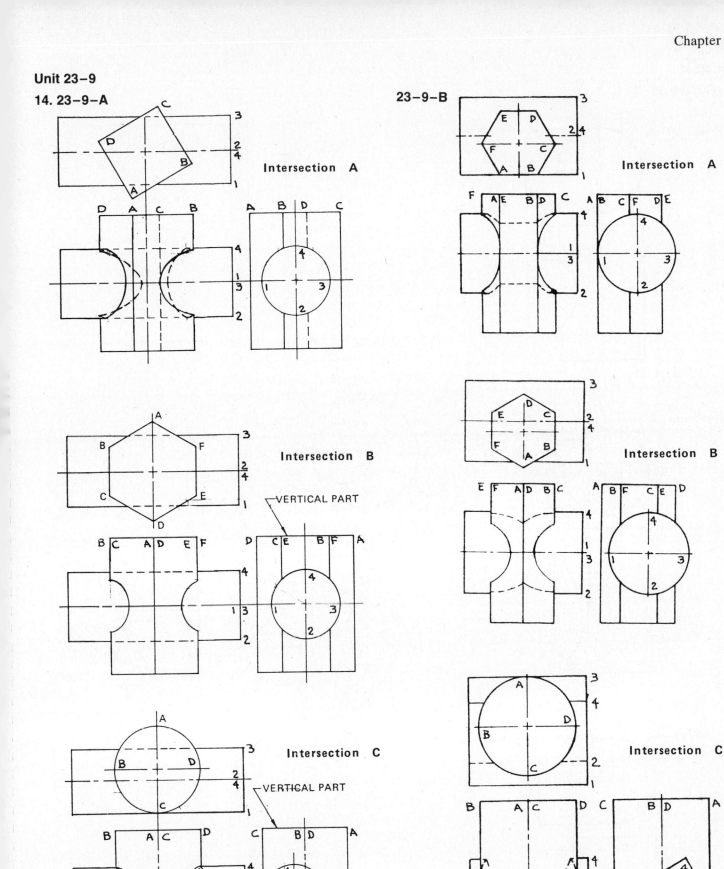

Intersection A

Intersection B

VERTICAL PART

Intersection C

VERTICAL PART

Intersection A

Intersection B

Intersection C

Unit 23-9
14. CONTD 23-9-C

Intersection A

Intersection B

Intersection C

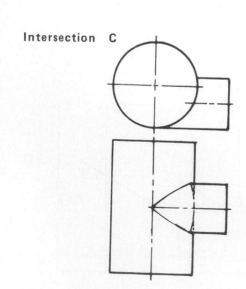

Unit 23-10
15. 23-10-A (A)

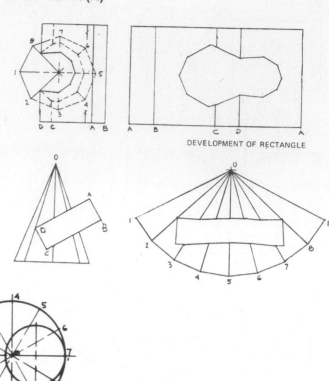

DEVELOPMENT OF RECTANGLE

23-10-A (B)

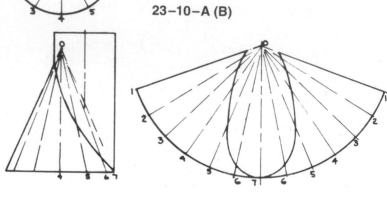

23-10-A (C)

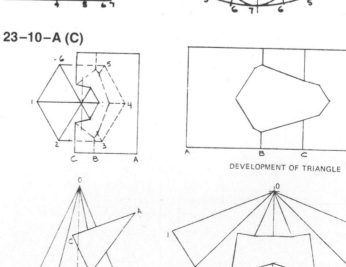

DEVELOPMENT OF TRIANGLE

Chapter 24 Pipe Drawings
Unit 24–1   1. 24–1–A

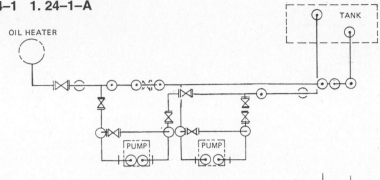

OIL HEATER

TANK

TOP VIEW

QTY	DESCRIPTION	SIZE	PT NO
7	VALVE – GLOBE	2.00	1
3	VALVE – RELIEF	2.00	2
2	VALVE – CHECK	2.00	3
22	ELBOW – 90°	2.00	4
10	TEE	2.00	5
4	TEE – REDUCING	2 x 2 x 1	6
1	LATERAL	2 x 2 x 2	7
38	FLANGE	2.00	8
1	CAP	2.00	9

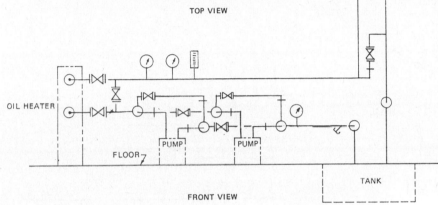

OIL HEATER

PUMP   PUMP

FLOOR

TANK

FRONT VIEW

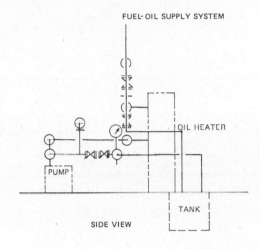

FUEL- OIL SUPPLY SYSTEM

OIL HEATER

PUMP

TANK

SIDE VIEW

2. 24–1–B

PRESSURE GAGE

TEMPERATURE GAGE

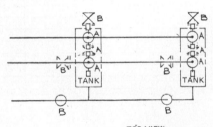

B   B
A   A

TANK   TANK

B   B

TOP VIEW

CODE	VALVES	SERVICE
A	GLOBE-REGRIND-RENEW	OIL LINES
B	GLOBE BEVEL SEAT	STEAM LINES
C	GATE	STEAM LINES

FUEL–OIL–STORAGE CONNECTIONS
WITH HEATING COILS

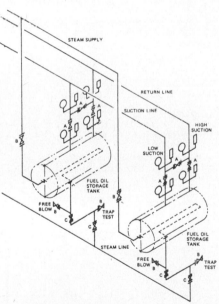

STEAM SUPPLY

RETURN LINE

SUCTION LINE

HIGH
SUCTION

LOW
SUCTION

FUEL OIL
STORAGE
TANK

FREE
BLOW   TRAP
TEST

STEAM LINE

FUEL OIL
STORAGE
TANK

FREE
BLOW   TRAP
TEST

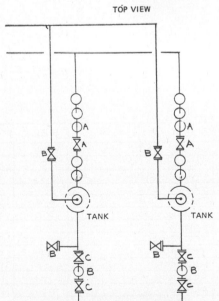

B   B
A   A

TANK   TANK

C   C
B   B
C   C

FRONT VIEW

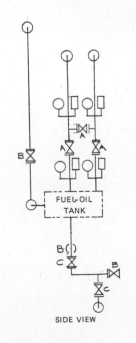

A

B   B

FUEL-OIL
TANK

B
C   B
C

SIDE VIEW

QTY	DESCRIPTION	SIZE	PT.
6	VALVE – GLOBE – REGRIND	1.50	1
6	VALVE – GLOBE – BEVEL SEAT	2.00	2
4	VALVE – GATE	2.00	3
18	ELBOW – 90°	1.50	4
8	ELBOW – 90°	2.00	5
6	TEE	1.50	6
6	TEE	2.00	7
8	CROSS – STRAIGHT	1.50	8
12	FLANGE	1.50	9
20	FLANGE	2.00	10

Unit 24-2

3. 24-2-A

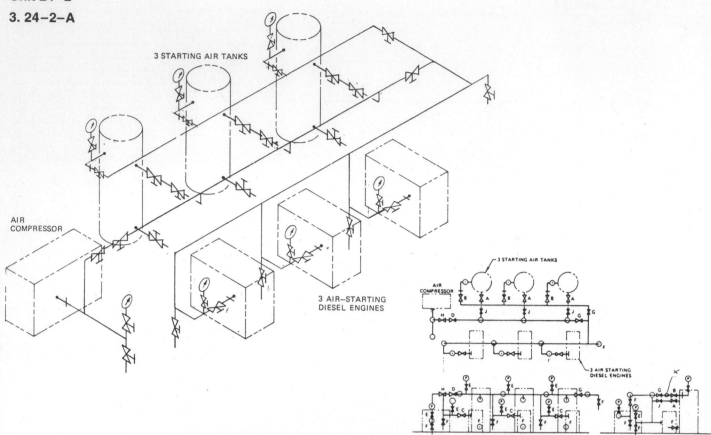

3 STARTING AIR TANKS

AIR
COMPRESSOR

3 AIR—STARTING
DIESEL ENGINES

NOTE-ALL FITTINGS ARE THREADED

QTY	DESCRIPTION	SIZE	PT NO
27	VALVE - BRONZE GLOBE	1.50	1
4	VALVE - BRONZE CHECK	1.50	2
2	VALVE - SPINDLE GATE	1.50	3
14	FLANGE	1.50	4
14	ELBOW	1.50	5
20	TEE	1.50	6

4. 24-2-B

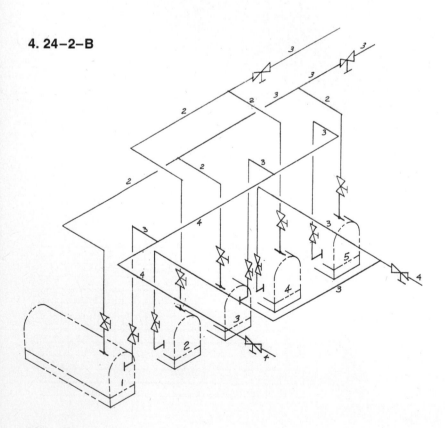

QTY	DESCRIPTION	SIZE	PT NO
2	VALVE - GATE	4.00	1
2	VALVE - GATE	3.00	2
5	VALVE - GLOBE	2.00	3
5	VALVE - CHECK	3.00	4
1	ELBOW - 90°	4.00	5
13	ELBOW - 90°	3.00	6
6	ELBOW - 90°	2.00	7
3	TEE - REDUCER	4×4×3	8
3	TEE - REDUCER	3×3×2	9
2	REDUCER	4×3	10
2	REDUCER	3×2	11
10	FLANGE	3.00	12
10	FLANGE	2.00	13

Unit 24–3

5. 24–3–A

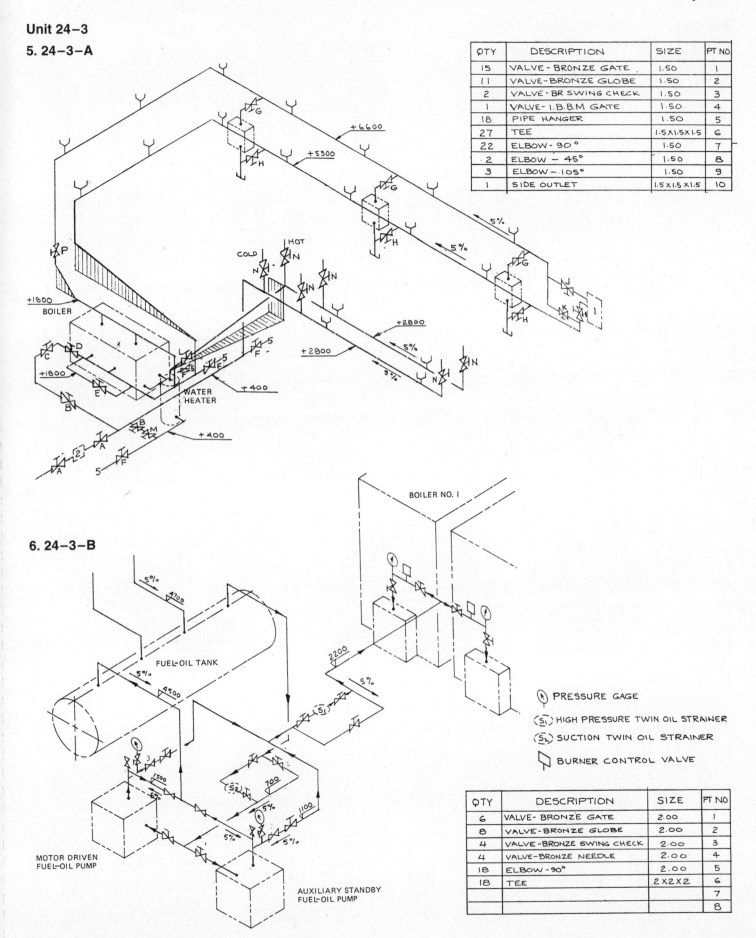

QTY	DESCRIPTION	SIZE	PT NO
15	VALVE – BRONZE GATE	1.50	1
11	VALVE–BRONZE GLOBE	1.50	2
2	VALVE – BR SWING CHECK	1.50	3
1	VALVE– I.B.B.M GATE	1.50	4
18	PIPE HANGER	1.50	5
27	TEE	1.5X1.5X1.5	6
22	ELBOW – 90°	1.50	7
2	ELBOW – 45°	1.50	8
3	ELBOW – 105°	1.50	9
1	SIDE OUTLET	1.5X1.5X1.5	10

6. 24–3–B

QTY	DESCRIPTION	SIZE	PT NO
6	VALVE– BRONZE GATE	2.00	1
8	VALVE–BRONZE GLOBE	2.00	2
4	VALVE –BRONZE SWING CHECK	2.00	3
4	VALVE–BRONZE NEEDLE	2.00	4
18	ELBOW – 90°	2.00	5
18	TEE	2 X2X2	6
			7
			8

Chapter 25 Structural Drafting
Unit 25–1,   1. 25–1–A

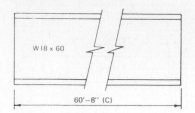 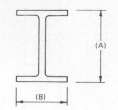 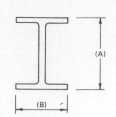

W18 x 60   60'–8" (C)   (A) (B)

DIM	NOMINAL	TOLERANCE		LIMITS	
				UPPER	LOWER
A	18 1/4	+ 1/8	– 1/8	18 3/8	18 1/8
B	7 1/2	+ 1/4	– 3/16	7 3/4	7 5/16
C	60'–8	+ 3/4	– 3/8	60'–8 3/4	60'–7 5/8

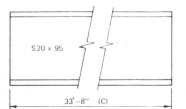 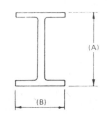 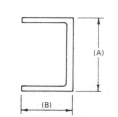

S20 x 95   33'–8" (C)   (A) (B)

DIM	NOMINAL	TOLERANCE		LIMITS	
				UPPER	LOWER
A	20	+ 3/16	– 1/8	20 3/16	19 7/8
B	7 1/4	+ 3/16	– 3/16	7 7/16	7 1/16
C	33'–8	+ 3/4	– 1/4	33'–8 3/4	33'–7 3/4

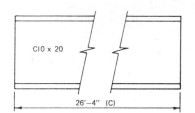 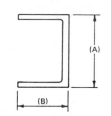 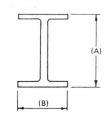

C10 x 20   26'–4" (C)   (A) (B)

DIM	NOMINAL	TOLERANCE		LIMITS	
				UPPER	LOWER
A	10	+ 1/8	– 1/16	10 1/8	9 15/16
B	2 3/4	+ 1/8	– 1/8	2 7/8	2 5/8
C	26'–4	+ 1/2	– 1/4	26'–4 1/2	26'–3 3/4

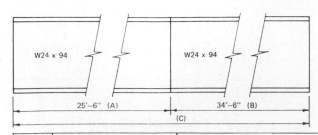

W24 x 94   W24 x 94   25'–6" (A)   34'–6" (B)   (C)

DIM	TOLERANCE		LIMITS	
			UPPER	LOWER
A	+ 3/8	– 3/8	25'–6 3/8	25'–5 5/8
B	+ 1/2	– 3/8	3'–6 1/2	34'–5 5/8
C			6'–0 7/8	59'–11 1/4

25–1–B

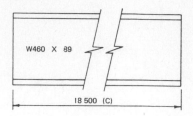

W460 X 89   18 500 (C)   (A) (B)

DIM	NOMINAL	TOLERANCE		LIMITS	
				UPPER	LOWER
A	464	+ 4	– 3	468	461
B	192	+ 6	– 5	198	187
C	18 500	+ 21	– 10	18 521	18 490

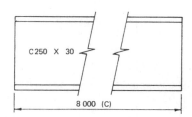

C250 X 30   8 000 (C)   (A) (B)

DIM	NOMINAL	TOLERANCE		LIMITS	
				UPPER	LOWER
A	254	+ 3	– 2	257	252
B	70	+ 3	– 4	73	66
C	8 000	+ 13	– 6	8 013	7 994

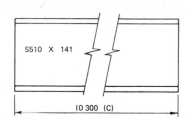

S510 X 141   10 300 (C)   (A) (B)

DIM	NOMINAL	TOLERANCE		LIMITS	
				UPPER	LOWER
A	508	+ 5	– 3	513	505
B	183	+ 5	– 5	188	178
C	10 300	+ 19	– 6	10 319	10 294

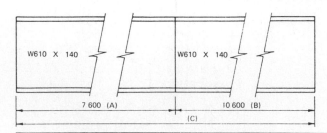

W610 X 140   W610 X 140   7 600 (A)   10 600 (B)   (C)

DIM	TOLERANCE		LIMITS	
			UPPER	LOWER
A	+ 10	– 10	7 610	7 590
B	+ 13	– 10	10 613	10 590
C			18 223	18 180

Unit 25–2

2. 25–2–A

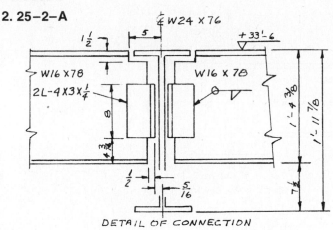

DETAIL OF CONNECTION

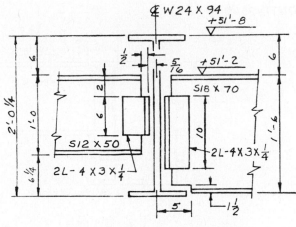

DETAIL OF CONNECTION

25–2–B

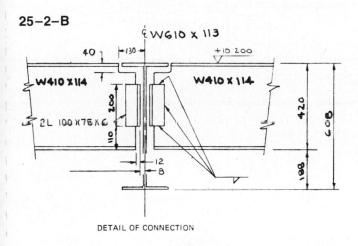

DETAIL OF CONNECTION

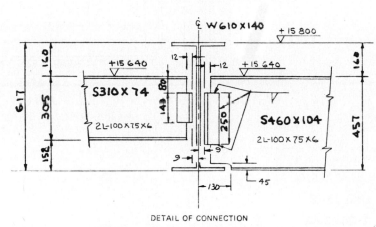

DETAIL OF CONNECTION

3. 25–2–C

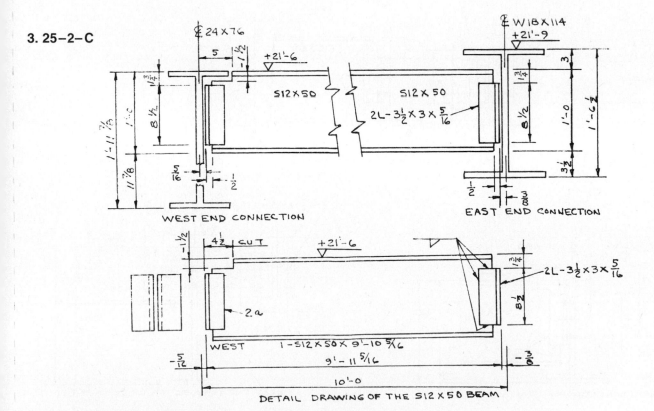

WEST END CONNECTION

EAST END CONNECTION

DETAIL DRAWING OF THE S12 X 50 BEAM

Unit 25-2

3. CONTD 25-2-D

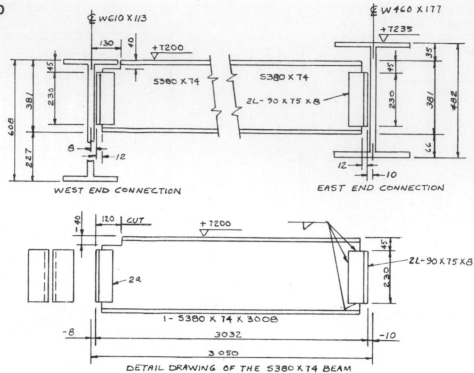

WEST END CONNECTION

EAST END CONNECTION

DETAIL DRAWING OF THE S380 X74 BEAM

Unit 25-3

4. 25-3-A

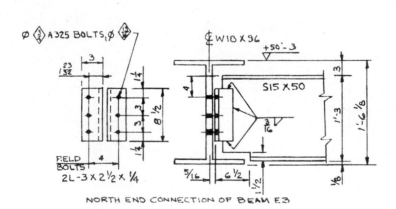

NORTH END CONNECTION OF BEAM E3

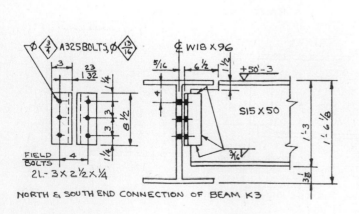

NORTH & SOUTH END CONNECTION OF BEAM K3

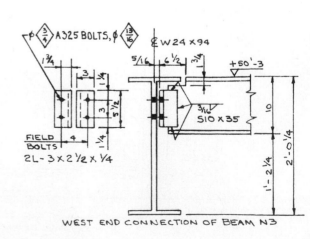

WEST END CONNECTION OF BEAM N3

Unit 25–3

4. CONTD 25–3–B

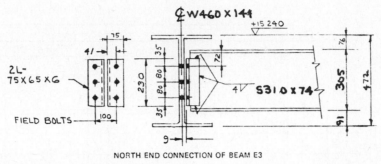

NORTH END CONNECTION OF BEAM E3

SOUTH END CONNECTION OF BEAM K3

WEST END CONNECTION OF BEAM N3

5. 25–3–C

	PROBLEM					
DIM	1	2	3	4	5	6
L	22'-6	17'-6 7/8	21'-6 3/4	24'-2 1/2	40'-10 3/4	20'-7
X	9'-1 1/8	6'-11 1/8	8'-8 3/4	6'-9 1/4	16'-9 1/2	8'-9 1/2
Y	11'-9	10'-6 3/8	12'-9 3/4	11'-7 1/4	24'-7 1/4	11'-9
Z	18'-7 1/2	14'-0 5/8	16'-7	17'-5	33'-2 3/4	15'-8

	PROBLEM		
DIM	1	2	3
L	14'-0 3/4	22'-8 1/2	31'-10 3/4
V	3'-9 1/4	4'-8 3/4	7'-9 3/4
W	2'-6 3/4	4'-6 1/2	6'-0 1/2
X	2'-3 1/2	3'-2	5'-11 1/4
Y	2'-11 1/4	4'-0 1/2	7'-11 3/4
Z	2'-6	4'-3 3/4	4'-2 1/2

Unit 25–4

6. 25–4–A

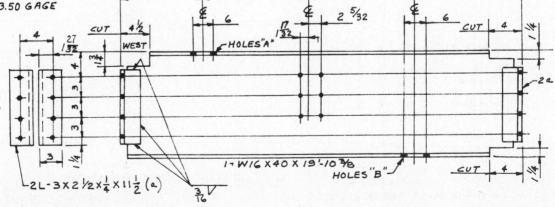
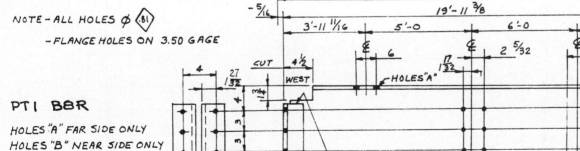
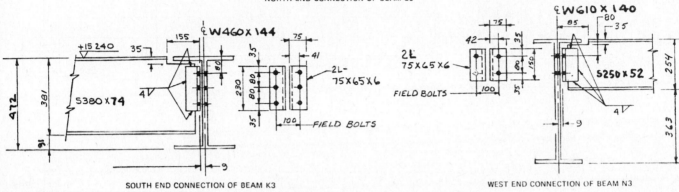

NOTE – ALL HOLES ∅ (81)
– FLANGE HOLES ON 3.50 GAGE

PT1 B8R

HOLES "A" FAR SIDE ONLY
HOLES "B" NEAR SIDE ONLY

PT2 B8L

HOLES "A" NEAR SIDE ONLY
HOLES "B" FAR SIDE ONLY

Unit 25–4

6. CONTD 25–4–B

NOTE – ALL HOLES ⌀ ⟨22⟩
– FLANGE HOLES ON 90 GAGE

PT 1 B8R

HOLES "A" FAR SIDE ONLY
HOLES "B" NEAR SIDE ONLY

PT 2 B8L

HOLES "A" NEAR SIDE ONLY
HOLES "B" FAR SIDE ONLY

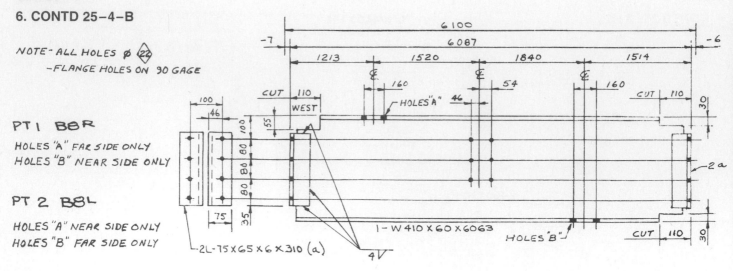

7. 25–4–A

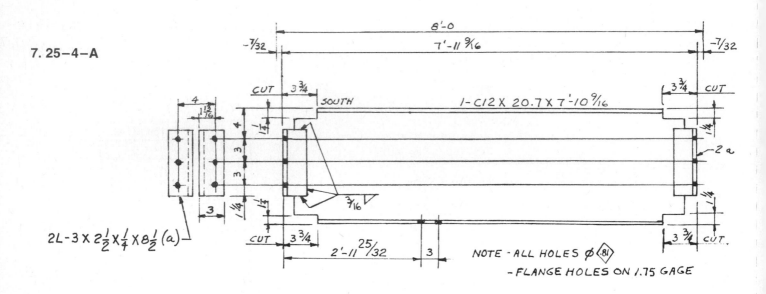

25–4–B

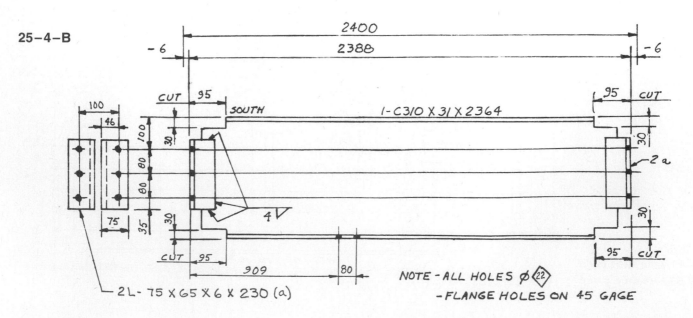

Unit 25-5

8. 25-5-A

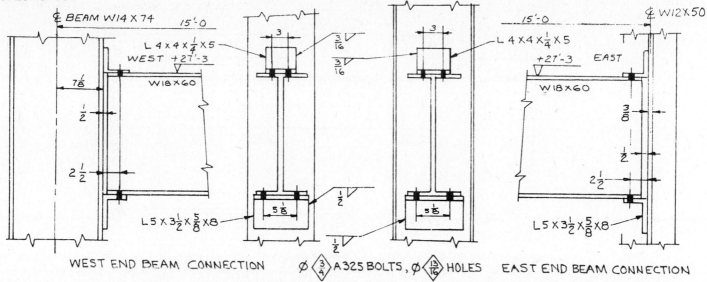

WEST END BEAM CONNECTION $\emptyset \langle \tfrac{3}{4} \rangle$ A325 BOLTS, $\emptyset \langle \tfrac{13}{16} \rangle$ HOLES EAST END BEAM CONNECTION

9. 25-5-A

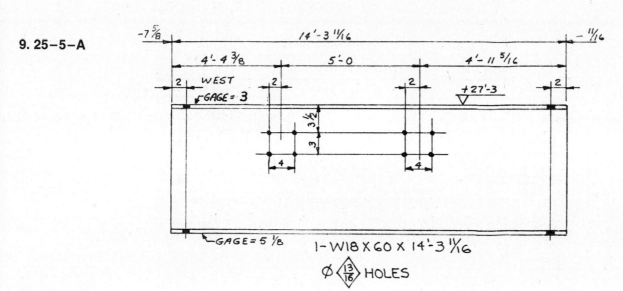

1-W18 X 60 X 14'-3 $\tfrac{11}{16}$

$\emptyset \langle \tfrac{13}{16} \rangle$ HOLES

10. 25-5-B

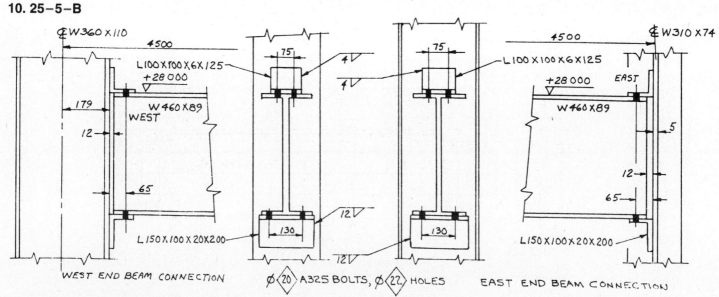

WEST END BEAM CONNECTION $\emptyset \langle 20 \rangle$ A325 BOLTS, $\emptyset \langle 22 \rangle$ HOLES EAST END BEAM CONNECTION

Unit 25−5

11. 25−5−B

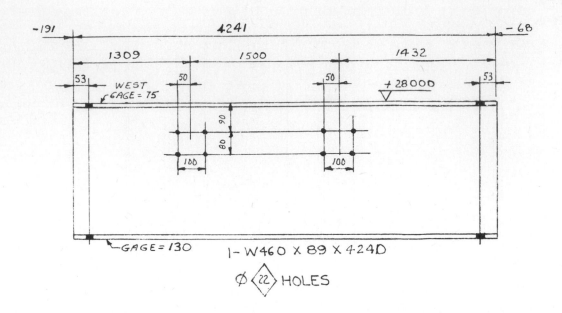

Unit 25−6

12. 25−6−A

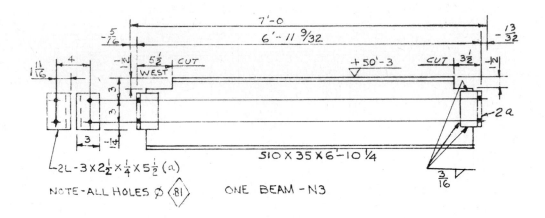

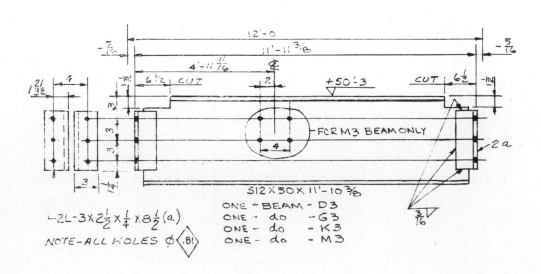

Unit 25–6

12. 25–6–A CONTD

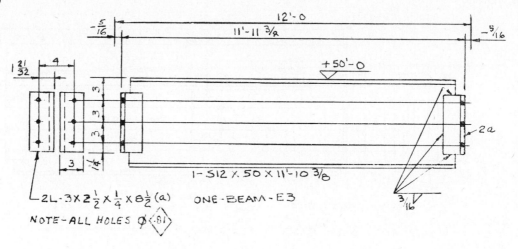

$-\frac{5}{16}$

12'-0

11'-11 $\frac{3}{8}$

$-\frac{5}{16}$

$1\frac{21}{32}$ 4

+50'-0

3

3

3

$\frac{1}{4}$

3

-2a

1-S12 X 50 X 11'-10 $\frac{3}{8}$

$\frac{3}{16}$

2L-3X2$\frac{1}{2}$X$\frac{1}{4}$X8$\frac{1}{2}$(a)   ONE-BEAM-E3

NOTE-ALL HOLES Ø ◇.81

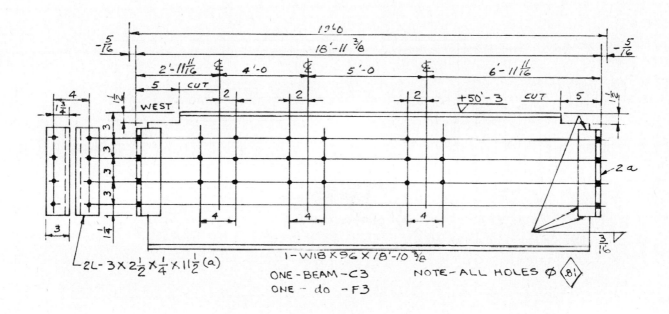

1960

18'-11 $\frac{3}{8}$

$-\frac{5}{16}$

$-\frac{5}{16}$

2'-11$\frac{11}{16}$ ₵   4'-0   ₵   5'-0   ₵   6'-11$\frac{11}{16}$

4   5   CUT   2   2   2   +50'-3   CUT   5

$1\frac{3}{4}$ WEST

3   3   3   3

3

$\frac{1}{4}$

-2a

4   4   4

2L-3X2$\frac{1}{2}$X$\frac{1}{4}$X11$\frac{1}{2}$(a)   1-W18X96X18'-10$\frac{3}{8}$

ONE-BEAM-C3   NOTE-ALL HOLES Ø ◇.81

ONE - do -F3

$\frac{3}{16}$

25–6–B

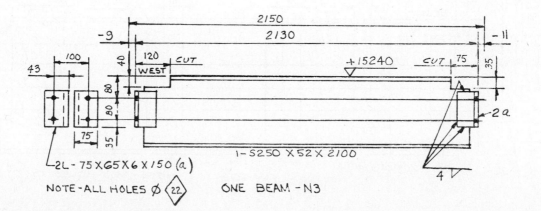

2150

2130

-9   -11

43   100   40   120   CUT   +15240   CUT   75   35

WEST

80

80

35

75

-2a

2L-75X65X6X150(a)   1-S250 X 52 X 2100

NOTE-ALL HOLES Ø ◇22   ONE BEAM - N3

4

Unit 25-6

12. 25-6-B CONTD

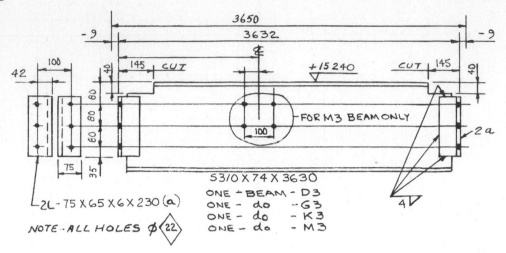

S 310 X 74 X 3630

ONE - BEAM - D3
ONE - do - G3
ONE - do - K3
ONE - do - M3

2L - 75 X 65 X 6 X 230 (a)

NOTE · ALL HOLES Ø ◇22◇

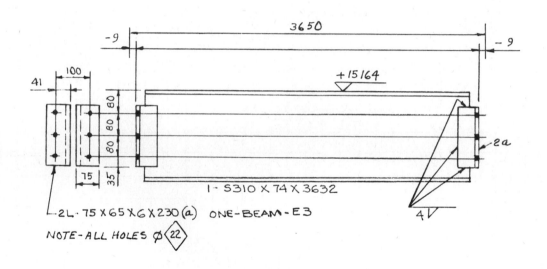

1 - S 310 X 74 X 3632   ONE - BEAM - E3

2L - 75 X 65 X 6 X 230 (a)

NOTE - ALL HOLES Ø ◇22◇

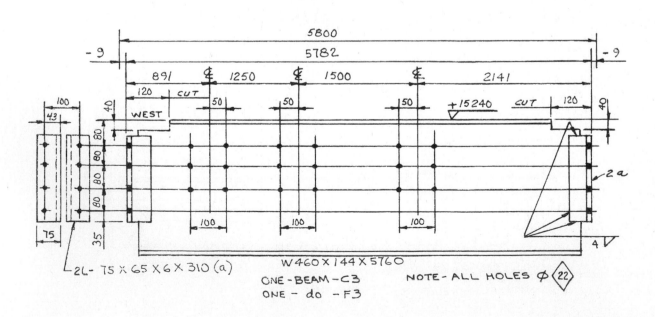

W 460 X 144 X 5760

ONE - BEAM - C3
ONE - do - F3

2L - 75 X 65 X 6 X 310 (a)

NOTE - ALL HOLES Ø ◇22◇

Unit 25–6

13. 25–6–A

BILL OF STRUCTURAL MATERIAL						SHIPPING LIST			
NO. OF PIECES	ASSEMBLY MARK	DESC.	SIZE	LENGTH	EST. WT	NO PCS	DESC.	MARK	SHIPPING RECORD
5		S	S12×50	11'-10⅜	2965	1	BEAM	D3	606
20	ⓐ	L	3×2½×¼	8½	64	1	do	E3	606
					‾‾‾‾ 3029	1	do	G3	606
						1	do	K3	606
						1	do	M3	606
2		W	W18×96	18'-0⅜	3062	1	BEAM	C3	1748
8	a	L	3×2½×¼	11½	34	1	do	F3	1748
					‾‾‾‾ 3496				
1		S	S10×35	6'-10¼	240	1	BEAM	N3	248
4	a	L	3×2½×¼	5½	8				
					‾‾‾‾ 248				

25–6–B

BILL OF STRUCTURAL MATERIAL						SHIPPING LIST			
NO. OF PIECES	ASSEMBLY MARK	DESC.	SIZE	LENGTH	EST. WT	NO PCS	DESC.	MARK	SHIPPING RECORD
5		S	S310×74	3630	1343	1	BEAM	D3	275
20	a	L	75×65×6	230	31	1	do	E3	275
					‾‾‾‾ 1374	1	do	G3	275
						1	do	K3	275
						1	do	M3	275
2		W	W460×144	5760	1659	1	BEAM	C3	838
8	a	L	75×65×6	310	17	1	do	F3	838
					‾‾‾‾ 1676				
1		S	S250×52	2100	109	1	BEAM	N3	113
4	a	L	75×65×6	150	4				
					‾‾‾‾ 113				

Chapter 26 Jigs and Fixtures
Unit 26–1

1. 26–1–A

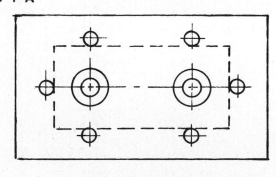

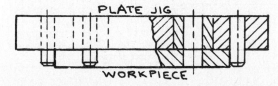

LOCKING PIN. NOT REQUIRED

26–1–B

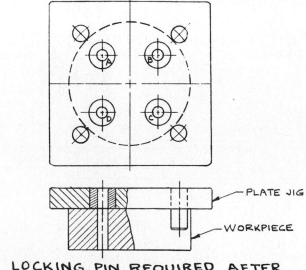

LOCKING PIN REQUIRED AFTER
FIRST HOLE IS DRILLED

Unit 26−1
1. CONTD 26−1−C

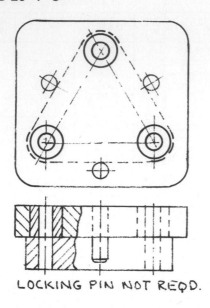

LOCKING PIN NOT REQD.

Unit 26−2
2. 26−2−A

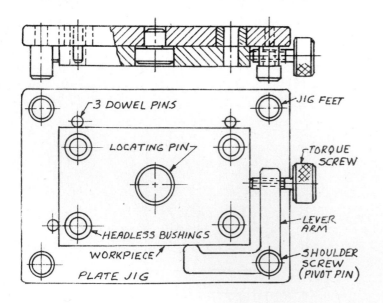

3 DOWEL PINS

JIG FEET

LOCATING PIN

TORQUE SCREW

HEADLESS BUSHINGS

WORKPIECE

LEVER ARM

SHOULDER SCREW (PIVOT PIN)

PLATE JIG

2. CONTD 26−2−B

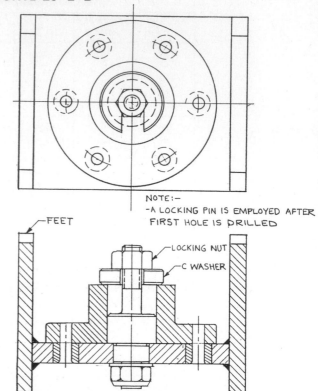

NOTE:−
−A LOCKING PIN IS EMPLOYED AFTER FIRST HOLE IS DRILLED

FEET

LOCKING NUT

C WASHER

3. 26−2−C

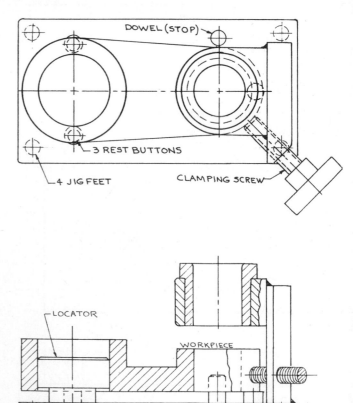

DOWEL (STOP)

3 REST BUTTONS

4 JIG FEET

CLAMPING SCREW

LOCATOR

WORKPIECE

Unit 26-2

4. 26-2-D

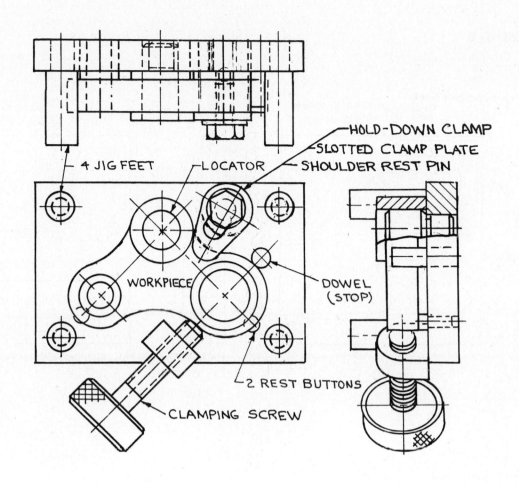

HOLD-DOWN CLAMP
SLOTTED CLAMP PLATE
SHOULDER REST PIN

4 JIG FEET — LOCATOR

DOWEL (STOP)

WORKPIECE

2 REST BUTTONS

CLAMPING SCREW

Unit 26-3

5. 26-3-A

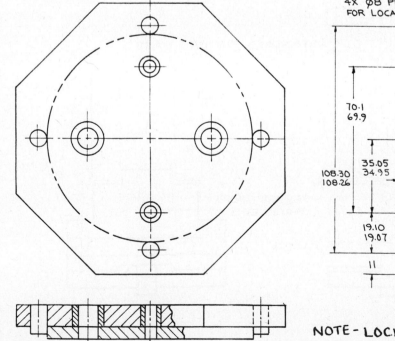

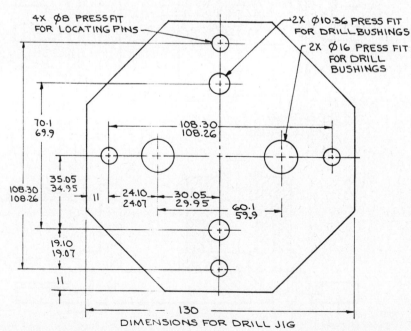

4X Ø8 PRESS FIT FOR LOCATING PINS

2X Ø10.36 PRESS FIT FOR DRILL BUSHINGS

2X Ø16 PRESS FIT FOR DRILL BUSHINGS

70.1
69.9

108.30
108.26

108.30
108.26

35.05
34.95

24.10
24.07

30.05
29.95

60.1
59.9

11

19.10
19.07

11

130

DIMENSIONS FOR DRILL JIG

NOTE- LOCKING PINS REQUIRED AFTER 1 & 2 HOLES

Unit 26–3

5. CONTD 26–3–B

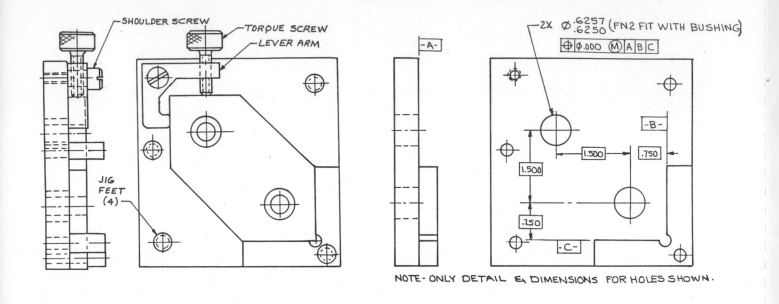

NOTE - ONLY DETAIL & DIMENSIONS FOR HOLES SHOWN.

Unit 26–4

6. 26–4–A

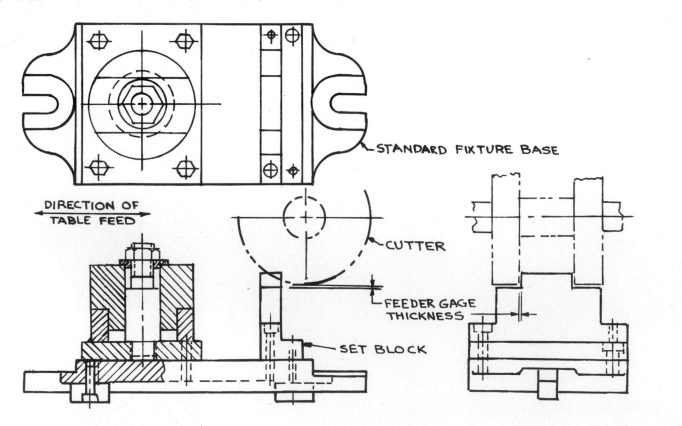

Unit 26-4

6. CONTD 26-4-B

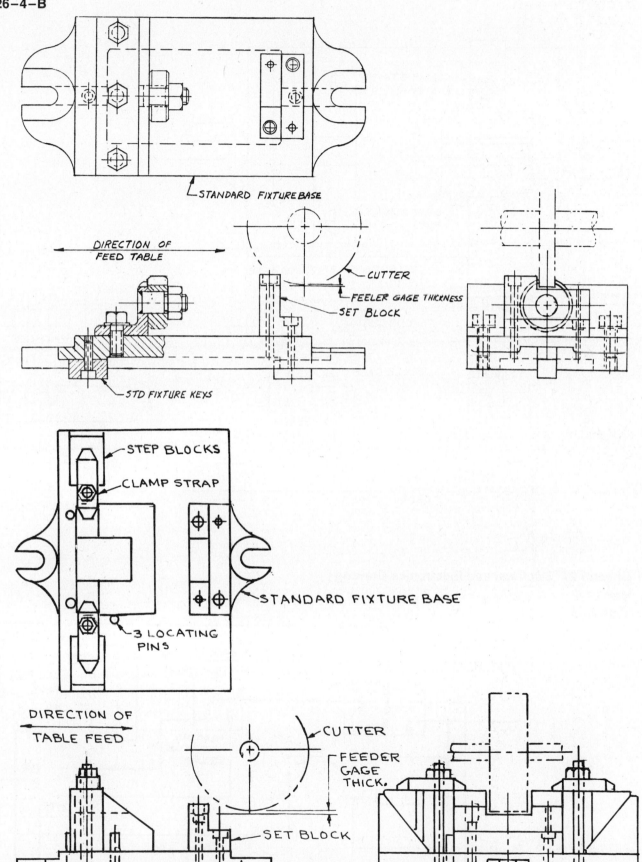

STANDARD FIXTURE BASE

DIRECTION OF
FEED TABLE

CUTTER

FEELER GAGE THICKNESS

SET BLOCK

STD FIXTURE KEYS

26-4-C

STEP BLOCKS

CLAMP STRAP

STANDARD FIXTURE BASE

3 LOCATING
PINS

DIRECTION OF
TABLE FEED

CUTTER

FEEDER
GAGE
THICK.

SET BLOCK

Unit 26-4
6. CONTD 26-4-D

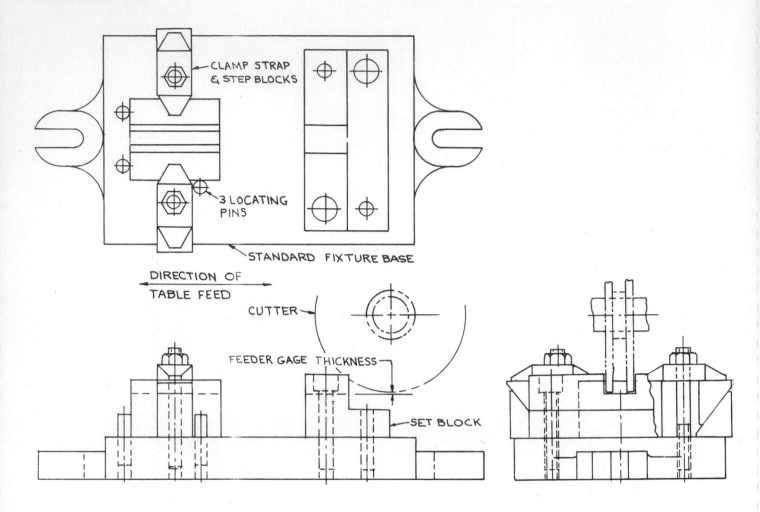

Chapter 27 Electrical and Electronics Drawing
Unit 27-2
1. 27-3-A

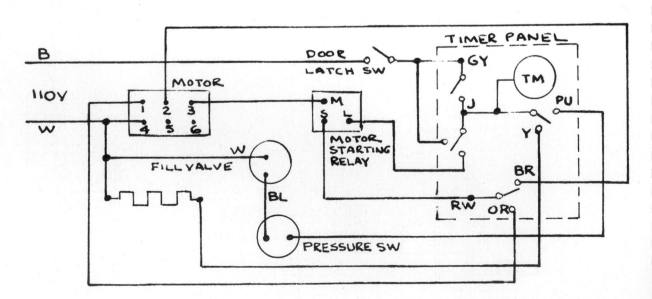

Unit 27–2

1. CONTD 27–3–B

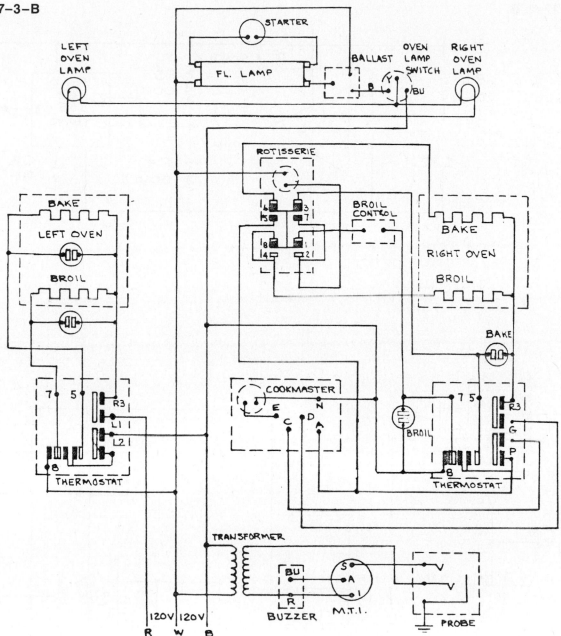

2. 27–2–A

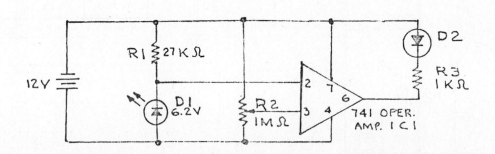

Unit 27–2

2. CONTD 27–2–B

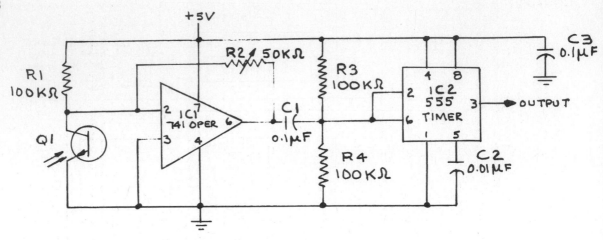

27–2–C

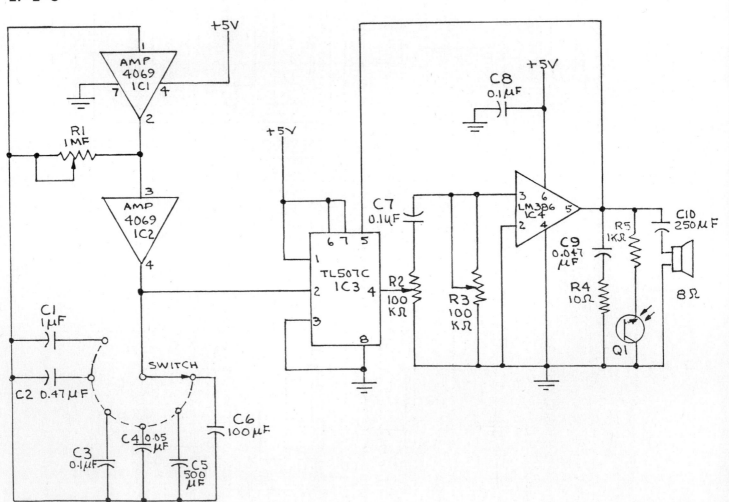

Unit 27–2

3. 27–2–D

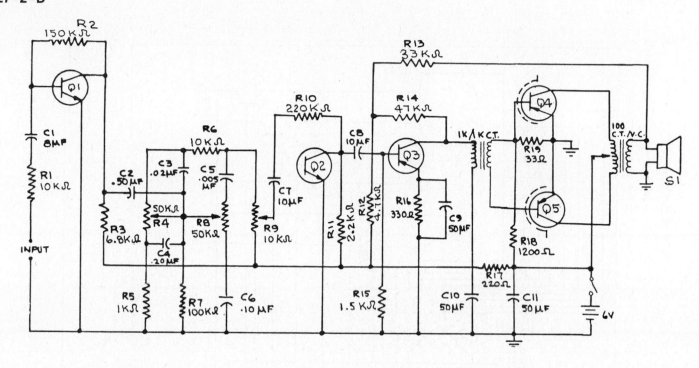

4. 27–2–E

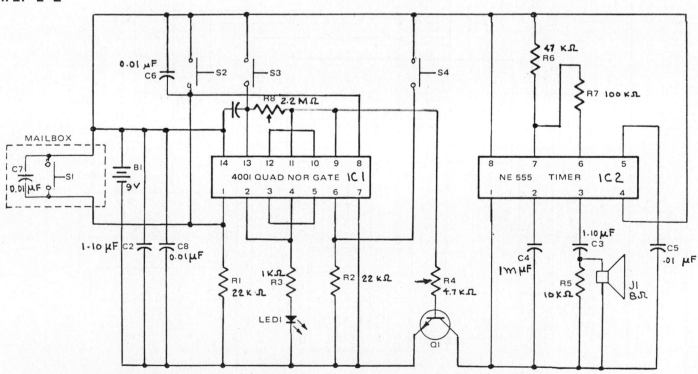

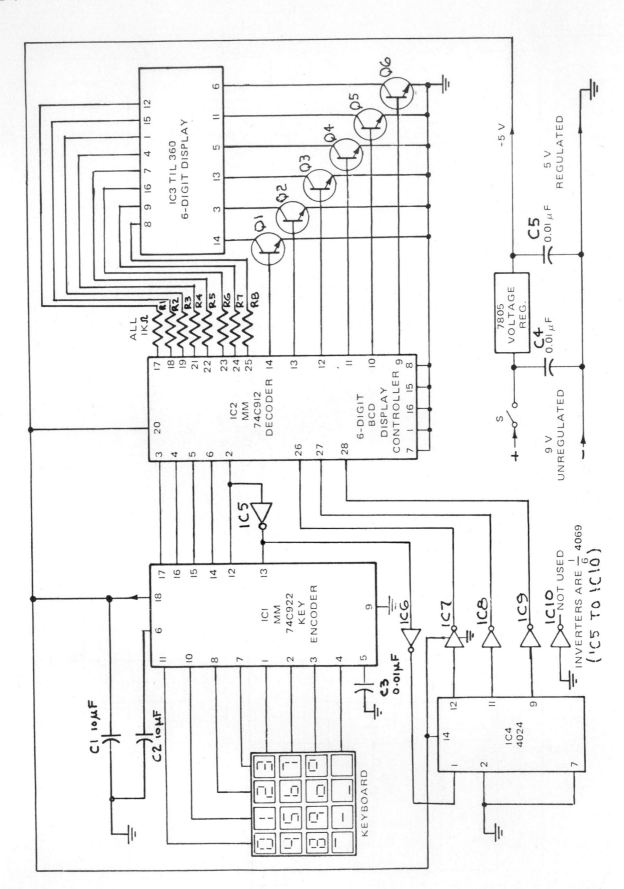

Unit 27–3

6. 27–3–1

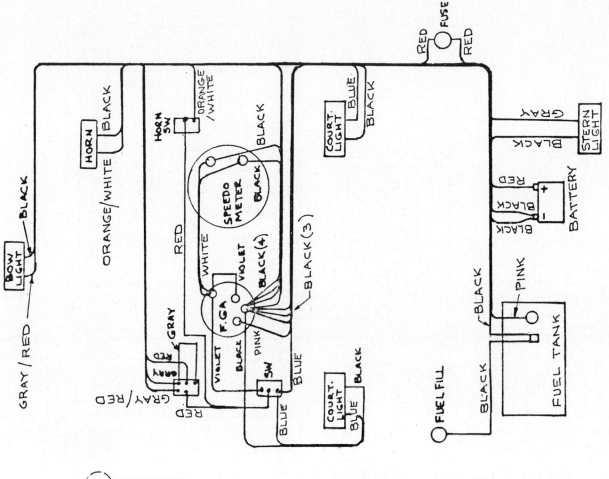

7. 27–3–2

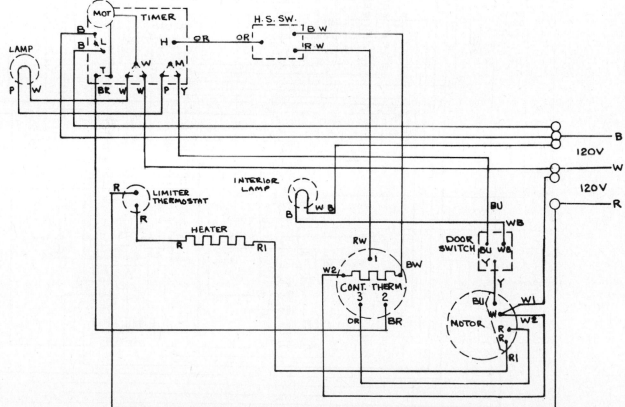

Unit 27–3

8. 27–3–A

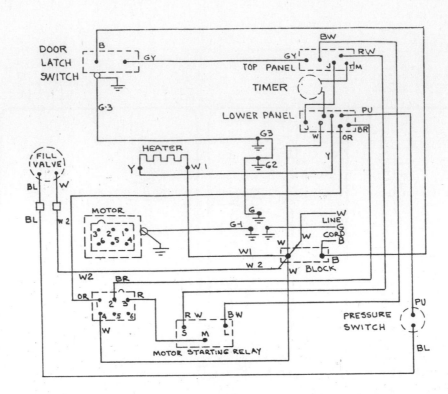

27–3–B

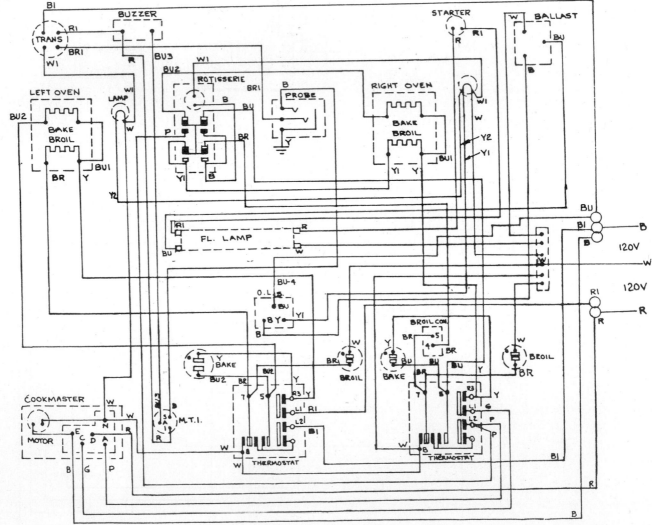

Unit 27–3

9. 27–3–2

COLOR	SYMBOL	LENGTH
BLACK	B1	12.5
	B2	10.5
WHITE	W1	26
	W2	18.5
	W3	8.5
RED	R1	35.5
	R2	15
GREEN	G	22.5
YELLOW	Y	10
BLUE	BU	28.5
BROWN	BR	40.5
ORANGE	OR	20
WHITE – BLACK	WB	28.5
BLACK – WHITE	BW	34.5
RED-WHITE	RW	37

NOTES:- DIMENSIONS IN INCHES
    - STRIP END OF WIRE FOR .50

27–3–A

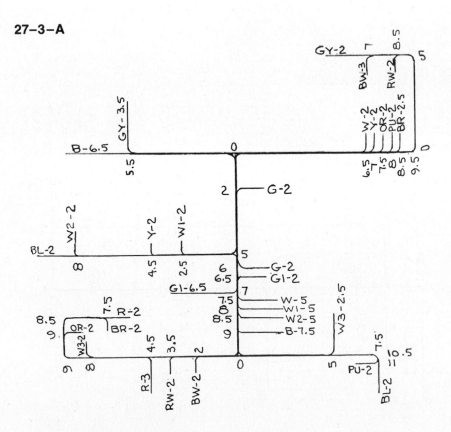

COLOR	SYMBOL	LENGTH
BLACK	B	28.5
WHITE	W	21
	W1	17.5
	W2	22
RED	R	13
GRAY	GY	28
YELLOW	Y	20.5
ORANGE	OR	32.5
PURPLE	PU	30.5
BROWN	BR	36
BLACK – WHITE	BW	34.5
RED – WHITE	RW	33.5
GREEN	G	8
	G1	9
BLUE	BL	25.5

NOTES: –DIMENSIONS IN INCHES
    – STRIP ENDS OF WIRE FOR .50

Unit 27–3

9. CONTD 27–3–B (7 Harnesses Required)

NOTES:- DIMENSIONS IN INCHES
- STRIP END OF WIRE FOR .50

COLOR	SYMBOL	LENGTH
BLACK	B	19.5
	BI	28.5
WHITE	W	20.5
	WI	19
RED	R	30
BROWN	BR	26
	BRI	26
BLUE	BUI	13
	BU2	36.5
	BU3	29
	BU4	14.5
YELLOW	Y	27
	YI	14
	Y2	35.5

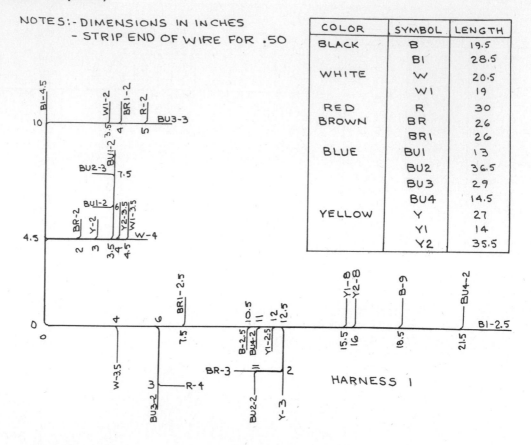

HARNESS 1

COLOR	SYMBOL	LENGTH
BLACK	B	37.5
	BI	26.5
WHITE	W	11
	WI	15
	W3	14
RED	R	33.5
	RI	26.5
GREEN	G	30
PURPLE	P	29
BLUE	BU	10.5

HARNESS 2

NOTES:- DIMENSIONS IN INCHES
- STRIP END OF WIRE FOR .50

Unit 27–3

9. 27–3–B CONTD

NOTES –DIMENSIONS IN INCHES
– STRIP ENDS OF WIRE FOR .50

COLOR	SYMBOL	LENGTH
BLUE	BU	23
	BU1	7.5
	BU2	19
WHITE	W	10.5
	W1	15.5
YELLOW	Y	16
	Y1	16.5
PURPLE	P	24.5
BROWN	BR	21.5

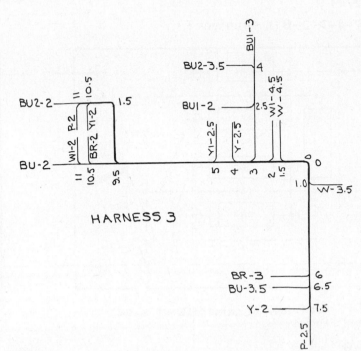

HARNESS 3

COLOR	SYMBOL	LENGTH
RED	R	16.5
	R1	29.5
BLUE	BU	27.5
WHITE	W	9

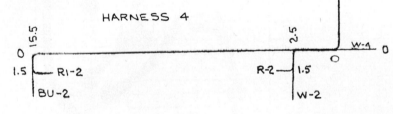
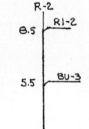

HARNESS 4

COLOR	SYMBOL	LENGTH
BLUE	BU	9.5
BROWN	BR	10
WHITE	W1	12.5

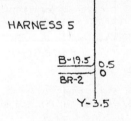
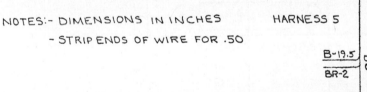

NOTES:– DIMENSIONS IN INCHES
– STRIP ENDS OF WIRE FOR .50

HARNESS 5

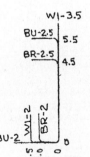

NOTES: –DIMENSIONS IN INCHES
– STRIP ENDS OF WIRE FOR .50   HARNESS 7   2 REQD

HARNESS 6

COLOR	SYMBOL	LENGTH
BLACK	B	28.5
BROWN	BR	11.5
YELLOW	Y	12.5

COLOR	SYMBOL	LENGTH
YELLOW	Y	6
BLUE	BU2	5.5

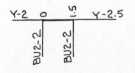

Unit 27–4
10. 27–4–A

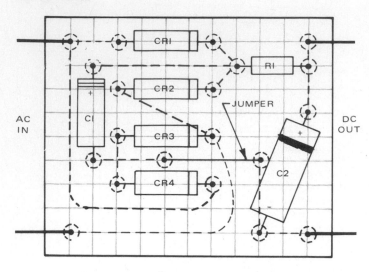

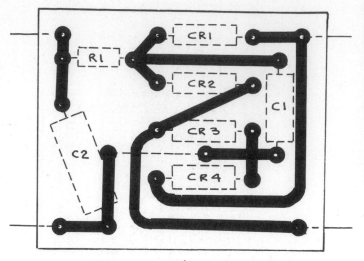

TOP VIEW (COMPONENT SIDE)

BOTTOM VIEW (CIRCUIT SIDE)

27–4–B

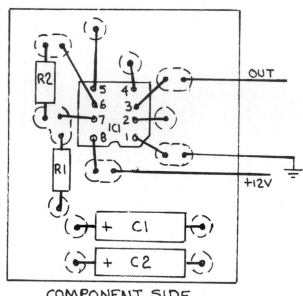

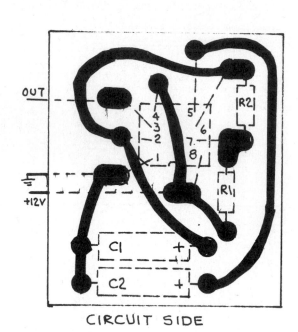

COMPONENT SIDE

CIRCUIT SIDE

11. 27–4–C

Unit 27–4
11. CONTD 27–4–D

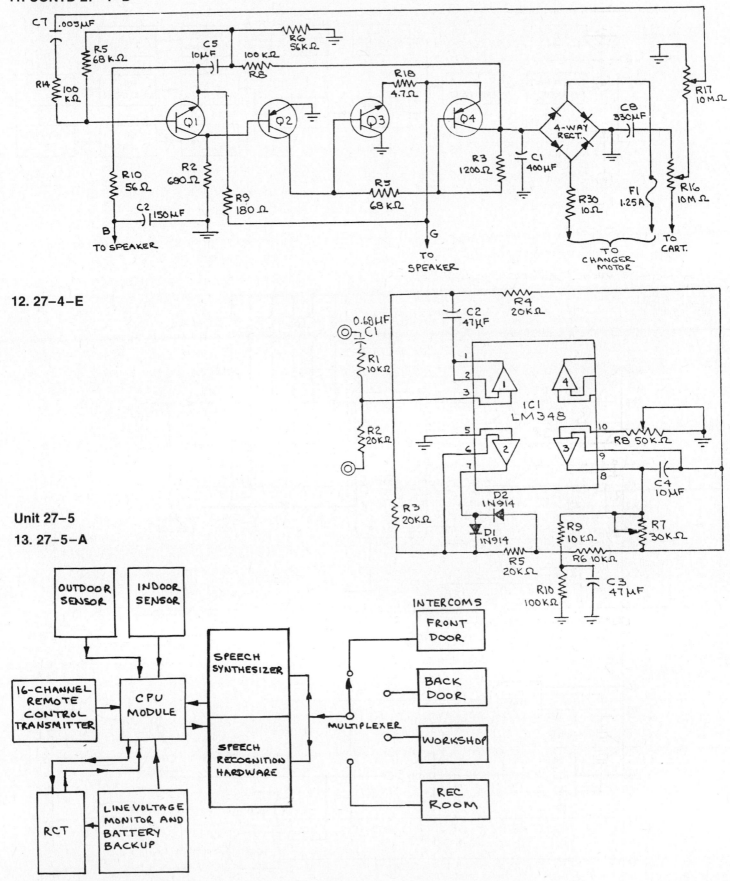

12. 27–4–E

Unit 27–5
13. 27–5–A

Unit 27–5
14. TV Block Diagram

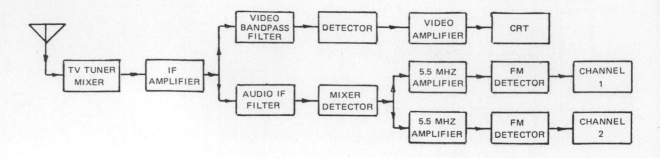

15. 27–5–B

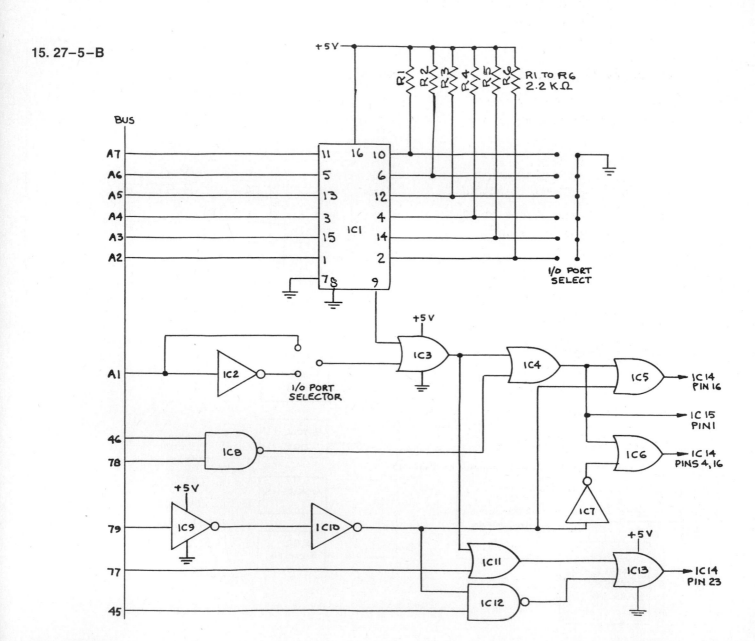

CONTENTS

Chapter Tests and Answers

Basic Drawing and Design

Chapter   1   Engineering Graphics as a Language 261
            2   Computer-Aided Drafting (CAD) 263
            3   Drawing Media, Filing, Storage, and Reproduction 265
            4   Basic Drafting Skills 267
            5   Applied Geometry 269
            6   Theory of Shape Description 271
            7   Auxiliary Views and Revolutions 273
            8   Basic Dimensioning 275
            9   Sections 277

Fasteners, Materials, and Forming Processes

Chapter 10   Threaded Fasteners 279
          11   Miscellaneous Types of Fasteners 281
          12   Manufacturing Materials 283
          13   Forming Processes 285

Working Drawings and Design

Chapter 14   Detail and Assembly Drawings 287
          15   Pictorial Drawings 289
          16   Geometric Dimensioning and Tolerancing 291
          17   Drawings for Numerical Control 293
          18   Welding Drawings 295
          19   Design Concepts 297

Power Transmissions

Chapter 20   Belts, Chains, and Gears 299
          21   Couplings, Bearings, and Seals 301
          22   Cams, Linkages, and Actuators 303

Special Fields of Drafting

Chapter 23   Developments and Intersections 305
          24   Pipe Drawings 307
          25   Structural Drafting 309
          26   Jigs and Fixtures 311
          27   Electrical and Electronics Drawing 313

Answers to Tests **315**

Test

Chapter 1

Multiple Choice   Circle the answer that *best* fits the question.

1. Drafting can be described as a _____ language.
   a. symbolic
   b. technical
   c. graphic
   d. written
   e. spoken

2. The acronym for the organization that establishes drafting standards in the United States is
   a. DIN.
   b. JIS.
   c. ISO.
   d. NBS.
   e. ANSI.

3. The most common definition for the acronym CAD is
   a. computer assisted design.
   b. computer aided design.
   c. computer aided drafting.
   d. computer assisted drafting.
   e. none of the above.

4. A drawing made without instruments is referred to as a
   a. sketch.
   b. picture.
   c. image.
   d. drawing.
   e. none of the above.

5. In a drafting department, checkers are
   a. responsible for identifying errors in drawings.
   b. experienced drafters.
   c. responsible for ensuring compliance with standards.
   d. all of the above.
   e. none of the above.

6. Triangles can be combined to form angles in multiples of
   a. 1°.
   b. 2°.
   c. 5°.
   d. 10°.
   e. 15°.

7. Dividers are used to
   a. measure.
   b. transfer distances.
   c. layout curves.
   d. draw circles.
   e. all of the above.

8. A graphite lead of hardness 2H would be
   a. hard.
   b. medium hard.
   c. medium.
   d. medium soft.
   e. soft.

9. The advantage of a drafting machine over a T-square is that a drafting machine
   a. is more accurate.
   b. can eliminate the need for individual triangles.
   c. can eliminate the need for separate scales.
   d. all of the above.
   e. none of the above.

10. A scale of 1:2 is usually referred to as
    a. half size.
    b. double size.
    c. It would depend on the units used.
    d. a metric dimension.
    e. none of the above.

11. A scale of 2:1 is usually referred to as
    a. half size.
    b. double size.
    c. It would depend on the units used.
    d. a metric dimension.
    e. none of the above.

12. Vellum is
    a. a technical term for drafting paper.
    b. translucent paper.
    c. a type of transparent plastic.
    d. a type of drafting film.
    e. none of the above.

13. A french curve is used to draw
    a. circles.
    b. polygons.
    c. irregular curves.
    d. large circles.
    e. accurate ellipses.

14. Drafting tape
    a. is no different than masking tape.
    b. is transparent.
    c. comes only in small rolls.
    d. is more tacky than masking tape.
    e. is less tacky than masking tape.

Test

Chapter 2

Multiple Choice   Circle the answer that *best* fits the question.

1. The smallest unit of information recognized by a computer is a
   a. byte.
   b. nibble.
   c. word.
   d. bit.
   e. all of the above.

2. In computer systems. permanent memory is referred to as
   a. ROM.
   b. RAM.
   c. storage.
   d. a hard drive.
   e. a floppy drive.

3. The most commonly used input device for computers is a
   a. digitizing tablet.
   b. mouse.
   c. function keypad.
   d. joystick.
   e. keyboard.

4. The most commonly used input device for a CAD system is a
   a. digitizing tablet.
   b. mouse.
   c. function keypad.
   d. joystick.
   e. keyboard.

5. The most common type of computer monitor is a(n)
   a. LCD screen.
   b. TV.
   c. CRT.
   d. vector tube.
   e. plasma screen.

6. A plotter is a(n) _____ output device and a laser printer is a(n) _____ output
   device.
   a. raster, vector
   b. vector, raster
   c. high quality, medium quality
   d. high speed, medium speed
   e. color, monochrome

7. CAM is an acronym for
   a. computer-aided machining.
   b. continuous automated manufacturing.
   c. computer analysis and modeling.
   d. computer automated manufacturing.
   e. computer-aided manufacturing.

8. How many bits are in a byte?
   a. 1
   b. 8
   c. 64
   d. 1024
   e. none

9. The correct units abbreviation for 1000 bytes is
   a. 1 Kb.
   b. 1 KB.
   c. 1 kb.
   d. 1 kB.
   e. any of the above.

10. A collection of computers that can communicate and operate as a group is referred
    to as
    a. a system.
    b. compatible.
    c. using one operating system.
    d. a computer laboratory.
    e. networked.

11. The general term used to describe a collection of data stored on a diskette is
    a. a program.
    b. software.
    c. a file.
    d. binary data.
    e. a database.

12. A graphic representation of a program command or function is referred to as
    a. an icon.
    b. a symbol.
    c. a menu.
    d. a device driver.
    e. a command line.

13. The three main categories of computer system elements are
    a. CPU, monitor, and hard drive.
    b. system, display, and printer.
    c. CPU, memory, and storage.
    d. processing and storage, input, and output.
    e. EGA, VGA, and SVGA.

14. Robotics are best used for applications involving
    a. high temperature environments.
    b. toxic or dangerous environments.
    c. heavy lifting.
    d. work with delicate parts.
    e. all of the above.

Multiple Choice

Circle the answer that *best* fits the question.

1. The term *drawing media* refers to
   a. any material on which a drawing is made.
   b. any printer or plotter output.
   c. paper, vellum, or film.
   d. duplicated or reproduced drawings.
   e. blueprints or whiteprints.

2. The principal function of the title block is to
   a. identify the drawing.
   b. identify the company.
   c. identify who produced and released the drawing.
   d. indicate the number of drawings in a set.
   e. all of the above.

3. For CAD based drawings the title block will always include
   a. what type of computer was used.
   b. what type of software was used.
   c. the name and location of the computer file.
   d. all of the above.
   e. none of the above.

4. A D size drawing would be (height x width)
   a. 8.5 x 11 in.
   b. 11 x 17 in.
   c. 17 x 22 in.
   d. 22 x 34 in.
   e. 24 x 36 in.

5. The nearest metric size for an A size drawing would be
   a. A0.
   b. A1.
   c. A2.
   d. A3.
   e. A4.

6. The space above the title block is reserved for
   a. notes.
   b. the item list.
   c. the revision list.
   d. copyright and patent notices.
   e. small detail drawings.

7. When using diskettes to store drawing files,
   a. always make a backup copy.
   b. label the diskette properly.
   c. store the diskette properly.
   d. keep them away from magnets.
   e. all of the above.

8. The arrows in the margins of a drawing form
   a. indicate how the paper is to be folded.
   b. help the drafter place the drawing in the center of the page.
   c. are used to load the preprinted from into a plotter.
   d. are used to locate the center of the drawing for microfilming.
   e. no longer serve any purpose.

9. Zoning refers to
   a. locating information in predefined locations of the drawing form.
   b. referencing an area of a drawing form by letters and numbers in the margins.
   c. breaking a large drawing into smaller drawings.
   d. the same thing as tiling.
   e. splitting a CAD file too large to fit into a diskette into several files.

10. On drawings larger than B size, letters are placed on the _____ margin.
   a. vertical
   b. horizontal
   c. top
   d. bottom
   e. the same

11. The diazo reproduction process is also referred to as
   a. photocopying.
   b. zerographic reproduction.
   c. photo duplication.
   d. a whiteprint.
   e. a blueprint.

12. The processing of copying an existing drawing into a CAD system is
   a. translation.
   b. inputting.
   c. digitizing.
   d. scanning.
   e. none of the above.

13. The main advantage of a CD-ROM is
   a. they are the same size a 5 1/4" diskettes .
   b. the data can be write protected.
   c. the amount of data that can be stored.
   d. they are more rugged.
   e. all of the above.

14. One of the problems in using a photocopier to reproduce engineering drawings is
   that
   a. the size of the image may change.
   b. the quality is too poor.
   c. it is too expensive.
   d. all of the above.
   e. none of the above.

Test

Chapter 4

Multiple Choice   Circle the answer that *best* fits the question.

1. The widths of lines used in engineering drawings are
    a. fat and skinny.
    b. thin, medium, and thick.
    c. thin and thick.
    d. 1mm and 2mm
    e. dependent on the drawing size.

2. Guidelines in drafting are
    a. based on national standards.
    b. established by each company.
    c. the same as grid lines.
    d. used to ensure uniform lettering.
    e. not always followed by the drafter.

3. A coordinate system based on a fixed 0,0 point is a(n)
    a. absolute coordinate system.
    b. relative coordinate system.
    c. variable coordinate system.
    d. rectangular coordinate system.
    e. cylindrical coordinate system.

4. In a polar coordinate system, the first element of a coordinate pair represents
    a. distance.
    b. angle.
    c. beta.
    d. theta.
    e. relative offset.

5. According to standards, the minimum text height for drawings is
    a. .10.
    b. .12.
    c. .1875.
    d. .25.
    e. based on the size of the drawing.

6. Lettering on mechanical drawings is usually
    a. bold and underlined.
    b. italics or slanted.
    c. upper- and lowercase.
    d. lowercase.
    e. uppercase.

7. A beam compass is used to draw
    a. very large arcs.
    b. very small arcs.
    c. irregular curves.
    d. curves when inking.
    e. very accurate curves.

8. An irregular curve, when drawn using a CAD system, is often referred to as a(n)
   a. polyline
   b. lofted curve.
   c. conic section.
   d. spline.
   e. biased curve.

9. Compass lead is usually _____ pencil lead.
   a. harder than
   b. softer than
   c. the same as.
   d. heavier than.
   e. lighter than.

10. Sketches
    a. are not used in engineering design.
    b. should be drawn with instruments.
    c. are usually unnecessary.
    d. are used to express ideas graphically.
    e. are obsolete now that engineers are using CAD.

11. Sketched lines should have a _____ appearance.
    a. feathered
    b. natural
    c. irregular.
    d. very dark.
    e. uniform.

12. A picturelike sketch is referred to as a _____ sketch.
    a. pictorial
    b. illustrative
    c. multiview
    d. isometric
    e. preliminary

13. Inked drawings
    a. are more permanent.
    b. make better copies.
    c. take longer to produce.
    d. are more difficult to change.
    e. all of the above.

14. The usual order of inking for drawings is
    a. dimensions, arcs, horizontal lines, and vertical lines.
    b. vertical lines, horizontal lines, arcs, and dimensions.
    c. horizontal lines, vertical lines, arcs, and dimensions.
    d. arcs, horizontal lines, vertical lines, and dimensions.
    e. left to right.

Test

Chapter 5

Multiple Choice   Circle the answer that *best* fits the question.

1. Geometry is
   a. a part of mathematics.
   b. the study of the size and shape of objects.
   c. a part of engineering.
   d. the theoretical study of form.
   e. all of the above.

2. Geometric construction refers to
   a. any boolean operation involving solids.
   b. the production of models.
   c. complex CAD operations and programs.
   d. the lines and points drawn to produce geometry.
   e. none of the above.

3. The shape formed by lines that extend from a common point is called a(n)
   a. intersection.
   b. triangle.
   c. tangency.
   d. vertex.
   e. angle.

4. A line that is bisected is divided into _____ parts.
   a. 2
   b. 4
   c. 8
   d. 16
   e. a specified number of

5. Two lines that, if extended, would never intersect are
   a. perpendicular.
   b. parallel.
   c. offset.
   d. coincidental.
   e. coplanar.

6. Two lines that intersect at 90° are
   a. perpendicular.
   b. parallel.
   c. offset.
   d. coincidental.
   e. coplanar.

7. A line that intersect a circle at one point is _____ to that circle.
   a. coincidental
   b. acute
   c. oblique
   d. tangent
   e. coplanar

Test

Chapter 5

8. In engineering graphics, a reverse curve is also know as a(n)
   a. s-curve.
   b. ogee.
   c. spline.
   d. double curve.
   e. all of the above.

9. A hexagon has _____ sides.
   a. 5
   b. 6
   c. 8
   d. 12
   e. a variable number of

10. A regular polygon has
    a. an even number of sides.
    b. an odd number of sides.
    c. obtuse interior angles.
    d. interior angles divisible by 15°.
    e. sides of even length.

11. The interior angles of an octagon are
    a. 15°.
    b. 30°.
    c. 37.5°.
    d. 45°.
    e. 60°.

12. The two focal points of an ellipse are called
    a. foci.
    b. fulcrums.
    c. center pairs.
    d. medial points.
    e. bifurcated points.

13. The fixed line of a parabola is referred to as
    a. an axis.
    b. the minor axis.
    c. the major axis.
    d. a lead.
    e. the directrix.

14. A helix is a curve generated by
    a. a conic section.
    b. sweeping a parabolic curve.
    c. a point revolving around a cone.
    d. a point revolving around a cylinder.
    e. none of the above.

Test _____ Chapter 6

Multiple Choice   Circle the answer that *best* fits the question.

1. In the United States and Canada, multiview drawings are based on
   a. first-angle projection.
   b. second-angle projection.
   c. third-angle projection.
   d. fourth-angle projection.
   e. the appropriate national standard.

2. The three standard views for multiview drawings in the United States, are
   a. top, back, profile.
   b. front, rear, side.
   c. front. top, side.
   d. front, top, left profile.
   e. front, top, right profile.

3. Features of a part that cannot be seen are represented in a drawing by
   a. phantom lines.
   b. hidden lines.
   c. object lines.
   d. thin lines.
   e. all of the above.

4. A view that is drawn to describe an inclined surface is known as a(n)
   a. auxiliary view.
   b. supplemental view.
   c. secondary view.
   d. helper view.
   e. adjunct view.

5. Center lines
   a. are drawn as thin, broken lines of long and short dashes.
   b. may be used as extension lines for dimensioning.
   c. extend beyond the object outline.
   d. do not extend between views.
   e. all of the above.

6. Oblique surfaces are
   a. perpendicular to one viewing plane.
   b. perpendicular to two viewing planes.
   c. perpendicular to three viewing planes.
   d. not perpendicular to any viewing plane.
   e. none of the above.

7. The number of views required to represent an object is
   a. at least three.
   b. the minimum required.
   c. dependent on the manufacturing process.
   d. often only one or two.
   e. all of the above.

8. When selecting the views for a drawing
   a. hidden lines should be minimized.
   b. the object should be fully described.
   c. the normal position of the object should be considered.
   d. the shape of the object should be considered.
   e. all of the above.

9. The intersection of fillets and rounds are
   a. shown as object lines.
   b. shown as hidden lines.
   c. shown as phantom lines.
   d. shown in an auxiliary projection.
   e. not shown.

10. Small details on a part are
    a. normally omitted.
    b. shown on a separate drawing.
    c. shown as an enlarged view.
    d. shown in an auxiliary view.
    e. shown simplified or by a note.

11. Repetitive parts are
    a. normally omitted.
    b. shown fully drawn for clarity.
    c. represented by corner "tick" marks.
    d. drawn using phantom lines.
    e. shown as a simple outline.

12. Long, simple parts are
    a. are drawn using conventional breaks.
    b. are fully detailed to avoid confusion.
    c. drawn to a smaller scale.
    d. referenced by a note.
    e. none of the above.

13. When a fillet or round blends into another surface, it is known as a(n)
    a. running intersection.
    b. runout.
    c. irregular intersection.
    d. any of the above.
    e. none of the above.

14. Repetitive details
    a. are fully drawn to avoid confusion.
    b. are often omitted.
    c. are simplified and referenced by a note.
    d. are partially drawn and then continued as hidden lines.
    e. are partially drawn and then continued as phantom lines.

Test

Chapter 7

Multiple Choice   Circle the answer that *best* fits the question.

1. Auxiliary views are drawn to
   a. describe the true shape of a feature.
   b. describe very complex features.
   c. describe distorted features.
   d. describe double-curved features.
   e. all of the above.

2. In auxiliary views hidden lines are
   a. usually drawn.
   b. usually omitted.
   c. drawn if required for clarity.
   d. drawn if the part is complex.
   e. drawn thinner than in principal views.

3. Auxiliary views are
   a. not dimensioned.
   b. dimensioned using the unidirectional system.
   c. dimensioned using the directional system.
   d. dimensioned using the aligned system.
   e. dimensioned using the multidirectional system.

4. A feature that is circular in an auxiliary view will appear _____ in a principal view.
   a. circular
   b. stretched
   c. elliptical
   d. squashed
   e. all of the above

5. A secondary auxiliary view would be required to determine the
   a. point view of a horizontal line.
   b. point view of a vertical line.
   c. true length of an inclined line.
   d. true length of an oblique line.
   e. all of the above.

6. In an auxiliary view
   a. only the true shape surface is drawn.
   b. the true shape surface and hidden lines are drawn.
   c. the true shape surface and perpendicular features are drawn.
   d. all of the object is drawn.
   e. none of the object is drawn.

7. A drawing with several auxiliary views is
   a. not standard practice.
   b. a complex drawing.
   c. usually avoided.
   d. a multi-auxiliary-view drawing.
   e. a multiple view auxiliary drawing.

8. A plane that appears as a line in two of the principal views is a
   a. normal plane.
   b. inclined plane.
   c. oblique plane.
   d. skewed plane.
   e. any of the above.

9. To determine the angle between two planes, you must
   a. first construct a point view of the vertex.
   b. determine the true length of the vertex.
   c. construct a true shape view of one of the planes.
   d. construct true shape views of both planes.
   e. construct an edge view of one plane.

10. To determine the true length of a line, you must first generate
   a. an auxiliary view
   b. a view of revolution.
   c. an edge view of the line.
   d. a point view of the line.
   e. any of the above.

11. Drawings created by revolution are
   a. not dimensioned.
   b. dimensioned using the unidirectional system.
   c. dimensioned using the directional system.
   d. dimensioned using the aligned system.
   e. dimensioned using the multidirectional system.

12. The true shape of an oblique surface can be found by
   a. an auxiliary view.
   b. successive revolutions.
   c. combining auxiliary and revolved views.
   d. determining the true length of a line on the plane.
   e. any of the above.

Test

Chapter 8

Multiple Choice Circle the answer that *best* fits the question.

1. Dimensioning is used to indicate the _____ and _____ of a feature.
   a. shape, size
   b. size, location
   c. shape, location
   d. extent, position
   e. orientation, shape

2. Dimension lines indicate the
   a. location of a dimension.
   b. size of a dimension.
   c. direction and orientation of a dimension.
   d. extent and direction of a dimension.
   e. shape of the feature.

3. The two types of notes used in dimensioning are
   a. general and local.
   b. global and local.
   c. general and special.
   d. general and limited.
   e. global and special.

4. Dual dimensions include
   a. decimal and fractional inch.
   b. inch and foot notations.
   c. decimal inches and millimeters.
   d. fractional inches and millimeters.
   e. English and metric units.

5. A dimension shown for information purposes only is a
   a. reference dimension.
   b. basic dimension.
   c. nominal dimension.
   d. secondary dimension.
   e. datum dimension.

6. Arcs are dimensioned by _____ and circles are dimensioned by _____.
   a. note, diameter
   b. radius, radius
   c. chord, diameter
   d. diameter, radius
   e. radius, diameter

7. Rounds and fillets are usually dimensioned by
   a. a local note.
   b. a general note.
   c. nominal diameters.
   d. radius and diameter.
   e. chordal lengths.

Test   Chapter 8

8. Which of the following is a correct chamfer specification?
   a. CHAM .06
   b. .06 CHAM
   c. .06 x 45 CHAM
   d. 0.6 x 45 CHAM
   e. 45 x .06 CHAM

9. Which of the following dimensioning methods permits the accumulation of error?
   a. chain dimensioning
   b. datum dimensioning
   c. ordinate dimensioning
   d. rectangular coordinate dimensioning
   e. tabular dimensioning

10. _____ size is the theoretical size of a feature while _____ is used for general identification.
   a. Design, actual
   b. Basic, actual
   c. Nominal, basic
   d. Tolerance, design
   e. Basic, nominal

11. $.500 \pm .004$ is a
   a. limit dimension.
   b. unilateral tolerance.
   c. bilateral tolerance.
   d. fixed range dimension.
   e. none of the above.

12. A fit that requires two parts to be forced together would be a(n)
   a. clearance fit.
   b. interference fit.
   c. transition fit.
   d. run-in fit.
   e. overlap fit.

13. The intentional difference in the size of mating parts is
   a. clearance.
   b. a fit.
   c. fundamental deviation.
   d. tolerance.
   e. allowance.

14. Surface texture is important for
   a. reducing friction.
   b. controlling wear.
   c. selecting the manufacturing process.
   d. all of the above.
   e. none of the above.

Test   Chapter 9

Multiple Choice   Circle the answer that *best* fits the question.

1. Section views are used to
   a. simplify complex parts.
   b. describe interior detail.
   c. reduce the number of required views.
   d. reduce the number of hidden lines.
   e. all of the above

2. A section view where half of the object is "removed" is a(n)
   a. half-section.
   b. full-section.
   c. phantom section.
   d. removed section.
   e. any of the above.

3. The arrows of the cutting-plane line point
   a. toward the viewer.
   b. away from the viewer.
   c. toward the top of the page.
   d. in the direction defined in the notes.
   e. none of the above.

4. For symmetrical objects, the most common type of section view is a(n)
   a. half-section.
   b. full-section.
   c. phantom section.
   d. removed section.
   e. any of the above.

5. A section view superimposed on a regular view is a
   a. half section.
   b. full-section.
   c. phantom section.
   d. removed section.
   e. any of the above.

6. In section view, hidden lines are usually
   a. included.
   b. omitted.
   c. shown if a full-section.
   d. included for special details.
   e. shown in removed sections.

7. Threads in section views are usually
   a. included.
   b. omitted.
   c. drawn as tap-drill diameter holes.
   d. drawn as simplified threads.
   e. drawn as schematic threads.

8. The direction of section lining
    a. affects the interpretation of the drawing.
    b. is used to indicate material type.
    c. does not affect the interpretation of a drawing.
    d. indicates different parts of the same material.
    e. is arbitrary.

9. Which of the following are not sectioned?
    a. shafts
    b. bolts
    c. pins
    d. keyseats
    e. all of the above

10. Ribs, holes, and lugs
    a. are aligned in section views.
    b. are not sectioned.
    c. are aligned and not sectioned.
    d. are shown in true projection.
    e. are shown by phantom section.

11. A section view drawn apart from the regular view is a
    a. half-section.
    b. full-section.
    c. phantom section.
    d. removed section.
    e. any of the above.

12. When only a small section of a part needs to be section, a(n) _____ is used.
    a. removed section.
    b. phantom section.
    c. broken-out section.
    d. offset section.
    e. revolved section.

13. General purpose section lining is drawn at
    a. 15°.
    b. 30°.
    c. 45°.
    d. 60°.
    e. any of the above.

14. For phantom sections, the section lining
    a. is drawn as thin broken lines.
    b. the same as regular section lining.
    c. is drawn at multiples of 17.5° only.
    d. is drawn as thin rather than thick lines.
    e. is drawn as hidden lines.

Test   Chapter 10

Multiple Choice   Circle the answer that *best* fits the question.

1. Knuckle, buttress, square, and acme are examples of thread
   a. series.
   b. classes.
   c. forms.
   d. divisions.
   e. pitches.

2. The most often used approach to thread representation in manual drafting is
   a. detailed.
   b. pictorial.
   c. minimal.
   d. schematic.
   e. simplified.

3. Most threaded fasteners are
   a. right-handed.
   b. left-handed.
   c. hybrid.
   d. sequential.
   e. nonsequential.

4. Multiple threads are used when _____ is desired.
   a. slow movement
   b. rapid movement
   c. additional strength
   d. shear resistance
   e. strain resistance

5. In thread class designations, A and B refer to
   a. internal and external.
   b. external and internal.
   c. coarse and fine.
   d. fine and coarse.
   e. none of the above.

6. The two most common head styles for bolts are
   a. machine and cap.
   b. flat and oval.
   c. square and hexagon.
   d. plain and dressed.
   e. all of the above.

7. Washers are used to
   a. maintain a spring-resistance pressure.
   b. guard against surface marring.
   c. provide a seal.
   d. all of the above.
   e. none of the above.

Test   Chapter 10

8. The term applied to a hole that permits the head of a fastener to sit below the surface
   is
   a. counterbored.
   b. countersunk.
   c. spot faced.
   d. recessed.
   e. tunneled.

9. The special fastener used to hold a gear, collar, or sheave onto a shaft is a
   a. stud screw.
   b. captive screw.
   c. cap screw.
   d. key screw.
   e. set screw.

10. A screw that is designed to work without a prepared hole is a
    a. cap screw.
    b. captive screw.
    c. tapping screw.
    d. set screw.
    e. machine screw.

11. Point types apply only to
    a. cap screws.
    b. captive screws.
    c. tapping screws.
    d. set screws.
    e. machine screws.

12. The clearance drill size for a threaded fastener is slightly greater than the
    a. tape drill size.
    b. pitch diameter.
    c. root diameter.
    d. major diameter.
    e. minor diameter.

13. In the following callout, .500 - 13 UNC - 2A X 4.00, "UNC" is the thread
    a. series.
    b. class.
    c. form.
    d. classification.
    e. pitch.

14. In the following callout, .500 - 13 UNC - 2A X 4.00, "2A" is the thread
    a. series.
    b. class.
    c. form.
    d. classification.
    e. pitch.

Test

Chapter 11

Multiple Choice   Circle the answer that *best* fits the question.

1. For keys, the groove in the shaft is referred to as a _____ and the groove in the hub is referred to as a _____.
   a. keyseat, keyslot
   b. keyslot, keyway
   c. keyway, keyseat
   d. keyseat, keyway
   e. none of the above

2. A type of key that has a head added for easy removal is a
   a. gib-head key.
   b. Pratt & Whitney key.
   c. Woodruff key.
   d. square-head key.
   e. all of the above.

3. Pins are best used in conditions of
   a. stress.
   b. shear.
   c. strain.
   d. vibration.
   e. all of the above.

4. A device used to lock components onto a shaft or in a bore housing is a
   a. splined shaft.
   b. retaining ring.
   c. serrated shaft.
   d. spring.
   e. all of the above.

5. A suspension would be an application for a
   a. controlled-action spring.
   b. variable-action spring.
   c. static spring.
   d. all of the above.
   e. none of the above.

6. A flat coil spring would be an example of a(n)
   a. compression spring.
   b. extension spring.
   c. flat spring.
   d. leaf spring.
   e. power spring.

7. The type of compression spring end that affords maximum stability is the
   a. plain open end.
   b. ground open end.
   c. plain closed end.
   d. ground closed end.
   e. all of the above.

8. Riveting is a popular method of fastening and joining because of its
   a. simplicity.
   b. dependability.
   c. low cost.
   d. all of the above.
   e. a and b only.

9. Pitch is the
   a. interval between center lines of adjacent rivets.
   b. interval between a row of rivets and the edge.
   c. offset interval between staggered, adjacent rivets.
   d. offset interval between rows of rivets.
   e. angled offset for staggered rows of rivets.

10. The length-to-diameter ratio of a rivet should not exceed
    a. 1.5:1.
    b. 2:1.
    c. 3:1.
    d. 4:1.
    e. 6:1.

11. The type of rivet used when both sides of the material cannot be accessed is a
    a. split rivet.
    b. composite rivet.
    c. blind rivet.
    d. shop rivet.
    e. field rivet.

12. Which of the following is the primary type of stress adhesive fasteners are subject to?
    a. tensile
    b. shear
    c. clearance
    d. peel
    e. all of the above

13. The most practical type of adhesive joint design for thin materials is the
    a. lap joint.
    b. angle joint.
    c. butt joint.
    d. corner joint.
    e. stiffener joint.

Test

Multiple Choice   Circle the answer that *best* fits the question.

1. White iron is produced by a process called
   a. quenching.
   b. chilling.
   c. alloying.
   d. super saturation.
   e. manganese diffusion.

2. Crankshafts would be a typical application for
   a. ductile iron.
   b. gray iron.
   c. white iron.
   d. high-alloy iron.
   e. malleable iron.

3. Automotive engine blocks would be a typical application for
   a. ductile iron.
   b. gray iron.
   c. white iron.
   d. high-alloy iron.
   e. malleable iron.

4. Rolling mill rolls would be a typical application for
   a. ductile iron.
   b. gray iron.
   c. white iron.
   d. high-alloy iron.
   e. malleable iron.

5. High-alloy cast steels contain at least _____ nickel and/or chromium.
   a. 2%
   b. 3%
   c. 5%
   d. 8%
   e. 10%

6. Which of the following additives does not increase the machinability of steel?
   a. phosphorus
   b. sulfur
   c. lead
   d. silicon
   e. none of the above

7. Copper is an example of a _____ metal.
   a. ferrous
   b. nonferrous
   c. transitional
   d. heavy
   e. light

8. Powder metallurgy forms parts by
   a. deep drawing.
   b. stamping.
   c. forging.
   d. extrusion.
   e. compaction.

9. The most common fabrication technique applied to plastics is
   a. extrusion.
   b. forming.
   c. casting.
   d. molding.
   e. machining.

10. Plastics that undergo an irreversible chemical reaction when heated are known as
    a. thermosetting plastics.
    b. thermoplastics.
    c. superpolymers.
    d. hybrid thermal plastics.
    e. conversion plastics.

11. Elastomers are
    a. composite materials.
    b. metalized plastics.
    c. soft metals.
    d. plasticlike substances.
    e. runnerlike substances.

12. The basic categories for rubber are
    a. opened and closed cell.
    b. natural and synthetic.
    c. inert and reactive.
    d. sponge and sheet.
    e. mechanical and cellular.

Test

Multiple Choice   Circle the answer that *best* fits the question.

1. The top part of a sand mold is known as a
   a. chaplet.
   b. cheek.
   c. drag.
   d. cope.
   e. core.

2. The part of a sand mold used to create an internal void or hollow is a
   a. chaplet.
   b. cheek.
   c. drag.
   d. cope.
   e. core.

3. The casting process in which the pattern is destroyed during casting is
   a. shell mold casting.
   b. investment casting.
   c. permanent mold casting.
   d. plaster mold casting.
   e. die casting.

4. The least expensive and most efficient casting process is
   a. shell mold casting.
   b. investment casting.
   c. permanent mold casting.
   d. plaster mold casting.
   e. die casting.

5. The slope given to the side walls of a pattern to facilitate easy removal from a mold, or a casting from a die, is known as
   a. cope.
   b. flash.
   c. draft.
   d. cheek.
   e. drag.

6. Small holes in castings are usually
   a. drilled.
   b. formed by cores.
   c. included in the pattern.
   d. formed by chaplets.
   e. bored.

7. The dimensional accuracy of castings is
   a. better than machining processes.
   b. worse than machining processes.
   c. the same as machining processes.
   d. of concern only for large castings.
   e. easily controlled through proper mold preparation.

8. In forging, excess metal produced by the process is referred to as
   a. runout.
   b. overflow.
   c. overrun.
   d. squeeze.
   e. flash.

9. Rough casting and forging outlines are shown on drawings by using
   a. hidden lines to describe the profile.
   b. phantom lines to describe the profile.
   c. object lines to describe the profile.
   d. phantom section views.
   e. regular section views.

10. Powder metallurgy forms parts by
    a. compacting powdered metal.
    b. casting powdered metal.
    c. using a flow of continuous powdered metal for casting.
    d. drop forging.
    e. none of the above.

11. The point at which plastic mold separates is known as a
    a. flash line.
    b. separation line.
    c. molding line.
    d. split line.
    e. ejection line.

12. Plastic molded parts
    a. have less dimensional accuracy than cast metal parts.
    b. require extensive post-mold processing.
    c. do not permit extensive interior cavities.
    d. all of the above.
    e. none of the above.

Test

Multiple Choice   Circle the answer that *best* fits the question.

1. A drawing that shows a product in its completed state is known as a(n)
   a. detail drawing.
   b. assembly drawing.
   c. release drawing.
   d. process drawing.
   e. working drawing.

2. Multiple detail drawings
   a. are usually combined onto one sheet.
   b. are usually drawn on separate sheets.
   c. may be combined or drawn on separate sheets.
   d. are usually not required.
   e. are not used for large complex products.

3. The drawing revision table is located _____ of a drawing.
   a. down the right side.
   b. across the bottom.
   c. across the top.
   d. a or b.
   e. b or c.

4. If a drawing is completely redrawn
   a. the original drawing date is retained.
   b. a new part number is used.
   c. no changes should be made.
   d. REDRAWN or REVISED appears in the revision table.
   e. REDRAWN or REVISED appears above the title block.

5. Item lists
   a. are always included on the first page of an assembly drawing.
   b. may be on a separate sheet for ease of handling and duplication.
   c. are usually prepared by the purchasing department.
   d. are no longer part of a CAD generated drawing.
   e. include the full name and addresses of all vendors.

6. General tolerances for a drawing are included in the
   a. title block.
   b. process notes.
   c. revision table.
   d. materials table.
   e. global notes.

7. The person who creates working drawings is known as a
   a. drafter.
   b. detailer.
   c. engineer.
   d. operator.
   e. checker.

8. On working drawings, tolerances should be
   a. uniform.
   b. shown as supplemental notes only.
   c. omitted.
   d. as liberal as the design will permit.
   e. as restrictive as the manufacturing process will permit.

9. CAD generated drawings should include
   a. the name of the software used to create the drawing.
   b. the data file name and location.
   c. a brief description of the system on which the drawing was created.
   d. the size of the data file.
   e. all of the above.

10. Detail drawings most often use _____ representation of threads.
    a. detailed
    b. schematic
    c. simplified
    d. pictorial
    e. none of the above

11. The most common time devices in use for manual drafting are
    a. copy machines.
    b. templates.
    c. different kinds of transfer type.
    d. technical pens.
    e. plastic media and pencils.

12. Simplified drafting
    a. increases design efficiency.
    b. accelerates the course of design.
    c. reduces the workload in the drafting office.
    d. enhances legibility of the drawings.
    e. all of the above.

13. The equivalent of templates for CAD is
    a. standard drawings.
    b. symbol libraries.
    c. prototype drawings.
    d. multiple copy commands.
    e. externally referenced drawings.

Test

Multiple Choice   Circle the answer that *best* fits the question.

1. Lines of sight are perpendicular to the plane of projection in _____ projection.
   a. axonometric
   b. linear
   c. oblique
   d. perspective
   e. pictorial

2. In isometric projection, the object is rotated at an angle of _____ to the horizon.
   a. 15°
   b. 30°
   c. 45°
   d. 60°
   e. 75°

3. The type of axonometric drawing in which two of the three principal axes are scaled the same is a _____ drawing.
   a. isometric
   b. dimetric
   c. trimetric
   d. cabinet
   e. cavalier

4. In axonometric drawings all measurements must be _____ to the principal axes.
   a. tagential
   b. skewed at 45°
   c. perpendicular
   d. parallel
   e. none of the above

5. In isometric drawings, circles normal to a principal plane would appear as
   a. circles.
   b. ovals.
   c. ellipses.
   d. all of the above.
   e. none of the above.

6. _____ dimensioning is the preferred method of dimensioning isometric drawings.
   a. Aligned
   b. Isodirectional
   c. Monodirectional
   d. Bidirectional
   e. Unidirectional

7. For manual drafting, the most efficient method for creating isometric circles is
   a. by the offset method.
   b. by the 4-point method.
   c. using a template.
   d. by sketching.
   e. by the grid method.

Test

8. Angles in isometric drawings can be constructed using
   a. isometric protractors.
   b. offset construction.
   c. grid construction.
   d. all of the above.
   e. none of the above.

9. The type of oblique projection where all lines are true length is
   a. isometric.
   b. dimetric.
   c. box.
   d. cabinet.
   e. cavalier.

10. Circles and arcs on oblique faces are usually drawn
    a. using templates.
    b. using a compass.
    c. using a grid.
    d. by the offset method.
    e. by the 4-point method.

11. In perspective drawings the lines of projection
    a. diverge.
    b. converge.
    c. radiate.
    d. coincide.
    e. are concurrent.

12. Which type of perspective drawing has three vanishing points?
    a. pastoral
    b. pictorial
    c. parallel
    d. angular
    e. oblique

13. An angular perspective drawing in which the viewer is looking up is known as a
    _____ grid.
    a. worm's eye
    b. bird's eye
    c. exterior
    d. interior
    e. none of the above

Test

Multiple Choice   Circle the answer that *best* fits the question.

1. A basic dimension is
   a. a geometric characteristic, the size of which is specified.
   b. the maximum permissible variation of form.
   c. the total permissible variation in the size of a dimension.
   d. a theoretical exact size, profile, orientation of location.
   e. the overall envelope of perfect form.

2. Geometric tolerance is
   a. a geometric characteristic, the size of which is specified.
   b. the maximum permissible variation of form.
   c. the total permissible variation in the size of a dimension.
   d. a theoretical exact size, profile, orientation of location.
   e. the overall envelope of perfect form.

3. The condition in which two or more cylindrical features are arranged with their axes on a straight line is
   a. circularity.
   b. coaxiality.
   c. concentricity.
   d. coplanarity.
   e. parallelism.

4. A _____ is a theoretically exact point, line, plane, or other geometric feature.
   a. datum
   b. feature
   c. runout
   d. profile
   e. none of the above

5. Geometric tolerances can be identified by the presence of
   a. a feature control frame.
   b. geometric symbols.
   c. very close tolerance.
   d. special notes and acronyms.
   e. none of the above.

6. Features of size have
   a. a reference to a datum.
   b. diameter or thickness.
   c. a virtual envelope.
   d. a correlative geometric condition.
   e. upper and lower deviations.

7. Datums are identified by
   a. feature control frames.
   b. geometric symbols.
   c. general notes.
   d. target symbols.
   e. all of the above.

8. A tertiary datum would be related to _____ other datums.
   a. one
   b. two
   c. three
   d. no
   e. several

9. In geometric tolerancing the space or area in which a feature could exist is known as a(n)
   a. profile zone.
   b. allowance zone.
   c. tolerance zone.
   d. all of the above.
   e. none of the above.

10. Coplanarity and concentricity are examples of _____ geometric tolerancing.
    a. positional
    b. virtual
    c. correlative
    d. all of the above
    e. none of the above

11. _____ is not applicable to runout tolerances.
    a. MMC
    b. LMC
    c. RFS
    d. FIM
    e. none of the above

12. A square tolerance zone is _____ restrictive when compared to a circular tolerance zone.
    a. as
    b. less
    c. more
    d. not
    e. none of the above

13. Form is used to control the _____ of features.
    a. location
    b. orientation
    c. shape
    d. true position
    e. relationship

14. Form of line includes
    a. straightness.
    b. angularity.
    c. parallelism.
    d. perpendicularity.
    e. flatness.

Test

Multiple Choice Circle the answer that *best* fits the question.

1. CAM is an acronym for
   a. computer-aided manufacturing.
   b. computer-assisted manufacturing.
   c. computer-automated manufacturing.
   d. computer-aided machining.
   e. computer-assisted machining.

2. The automated control of a lathe or mill is known as
   a. CAD/CAM.
   b. CIM.
   c. numerical control.
   d. computer numerical control.
   e. direct numerical control.

3. The total computer-based automation of the entire manufacturing process is known as
   a. CAD/CAM.
   b. CIM.
   c. numerical control.
   d. computer numerical control.
   e. direct numerical control.

4. _____ dimensioning is most often used in preparation of drawings for NC.
   a. Tabular
   b. Ordinate
   c. Polar
   d. Geometric
   e. Baseline

5. The system of programming NC machines that always references the last programmed point is known as _____ coordinate programming.
   a. absolute
   b. relative
   c. offset
   d. fixed origin
   e. floating origin

6. A known point or location on a part used to reference the origin is a(n)
   a. origin-point.
   b. initial-point.
   c. startup-point.
   d. setup-point.
   e. set-point.

7. The up-down motion of a turret drill corresponds to the _____ axis in an NC program.
   a. X
   b. Y
   c. Z
   d. W
   e. U

8. _____ tolerancing should be used for NC drawings.
   a. Limit
   b. Lower limit
   c. Upper limit
   d. Unilateral
   e. Bilateral

9. The position where the X and Y axes intersect is known as the
   a. origin.
   b. fulcrum.
   c. start-point.
   d. setup-point.
   e. offset-point.

10. On most NC machines, the Z zero plane is _____ the workpiece.
    a. on
    b. to the left of
    c. to the right of
    d. below
    e. above

11. The specified tolerance range for NC drawings should be _____ for drawings for manual production.
    a. more restrictive than
    b. less restrictive than
    c. the same as
    d. more restrictive for larger parts than
    e. less restrictive for smaller parts than

12. Parabolas or similar curves are _____ in NC machining.
    a. to be avoided
    b. approximated by curves
    c. defined by notes such as BLEND SMOOTHLY
    d. defined by mathematical equation
    e. approximated by linear interpolation

Test

Chapter 18

Multiple Choice

Circle the answer that *best* fits the question.

1. Metal that is added to a weld is referred to as _____ metal.
   a. adder
   b. flux
   c. brazing
   d. filler
   e. none of the above

2. A _____ is used to indicate the type of weld to be used.
   a. welding symbol
   b. weld symbol
   c. callout
   d. local note
   e. feature control frame

3. The most common form of gas welding is _____ welding.
   a. gas tungsten
   b. oxyhydrogen
   c. oxyacetylene
   d. pressure gas
   e. inert gas

4. Gas welding is normally used for
   a. automated welding systems.
   b. structural welding.
   c. field welds.
   d. mass production.
   e. repair and maintenance.

5. One of the commonly used welds is the _____ weld.
   a. spot
   b. backing
   c. groove
   d. fillet
   e. seam

6. A specific welding or allied process may be specified in
   a. the body of the welding symbol.
   b. the tail of the welding symbol.
   c. a general note.
   d. a local note.
   e. a process note.

7. The terms *far side* and *near side*
   a. are no longer used.
   b. indicate weld location.
   c. are synonymous with *arrow side* and *other side.*
   d. are used in local notes only.
   e. are used in specifying field gas welds.

Test

8. Finishing of welds, other than cleaning, is indicated
   a. by a contour symbol.
   b. by a finish symbol.
   c. by a callout.
   d. in the body of the welding symbol.
   e. in the tail of the welding symbol.

9. The center-to-center spacing of intermittent fillet welding is referred to as
   a. jump.
   b. pitch.
   c. spacing.
   d. gap.
   e. offset.

10. The root opening in groove weld design is to permit or allow
    a. the flow of metal.
    b. expansion during the welding process.
    c. the escape of accumulated gases.
    d. access by the welding electrode.
    e. easier access for inspection.

11. The quality of a weld is affected by the
    a. type of weld.
    b. design of the weld.
    c. preparation of the weld area.
    d. all of the above.
    e. none of the above.

12. A field weld is
    a. not made at the place of initial construction or fabrication.
    b. made at the place of initial construction or fabrication.
    c. made in an open fabrication area.
    d. specified on site to adjust for local conditions.
    e. a partial weld for repair purposes.

Test   Chapter 19

Multiple Choice   Circle the answer that *best* fits the question.

1. The purpose of design is to create products that meet _____ requirements.
   a. functional
   b. economic
   c. aesthetic
   d. all of the above
   e. none of the above

2. The first step in design is to
   a. create rough sketches.
   b. define end use requirements.
   c. write meaningful specifications.
   d. construct a prototype of mock-up.
   e. create detail drawings of the proposal.

3. Noncritical dimensions on design drawings should be
   a. marked as reference dimensions.
   b. marked as nominal dimensions.
   c. not toleranced.
   d. not dimensioned.
   e. shown as limit dimensions.

4. An assembly is a(n)
   a. single, completed part.
   b. two or more parts that are joined.
   c. two or more parts that are part of a more complex product.
   d. any of the above.
   e. none of the above.

5. A prototype is a
   a. sample product.
   b. half-scale mock-up.
   c. model.
   d. computer simulation and rendering of a product.
   e. none of the above.

6. Determination of end-use requirements for a new product is ascertained through
   a. market analysis.
   b. surveys.
   c. examination of competitive products.
   d. testing.
   e. all of the above.

7. If dimensional tolerances are too tight
   a. assembly problems can occur.
   b. the manufacturing processes will be difficult to control.
   c. tooling costs will be too high.
   d. the product may not be economically competitive.
   e. all of the above.

8. An example of semipermanent attachment would be
   a. riveting.
   b. crimping.
   c. studs.
   d. soft soldering.
   e. none of the above.

9. Individual fasteners in assemblies should be
   a. designed out of the product.
   b. replaced by other processes.
   c. avoided if possible.
   d. kept to a minimum.
   e. all of the above.

10. To prevent loosening of fasteners
    a. always use slotted nuts.
    b. always use cotter pins.
    c. use blind-rivets if possible.
    d. all of the above.
    e. none of the above.

11. The choice of material is
    a. seldom important to the design process.
    b. a very important factor in the design process.
    c. determined by the manufacturing process.
    d. determined by the tolerances specified.
    e. usually an aesthetic or visual appearance decision.

12. The consumer is
    a. usually involved in the product design process.
    b. is very useful in testing the manufactured product.
    c. does not have sufficient knowledge to be involved in the design process.
    d. is most concerned with product appearance.
    e. is too influenced by advertising to provide useful comments on a design.

Test

Chapter 20

Multiple Choice   Circle the answer that *best* fits the question.

1. Flat belt drives offer
   a. low cost.
   b. shock absorption.
   c. resistance to abrasive environments.
   d. efficient power transmission at high speeds.
   e. all the above.

2. _____ belts are the basic workhorse of industry.
   a. Flat
   b. V-
   c. Grooved
   d. Positive-drive
   e. Crowned

3. The grooved wheels of pulleys are referred to as
   a. power wheels.
   b. sheaves.
   c. idlers.
   d. transmission wheels.
   e. shunts.

4. The toothed drive wheel used in chain drives is known as a
   a. pintle.
   b. sprocket.
   c. pinion.
   d. roller.
   e. prawl.

5. Chains are used in applications where _____ compared to the application of V-belts.
   a. slower speeds are needed
   b. faster speeds are needed
   c. less efficient power transmission is needed
   d. more efficient power transmission is needed
   e. there are severe environmental conditions

6. Chain sag should be equivalent to approximately _____ percent of the center distance.
   a. 1
   b. 2
   c. 3
   d. 5
   e. 7

7. _____ gears connect shafts whose axes intersect.
   a. Bevel
   b. Worm
   c. Spur
   d. Pinion
   e. Miter

Test _____   Chapter 20

8. The smaller of two gears in mesh is called the _____ gear.
   a. bevel
   b. worm
   c. spur
   d. pinion
   e. miter

9. The first factor to be considered in a selecting a spur gear drive is the
   a. class of service.
   b. required horsepower and service factor.
   c. size of the spur gear pinion.
   d. size of the drive spur gear.
   e. any of the above.

10. The gears used to transfer power between perpendicular shafts without a change in revolution are _____ gears.
    a. bevel
    b. worm
    c. spur
    d. pinion
    e. miter

11. The revolutions of a gear can be transferred into linear motion by a
    a. prawl.
    b. idler.
    c. spur gear.
    d. pinion.
    e. rack.

12. Pinion gears are often made of stronger material because they
    a. are smaller.
    b. are larger.
    c. have more tooth contact.
    d. have less tooth contact.
    e. run at variable speeds.

13. Detailed representation of gear teeth may be used for gears shown in _____ drawings.
    a. detail
    b. assembly
    c. fabrication
    d. any of the above
    e. none of the above

14. The maximum speed for the operation of a gear drive is usually _____ a chain drive.
    a. lower than
    b. higher than
    c. equal to
    d. not compared to
    e. variable when compared to

Test

Multiple Choice   Circle the answer that *best* fits the question.

1. The two main types of couplings are
   a. permanent and semi-permanent.
   b. fixed and flexible.
   c. semi-permanent and clutches.
   d. closed and open faced.
   e. permanent and clutches.

2. _____ are used to transmit power around corners.
   a. Solid couplings
   b. Flexible couplings
   c. Universal couplings
   d. Flexible shafts
   e. None of the above

3. The type of universal coupling used in automotive systems is known as a
   a. Hotchkiss drive.
   b. Hook's joint.
   c. Peabody joint.
   d. ball drive.
   e. Harrison joint.

4. A plain bearing works by
   a. sliding action.
   b. using spherical ball bearings.
   c. using cylindrical ball bearings.
   d. using tapered or conical ball bearings.
   e. hydrostatic action.

5. The type of bearing used to support axial loads is
   a. journal.
   b. thrust.
   c. plain.
   d. self-aligning.
   e. sleeve.

6. Ball, roller, and needle bearings are classified as _____ bearings.
   a. antifriction
   b. combined load
   c. contact
   d. thrust
   e. self-aligning

7. When a bearing is ordered, the _____ are specified
   a. standard part number and manufacturer
   b. classification, rating, and size
   c. classification, series, and size
   d. series, classification, and rating
   e. series, type, and size

8. The function of a seal is to
   a. protect the bearing against contamination.
   b. retain lubricants.
   c. avoid the drying or evaporation of the lubricant.
   d. all of the above.
   e. none of the above.

9. A _____ is used to create and maintain a tight seal between separate members
   of a mechanical assembly.
   a. sleeve
   b. gasket
   c. seal
   d. retaining ring
   e. any of the above

10. _____ is a semisolid used a lubricant.
    a. Gel
    b. Bearing paste
    c. Grease
    d. Hydrodynamic gel
    e. Oil

11. Pictorial representations of bearings are
    a. used chiefly in catalogs.
    b. restricted to assembly drawings.
    c. used to clarify the type of bearing specified.
    d. used more frequently than schematic representations.
    e. used more frequently than simplified representations.

12. An example of an application of a split ring seal would be a
    a. shaft and sleeve.
    b. high-speed motor.
    c. clutch assembly.
    d. rotating plate.
    e. piston ring.

13. An O-ring is an example of a(n)
    a. shaft sealing element.
    b. bellows-type element.
    c. elimination element.
    d. exclusion element.
    e. none of the above.

Test

Chapter 22

Multiple Choice

Circle the answer that *best* fits the question.

1. The shape of a cam is determined by the
   a. required motion of the follower.
   b. rotational velocity of the cam.
   c. type of cam used.
   d. four-bar linkage.
   e. type of follower used.

2. The most popular type of cam is the
   a. plate cam.
   b. face cam.
   c. drum cam.
   d. conjugate cam.
   e. index cam.

3. The area of a cam displacement diagram where the follower does not rise or fall is called the _____.
   a. straight-line motion.
   b. dwell.
   c. idle.
   d. retarded motion.
   e. none of the above.

4. The follower motion that is commonly referred to as constant acceleration is
   a. parabolic.
   b. harmonic.
   c. cycloidal.
   d. modified trapezoidal.
   c. modified sine.

5. _____ motion is also referred to as crank motion.
   a. Parabolic
   b. Harmonic
   c. Cycloidal
   d. Modified trapezoidal
   e. Modified sine

6. _____ motion is noted for jerk-free motion.
   a. Parabolic
   b. Harmonic
   c. Cycloidal
   d. Modified trapezoidal
   e. Modified sine

7. _____ motion is also known as retarded motion.
   a. Parabolic
   b. Harmonic
   c. Cycloidal
   d. Modified trapezoidal
   e. Modified sine

Test _____   Chapter 22

8. Most cam displacement diagrams have cam displacements of how many degrees?
   a. 90°
   b. 180°
   c. 270°
   d. 360°
   e. 5° increments

9. The smallest circle that can be drawn on a plate cam profile is called
   a. displacement.
   b. prime.
   c. pitch.
   d. base.
   e. trace.

10. The _____ cam is used to produce an indexing motion.
    a. Geneva
    b. conjugate
    c. yoke-type
    d. Hopkins
    e. positive-motion

11. Ratchet wheels are usually used in combination with
    a. followers.
    b. clutches.
    c. indexing yokes.
    d. prawls.
    e. linkages.

12. The groove in a face cam is developed
    a. as a surface development.
    b. using a linkage diagram.
    c. using a displacement diagram.
    d. any of the above.
    e. none of the above.

13. An actuator usually produces a _____ motion.
    a. straight-line
    b. angular
    c. compound
    d. parabolic
    e. curved

14. The relationship of the movements of various machine members can be
    conveniently documented by
    a. showing the motion path of the mechanism.
    b. drawing the extreme positions of the mechanism.
    c. calculating the indicator path of the combined motion.
    d. constructing a displacement diagram.
    e. constructing a timing diagram.

Test

Chapter 23

Multiple Choice

Circle the answer that *best* fits the question.

1. A surface development drawing is sometimes referred to as a
   a. sheet-metal drawing.
   b. pattern drawing.
   c. layout drawing.
   d. box drawing.
   e. all of the above.

2. The line common to both surfaces when two surfaces meet is known as
   a. an edge line.
   b. a junction line.
   c. the line of intersection.
   d. the line of development.
   e. the bend line.

3. Sheet metal thickness for thin sheets of less than 6 mm is referred to by
   a. whole number increments.
   b. decimal inches.
   c. millimeters.
   d. gage numbers.
   e. gage letters.

4. If a thin piece of material can be wrapped smoothly about the surface of an object, it is said to be
   a. approximated.
   b. warped.
   c. uniform.
   d. developable.
   e. none of the above.

5. _____ is the development of an object that has surfaces on a flat plane of projection.
   a. Pattern development
   b. Parallel line development
   c. Radial line development
   d. Straight-line development
   e. Development by triangulation

6. _____ would be used to develop the surface of a cylindrically shaped object.
   a. Pattern development
   b. Parallel line development
   c. Radial line development
   d. Straight-line development
   e. Development by triangulation

7. _____ would be used to develop the surface of a cone.
   a. Pattern development
   b. Parallel line development
   c. Radial line development
   d. Straight-line development
   e. Development by triangulation

8. Nondevelopable surfaces can be approximated by a series of
   a. rectangular surfaces.
   b. triangular surfaces.
   c. alternating rectangular and triangular surfaces.
   d. mixed polygons.
   e. any of the above.

9. An example of a nondevelopable surface would be
   a. either element of an intersecting cylinder and a rectangular form.
   b. an element of two intersecting cylinders.
   c. an elliptical cylinder.
   d. a truncated oblique cone.
   e. a sphere.

10. Lines of intersection are established by
    a. interpolation.
    b. approximation.
    c. triangulation.
    d. projection.
    e. none of the above.

11. The name given to a piece whose shape changes from one basic form to another
    (e.g., square to round) is
    a. transformation.
    b. transition.
    c. morphed.
    d. nondevelopable.
    e. transient.

Test

Multiple Choice   Circle the answer that *best* fits the question.

1. Parts used to join pipes are referred to as
   a. joints.
   b. connectors.
   c. vavles.
   d. fittings.
   e. any of the above.

2. A pipe's schedule number indicates the pipe's
   a. weight per linear foot.
   b. wall thickness.
   c. outer diameter.
   d. inner diameter.
   e. sequence in the installation.

3. Flanged fittings are connected by
   a. nuts and bolts.
   b. being screwed together.
   c. welding.
   d. spring clamps.
   e. twist-lock mechanisms.

4. The material in a pipe that would be used when there is an expectation of vibration and misalignment is
   a. aluminum.
   b. wrought iron.
   c. plastic.
   d. seamless brass.
   e. copper.

5. A high-pressure pipe would have _____ fittings.
   a. flanged
   b. screwed
   c. welded
   d. forced fit
   e. impregnated

6. The major advantage of flanges is
   a. quick disassembly.
   b. quick assembly.
   c. resistance to high pressure.
   d. resistance to vibration.
   e. reduced space requirements.

7. The type of valve designed to permit flow in one direction only is
   a. globe.
   b. check.
   c. gate.
   d. restriction.
   e. control.

Test _____   Chapter 24

8. Most pipe drawings are
   a. single-line.
   b. double-line.
   c. isometric.
   d. detailed.
   e. schematic.

9. A detachable connection would be shown as a _____ on a pipe drawing.
   a. heavy dot
   b. thin cross
   c. thick cross
   d. thin line
   e. thick line

10. Adjoining apparatus such as tanks are
    a. shown as a dashed line outline.
    b. shown as a phantom line outline.
    c. shown as a schematic symbol.
    d. indicated by their centerlines.
    e. not shown.

11. On pipe drawings, pipe lengths are
    a. shown in inches and decimal inches.
    b. shown in feet and fraction inches.
    c. dimensioned and toleranced.
    d. not shown.
    e. based on standard lengths.

12. On isometric pipe drawings, dimensioning is
    a. omitted.
    b. aligned.
    c. unidirectional.
    d. indicated in the item list.
    e. none of the above.

13. The direction of flow in a pipe in a pipe drawing is
    a. indicated by an arrow adjacent to the line.
    b. indicated by an arrowhead on the line.
    c. labeled as to destination ID.
    d. labeled as from-to.
    e. labeled as to-from.

14. Flanged pipes are
    a. easier to insulate.
    b. have less mass.
    c. resistant to high temperatures and pressures.
    d. all the above.
    e. none of the above.

Test   Chapter 25

Multiple Choice   Circle the answer that *best* fits the question.

1. Steel, as shipped from the mill, is referred to as
   a. plain material.
   b. base material.
   c. shapes.
   d. coils.
   e. sections.

2. The tolerance that indicates permissible deviation from published contours and dimensions is
   a. shop.
   b. field.
   c. finished.
   d. mill.
   e. fabrication.

3. Detail drawings depicting individual steel members are called
   a. field drawings.
   b. fabrication drawings.
   c. shop drawings.
   d. setup drawings.
   e. installation drawings.

4. _____ dimensioning is used on structural drawings.
   a. Unidirectional
   b. Tabular
   c. Baseline
   d. Aligned
   e. None of the above

5. Elevation detail dimensions are known as
   a. levels.
   b. heights.
   c. details.
   d. setups.
   e. shop callouts.

6. A _____ is a convenient erection unit.
   a. beam
   b. column
   c. member
   d. support
   e. bracket

7. Bolt pitch is the
   a. stagger offset.
   b. edge to bolt row distance.
   c. distance between bolt holes.
   d. distance between bolt rows.
   e. row to row distance.

Test

Chapter 25

8. A member's attachment shape and means of fastening is called
   a. the connection plate.
   b. the attachment plate.
   c. the seat.
   d. stringer.
   e. none of the above.

9. In structural drafting detail drawings, fillets are
   a. shown and dimensioned.
   b. shown and dimensioned by local note.
   c. shown but not dimensioned.
   d. dimensioned by general note but not shown.
   e. usually omitted.

10. Section views in structural drafting
    a. include hidden lines in the view.
    b. use intermittent hatching.
    c. never indicate the material by using a pattern hatch.
    d. omit the cross hatching.
    e. none of the above.

11. Dimensions of beams in structural drawings are
    a. centerline to centerline.
    b. edge to edge.
    c. edge to centerline.
    d. omitted.
    e. none of the above.

12. The calculated mass of pieces is used to
    a. calculate the load on erection equipment.
    b. form the basis for making up loads.
    c. form the basis of payment for shipping.
    d. determine the type of transport equipment required.
    e. all of the above.

Test

Multiple Choice   Circle the answer that *best* fits the question.

1. A device that holds the workpiece and locates the path of the tool is the
   a. drill template.
   b. fixture.
   c. workpiece support.
   d. jig.
   e. any of the above.

2. The simplest tools used to locate holes for drilling are
   a. drill bushings.
   b. setup fixtures.
   c. alignment tools.
   d. open jigs.
   e. press-fit bushings.

3. The life of the average drill bushing is _____ pieces.
   a. less than 500
   b. 500–1,000
   c. 5,000–10,000
   d. 50,000–100,000
   e. 500,000–1,000,000

4. The small pieces of metal ejected as a result of a machining process are called
   a. burrs.
   b. chips.
   c. shards.
   d. chaff.
   e. none of the above.

5. Stop pins are also known as
   a. dowels.
   b. stop pads.
   c. plugs.
   d. all of the above.
   e. none of the above.

6. The preferred head type for cab screws used in jigs is
   a. flat.
   b. oval.
   c. round.
   d. fillister.
   e. hexagon.

7. The most common type of fixture used is a _____ fixture.
   a. milling
   b. drilling
   c. reaming
   d. broaching
   e. deburring

Test

8. When more than one hole is to be drilled in a part, a _____ is inserted after the first hole is drilled.
   a. locking pin
   b. alignment pin
   c. dowel
   d. liner
   e. shoulder pin

9. Rapid methods of clamping usually involve
   a. screw mechanisms.
   b. cam mechanisms.
   c. a pneumatic actuator.
   d. a hydraulic actuator.
   e. an interlock mechanism.

10. Drawings of jigs are usually based on a(n) _____ drawing of the workpiece.
    a. detail
    b. assembly
    c. process
    d. production
    e. any of the above

11. The first factor to consider in the design of a jig is the
    a. machining operation.
    b. number of parts to be produced.
    c. accuracy required.
    d. stage of the component.
    e. none of the above.

12. The frame that holds the parts of a jig assembly is referred to as a jig
    a. base.
    b. block.
    c. support.
    d. frame.
    e. body.

13. When laying out a milling fixture, the most important point to remember is to
    a. allow for chip and burr clearance.
    b. not interfere with the milling arbor.
    c. provide rapid clamping devices.
    d. make sure there is only one way to load the workpiece.
    e. maintain adequate coolant flow.

Test

Multiple Choice   Circle the answer that *best* fits the question.

1. The most fundamental type of electronics drawing is a
   a. block diagram.
   b. logic diagram.
   c. schematic diagram.
   d. connection diagram.
   e. wiring diagram.

2. Schematic diagrams are also knows as
   a. integrated diagrams.
   b. base diagrams.
   c. elementary diagrams.
   d. flow diagrams.
   e. none of the above.

3. The identifying element of graphic symbols in schematic diagrams is referred to as a(n)
   a. numerical value.
   b. local note.
   c. tag.
   d. identifier.
   e. reference designator.

4. The units part of a component value may be omitted if it is
   a. inferred from the component type.
   b. repetitive.
   c. indicated in a local note.
   d. indicated in a general note.
   e. It is never omitted.

5. Several wires close together are referred to as a
   a. cable.
   b. conduit.
   c. bundlc.
   d. highway.
   e. network.

6. If a component is inserted into a socket, its identifier is prefixed by a(n)
   a. /
   b. *
   c. S
   d. W
   e. X

7. The piece of copper that connects lands on a printed circuit board is know as a
   a. trace.
   b. lead.
   c. wire.
   d. flow line.
   e. all of the above.

Test

8. Most electronic drawings are produced
   a. using CAD.
   b. on polyester film.
   c. at an enlarged scale.
   d. using color.
   e. all of the above.

9. The point-to-point connection of lands on a PCB when using a CAD system is referred to as a
   a. connection diagram.
   b. crossover diagram.
   c. bundle.
   d. rat's nest.
   e. none of the above.

10. Schematic drawings
    a. are drawn half scale.
    b. are drawn double scale.
    c. are drawn full scale.
    d. are not drawn to scale.
    e. have no scale.

11. The shape of symbols in schematic diagrams indicate
    a. their size.
    b. their function.
    c. their power consumption.
    d. the type of component packaging.
    e. all of the above.

12. Blocks in block diagrams that are for alternative or future components should be drawn with
    a. broken lines.
    b. phantom lines.
    c. hidden lines.
    d. a type of drafting film.
    e. any of the above.

Chapter 1	Chapter 4	Chapter 7
1. c	1. c	1. a
2. e	2. d	2. c
3. b	3. a	3. b
4. a	4. a	4. c
5. d	5. b	5. d
6. e	6. e	6. a
7. b	7. a	7. d
8. c	8. d	8. a
9. d	9. b	9. a
10. a	10. d	10. d
11. b	11. b	11. b
12. b	12. a	12. b
13. c	13. e	
14. e	14. d	

Chapter 2	Chapter 5	Chapter 8
1. d	1. b	1. b
2. a	2. d	2. d
3. e	3. e	3. a
4. b	4. a	4. c
5. c	5. b	5. a
6. b	6. a	6. e
7. e	7. d	7. b
8. b	8. b	8. c
9. b	9. c	9. a
10. e	10. e	10. e
11. c	11. d	11. c
12. a	12. a	12. b
13. d	13. e	13. e
14. e	14. d	14. d

Chapter 3	Chapter 6	Chapter 9
1. a	1. c	1. e
2. e	2. e	2. b
3. c	3. b	3. b
4. d	4. a	4. a
5. e	5. e	5. c
6. b	6. d	6. b
7. e	7. b	7. e
8. d	8. e	8. d
9. b	9. e	9. e
10. a	10. c	10. a
11. d	11. d	11. d
12. b	12. a	12. c
13. c	13. b	13. c
14. a	14. e	14. a

Chapter 10
1. c
2. e
3. a
4. b
5. b
6. c
7. d
8. a
9. e
10. c
11. d
12. d
13. a
14. b

Chapter 11
1. d
2. a
3. b
4. b
5. b
6. e
7. d
8. d
9. a
10. e
11. c
12. e
13. a

Chapter 12
1. b
2. a
3. b
4. c
5. d
6. d
7. b
8. e
9. a
10. d
11. e
12. b

Chapter 13
1. d
2. e
3. b
4. d
5. c
6. a
7. b
8. e
9. b
10. a
11. a
12. e

Chapter 14
1. b
2. c
3. e
4. d
5. b
6. a
7. b
8. d
9. b
10. c
11. b
12. e
13. b

Chapter 15
1. a
2. c
3. b
4. d
5. c
6. e
7. c
8. d
9. e
10. d
11. b
12. d
13. a

Chapter 16
1. d
2. b
3. b
4. a
5. a
6. b
7. d
8. b
9. c
10. c
11. a
12. b
13. c
14. a

Chapter 17
1. a
2. c
3. b
4. e
5. b
6. d
7. c
8. e
9. a
10. e
11. c
12. e

Chapter 18
1. d
2. b
3. c
4. e
5. d
6. b
7. a
8. a
9. b
10. d
11. d
12. a

Chapter 19
1. d
2. b
3. a
4. b
5. a
6. e
7. e
8. c
9. e
10. e
11. b
12. a

Chapter 20
1. e
2. b
3. b
4. b
5. a
6. b
7. a
8. e
9. a
10. d
11. e
12. c
13. b
14. a

Chapter 21
1. e
2. d
3. a
4. a
5. b
6. a
7. e
8. d
9. b
10. c
11. a
12. e
13. a

Chapter 22
1. a
2. a
3. b
4. a
5. b
6. c
7. a
8. d
9. d
10. a
11. d
12. c
13. a
14. e

Chapter 23
1. b
2. c
3. d
4. d
5. d
6. b
7. c
8. b
9. e
10. d
11. b

Chapter 24
1. d
2. b
3. a
4. e
5. c
6. a
7. b
8. a
9. e
10. b
11. d
12. c
13. b
14. e

Chapter 25
1. a
2. d
3. c
4. d
5. a
6. a
7. c
8. a
9. e
10. d
11. a
12. e

Chapter 26
1. d
2. d
3. c
4. b
5. a
6. d
7. a
8. a
9. b
10. a
11. a
12. e
13. b

Chapter 27

1. a
2. c
3. e
4. d
5. d
6. e
7. a
8. a
9. d
10. e
11. b
12. a

CONTENTS _____

Transparency Masters

Basic Drawing and Design

Chapter 1	1	Figure 1-1-2	327
	2	Figure 1-2-2	328
	3	Figure 1-4-8	329
	4	Figure 1-4-13	330
	5	Basic Equipment	331
Chapter 2	1	Figure 2-1-1	332
	2	Figure 2-4-2	333
Chapter 3	1	Figure 3-1-3	334
	2	Figure 3-1-5	335
	3	Figures 3-1-6, 3-1-9	336
Chapter 4	1	Figure 4-1-1	337
	2	Figure 4-1-1	338
	3	Figure 4-1-2	339
	4	Figures 4-1-7, 4-1-9	340
	5	Figure 4-2-1	341
	6	Figure 4-4-4	342
	7	Figure 4-4-11	343
	8	Figure 4-4-16	344
Chapter 5	1	Figure 5-1-1	345
	2	Figures 5-1-4, 5-1-5, 5-1-6	346
	3	Figures 5-2-1, 5-2-2	347
	4	Figure 5-2-3	348
	5	Figures 5-2-4, 5-2-5	349
	6	Figure 5-2-7	350
	7	Figures 5-3-6, 5-4-2	351
Chapter 6	1	Figure 6-1-1	352
	2	Figure 6-1-2	353
	3	Figure 6-1-3	354
	4	Figure 6-1-4	355
	5	Figure 6-1-11	356
	6	Figure 6-2-2	357
	7	Figure 6-4-1, 6-5-1	358
	8	Figure 6-7-1	359
	9	Figure 6-10-1, 6-11-1	360
	10	Figure 6-14-1	361
	11	Figure 6-15-2	362

CONTENTS

Chapter 7

1	Figures 7-1-1, 7-1-2	363
2	Figure 7-1-5	364
3	Figures 7-2-1, 7-2-2	365
4	Figure 7-3-2	366
5	Figure 7-4-1	367
6	Figure 7-5-2	368
7	Figure 7-5-7	369
8	Figure 7-6-2	370
9	Figure 7-6-3	371
10	Figures 7-6-4, 7-6-5	372
11	Figure 7-7-1	373
12	Figure 7-8-1	374
13	Figure 7-9-1	375
14	Figure 7-10-1	376
15	Figure 7-11-2	377

Chapter 8

1	Figures 8-1-1, 8-1-2	378
2	Figure 8-1-3	379
3	Figure 8-1-5	380
4	Figure 8-1-10	381
5	Figure 8-1-12	382
6	Figure 8-2-1	383
7	Figure 8-2-3	384
8	Figure 8-2-8	385
9	Figure 8-2-11, 8-2-12	386
10	Figure 8-3-1	387
11	Figures 8-3-2, 8-3-3, 8-3-4	388
12	Figures 8-4-1, 8-4-4, 8-4-5	389
13	Figure 8-4-10	390
14	Figure 8-5-3	391
15	Figure 8-5-8	392
16	Figure 8-6-1	393
17	Figure 8-6-3	394
18	Figure 8-6-6	395
19	Figure 8-7-1	396
20	Figure 8-7-3	397
21	Figure 8-7-4	398
22	Figure 8-7-8	399
23	Figure 8-7-13	400

CONTENTS

Chapter 9	1	Figure 9-1-1	.401
	2	Figure 9-1-6	.402
	3	Figure 9-2-1	.403
	4	Figure 9-3-1	.404
	5	Figure 9-4-1	.405
	6	Figure 9-6-1	.406
	7	Figure 9-7-1	.407
	8	Figure 9-7-3	.408
	9	Figure 9-8-1	.409
	10	Figures 9-9-1, 9-10-1	.410

Fasteners, Materials, and Forming Processes

Chapter 10	1	Figures 10-1-4, 10-1-5	.411
	2	Figures 10-1-7, 10-1-8, 10-1-9	.412
	3	Figure 10-1-10	.413
	4	Figures 10-1-12, 10-1-13	.414
	5	Figure 10-2-1	.415
	6	Figure 10-3-1	.416
	7	Figure 10-3-2	.417
	8	Figures 10-3-3, 10-3-4	.418
	9	Figure 10-3-12	.419
	10	Figure 10-3-18	.420
Chapter 11	1	Figure 11-1-1	.421
	2	Figures 11-1-3, 11-1-4	.422
	3	Figure 11-1-7	.423
	4	Figure 11-3-1	.424
	5	Figure 11-4-1	.425
	6	Figures 11-4-2, 11-4-8	.426
	7	Figures 11-5-2, 11-5-3	.427
	8	Figure 11-6-1	.428
	9	Figure 11-7-1	.429
Chapter 12	1	Figure 12-1-1	.430
	2	Figure 12-2-1	.431
	3	Figure 12-2-3	.432
	4	Figure 12-2-8	.433
	5	Figure 12-3-1	.434
	6	Figure 12-3-2	.435
	7	Figure Common terms and definitions	.436

CONTENTS

Chapter 13 1   Figure 13-1-2 437

             2   Figure 13-1-5 438

             3   Figure 13-1-7 439

             4   Figure 13-1-16 440

             5   Figures 13-1-17, 13-1-19 441

             6   Figure 13-1-20 442

             7   Figure 13-2-3 443

             8   Figure 13-2-8 444

             9   Figures 13-2-10, 13-2-11 445

            10   Figure 13-3-1 446

Working Drawings and Design

Chapter 14 1   Figures 14-2-5, 14-2-6 447

             2   Figures 14-2-7, 14-2-8, 14-2-9 448

             3   Figure 14-3-1 449

             4   Figure 14-3-2 450

             5   Figure 14-3-2 451

             6   Figure 14-4-1 452

             7   Figure 14-6-5 453

             8   Figure 14-7-1(A) 454

             9   Figure 14-8-1 455

Chapter 15 1   Figure 15-1-1 456

             2   Figures 15-1-2, 15-1-3 457

             3   Figure 15-1-4 458

             4   Figure 15-1-6 459

             5   Figure 15-1-9 460

             6   Figures 15-2-2, 15-2-4 461

             7   Figure 15-3-2 462

             8   Figures 15-4-3, 15-4-4 463

             9   Figure 15-5-3 464

            10   Figure 15-6-4 465

            11   Figure 15-6-6 466

            12   Figure 15-7-2 467

            13   Figure 15-7-4 468

            14   Figure 15-7-5 469

CONTENTS

Chapter 16 1 Figures 16-1-2, 16-1-3 470

2 Figures 16-1-4, 16-1-5, 16-1-6 471

3 Figures 16-1-7, 16-1-8, 16-1-9 472

4 Figure 16-1-10 473

5 Figure 16-1-14 474

6 Figure 16-2-2 475

7 Figure 16-2-4 476

8 Figure 16-4-5 477

9 Figure 16-5-2 478

10 Figure 16-5-3 479

11 Figures 16-5-4, 16-5-6 480

12 Figure 16-6-2 481

13 Figure 16-7-1 482

14 Figure 16-7-9 483

15 Figure 16-9-1 484

16 Figure 16-9-2 485

17 Figure 16-9-7 486

18 Figure 16-10-1 487

19 Figures 16-11-1, 16-11-2 488

20 Figure 16-11-13 489

21 Figures 16-13-2, 16-13-3 490

22 Figure 16-13-16 491

23 Figure 16-14-19 492

24 Figure 16-15-3 493

25 Figure 16-17-3 494

Chapter 17 1 Figure 17-1-2 495

2 Figures 17-1-3, 17-1-4 496

3 Figure 17-1-5 497

4 Figure 17-2-1 498

5 Figures 17-2-4, 17-2-5 499

Chapter 18 1 Figure 18-1-4 500

2 Figure 18-2-1 501

3 Figure 18-2-2 502

4 Figure 18-2-3 503

5 Figure 18-2-7 504

6 Figures 18-2-11, 18-2-12 505

7 Figure 18-3-5 506

8 Figure 18-4-6 507

9 Figure 18-5-1 508

10 Figure 18-5-3 509

CONTENTS

Chapter 19   1   Figure 19-2-1 .... .510

Power Transmissions

Chapter 20   1   Figure 20-1-1 .... .511
            2   Figures 20-1-3, 20-1-5 .... .512
            3   Figure 20-1-10 .... .513
            4   Figure 20-2-3 .... .514
            5   Figure 20-2-7(A) .... .515
            6   Figures 20-3-3, 20-3-4 .... .516
            7   Figure 20-3-8 .... .517
            8   Figures 20-6-2, 20-6-3 .... .518
            9   Figure 20-6-4 .... .519
          10   Figure 20-7-4 .... .520
          11   Figure 20-7-6 .... .521

Chapter 21   1   Figures 21-2-1, 21-2-2 .... .522
            2   Figures 21-3-1, 21-3-2 .... .523
            3   Figure 21-3-8 .... .524
            4   Figure 21-3-9 .... .525
            5   Figure 21-5-4 .... .526
            6   Figures 21-5-7, 21-5-8 .... .527
            7   Figures 21-6-1, 21-6-2 .... .528

Chapter 22   1   Figures 22-1-3, 22-1-4 .... .529
            2   Figure 22-1-5 .... .530
            3   Figure 22-1-8 .... .531
            4   Figures 22-1-9, 22-1-10 .... .532
            5   Figure 22-2-1 .... .533
            6   Figure 22-2-2 .... .534
            7   Figure 22-2-6 .... .535
            8   Figure 22-2-9 .... .536
            9   Figure 22-4-2 .... .537
          10   Figure 22-7-1 .... .538

CONTENTS

Special Fields of Drafting

Chapter 23	1	Figures 23-1-1, 23-1-2	.539
	2	Figure 23-2-4	.540
	3	Figure 23-3-1	.541
	4	Figure 23-3-4	.542
	5	Figure 23-4-3	.543
	6	Figure 23-5-2	.544
	7	Figure 23-6-4	.545
	8	Figure 23-8-3	.546
	9	Figure 23-9-1	.547
	10	Figure 23-9-2	.548
Chapter 24	1	Figure 24-1-2	.549
	2	Figure 24-1-6	.550
	3	Figure 24-1-11	.551
	4	Figure 24-1-12	.552
	5	Figure 24-2-2	.553
	6	Figure 24-2-6	.554
	7	Figures 24-3-3, 24-3-4	.555
Chapter 25	1	Figure 25-1-2	.556
	2	Figure 25-1-6	.557
	3	Figures 25-1-15, 25-1-16	.558
	4	Figure 25-2-5	.559
	5	Figure 25-3-11	.560
	6	Figures 25-3-13, 25-3-14	.561
	7	Figures 25-6-1, 25-6-2, 25-6-3	.562
Chapter 26	1	Figure 26-1-3	.563
	2	Figures 26-2-1, 26-2-3	.564
	3	Figure 26-2-10	.565
	4	Figure 26-3-1	.566
Chapter 27	1	Figure 27-2-1	.567
	2	Figure 27-2-4	.568
	3	Figure 27-3-1	.569
	4	Figures 27-4-1, 27-4-2, 27-4-3	.570
	5	Figure 27-5-1, 27-5-2	.571

FIGURE 1-1-2 Various fields of drafting.

TYPICAL BRANCHES OF ENGINEERING GRAPHICS	ACTIVITIES	PRODUCTS	SPECIALIZED AREAS
MECHANICAL	DESIGNING TESTING MANUFACTURING MAINTENANCE CONSTRUCTION	MATERIALS MACHINES DEVICES	POWER GENERATION TRANSPORTATION MANUFACTURING POWER SERVICES ATOMIC ENERGY MARINE VESSELS
ARCHITECTURAL	PLANNING DESIGNING SUPERVISING	BUILDINGS ENVIRONMENT LANDSCAPE	COMMERCIAL BUILDINGS RESIDENTIAL BUILDINGS INSTITUTIONAL BUILDINGS ENVIRONMENTAL SPACE FORMS
ELECTRICAL	DESIGNING DEVELOPING SUPERVISING PROGRAMMING	COMPUTERS ELECTRONICS POWER ELECTRICAL	POWER GENERATION POWER APPLICATION TRANSPORTATION ILLUMINATION INDUSTRIAL ELECTRONICS COMMUNICATIONS INSTRUMENTATION MILITARY ELECTRONICS
AEROSPACE	PLANNING DESIGNING TESTING	MISSILES PLANES SATELLITES ROCKETS	AERODYNAMICS STRUCTURAL DESIGN INSTRUMENTATION PROPULSION SYSTEMS MATERIALS RELIABILITY TESTING PRODUCTION METHODS
PIPING	DESIGNING TESTING MANUFACTURING MAINTENANCE CONSTRUCTION	BUILDINGS HYDRAULICS PNEUMATICS PIPE LINES	LIQUID TRANSPORTATION MANUFACTURING POWER SERVICES HYDRAULICS PNEUMATICS
STRUCTURAL	PLANNING DESIGNING MANUFACTURING CONSTRUCTION	MATERIALS BUILDINGS MACHINES VEHICLES BRIDGES	STRUCTURAL DESIGNS BUILDINGS PLANES SHIPS AUTOMOBILES BRIDGES
TECHNICAL ILLUSTRATING	PROMOTION DESIGNING ILLUSTRATING	CATALOGS MAGAZINES DISPLAYS	NEW PRODUCTS ASSEMBLY INSTRUCTIONS PRESENTATIONS COMMUNITY PROJECTS RENEWAL PROGRAMS

FIGURE 1-2-2 Positions within the drafting office.

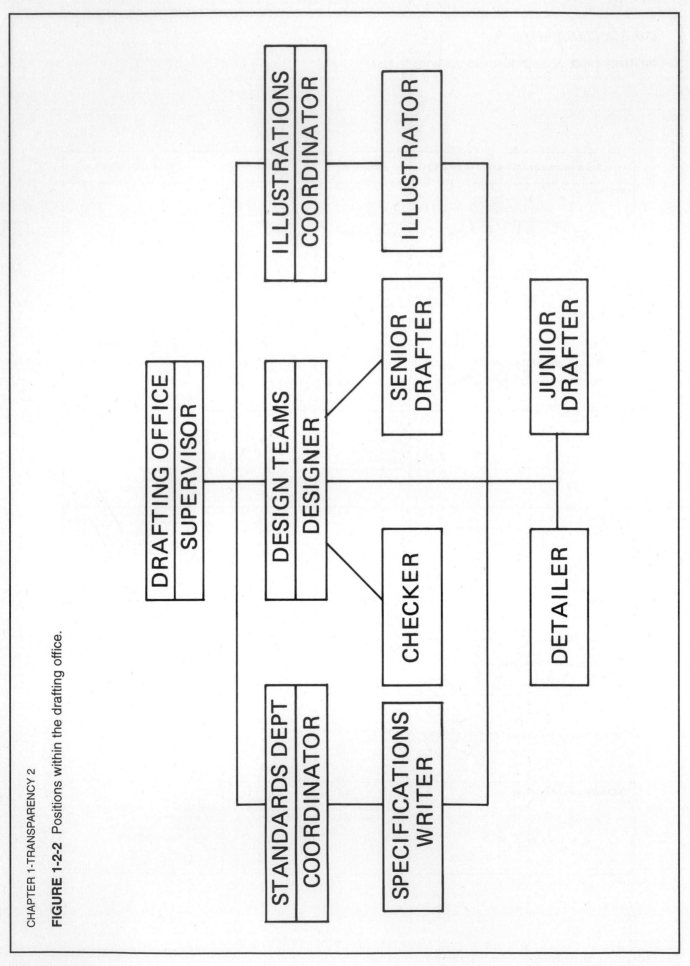

FIGURE 1-4-8 Triangles.

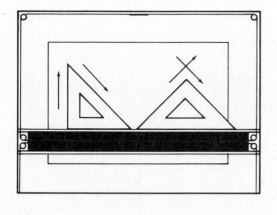

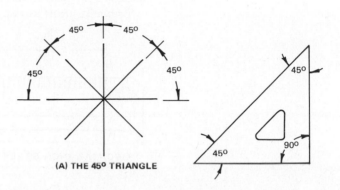

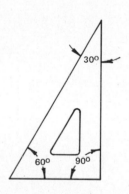

(A) THE 45° TRIANGLE

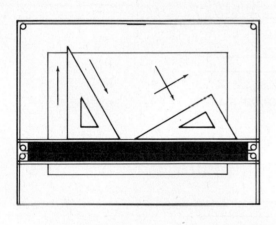

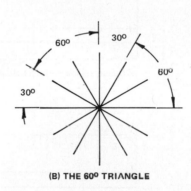

(B) THE 60° TRIANGLE

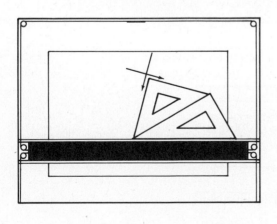

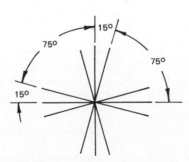

(C) THE TRIANGLES IN COMBINATION

FIGURE 1-4-13 Inch scales.

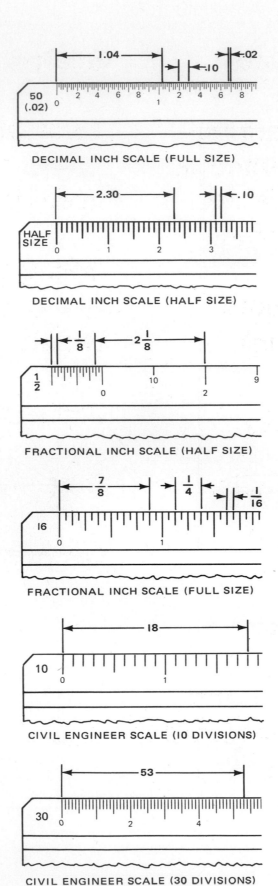

Basic Equipment.

Drawing board

T square, parallel-ruling straightedge (parallel slide), or drafting machine

Drawing sheets (paper or film)

Drafting tape

Drafting pencils

Erasers

Erasing shield

Triangles, 45° and 30/60° (not required with drafting machines)

Scales

Templates

Irregular curves

Inking pen

Brush

Protractor

Cleaning powder

Calculator

FIGURE 2-1-1 Operational flowchart of a CAD system.

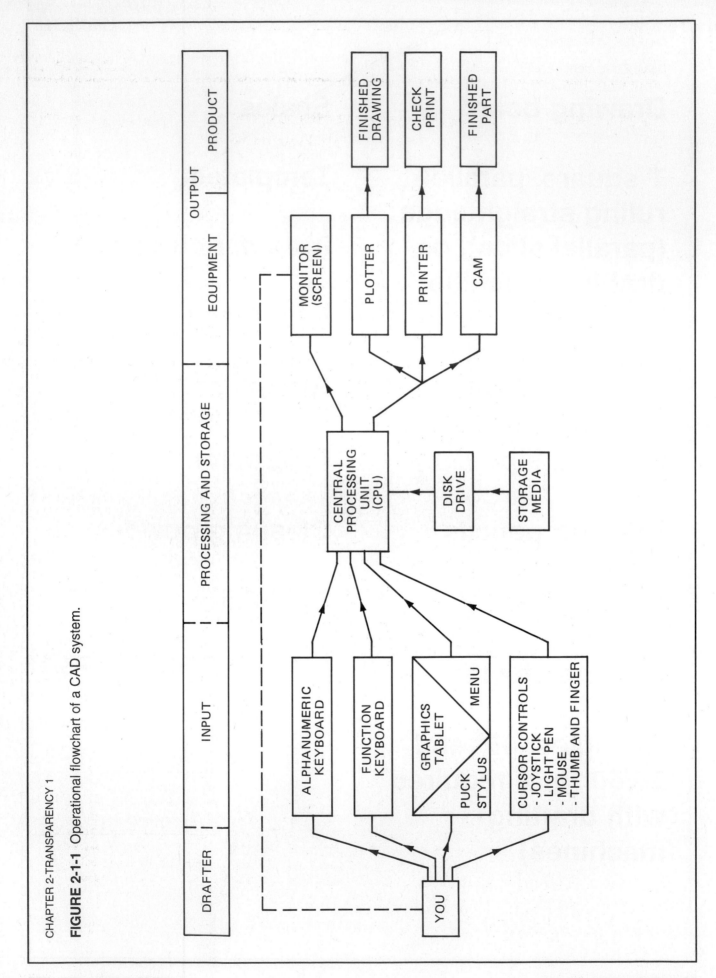

FIGURE 2-4-2 Raster display on a CRT.

FIGURE 3-1-3 Standard drawing sizes.

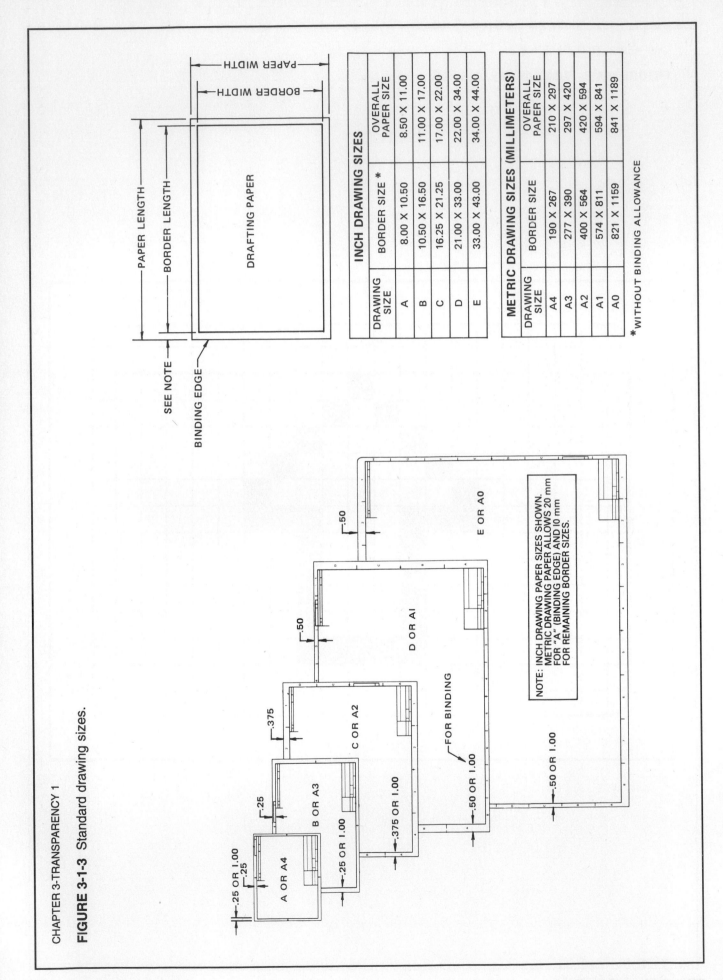

INCH DRAWING SIZES

DRAWING SIZE	BORDER SIZE *	OVERALL PAPER SIZE
A	8.00 X 10.50	8.50 X 11.00
B	10.50 X 16.50	11.00 X 17.00
C	16.25 X 21.25	17.00 X 22.00
D	21.00 X 33.00	22.00 X 34.00
E	33.00 X 43.00	34.00 X 44.00

METRIC DRAWING SIZES (MILLIMETERS)

DRAWING SIZE	BORDER SIZE	OVERALL PAPER SIZE
A4	190 X 267	210 X 297
A3	277 X 390	297 X 420
A2	400 X 564	420 X 594
A1	574 X 811	594 X 841
A0	821 X 1159	841 X 1189

*WITHOUT BINDING ALLOWANCE

NOTE: INCH DRAWING PAPER SIZES SHOWN. METRIC DRAWING PAPER ALLOWS 20 mm FOR "A" (BINDING EDGE) AND 10 mm FOR REMAINING BORDER SIZES.

FIGURE 3-1-5 Drawing format.

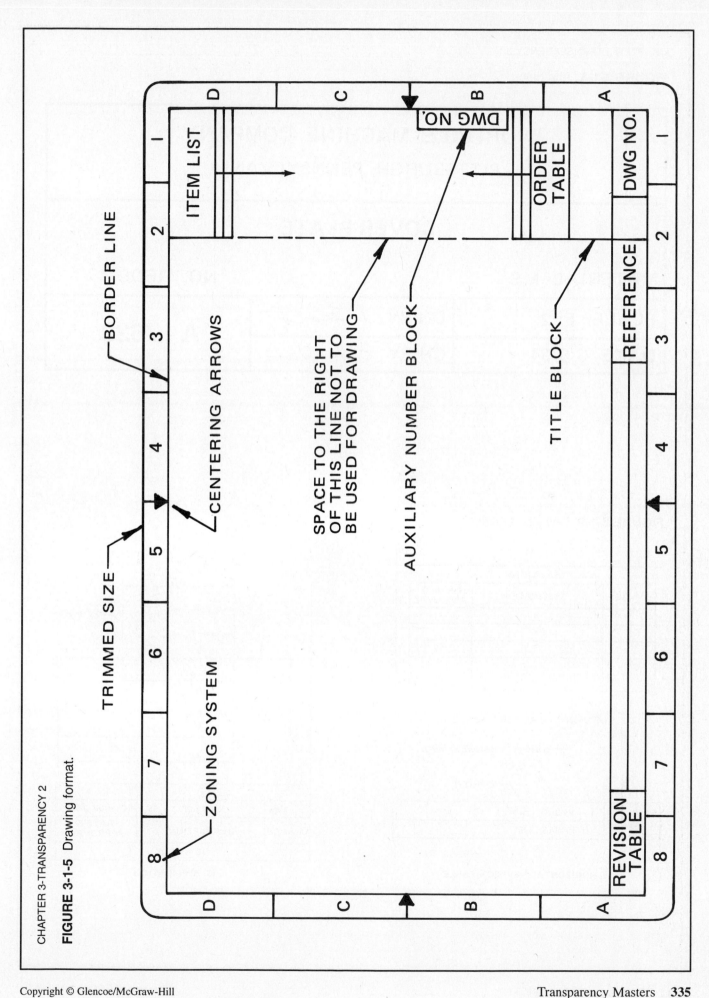

FIGURE 3-1-6 Title block.

NORDALE MACHINE COMPANY		
PITTSBURGH, PENNSYLVANIA		
COVER PLATE		
MATERIAL- MS		NO. REQD-4
SCALE- 1 : 2	DN BY *D Scott*	A - 7628
DATE- 3/6/94	CH BY *B Jensen*	

FIGURE 3-1-9 Revision tables.

REVISIONS		
SYMBOL	DESCRIPTION	DATE & APPROVAL

(A) VERTICAL REVISION TABLE

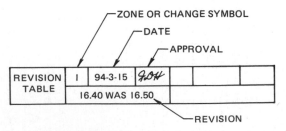

ZONE OR CHANGE SYMBOL
DATE
APPROVAL

REVISION TABLE	I	94-3-15	*HDH*			
	16.40 WAS 16.50					

REVISION

(B) HORIZONTAL REVISION TABLE

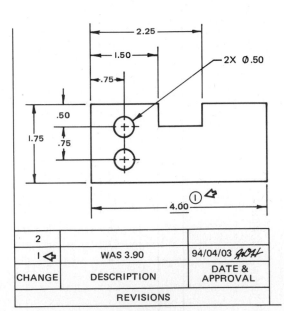

2		
I ◁	WAS 3.90	94/04/03 *HDH*
CHANGE	DESCRIPTION	DATE & APPROVAL
REVISIONS		

(C) APPLICATION

CHAPTER 4-TRANSPARENCY 1

FIGURE 4-1-1 Types of lines.

TYPE OF LINE	APPLICATION	DESCRIPTION
HIDDEN LINE THIN	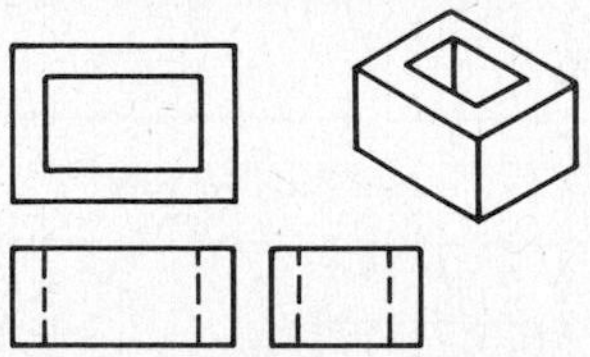	THE HIDDEN OBJECT LINE IS USED TO SHOW SURFACES, EDGES, OR CORNERS OF AN OBJECT THAT ARE HIDDEN FROM VIEW.
CENTER LINE THIN ALTERNATE LINE AND SHORT DASHES	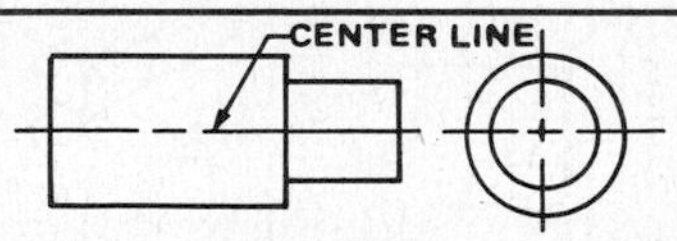	CENTER LINES ARE USED TO SHOW THE CENTER OF HOLES AND SYMMETRICAL FEATURES.
SYMMETRY LINE CENTER LINE THICK SHORT LINES	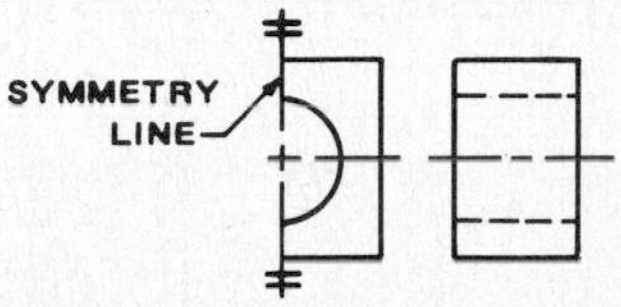	SYMMETRY LINES ARE USED WHEN PARTIAL VIEWS OF SYMMETRICAL PARTS ARE DRAWN. IT IS A CENTER LINE WITH TWO THICK SHORT PARALLEL LINES DRAWN AT RIGHT ANGLES TO IT AT BOTH ENDS.
EXTENSION AND DIMENSION LINES THIN DIMENSION LINE EXTENSION LINE	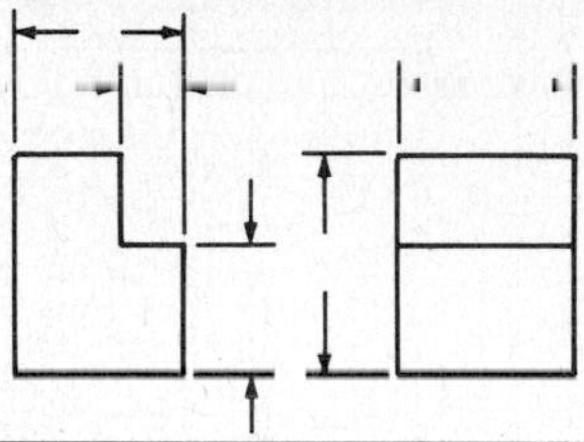	EXTENSION AND DIMENSION LINES ARE USED WHEN DIMENSIONING AN OBJECT.
LEADERS ARROW DOT THIN	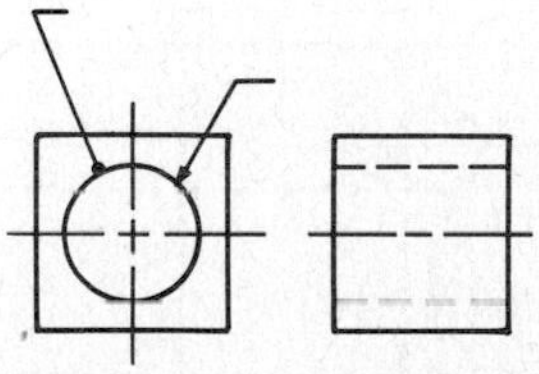	LEADERS ARE USED TO INDICATE THE PART OF THE DRAWING TO WHICH A NOTE REFERS. ARROWHEADS TOUCH THE OBJECT LINES WHILE THE DOT RESTS ON A SURFACE.
BREAK LINES THIN LONG BREAK THICK SHORT BREAK	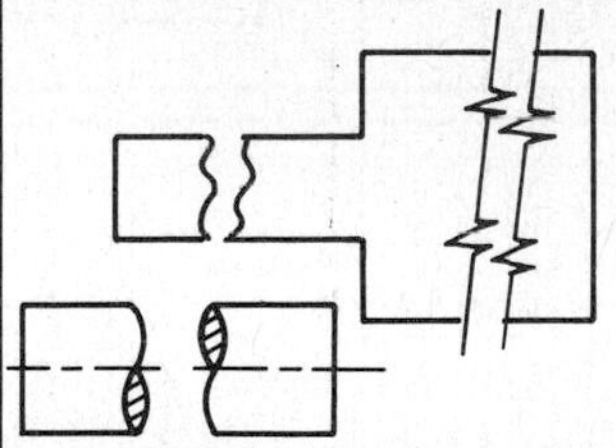	BREAK LINES ARE USED WHEN IT IS DESIRABLE TO SHORTEN THE VIEW OF A LONG PART.
CUTTING-PLANE LINE THICK OR	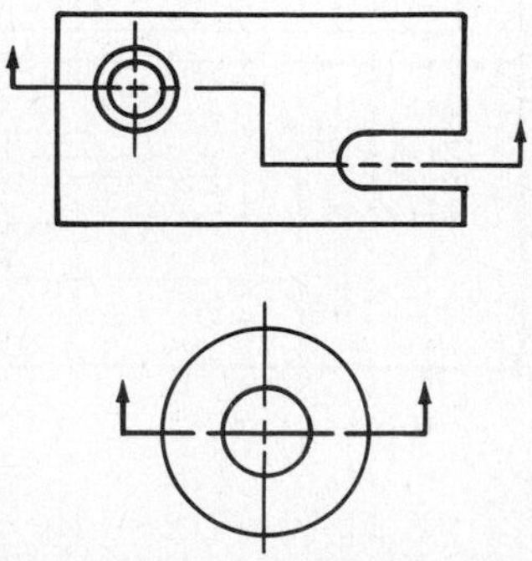	THE CUTTING-PLANE LINE IS USED TO DESIGNATE WHERE AN IMAGINARY CUTTING TOOK PLACE.

FIGURE 4-1-1 Types of lines (continued).

TYPE OF LINE	APPLICATION	DESCRIPTION
VISIBLE LINE THICK SECTION LINES THIN LINES		THE VISIBLE LINE IS USED TO INDICATE ALL VISIBLE EDGES OF AN OBJECT. THEY SHOULD STAND OUT CLEARLY IN CONTRAST TO OTHER LINES SO THAT THE SHAPE OF AN OBJECT IS APPARENT TO THE EYE. SECTION LINING IS USED TO INDICATE THE SURFACE IN THE SECTION VIEW IMAGINED TO HAVE BEEN CUT ALONG THE CUTTING-PLANE LINE.
VIEWING-PLANE LINE THICK OR		THE VIEWING-PLANE LINE IS USED TO INDICATE DIRECTION OF SIGHT WHEN A PARTIAL VIEW IS USED.
PHANTOM LINE THIN		PHANTOM LINES ARE USED TO INDICATE ALTERNATE POSITION OF MOVING PARTS, ADJACENT POSITION OF MOVING PARTS, ADJACENT POSITION OF RELATED PARTS, AND REPETITIVE DETAIL.
STITCH LINE THIN OR SMALL DOTS		STITCH LINES ARE USED FOR INDICATING A SEWING OR STITCHING PROCESS.
CHAIN LINE THICK		CHAIN LINES ARE USED TO INDICATE THAT A SURFACE OR ZONE IS TO RECEIVE ADDITIONAL TREATMENT OR CONSIDERATIONS.

FIGURE 4-1-2 Application of lines. *(ANSI Y14.2M, 1982)*

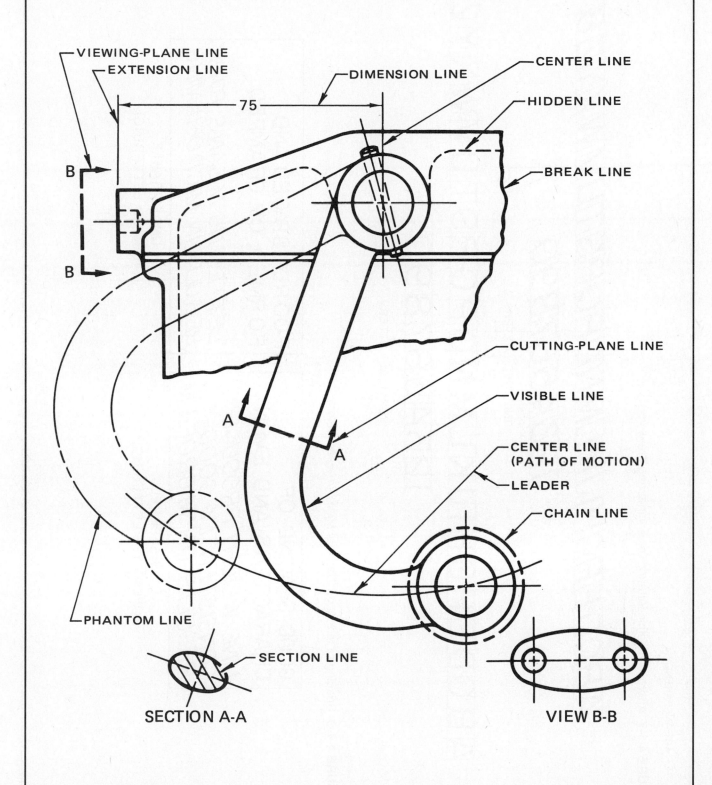

SECTION A-A

VIEW B-B

CHAPTER 4-TRANSPARENCY 4

FIGURE 4-1-7 Approved Gothic lettering for engineering drawings.

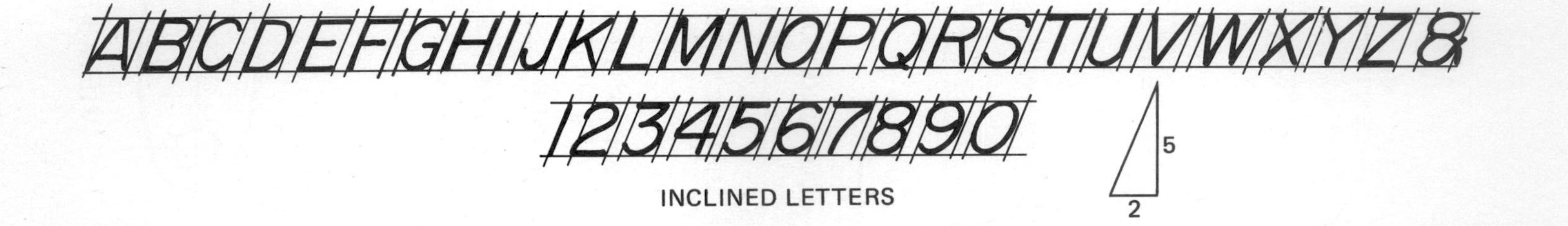
ABCDEFGHIJKLMNOPQRSTUVWXYZ&
1234567890
5
2
INCLINED LETTERS
ABCDEFGHIJKLMNOPQRSTUVWXYZ&
1234567890
VERTICAL LETTERS

FIGURE 4-1-9 Spacing of lettering. *(National Microfilm Association)*

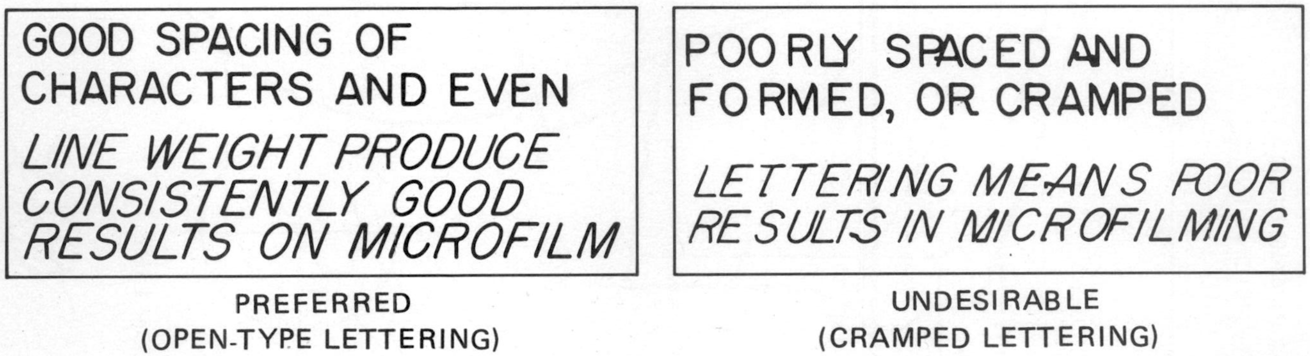
GOOD SPACING OF
CHARACTERS AND EVEN
LINE WEIGHT PRODUCE
CONSISTENTLY GOOD
RESULTS ON MICROFILM
PREFERRED
(OPEN-TYPE LETTERING)
POORLY SPACED AND
FORMED, OR CRAMPED
LETTERING MEANS POOR
RESULTS IN MICROFILMING
UNDESIRABLE
(CRAMPED LETTERING)

FIGURE 4-2-1 Center line technique.

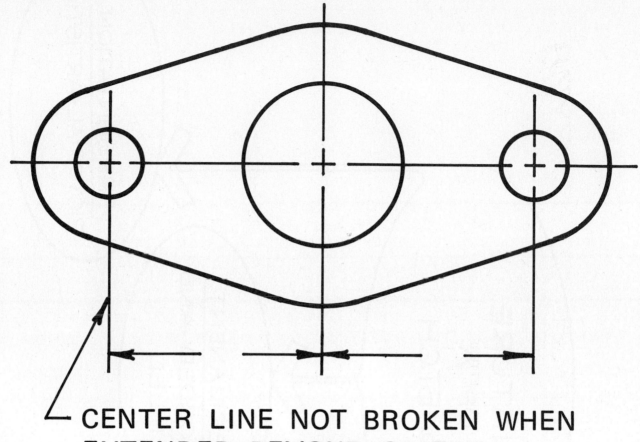

CENTER LINE NOT BROKEN WHEN
EXTENDED BEYOND OBJECT

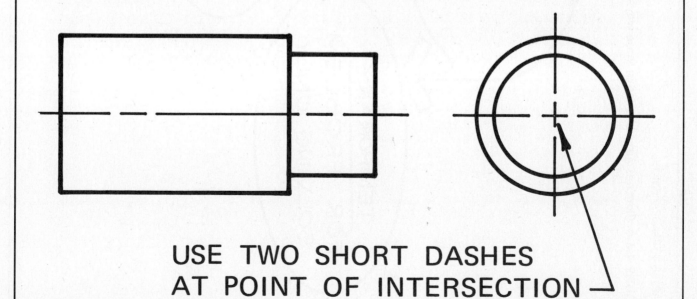

USE TWO SHORT DASHES
AT POINT OF INTERSECTION

FIGURE 4-4-4 Sketches are classified in three ways.

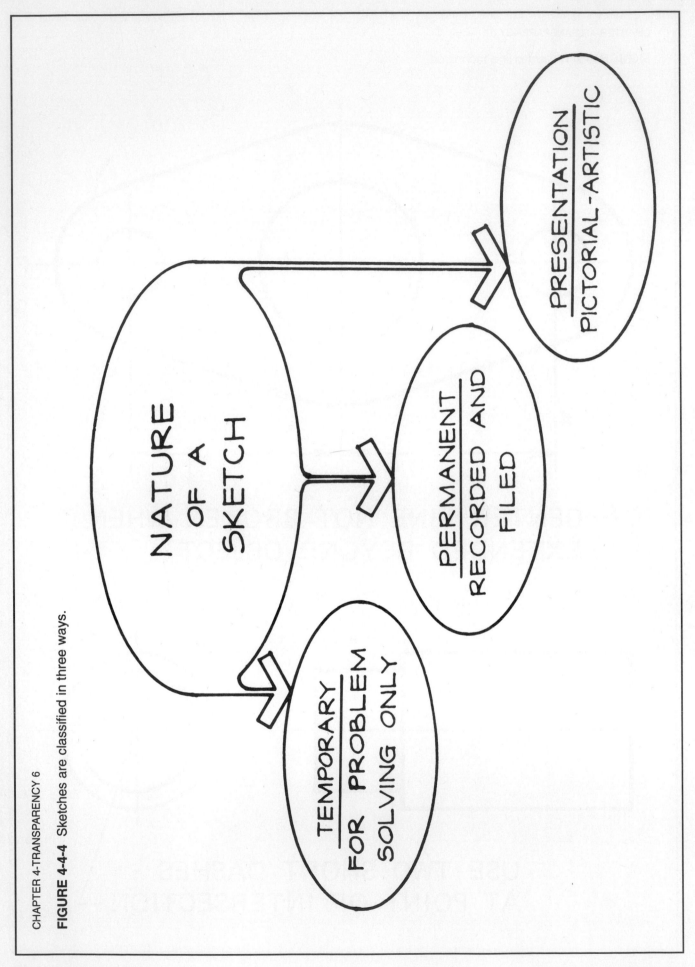

FIGURE 4-4-11 Sketching circles.

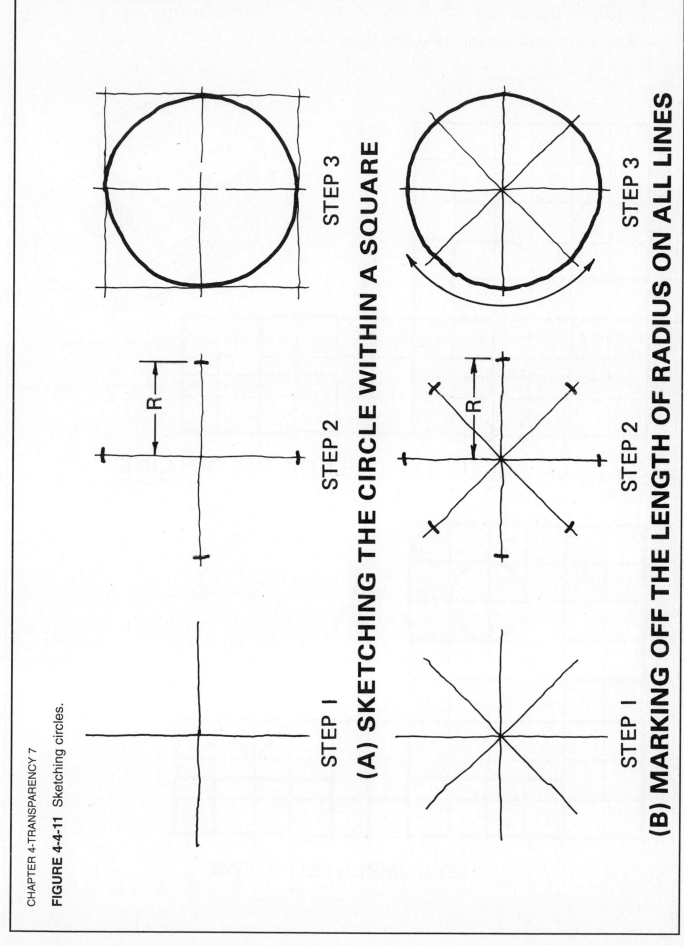

STEP 1   STEP 2   STEP 3

(A) SKETCHING THE CIRCLE WITHIN A SQUARE

STEP 1   STEP 2   STEP 3

(B) MARKING OFF THE LENGTH OF RADIUS ON ALL LINES

FIGURE 4-4-16 Usual procedure for sketching three views.

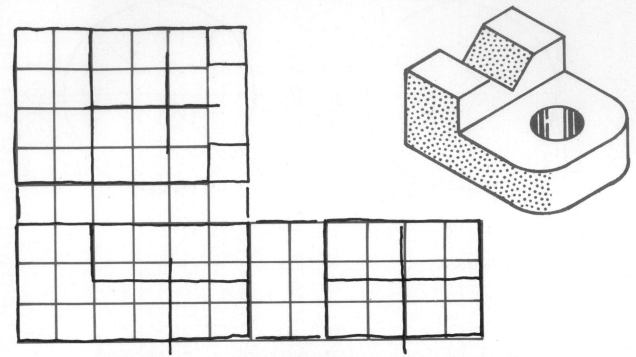

(A) SKETCHING PART OF THE OUTLINE

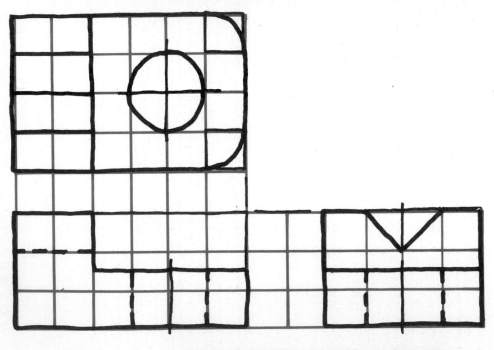

(B) COMPLETED VIEWS

FIGURE 5-1-1 Dictionary of drafting geometry.

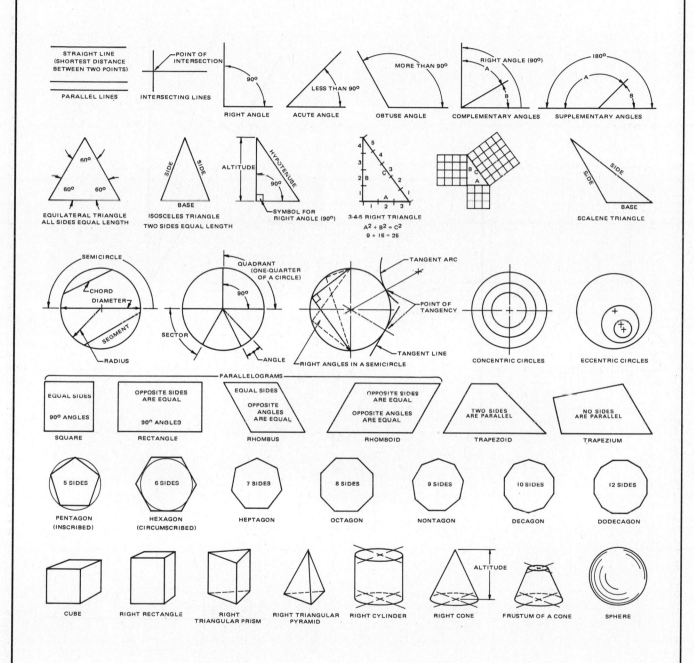

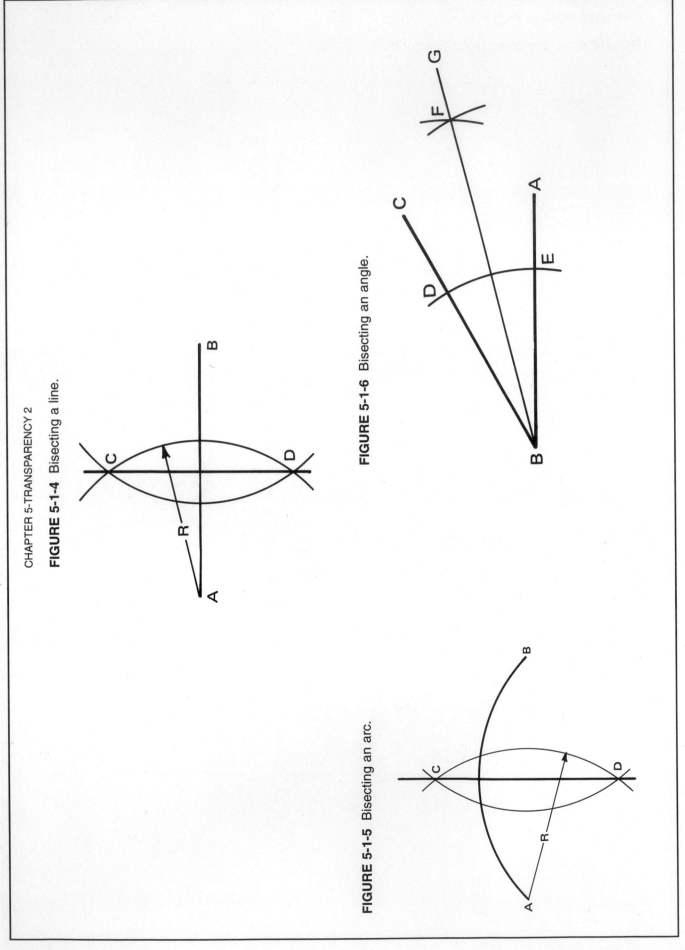

FIGURE 5-1-4 Bisecting a line.

FIGURE 5-1-5 Bisecting an arc.

FIGURE 5-1-6 Bisecting an angle.

FIGURE 5-2-1 Arc tangent to two lines at right angles to each other.

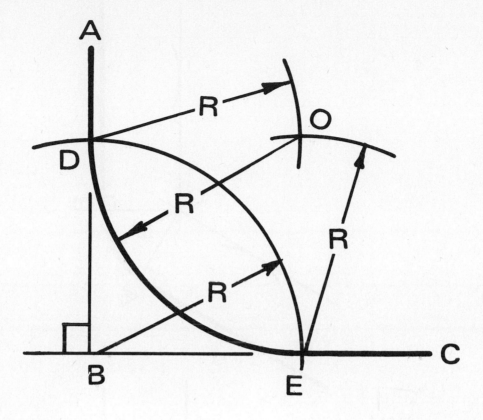

FIGURE 5-2-2 Drawing an arc tangent to the sides of an acute.

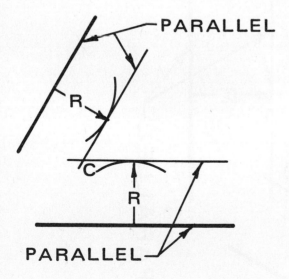

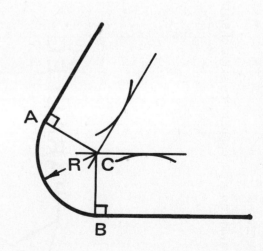

FIGURE 5-2-3 Drawing an arc tangent to the sides of an obtuse angle.

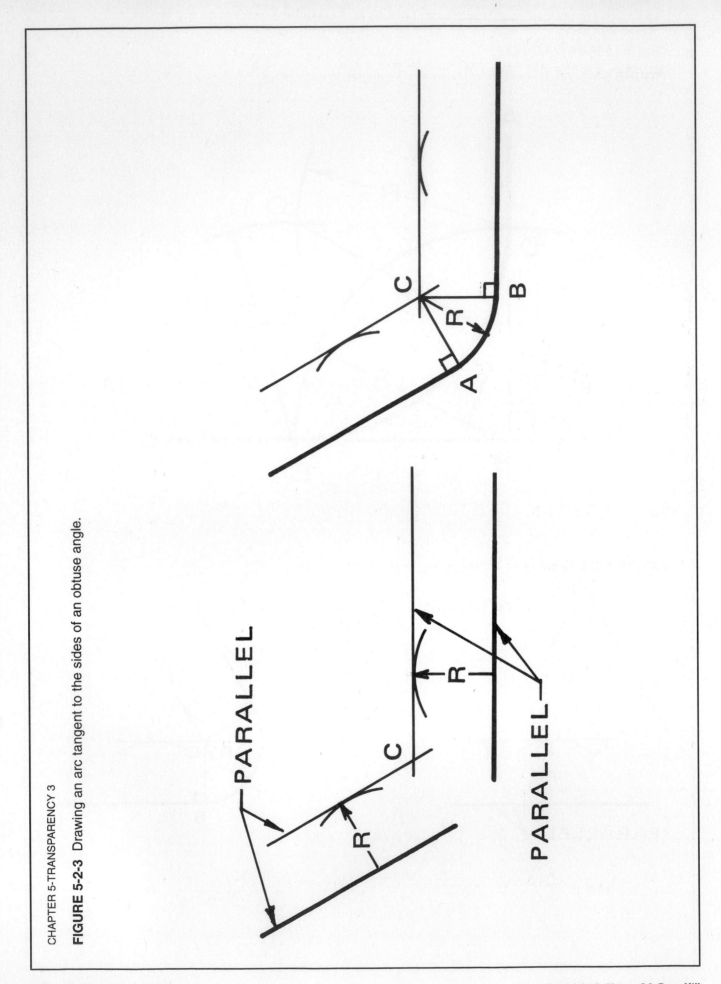

FIGURE 5-2-4 Drawing a circle on a regular polygon.

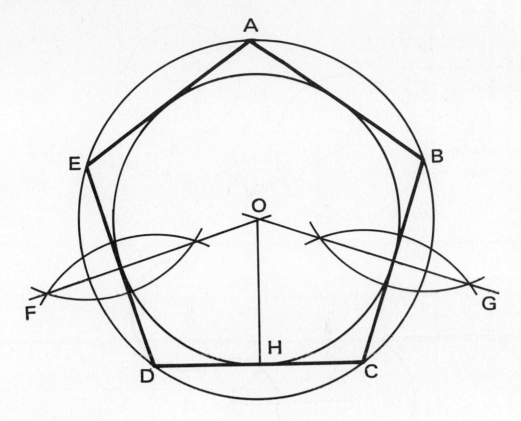

FIGURE 5-2-5 Drawing a reverse (ogee) curve connecting two parallel lines.

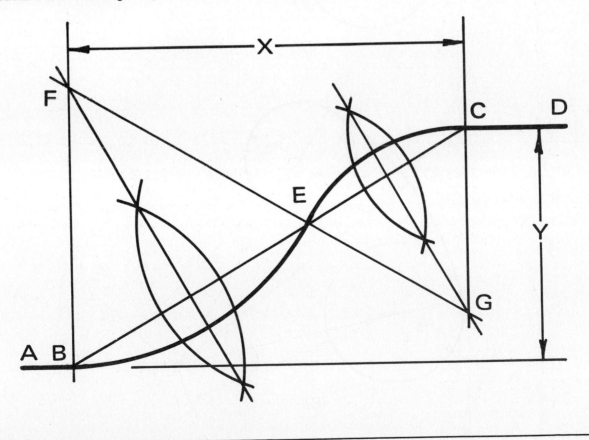

FIGURE 5-2-7 Drawing an arc tangent to two circles.

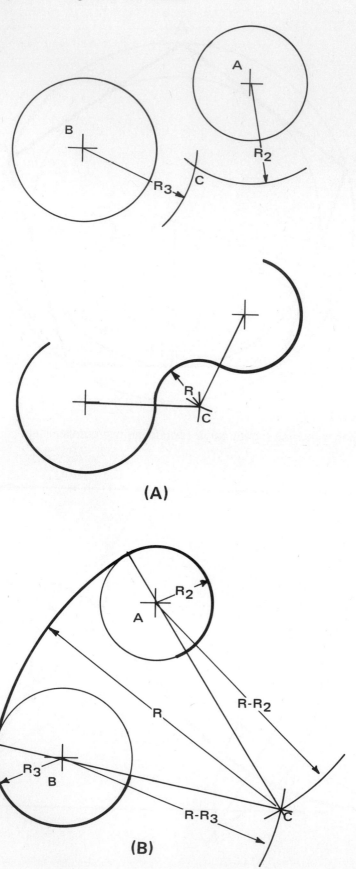

(A)

(B)

FIGURE 5-3-6 Inscribing a regular pentagon in a given circle.

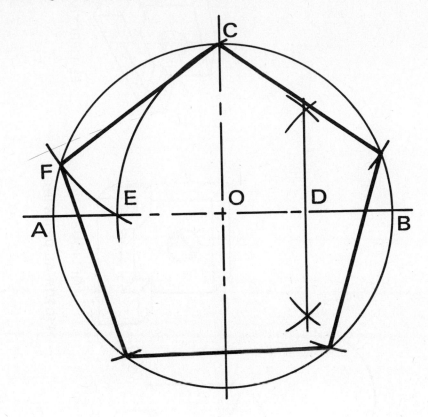

FIGURE 5-4-2 Drawing an ellipse—four-center method.

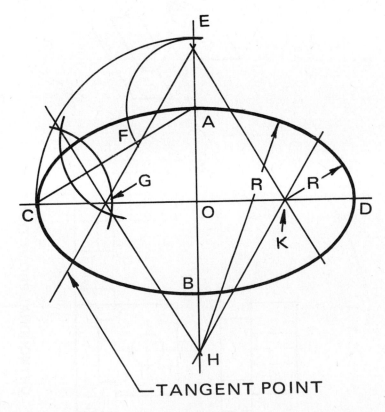

CHAPTER 6-TRANSPARENCY 1

FIGURE 6-1-1 Types of projection used in drafting.

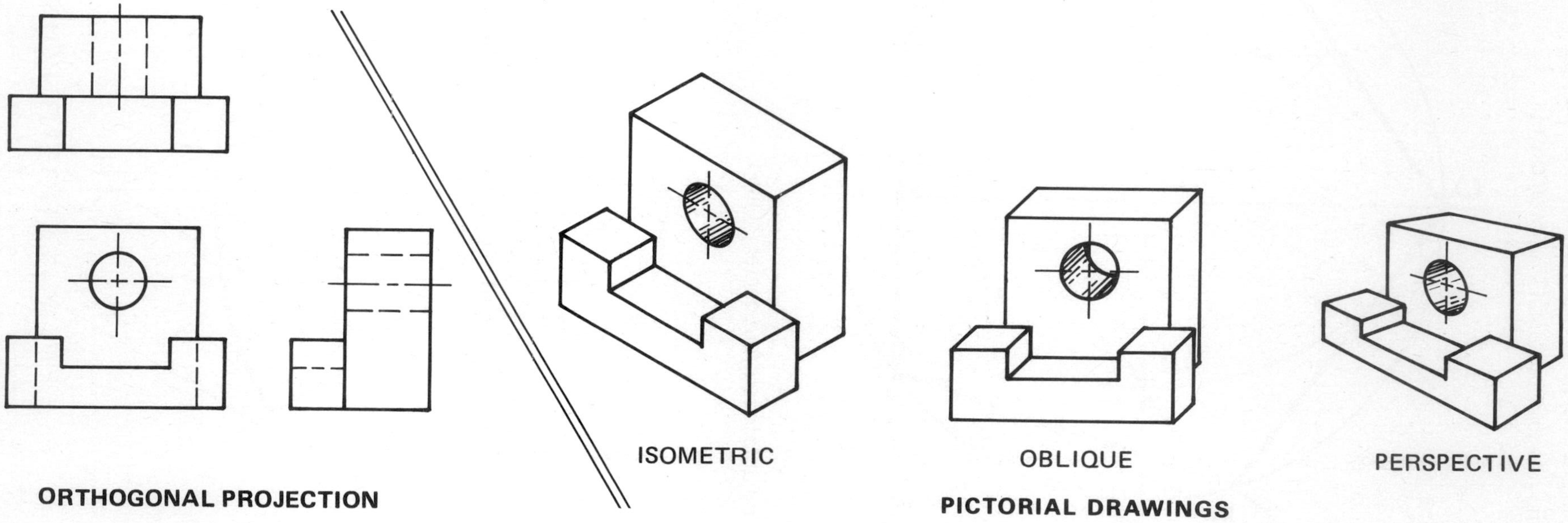

ORTHOGONAL PROJECTION
ISOMETRIC
OBLIQUE
PERSPECTIVE
PICTORIAL DRAWINGS

FIGURE 6-1-2 Designation of views.

DIRECTION OF OBSERVATION		DESIGNATION OF VIEW
VIEW IN DIRECTION	**VIEW FROM**	
a	THE FRONT	A
b	ABOVE	B
c	THE LEFT	C
d	THE RIGHT	D
e	BELOW	E
f	THE REAR	F

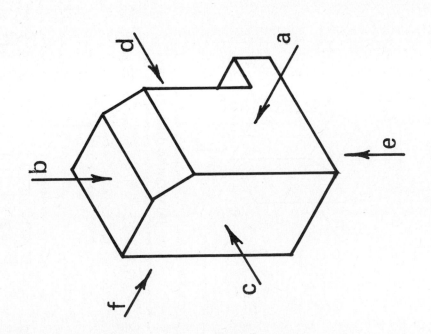

FIGURE 6-1-3 Third-angle projection.

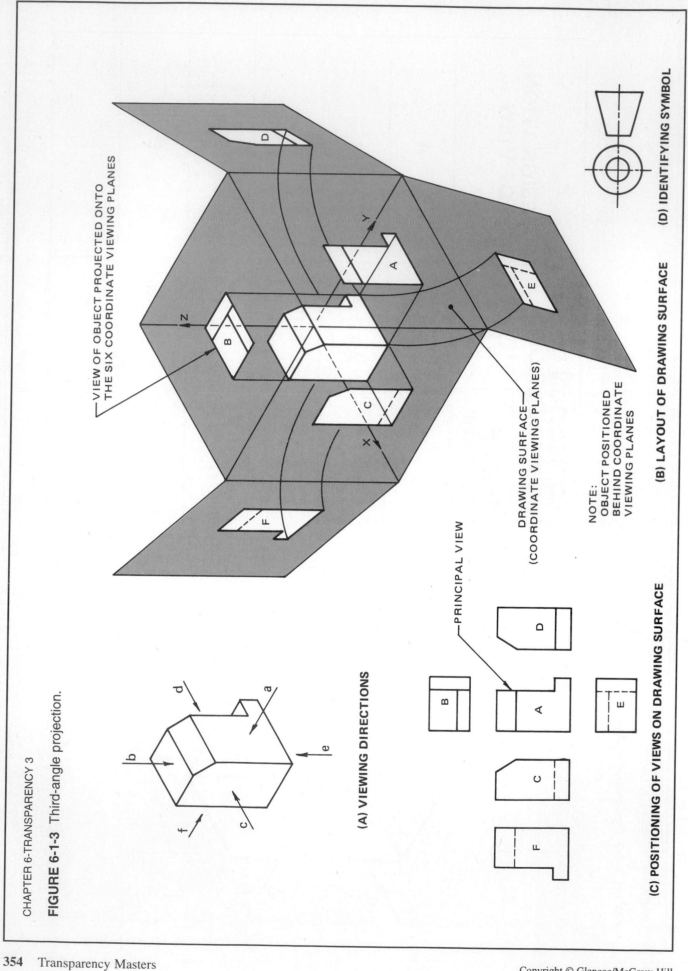

VIEW OF OBJECT PROJECTED ONTO
THE SIX COORDINATE VIEWING PLANES

NOTE:
OBJECT POSITIONED
BEHIND COORDINATE
VIEWING PLANES

DRAWING SURFACE
(COORDINATE VIEWING PLANES)

(B) LAYOUT OF DRAWING SURFACE

(D) IDENTIFYING SYMBOL

(A) VIEWING DIRECTIONS

PRINCIPAL VIEW

(C) POSITIONING OF VIEWS ON DRAWING SURFACE

FIGURE 6-1-4 First-angle projection.

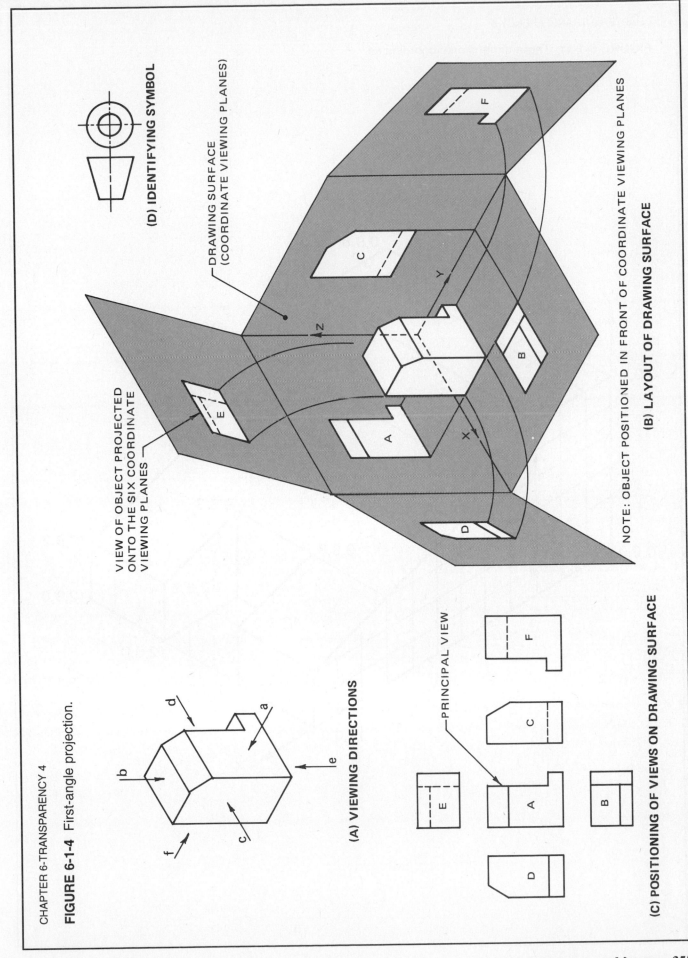

(A) VIEWING DIRECTIONS

(B) LAYOUT OF DRAWING SURFACE

NOTE: OBJECT POSITIONED IN FRONT OF COORDINATE VIEWING PLANES

VIEW OF OBJECT PROJECTED ONTO THE SIX COORDINATE VIEWING PLANES

DRAWING SURFACE (COORDINATE VIEWING PLANES)

(C) POSITIONING OF VIEWS ON DRAWING SURFACE

PRINCIPAL VIEW

(D) IDENTIFYING SYMBOL

FIGURE 6-1-11 Three-dimensional coordinates.

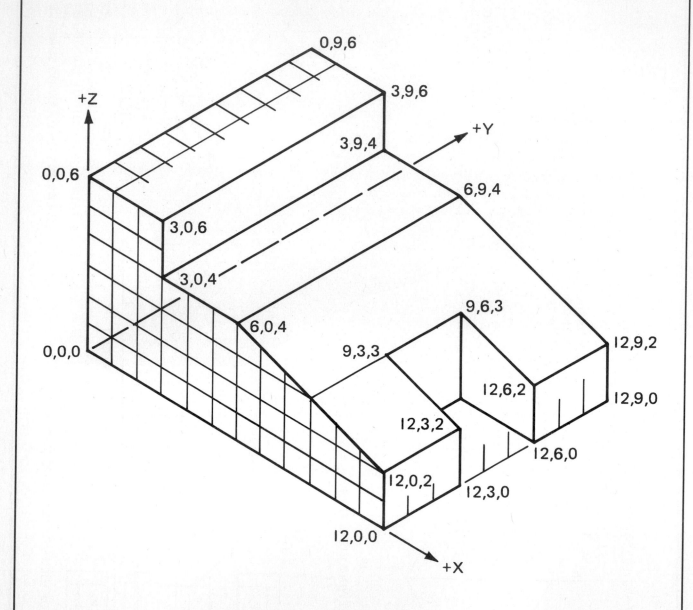

FIGURE 6-2-2 Use of a miter line.

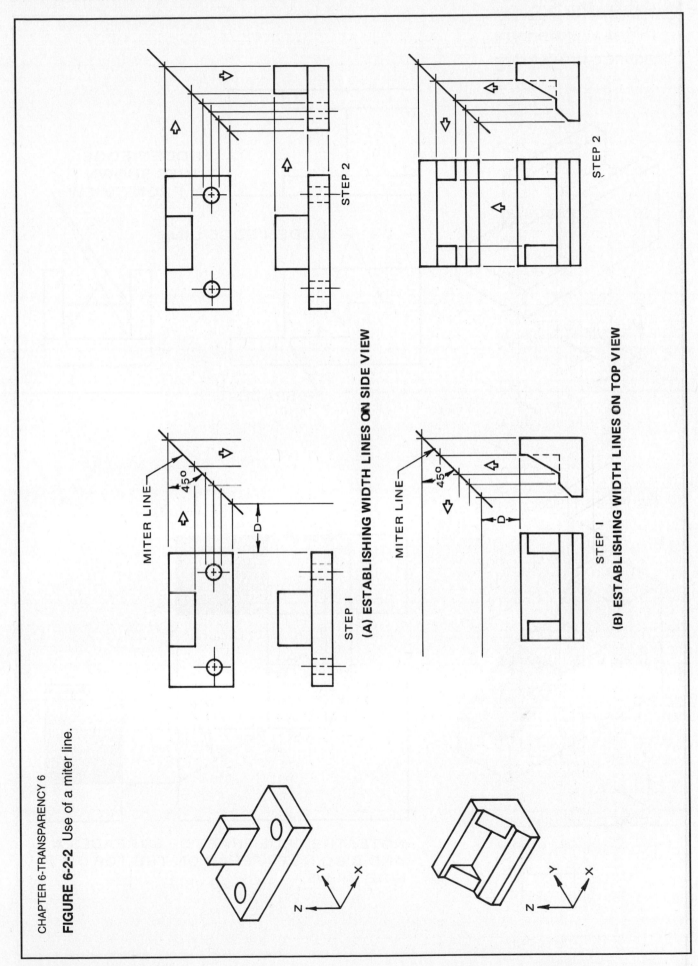

(A) ESTABLISHING WIDTH LINES ON SIDE VIEW

(B) ESTABLISHING WIDTH LINES ON TOP VIEW

FIGURE 6-4-1 Hidden lines.

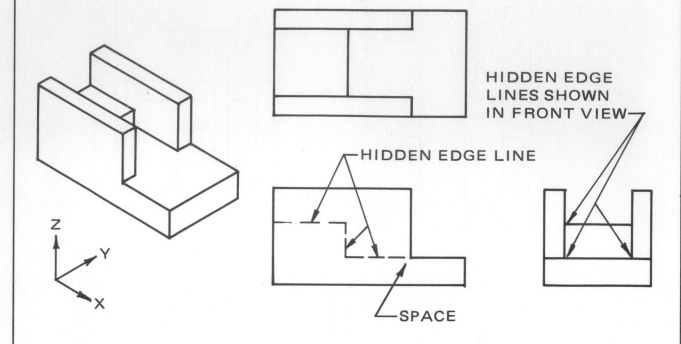

HIDDEN EDGE
LINES SHOWN
IN FRONT VIEW

HIDDEN EDGE LINE

SPACE

FIGURE 6-5-1 Sloping surfaces.

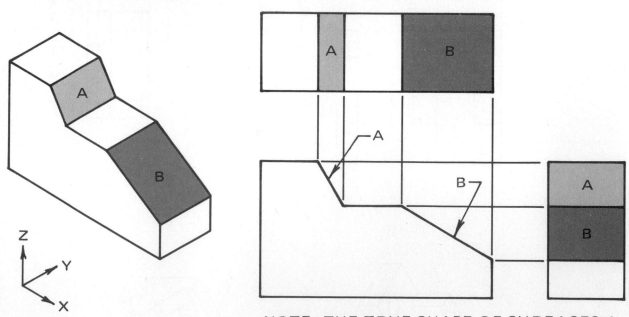

NOTE: THE TRUE SHAPE OF SURFACES A
AND B DO NOT APPEAR ON THE TOP OR
SIDE VIEWS.

FIGURE 6-7-1 Oblique surface is not its true shape in any of the three views.

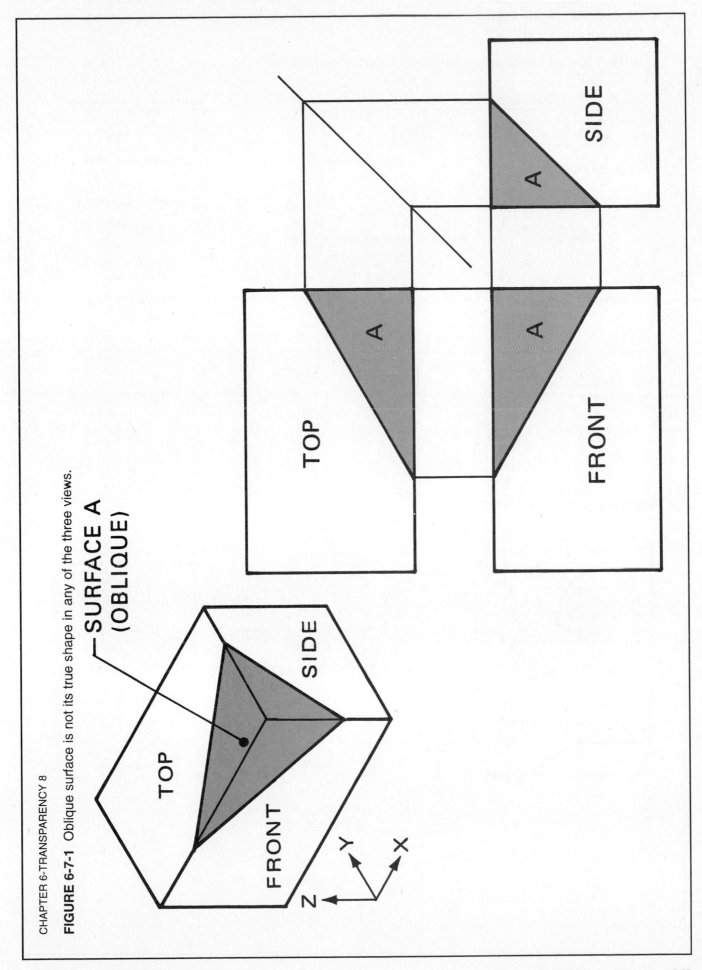

FIGURE 6-10-1 Conventional representation of common features.

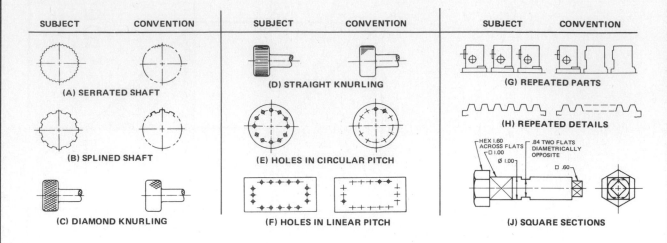

FIGURE 6-11-1 Conventional breaks.

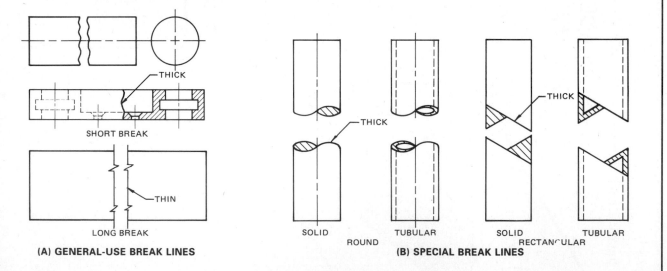

FIGURE 6-14-1 Alignment of parts and holes to show true relationship.

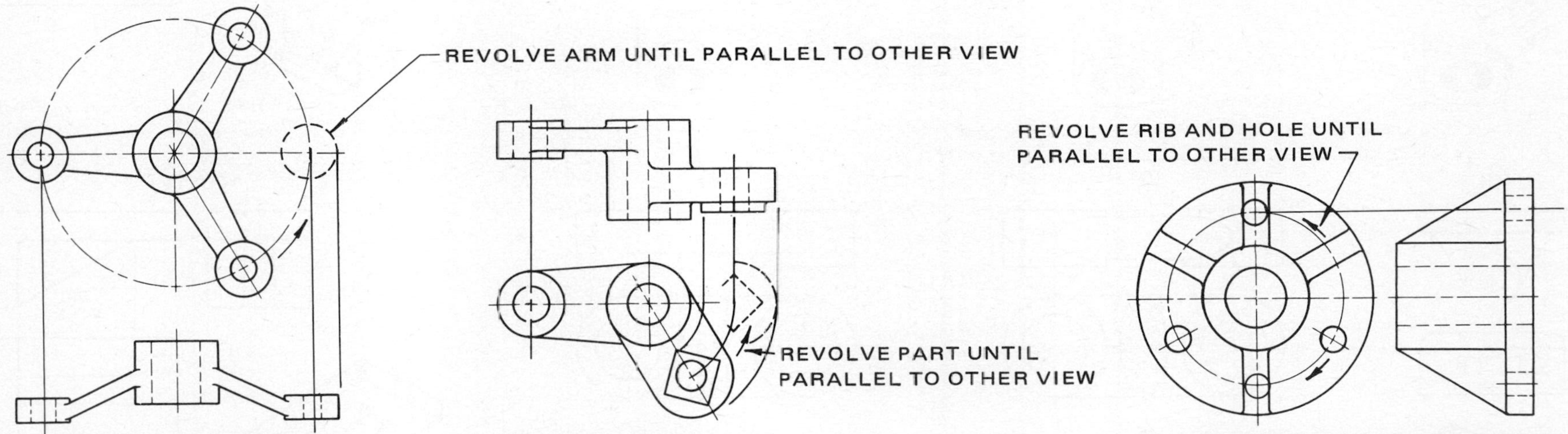

REVOLVE RIB AND HOLE UNTIL PARALLEL TO OTHER VIEW

(C) ALIGNMENT OF RIB AND HOLES

REVOLVE ARM UNTIL PARALLEL TO OTHER VIEW

REVOLVE PART UNTIL PARALLEL TO OTHER VIEW

(B) ALIGNMENT OF PART

REVOLVE ARM UNTIL PARALLEL TO OTHER VIEW

(A) ALIGNMENT OF ARM

FIGURE 6-15-2 Conventional representation of runouts.

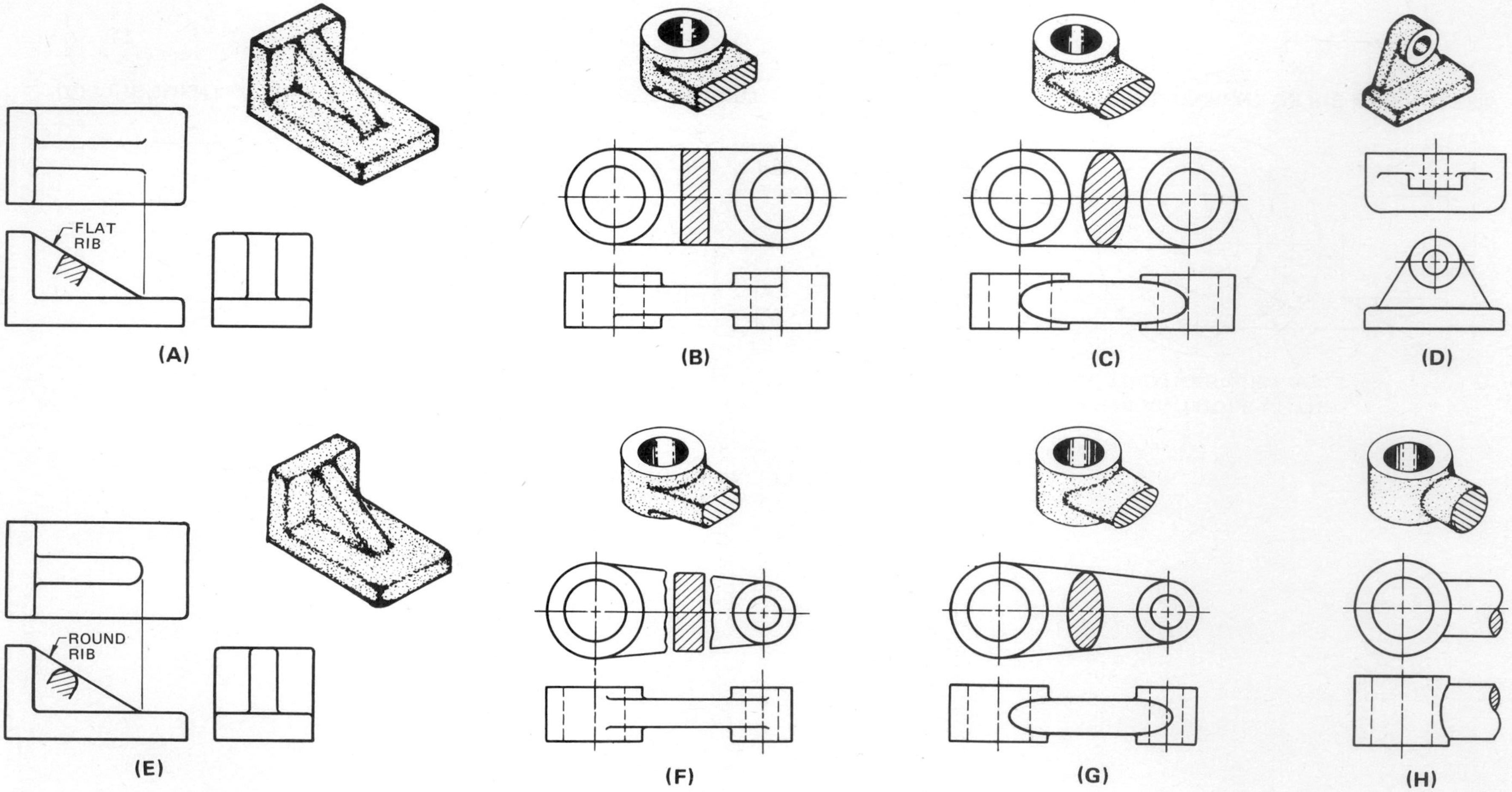

(A)

(B)

(C)

(D)

(E)

(F)

(G)

(H)

FLAT RIB

ROUND RIB

FIGURE 7-1-1 Relationship of the auxiliary plane to the three principal planes.

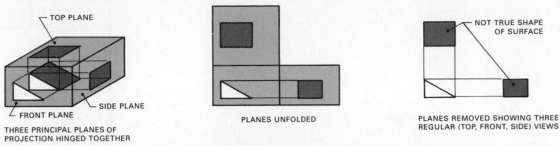

THREE PRINCIPAL PLANES OF PROJECTION HINGED TOGETHER

PLANES UNFOLDED

PLANES REMOVED SHOWING THREE REGULAR (TOP, FRONT, SIDE) VIEWS

NOTE: IN NONE OF THESE VIEWS DOES THE SLANTED (COLORED) SURFACE APPEAR IN ITS TRUE SHAPE.

(A) WEDGED BLOCK SHOWN IN THREE REGULAR VIEWS

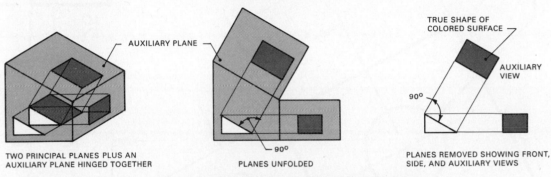

TWO PRINCIPAL PLANES PLUS AN AUXILIARY PLANE HINGED TOGETHER

PLANES UNFOLDED

PLANES REMOVED SHOWING FRONT, SIDE, AND AUXILIARY VIEWS

NOTE: IN THIS EXAMPLE THE AUXILIARY PLANE REPLACED THE TOP PLANE IN ORDER THAT THE SLANTED (COLORED) SURFACE MAY BE SHOWN IN ITS TRUE SHAPE.

(B) REPLACING THE TOP PLANE WITH AN AUXILIARY PLANE

FIGURE 7-1-2 Auxiliary views replacing regular views.

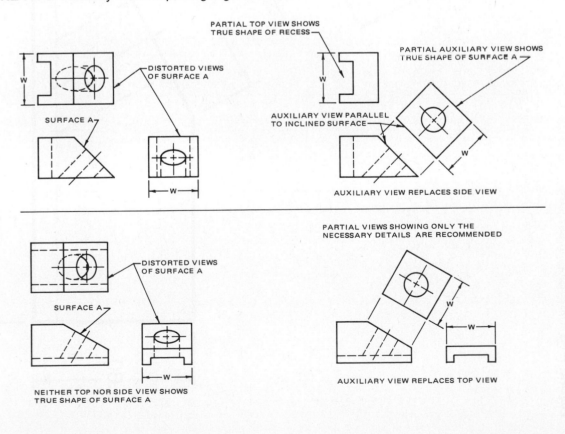

PARTIAL TOP VIEW SHOWS TRUE SHAPE OF RECESS

DISTORTED VIEWS OF SURFACE A

SURFACE A

PARTIAL AUXILIARY VIEW SHOWS TRUE SHAPE OF SURFACE A

AUXILIARY VIEW PARALLEL TO INCLINED SURFACE

AUXILIARY VIEW REPLACES SIDE VIEW

PARTIAL VIEWS SHOWING ONLY THE NECESSARY DETAILS ARE RECOMMENDED

DISTORTED VIEWS OF SURFACE A

SURFACE A

NEITHER TOP NOR SIDE VIEW SHOWS TRUE SHAPE OF SURFACE A

AUXILIARY VIEW REPLACES TOP VIEW

FIGURE 7-1-5 Dimensioning auxiliary.

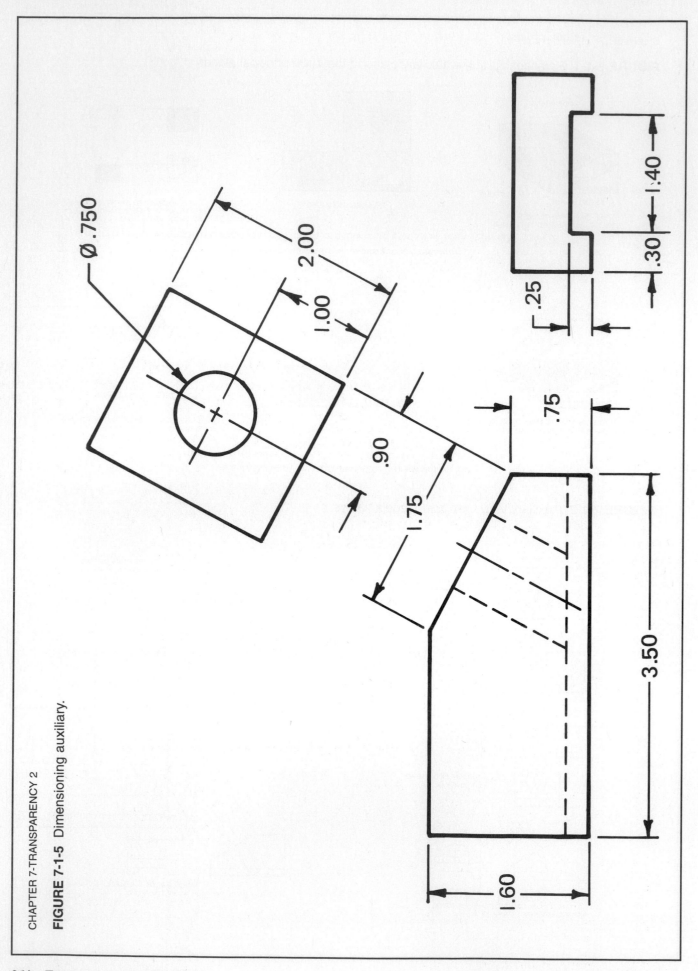

FIGURE 7-2-1 Establishing true shape of truncated cylinder.

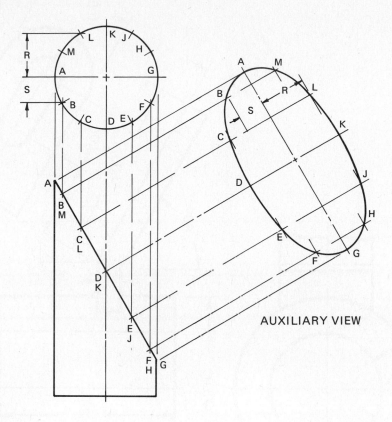

AUXILIARY VIEW

FIGURE 7-2-2 Constructing the true shape of a curved surface by plotting method.

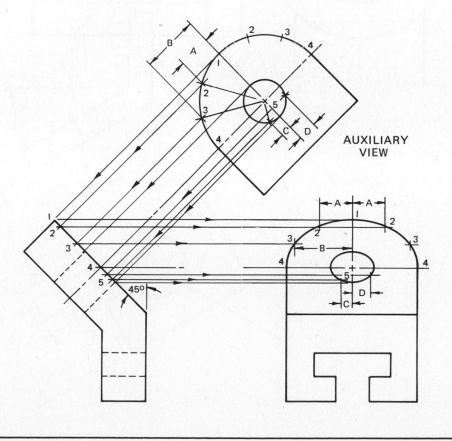

AUXILIARY VIEW

FIGURE 7-3-2 Dimensioning a multi-auxiliary-view drawing.

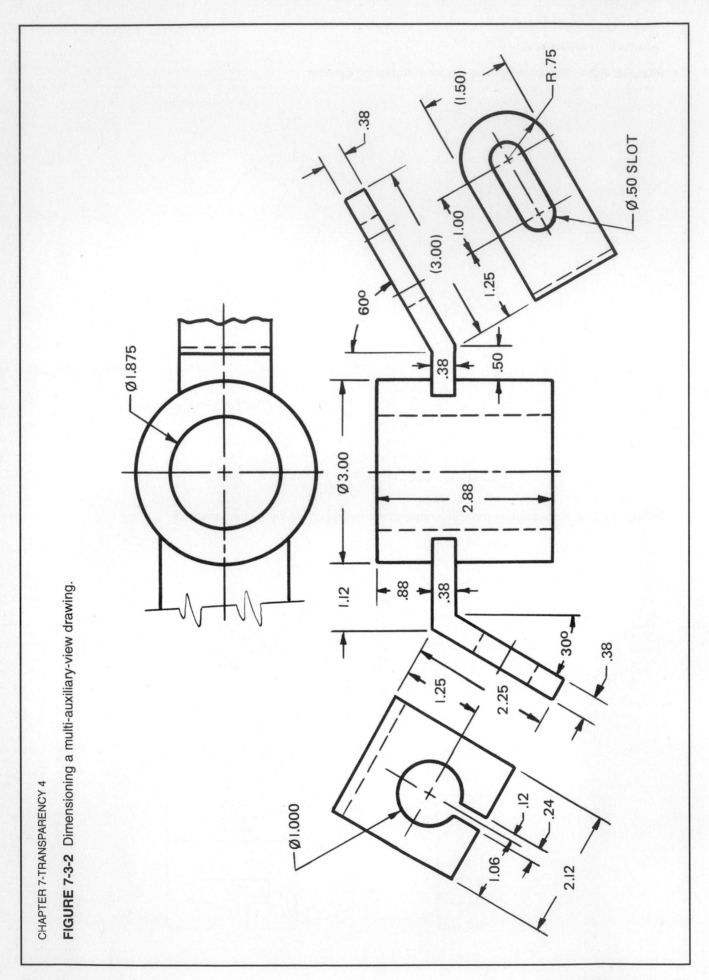

FIGURE 7-4-1 Steps in drawing a secondary auxiliary view.

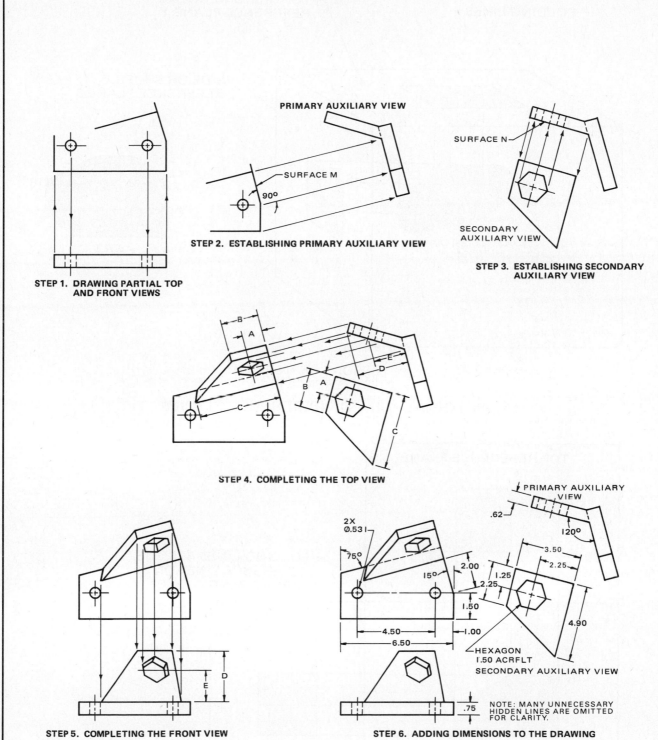

PRIMARY AUXILIARY VIEW

SURFACE M

90°

STEP 2. ESTABLISHING PRIMARY AUXILIARY VIEW

STEP 1. DRAWING PARTIAL TOP
AND FRONT VIEWS

SURFACE N

SECONDARY
AUXILIARY VIEW

STEP 3. ESTABLISHING SECONDARY
AUXILIARY VIEW

STEP 4. COMPLETING THE TOP VIEW

STEP 5. COMPLETING THE FRONT VIEW

PRIMARY AUXILIARY
VIEW

.62   120°

2X
Ø.531

75°

15°

2.00

1.25
2.25

3.50
2.25

1.50

4.50

1.00

6.50

4.90

HEXAGON
1.50 ACRFLT
SECONDARY AUXILIARY VIEW

.75

NOTE: MANY UNNECESSARY
HIDDEN LINES ARE OMITTED
FOR CLARITY.

STEP 6. ADDING DIMENSIONS TO THE DRAWING

FIGURE 7-5-2 Reference lines.

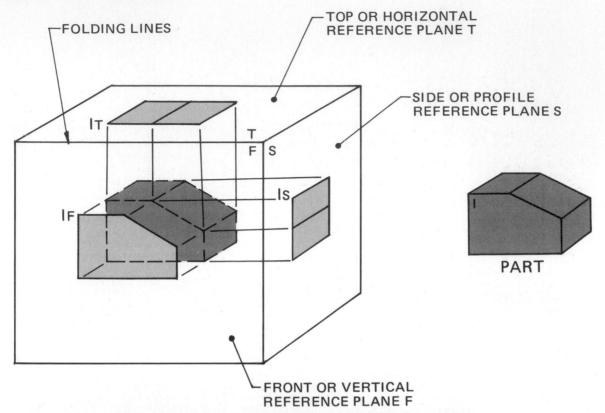

FOLDING LINES

TOP OR HORIZONTAL
REFERENCE PLANE T

SIDE OR PROFILE
REFERENCE PLANE S

IT

T
F S

Is

IF

PART

FRONT OR VERTICAL
REFERENCE PLANE F

(A) PICTORIAL VIEW OF REFERENCE PLANES

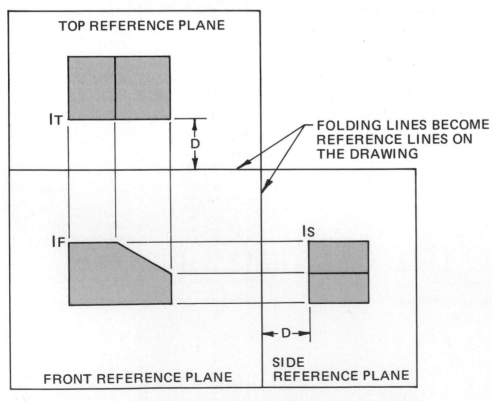

TOP REFERENCE PLANE

IT

D

FOLDING LINES BECOME
REFERENCE LINES ON
THE DRAWING

IF

Is

D

FRONT REFERENCE PLANE

SIDE
REFERENCE PLANE

(B) UNFOLDING OF THE THREE REFERENCE PLANES

FIGURE 7-5-7 The true shape of surface 1-2-3-4 obtained by successive revolutions.

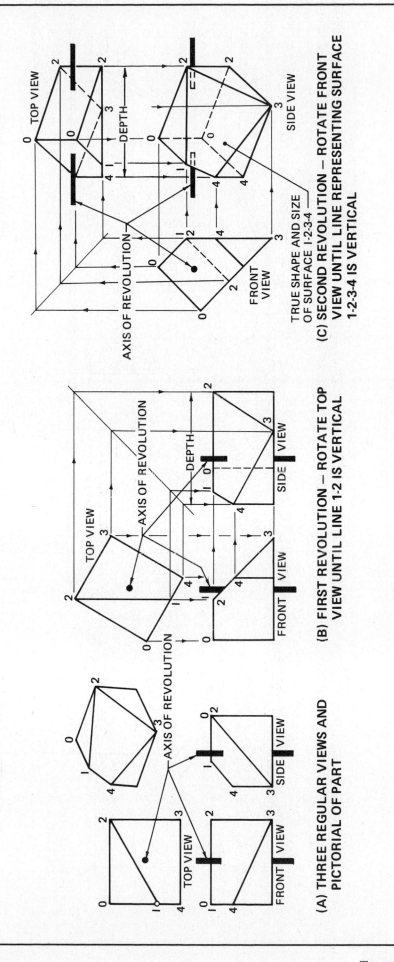

(A) THREE REGULAR VIEWS AND PICTORIAL OF PART

(B) FIRST REVOLUTION – ROTATE TOP VIEW UNTIL LINE 1-2 IS VERTICAL

(C) SECOND REVOLUTION – ROTATE FRONT VIEW UNTIL LINE REPRESENTING SURFACE 1-2-3-4 IS VERTICAL

FIGURE 7-6-2 Lines in space.

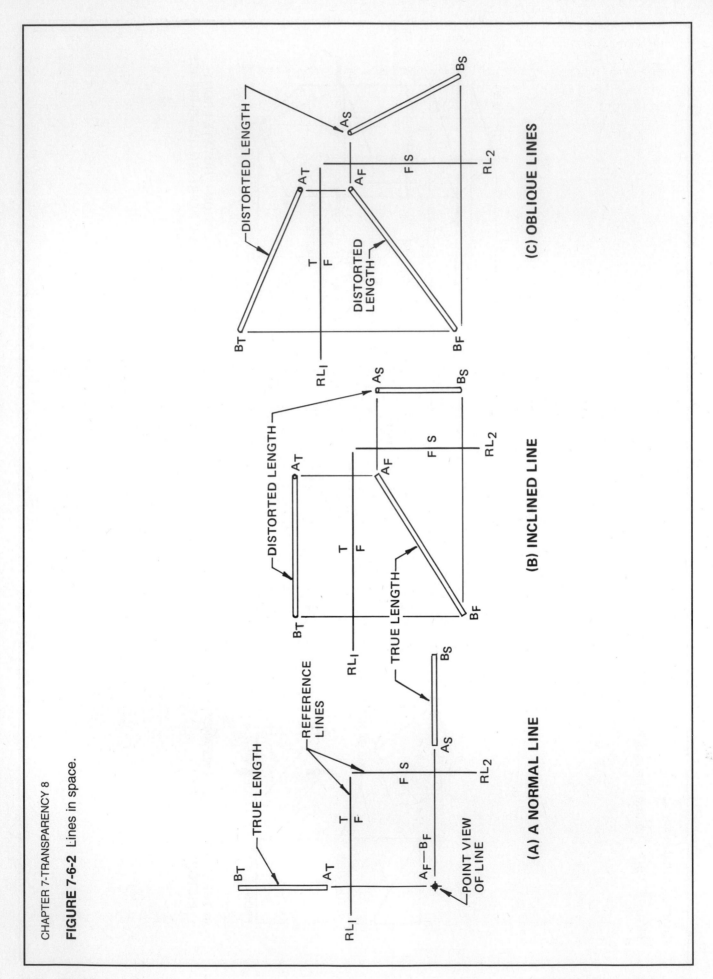

(A) A NORMAL LINE

(B) INCLINED LINE

(C) OBLIQUE LINES

FIGURE 7-6-3 The length of an oblique line by auxiliary view projection.

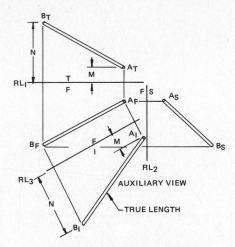

(A) REFERENCE LINE RL₃ PLACED PARALLEL TO FRONT VIEW

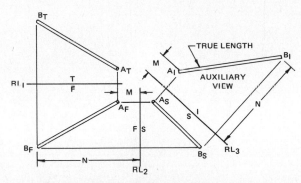

(B) REFERENCE LINE RL₃ PLACED PARALLEL TO SIDE VIEW

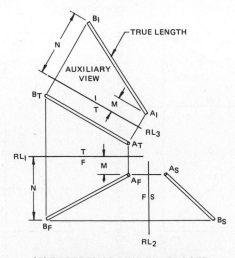

(C) REFERENCE LINE RL₃ PLACED PARALLEL TO TOP VIEW

FIGURE 7-6-4 Point on a line.

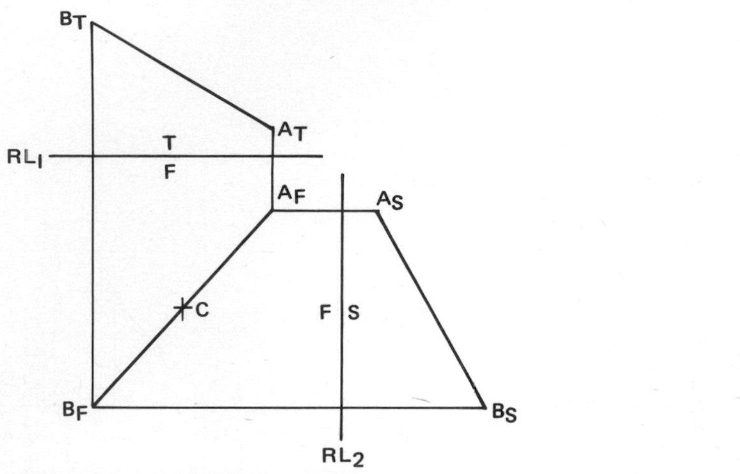

(A) PROBLEM—TO LOCATE POINT C ON LINE A-B IN OTHER VIEWS

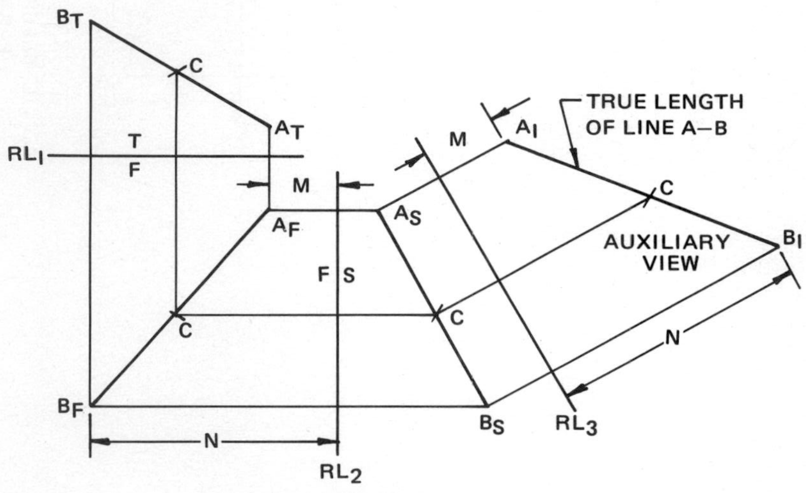

(B) SOLUTION

FIGURE 7-6-5 Point-on-point view of a line.

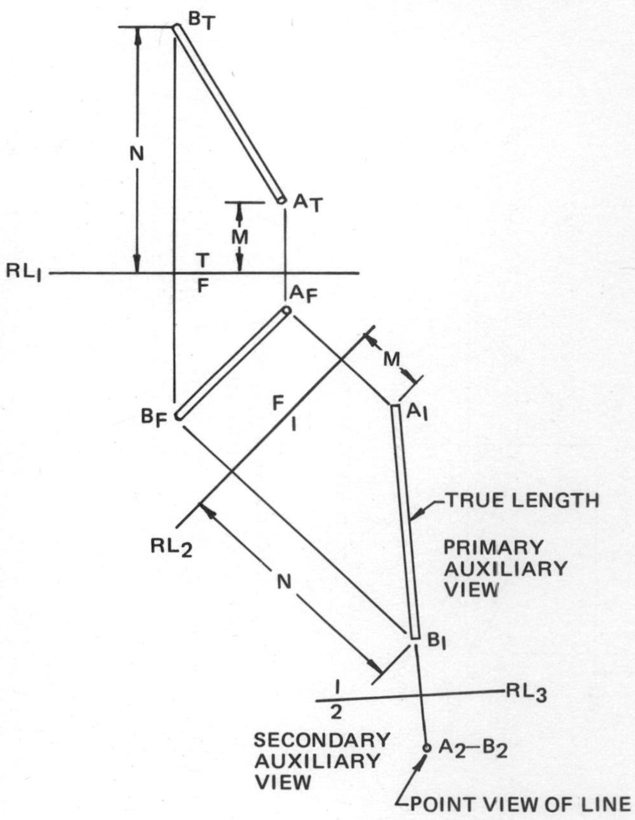

FIGURE 7-7-1 Planes in space.

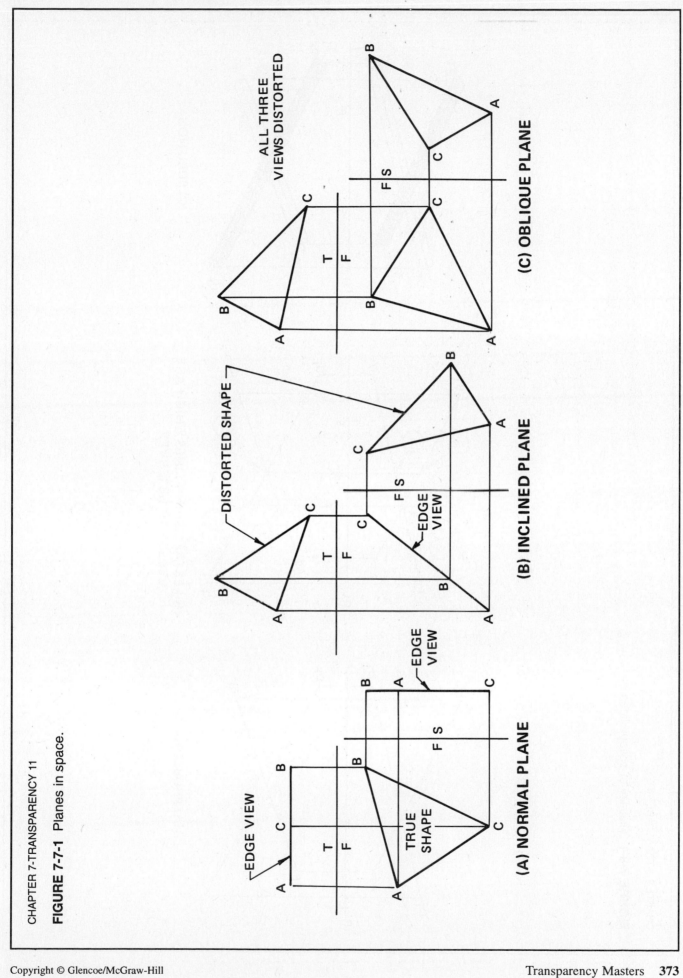

(A) NORMAL PLANE

(B) INCLINED PLANE

(C) OBLIQUE PLANE

CHAPTER 7-TRANSPARENCY 12

FIGURE 7-8-1 Visibility of oblique lines by testing.

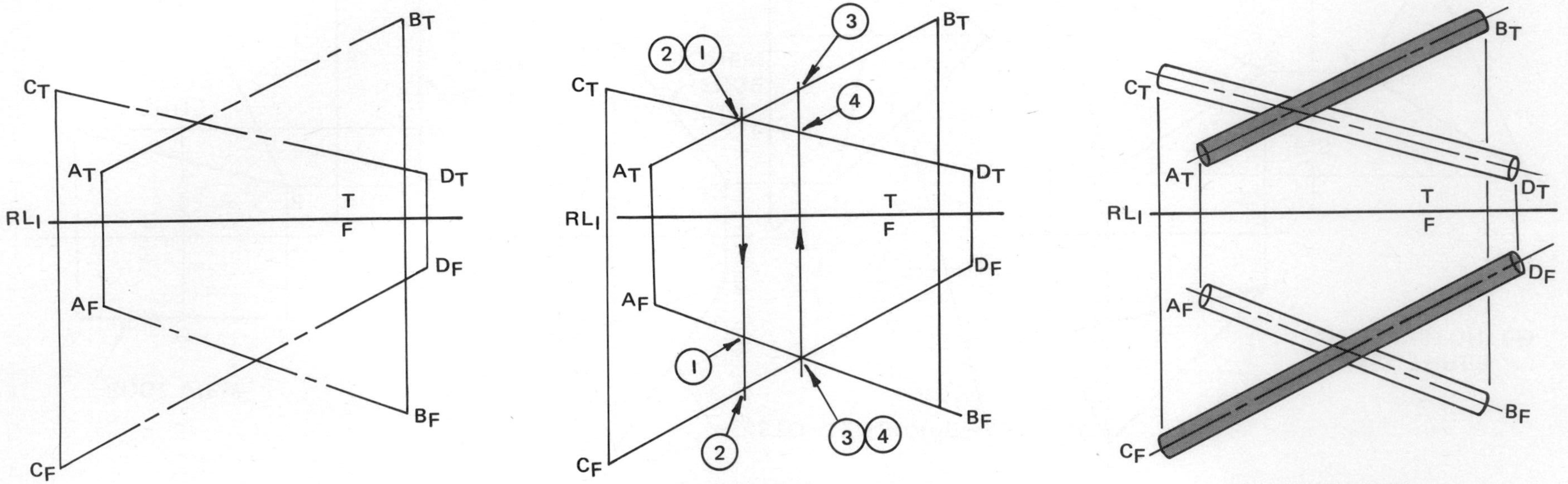

(A) PROBLEM

(B) ESTABLISHING LINES WHICH ARE CLOSER TO OBSERVER

(C) SOLUTION

FIGURE 7-9-1 Distance from a point to a line.

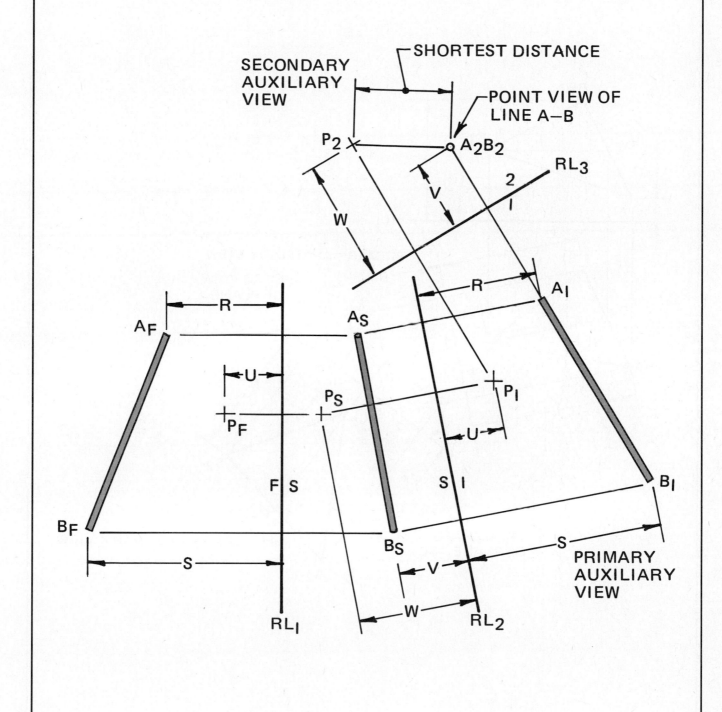

FIGURE 7-10-1 True views of a plane.

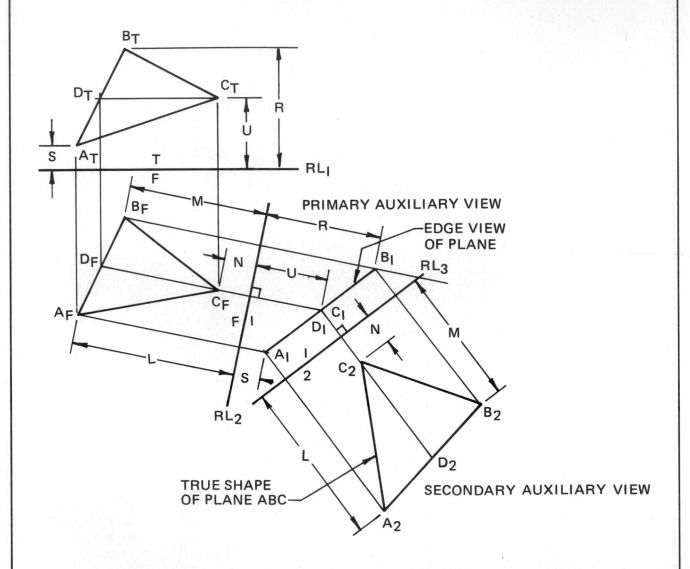

PRIMARY AUXILIARY VIEW

EDGE VIEW OF PLANE

SECONDARY AUXILIARY VIEW

TRUE SHAPE OF PLANE ABC

FIGURE 7-11-2 Edge lines of two planes.

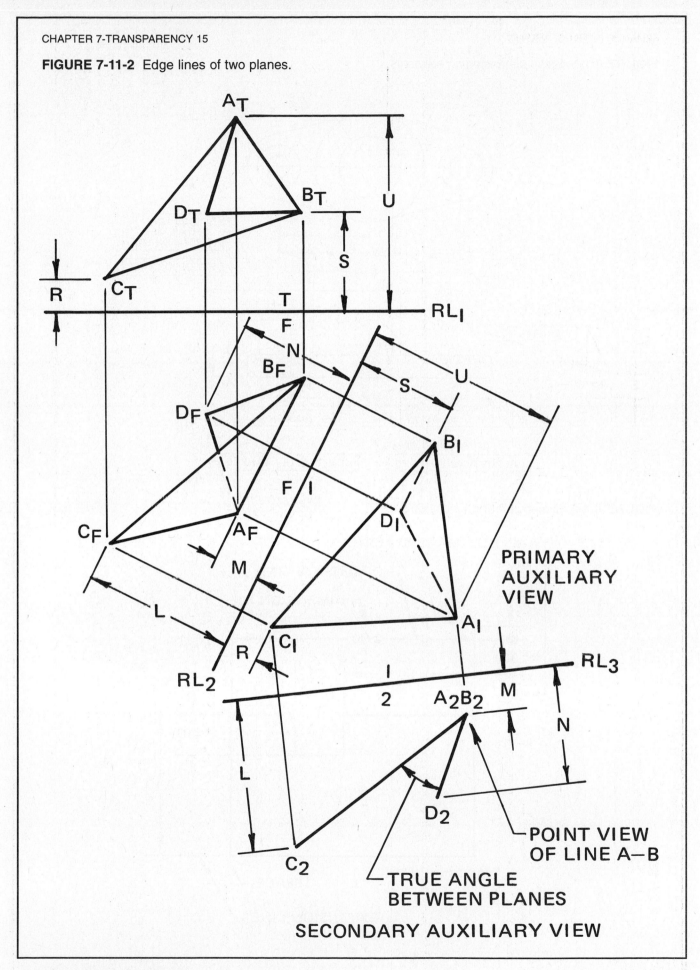

PRIMARY
AUXILIARY
VIEW

POINT VIEW
OF LINE A—B

TRUE ANGLE
BETWEEN PLANES

SECONDARY AUXILIARY VIEW

FIGURE 8-1-1 Basic dimensioning elements.

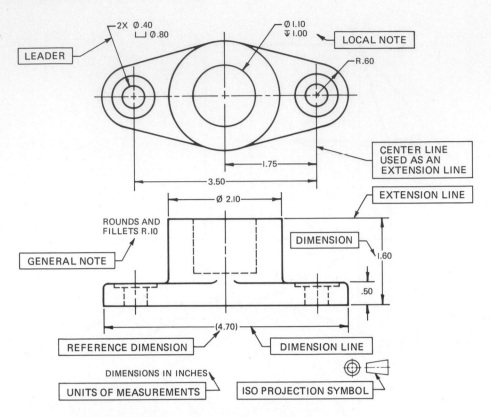

FIGURE 8-1-2 Dimension and extension lines.

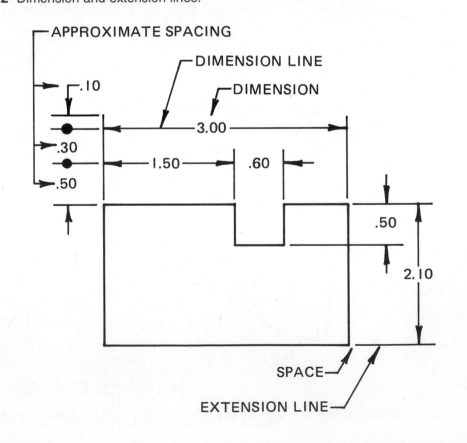

FIGURE 8-1-3 Dimensioning linear features.

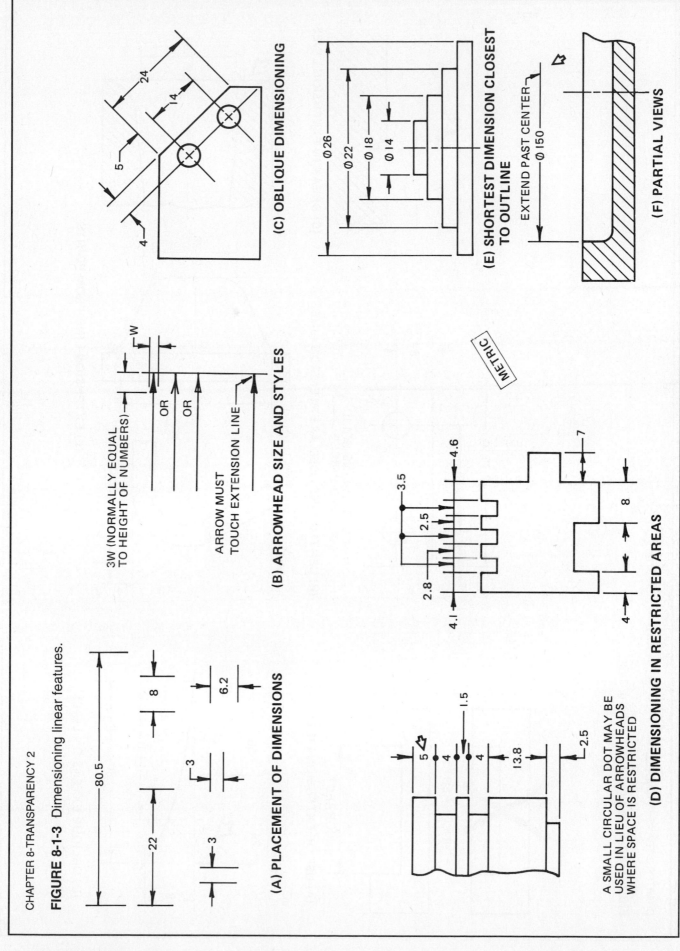

(A) PLACEMENT OF DIMENSIONS

(B) ARROWHEAD SIZE AND STYLES

(C) OBLIQUE DIMENSIONING

(D) DIMENSIONING IN RESTRICTED AREAS

(E) SHORTEST DIMENSION CLOSEST TO OUTLINE

(F) PARTIAL VIEWS

3W (NORMALLY EQUAL TO HEIGHT OF NUMBERS)

ARROW MUST TOUCH EXTENSION LINE

A SMALL CIRCULAR DOT MAY BE USED IN LIEU OF ARROWHEADS WHERE SPACE IS RESTRICTED

EXTEND PAST CENTER

METRIC

FIGURE 8-1-5 Extension (projection) lines.

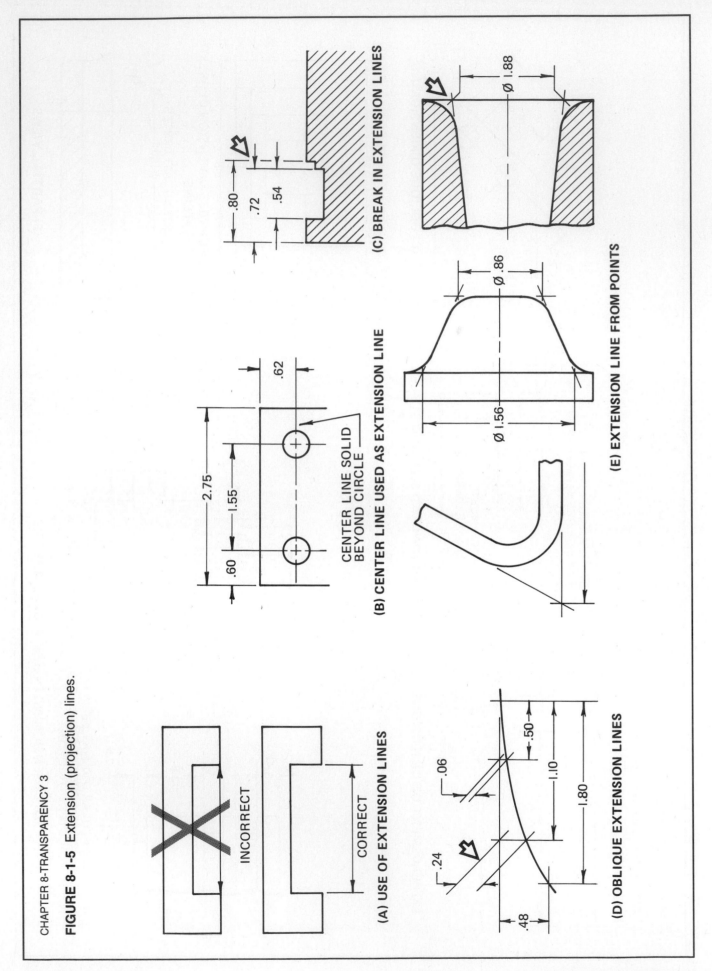

(A) USE OF EXTENSION LINES

INCORRECT

CORRECT

(B) CENTER LINE USED AS EXTENSION LINE

CENTER LINE SOLID
BEYOND CIRCLE

.62

2.75

1.55

.60

(C) BREAK IN EXTENSION LINES

.80

.72

.54

Ø 1.88

(D) OBLIQUE EXTENSION LINES

.06

.50

.24

1.10

1.80

.48

(E) EXTENSION LINE FROM POINTS

Ø .86

1.56

FIGURE 8-1-10 Angular units.

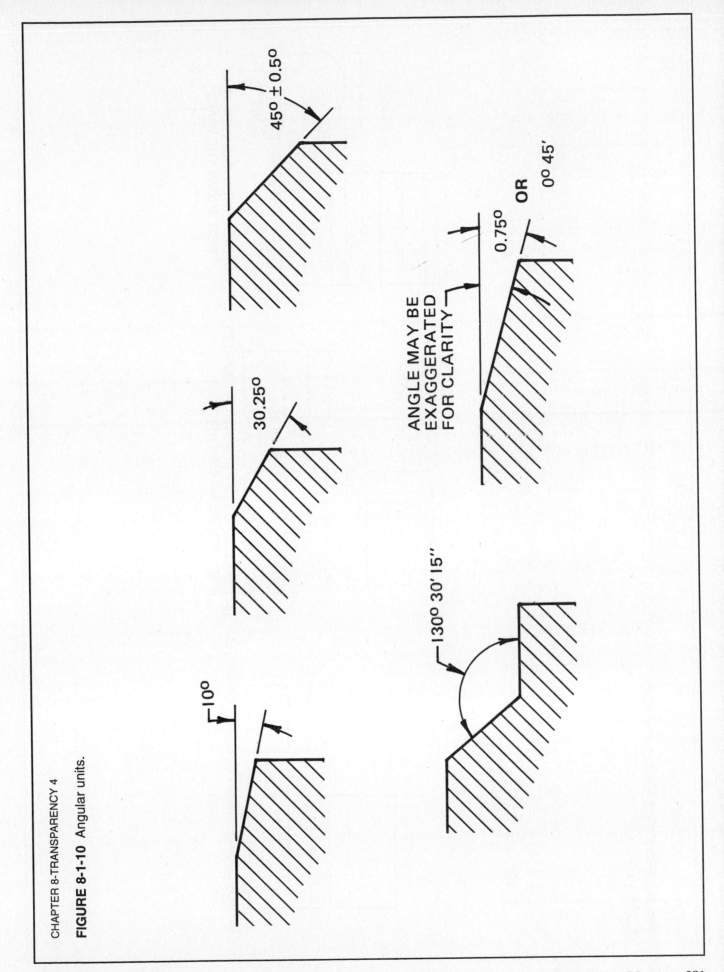

FIGURE 8-1-12 Basic dimensioning rules.

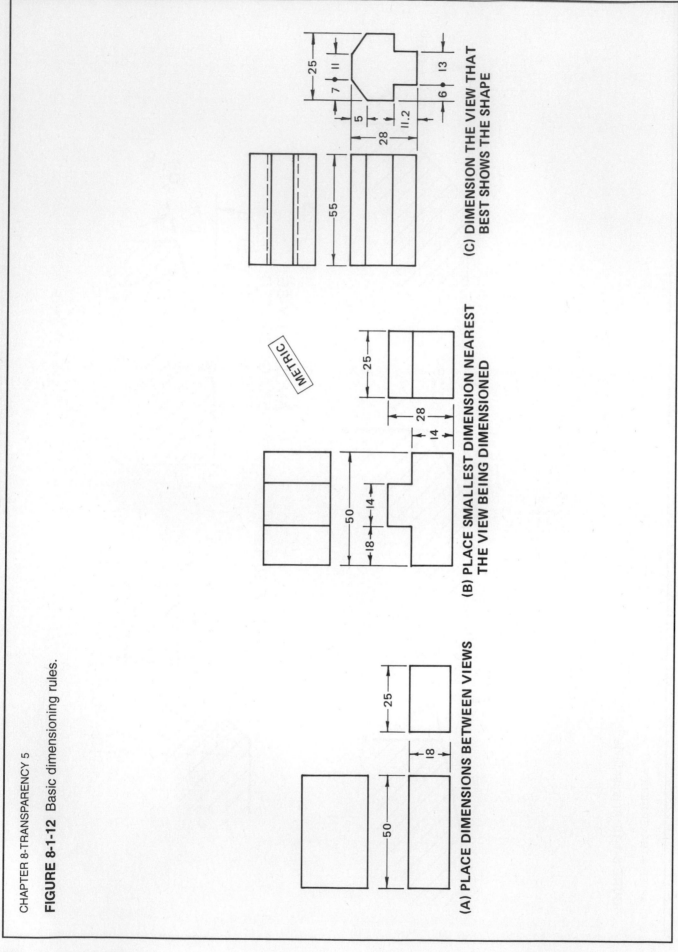

(A) PLACE DIMENSIONS BETWEEN VIEWS

(B) PLACE SMALLEST DIMENSION NEAREST THE VIEW BEING DIMENSIONED

(C) DIMENSION THE VIEW THAT BEST SHOWS THE SHAPE

METRIC

FIGURE 8-2-1 Dimensioning diameters.

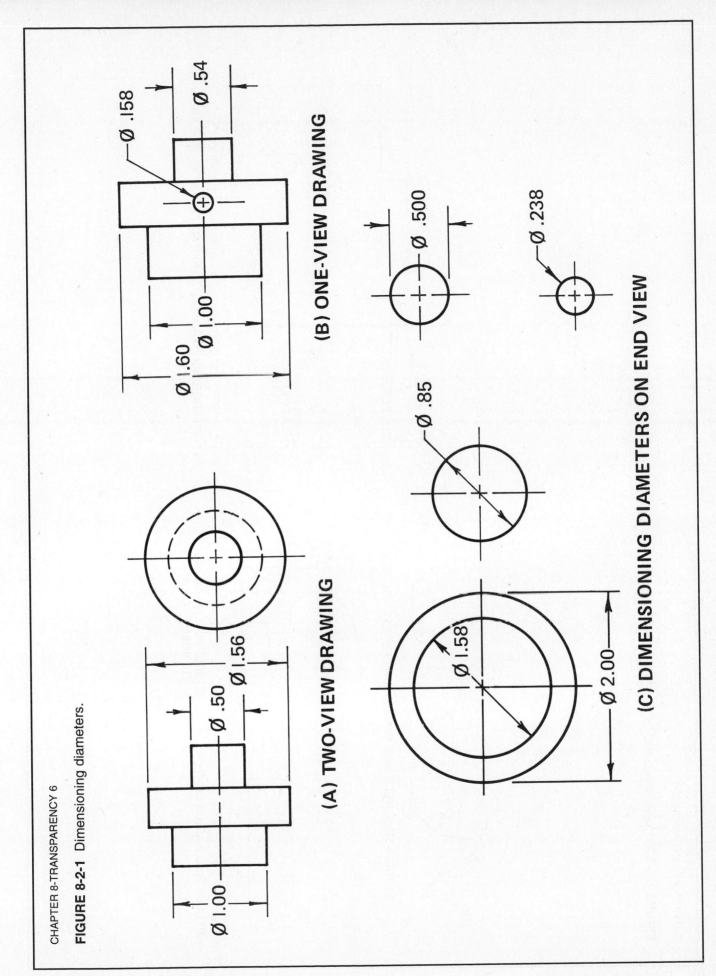

(A) TWO-VIEW DRAWING

(B) ONE-VIEW DRAWING

(C) DIMENSIONING DIAMETERS ON END VIEW

FIGURE 8-2-3 Dimensioning radii.

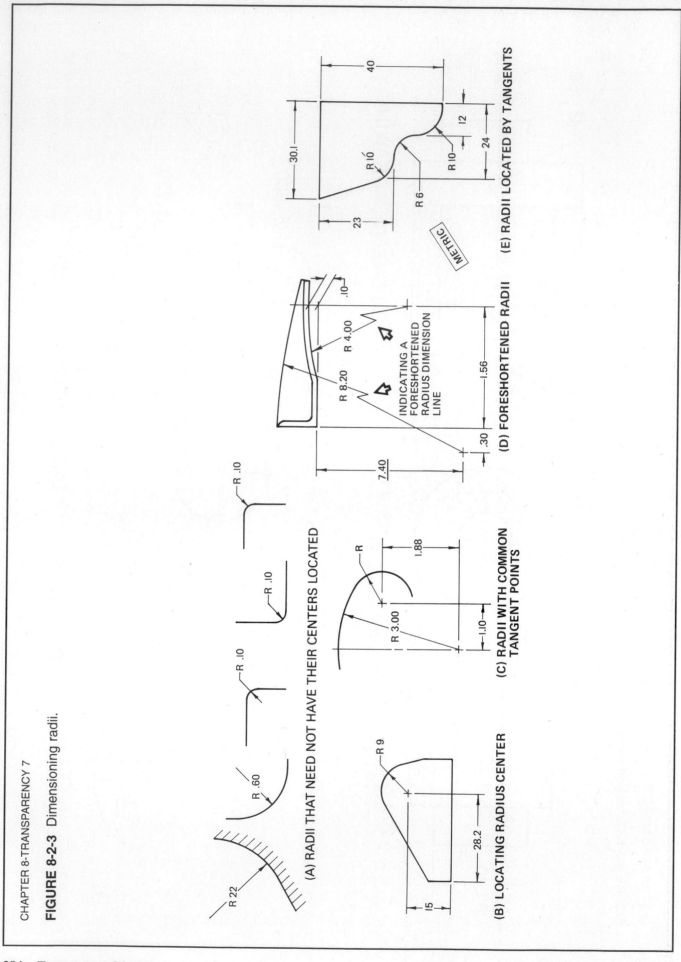

(A) RADII THAT NEED NOT HAVE THEIR CENTERS LOCATED

(B) LOCATING RADIUS CENTER

(C) RADII WITH COMMON TANGENT POINTS

(D) FORESHORTENED RADII

(E) RADII LOCATED BY TANGENTS

FIGURE 8-2-8 Dimensioning cylindrical holes.

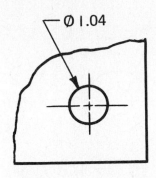

Ø1.04

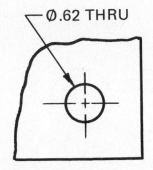

Ø.62 THRU

(B) ADDING THE WORD "THRU" WHEN IT IS NOT CLEAR THE HOLE GOES THROUGH

(A) DIMENSIONING ONE HOLE

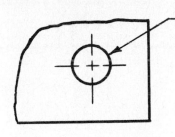

Ø.75
↧1.00

TOP VIEW NOT SHOWN

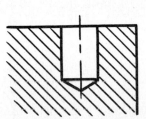

OR

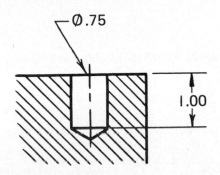

Ø.75

1.00

(C) DIMENSIONING A BLIND HOLE

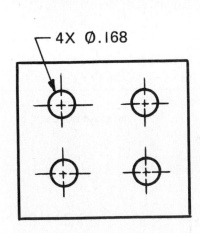

4X Ø.168

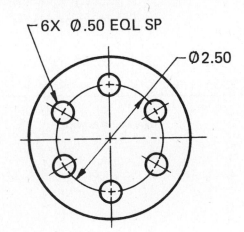

6X Ø.50 EQL SP

Ø2.50

NOTE: SEE UNIT 8–3 FOR DIMENSIONING REPETITIVE FEATURES

(D) DIMENSIONING A GROUP OF HOLES

FIGURE 8-2-11 Counterbored and spotfaced holes.

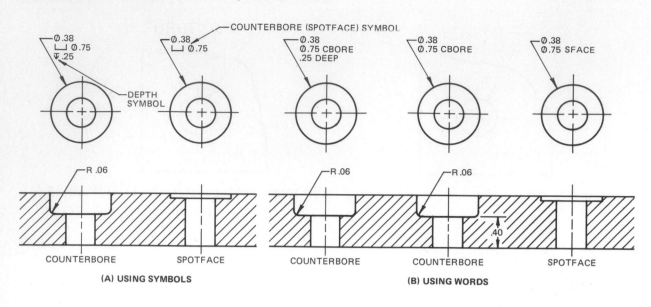

(A) USING SYMBOLS (B) USING WORDS

FIGURE 8-2-12 Countersunk and counterdrilled holes.

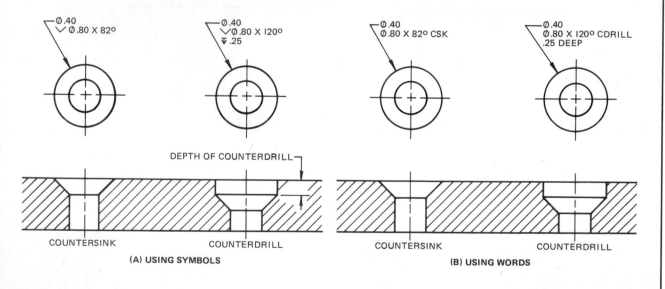

(A) USING SYMBOLS (B) USING WORDS

FIGURE 8-3-1 Dimensioning repetetive detail.

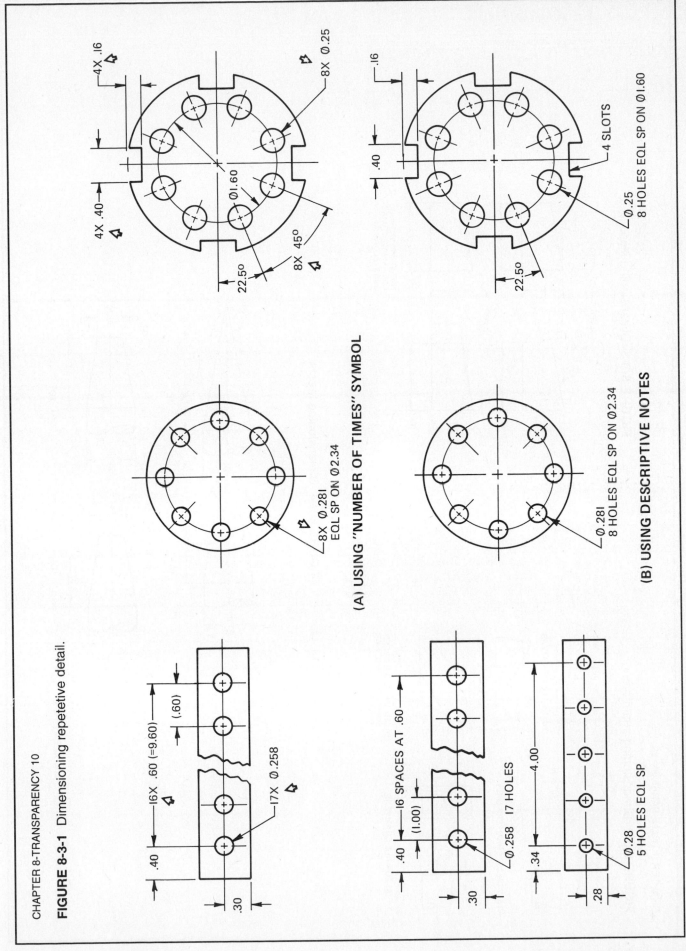

(A) USING "NUMBER OF TIMES" SYMBOL

(B) USING DESCRIPTIVE NOTES

FIGURE 8-3-2 Dimensioning chamfers.

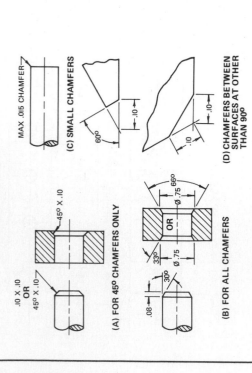

(A) FOR 45° CHAMFERS ONLY

(B) FOR ALL CHAMFERS

(C) SMALL CHAMFERS

(D) CHAMFERS BETWEEN SURFACES AT OTHER THAN 90°

FIGURE 8-3-3 Dimensioning slopes.

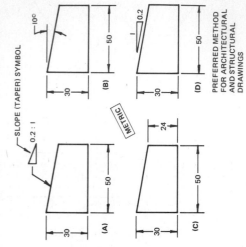

SLOPE (TAPER) SYMBOL

METRIC

(A)

(B)

(C)

(D)

PREFERRED METHOD FOR ARCHITECTURAL AND STRUCTURAL DRAWINGS

FIGURE 8-3-4 Dimensioning tapers.

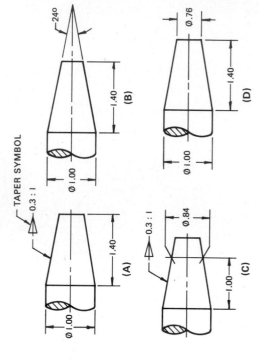

TAPER SYMBOL

(A)

(B)

(C)

(D)

FIGURE 8-4-1 Rectangular coordinate dimensioning.

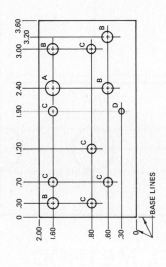

HOLE SYMBOL	HOLE SIZE
A	.246
B	.189
C	.154
D	.125

FIGURE 8-4-4 Rectangular coordinate dimensioning without dimension lines (arrowless dimensioning).

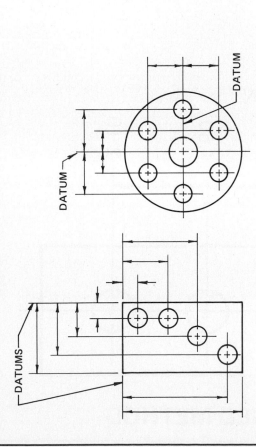

FIGURE 8-4-5 Tabular dimensioning.

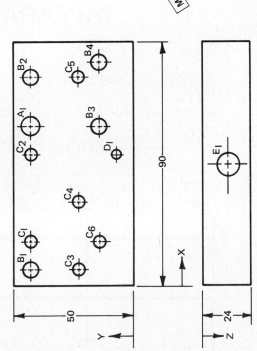

HOLE DIA	HOLE SYMBOL	LOCATION		
		X	Y	Z
5.6	A1	60	40	18
4.8	B1	10	40	THRU
	B2	75	40	THRU
	B3	60	16	THRU
	B4	80	16	THRU
4	C1	18	40	THRU
	C2	55	40	THRU
	C3	10	20	THRU
	C4	30	20	THRU
	C5	75	20	THRU
	C6	18	16	THRU
3.2	D1	55	8	12
8.1	E1	42	20	12

FIGURE 8-4-10 Common-point (baseline) dimensioning.

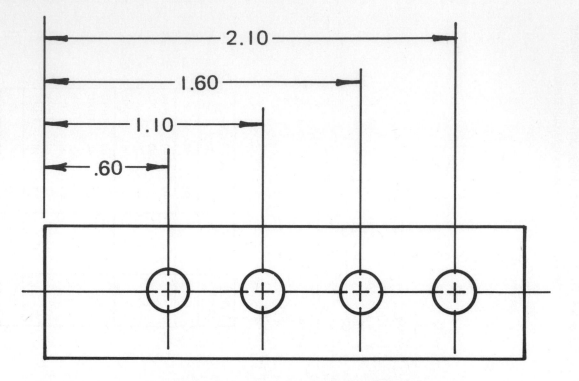

(A) PARALLEL METHOD

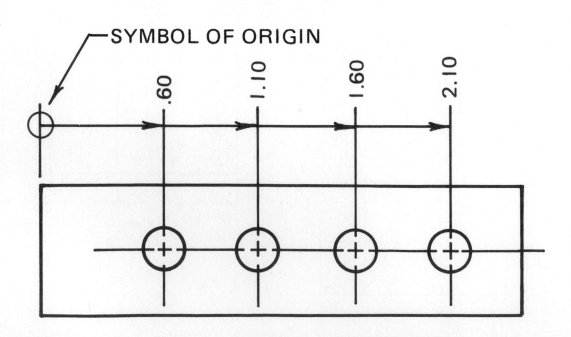

(B) SUPERIMPOSED METHOD

FIGURE 8-5-3 Methods of indicating tolerance on drawing.

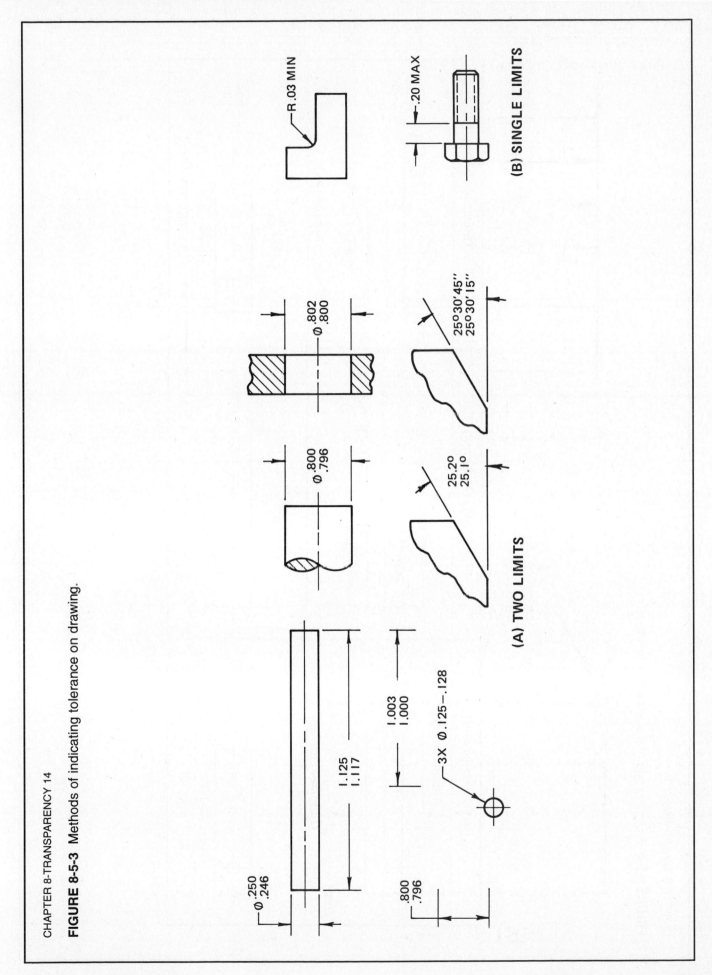

(A) TWO LIMITS

(B) SINGLE LIMITS

FIGURE 8-5-8 A comparison of the tolerance methods.

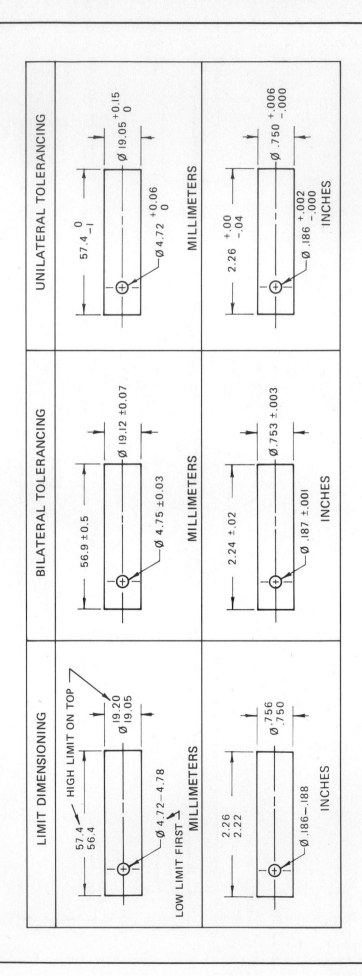

FIGURE 8-6-1 Illustration of definitions.

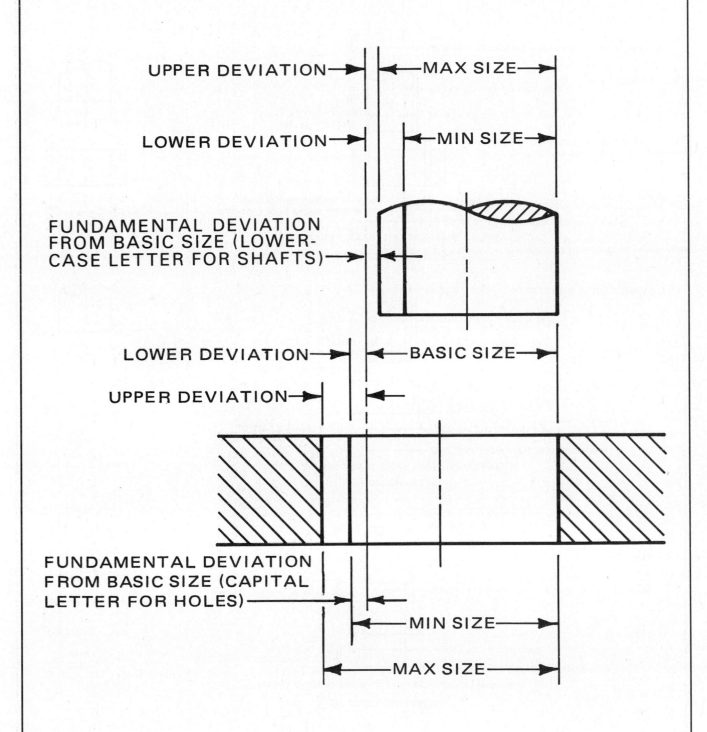

FIGURE 8-6-3 Types of inch fits.

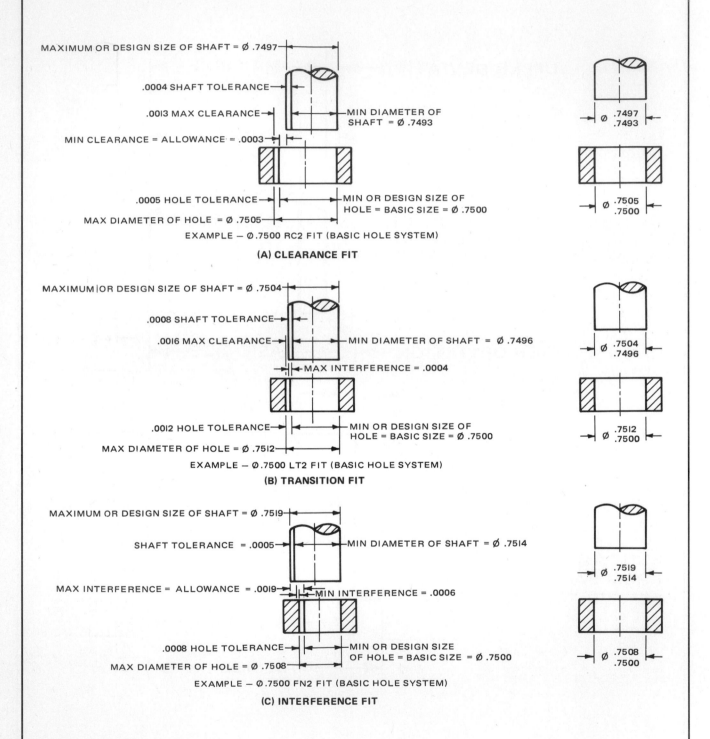

MAXIMUM OR DESIGN SIZE OF SHAFT = Ø .7497

.0004 SHAFT TOLERANCE

.0013 MAX CLEARANCE

MIN DIAMETER OF SHAFT = Ø .7493

MIN CLEARANCE = ALLOWANCE = .0003

.0005 HOLE TOLERANCE

MIN OR DESIGN SIZE OF HOLE = BASIC SIZE = Ø .7500

MAX DIAMETER OF HOLE = Ø .7505

Ø .7497 / .7493

Ø .7505 / .7500

EXAMPLE — Ø.7500 RC2 FIT (BASIC HOLE SYSTEM)

(A) CLEARANCE FIT

MAXIMUM OR DESIGN SIZE OF SHAFT = Ø .7504

.0008 SHAFT TOLERANCE

.0016 MAX CLEARANCE

MIN DIAMETER OF SHAFT = Ø .7496

MAX INTERFERENCE = .0004

.0012 HOLE TOLERANCE

MIN OR DESIGN SIZE OF HOLE = BASIC SIZE = Ø .7500

MAX DIAMETER OF HOLE = Ø .7512

Ø .7504 / .7496

Ø .7512 / .7500

EXAMPLE — Ø.7500 LT2 FIT (BASIC HOLE SYSTEM)

(B) TRANSITION FIT

MAXIMUM OR DESIGN SIZE OF SHAFT = Ø .7519

SHAFT TOLERANCE = .0005

MIN DIAMETER OF SHAFT = Ø .7514

MAX INTERFERENCE = ALLOWANCE = .0019

MIN INTERFERENCE = .0006

.0008 HOLE TOLERANCE

MIN OR DESIGN SIZE OF HOLE = BASIC SIZE = Ø .7500

MAX DIAMETER OF HOLE = Ø .7508

Ø .7519 / .7514

Ø .7508 / .7500

EXAMPLE — Ø.7500 FN2 FIT (BASIC HOLE SYSTEM)

(C) INTERFERENCE FIT

FIGURE 8-6-6 Types of metric fits.

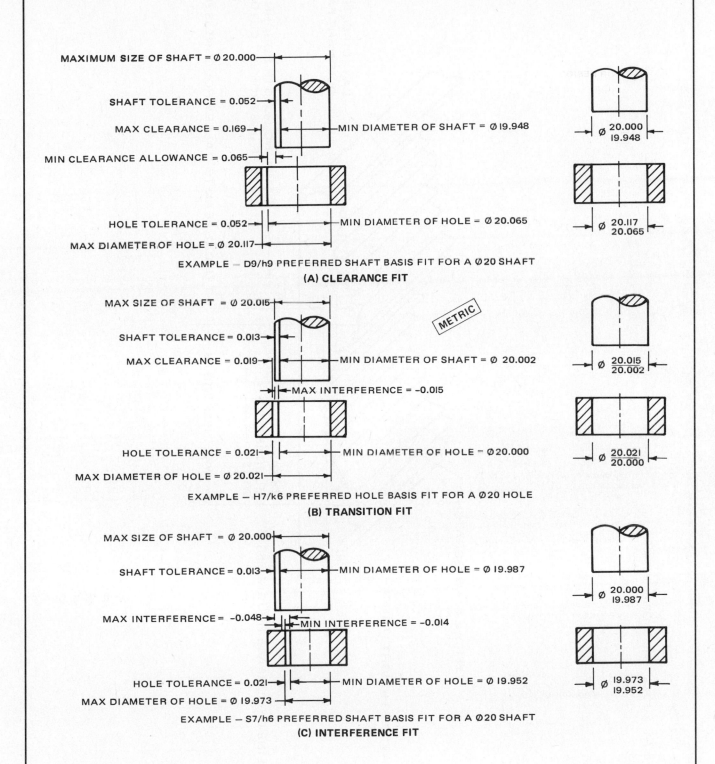

MAXIMUM SIZE OF SHAFT = Ø 20.000

SHAFT TOLERANCE = 0.052

MAX CLEARANCE = 0.169 — MIN DIAMETER OF SHAFT = Ø 19.948

MIN CLEARANCE ALLOWANCE = 0.065

HOLE TOLERANCE = 0.052 — MIN DIAMETER OF HOLE = Ø 20.065

MAX DIAMETER OF HOLE = Ø 20.117

Ø 20.000 / 19.948

Ø 20.117 / 20.065

EXAMPLE — D9/h9 PREFERRED SHAFT BASIS FIT FOR A Ø20 SHAFT

(A) CLEARANCE FIT

MAX SIZE OF SHAFT = Ø 20.015

SHAFT TOLERANCE = 0.013

MAX CLEARANCE = 0.019 — MIN DIAMETER OF SHAFT = Ø 20.002

MAX INTERFERENCE = –0.015

HOLE TOLERANCE = 0.021 — MIN DIAMETER OF HOLE – Ø 20.000

MAX DIAMETER OF HOLE = Ø 20.021

METRIC

Ø 20.015 / 20.002

Ø 20.021 / 20.000

EXAMPLE — H7/k6 PREFERRED HOLE BASIS FIT FOR A Ø20 HOLE

(B) TRANSITION FIT

MAX SIZE OF SHAFT = Ø 20.000

SHAFT TOLERANCE = 0.013 — MIN DIAMETER OF HOLE = Ø 19.987

MAX INTERFERENCE = –0.048 — MIN INTERFERENCE = –0.014

HOLE TOLERANCE = 0.021 — MIN DIAMETER OF HOLE = Ø 19.952

MAX DIAMETER OF HOLE = Ø 19.973

Ø 20.000 / 19.987

Ø 19.973 / 19.952

EXAMPLE — S7/h6 PREFERRED SHAFT BASIS FIT FOR A Ø20 SHAFT

(C) INTERFERENCE FIT

FIGURE 8-7-1 Surface texture characteristics.

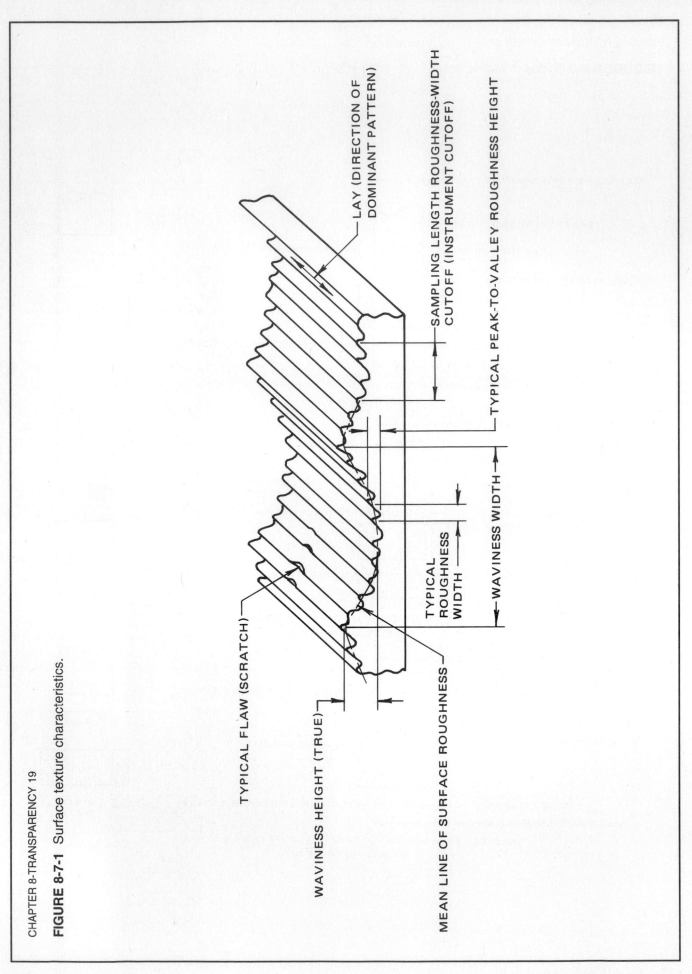

FIGURE 8-7-3 Location of notes and symbols on surface texture symbols.

PRESENT SYMBOLS VALUES SHOWN IN CUSTOMARY OR METRIC		FORMER SYMBOLS VALUES SHOWN IN MICROINCHES AND INCHES	
BASIC SURFACE TEXTURE SYMBOL	√	√	BASIC SURFACE TEXTURE SYMBOL
ROUGHNESS-HEIGHT RATING IN MICROINCHES OR MICROMETERS AND N SERIES ROUGHNESS NUMBERS	63 √ N8 √	63 √	ROUGHNESS-HEIGHT RATING IN MICROINCHES
MAXIMUM AND MINIMUM ROUGHNESS HEIGHT IN MICROINCHES OR MICROMETERS	63 32 √	63 32 √	MAXIMUM AND MINIMUM ROUGHNESS-HEIGHT RATINGS IN MICROINCHES
WAVINESS HEIGHT IN INCHES OR MILLIMETERS (F)	63 32 F √	63 32 .002 √	WAVINESS HEIGHT IN INCHES
WAVINESS SPACING IN INCHES OR MILLIMETERS (G)	63 32 F–G √	63 32 .002–I √	WAVINESS WIDTH IN INCHES
LAY SYMBOL (D)	63 32 √⊥	63 32 √⊥	LAY SYMBOL
MAXIMUM ROUGHNESS SPACING IN INCHES OR MILLIMETERS (B)	63 32 √⊥ B	.002–I √⊥ .008	SURFACE ROUGHNESS WIDTH IN INCHES
ROUGHNESS SAMPLING LENGTH OR CUTOFF RATING IN INCHES OR MILLIMETERS (C)	63 32 C √⊥	.002–I .030 √⊥ .008	ROUGHNESS WIDTH CUTOFF IN INCHES

FIGURE 8-7-4 Surface roughness range for common production.

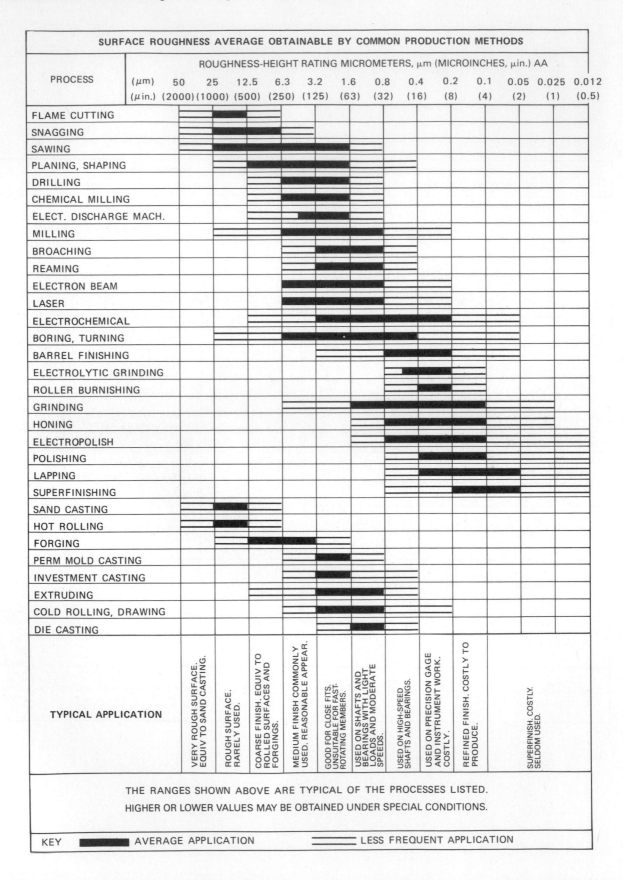

FIGURE 8-7-8 Lay symbols.

SYMBOL	DESIGNATION	EXAMPLE
=	LAY PARALLEL TO THE LINE REPRESENTING THE SURFACE TO WHICH THE SYMBOL IS APPLIED.	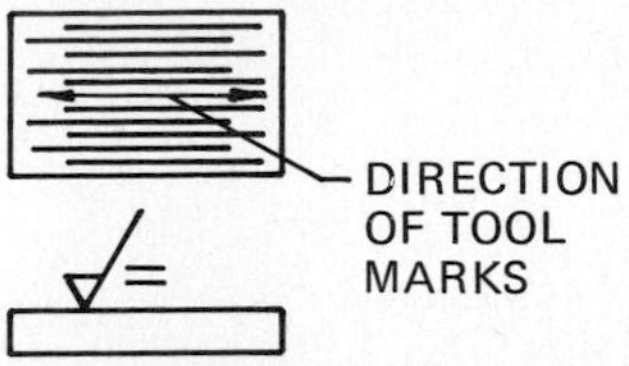 DIRECTION OF TOOL MARKS
⊥	LAY PERPENDICULAR TO THE LINE REPRESENTING THE SURFACE TO WHICH THE SYMBOL IS APPLIED.	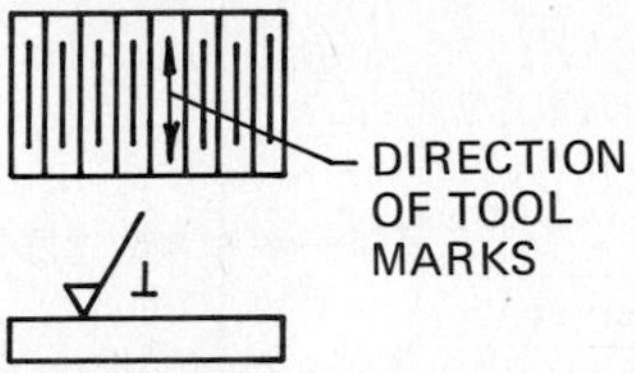 DIRECTION OF TOOL MARKS
X	LAY ANGULAR IN BOTH DIRECTIONS TO LINE REPRESENTING THE SURFACE TO WHICH SYMBOL IS APPLIED.	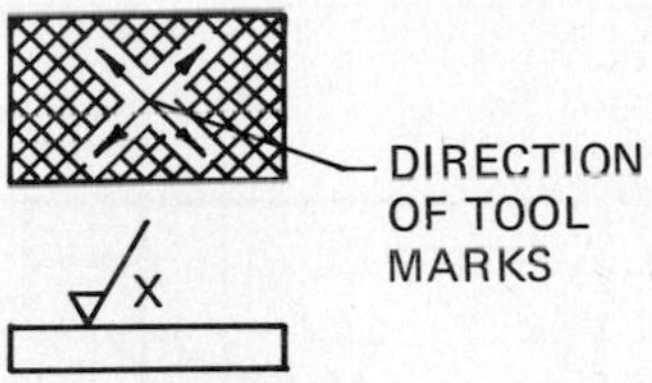 DIRECTION OF TOOL MARKS
M	LAY MULTIDIRECTIONAL.	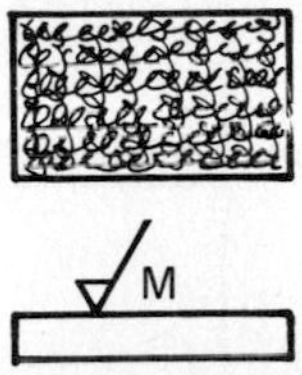
C	LAY APPROXIMATELY CIRCULAR RELATIVE TO THE CENTER OF THE SURFACE TO WHICH THE SYMBOL IS APPLIED.	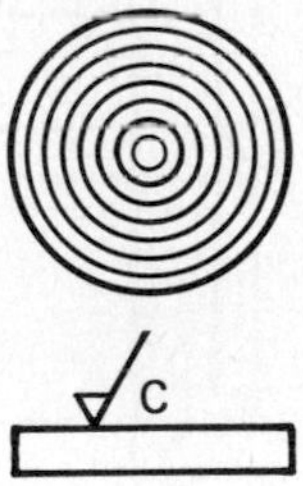
R	LAY APPROXIMATELY RADIAL RELATIVE TO THE CENTER OF THE SURFACE TO WHICH THE SYMBOL IS APPLIED.	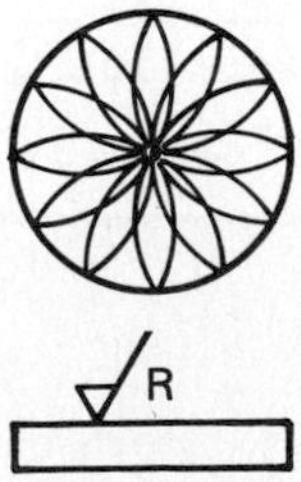
P	LAY NONDIRECTIONAL, PITTED, OR PROTUBERANT.	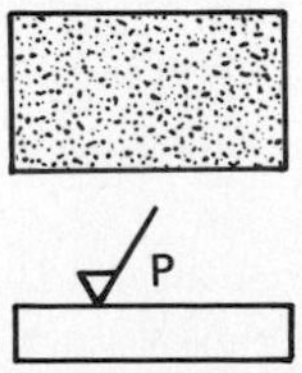

FIGURE 8-7-13 Extra metal allowance for machined surfaces.

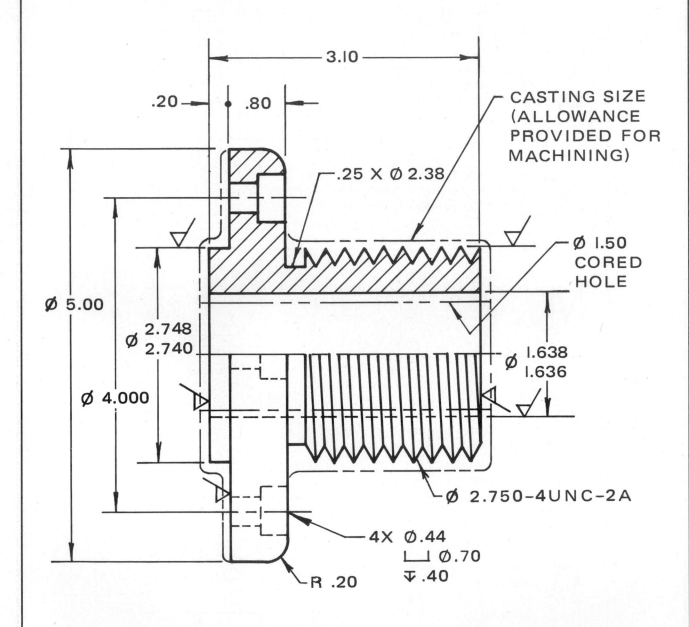

FIGURE 9-1-1 A full-section drawing.

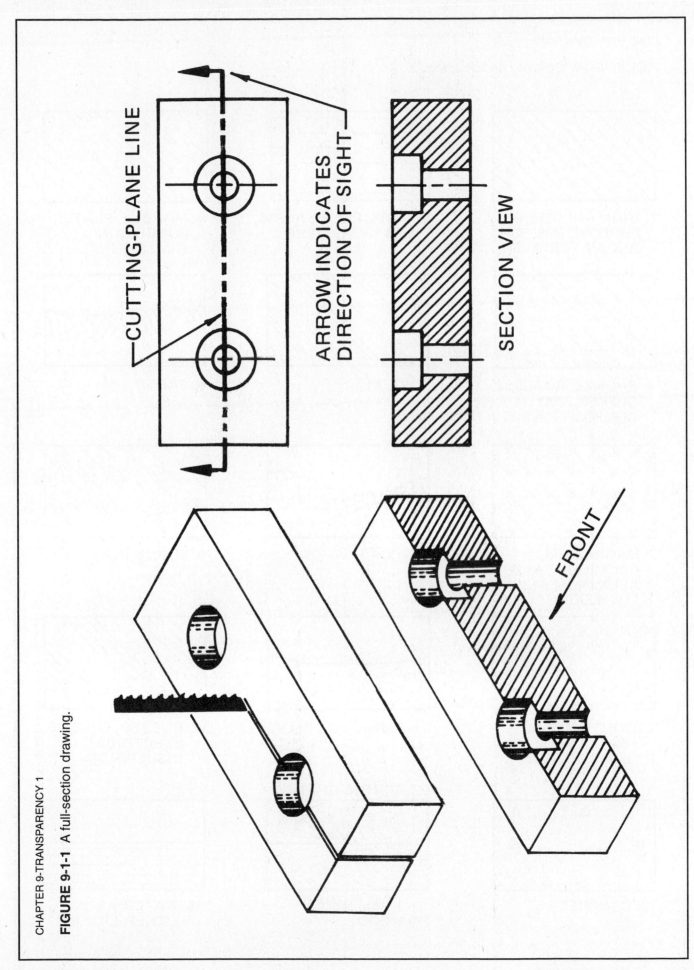

CUTTING-PLANE LINE

ARROW INDICATES DIRECTION OF SIGHT

SECTION VIEW

FRONT

FIGURE 9-1-6 Symbolic section lining.

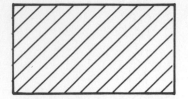

1. IRON AND GENERAL—PURPOSE USE FOR ALL MATERIALS

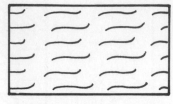

2. CORK, FELT, FABRIC, LEATHER, FIBER

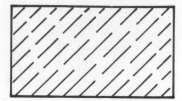

3. MARBLE, SLATE, PORCELAIN, GLASS, ETC.

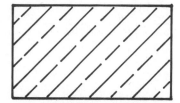

4. BRONZE, BRASS, COPPER, AND COMPOSITIONS

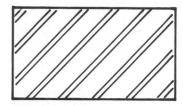

5. STEEL

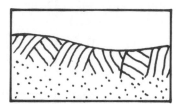

6. EARTH

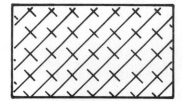

7. MAGNESIUM, ALUMINUM, AND ALUMINUM ALLOYS

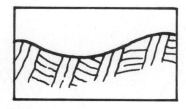

8. ROCK

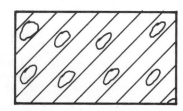

9. SOUND INSULATION

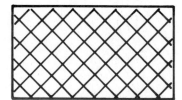

10. WHITE METAL, ZINC, LEAD, BABBITT, AND ALLOYS

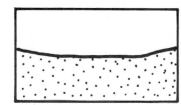

11. SAND

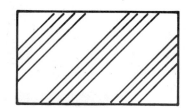

12. RUBBER, PLASTIC, ELECTRICAL INSULATION

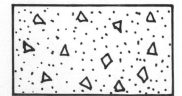

13. CONCRETE

ACROSS GRAIN

WITH GRAIN
14. WOOD

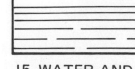

15. WATER AND OTHER LIQUIDS

FIGURE 9-2-1 Detail drawing having two sectional views.

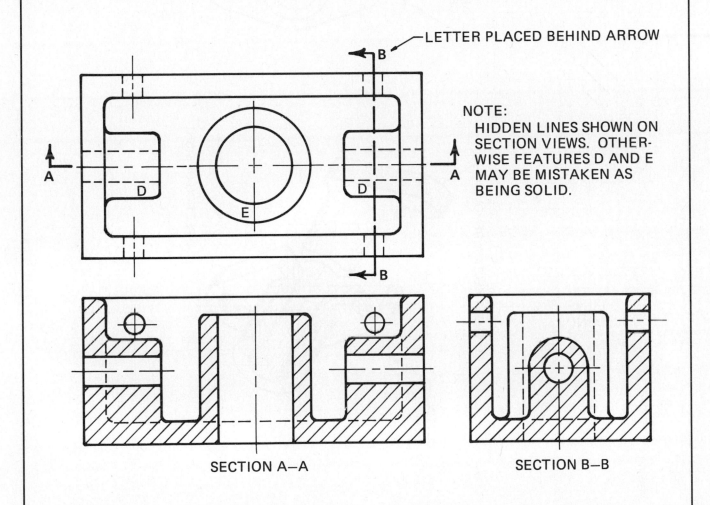

LETTER PLACED BEHIND ARROW

NOTE:
HIDDEN LINES SHOWN ON SECTION VIEWS. OTHER-WISE FEATURES D AND E MAY BE MISTAKEN AS BEING SOLID.

SECTION A—A

SECTION B—B

FIGURE 9-3-1 Half-section drawing.

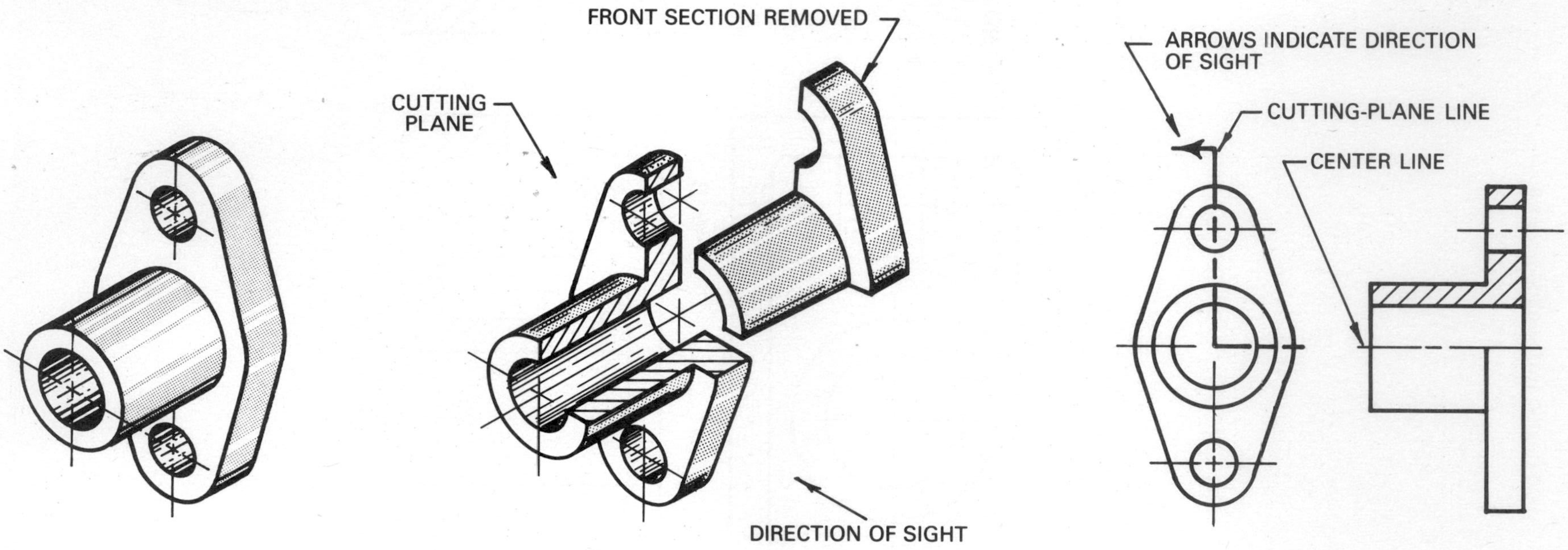

ARROWS INDICATE DIRECTION OF SIGHT

CUTTING-PLANE LINE

CENTER LINE

FRONT SECTION REMOVED

DIRECTION OF SIGHT

CUTTING PLANE

CHAPTER 9-TRANSPARENCY 5

FIGURE 9-4-1 Threads in section.

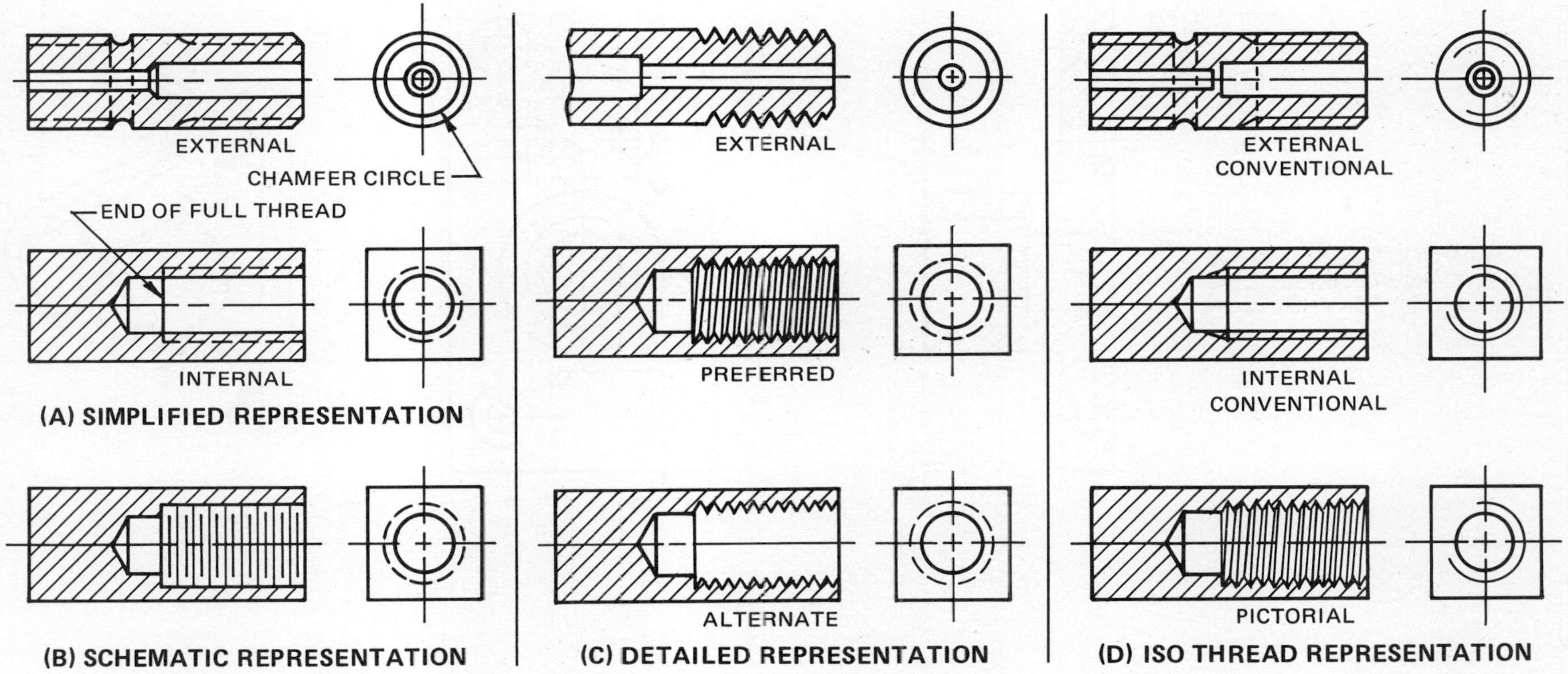
EXTERNAL
CHAMFER CIRCLE
END OF FULL THREAD
INTERNAL
(A) SIMPLIFIED REPRESENTATION
(B) SCHEMATIC REPRESENTATION
EXTERNAL
PREFERRED
ALTERNATE
(C) DETAILED REPRESENTATION
EXTERNAL CONVENTIONAL
INTERNAL CONVENTIONAL
PICTORIAL
(D) ISO THREAD REPRESENTATION

FIGURE 9-6-1 An offset section.

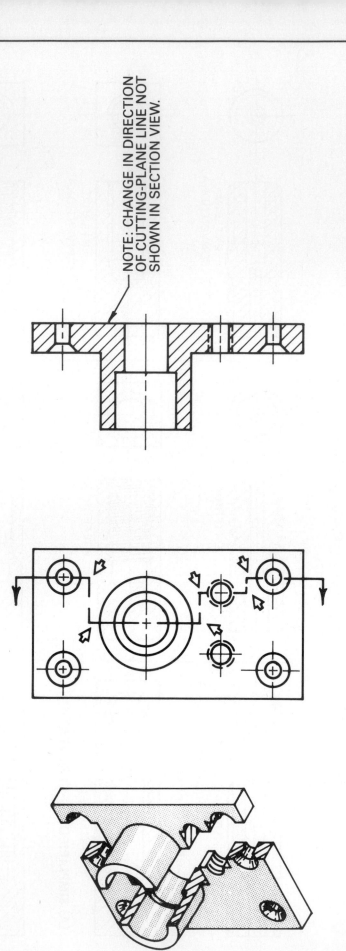

NOTE: CHANGE IN DIRECTION OF CUTTING-PLANE LINE NOT SHOWN IN SECTION VIEW.

FIGURE 9-7-1 Preferred and true projection through ribs and holes.

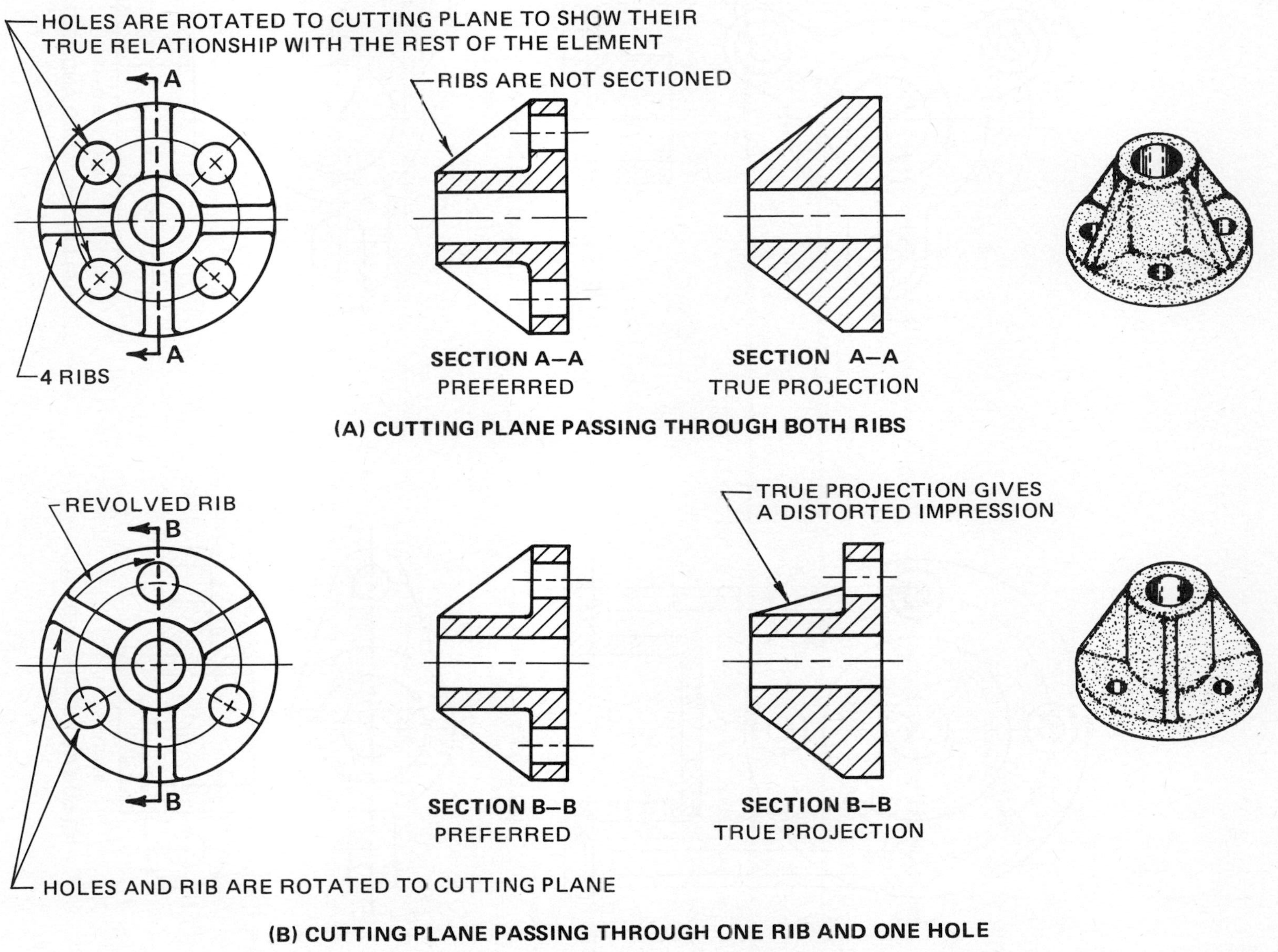

HOLES ARE ROTATED TO CUTTING PLANE TO SHOW THEIR
TRUE RELATIONSHIP WITH THE REST OF THE ELEMENT

RIBS ARE NOT SECTIONED

SECTION A–A
TRUE PROJECTION

SECTION A–A
PREFERRED

4 RIBS

(A) CUTTING PLANE PASSING THROUGH BOTH RIBS

TRUE PROJECTION GIVES
A DISTORTED IMPRESSION

SECTION B–B
TRUE PROJECTION

SECTION B–B
PREFERRED

REVOLVED RIB

HOLES AND RIB ARE ROTATED TO CUTTING PLANE

(B) CUTTING PLANE PASSING THROUGH ONE RIB AND ONE HOLE

FIGURE 9-7-3 Lugs in section.

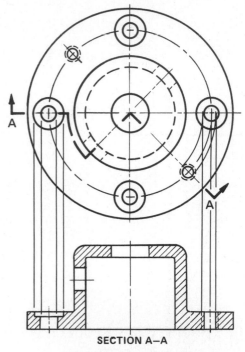

SECTION A—A

(A) HOLES ALIGNED

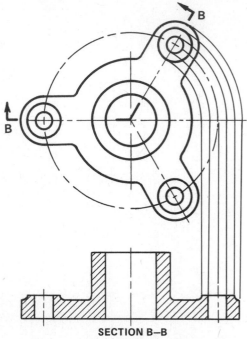

SECTION B—B

(B) LUGS ALIGNED AND SECTIONED

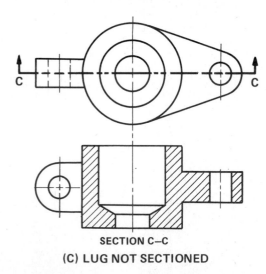

SECTION C—C

(C) LUG NOT SECTIONED

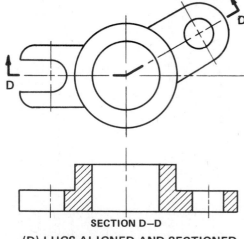

SECTION D—D

(D) LUGS ALIGNED AND SECTIONED

CHAPTER 9-TRANSPARENCY 9

FIGURE 9-8-1 Revolved sections.

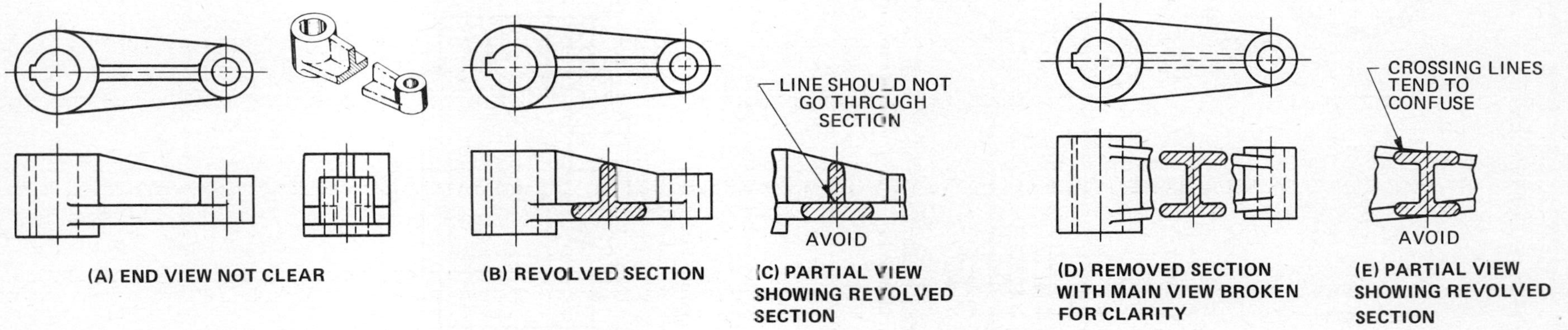
LINE SHOULD NOT GO THROUGH SECTION
CROSSING LINES TEND TO CONFUSE
AVOID
AVOID
(A) END VIEW NOT CLEAR
(B) REVOLVED SECTION
(C) PARTIAL VIEW SHOWING REVOLVED SECTION
(D) REMOVED SECTION WITH MAIN VIEW BROKEN FOR CLARITY
(E) PARTIAL VIEW SHOWING REVOLVED SECTION

FIGURE 9-9-1 Preferred and true projection of spokes.

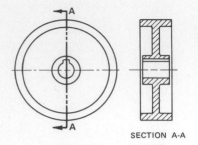

SECTION A-A

(A) FLAT PULLEY WITH WEB

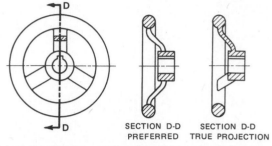

SECTION B-B
PREFERRED

SECTION B-B
TRUE PROJECTION

(B) HANDWHEEL WITH EVEN NUMBER OF SPOKES

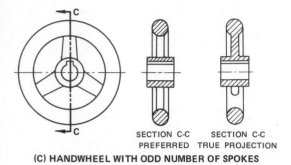

SECTION C-C
PREFERRED

SECTION C-C
TRUE PROJECTION

(C) HANDWHEEL WITH ODD NUMBER OF SPOKES

SECTION D-D
PREFERRED

SECTION D-D
TRUE PROJECTION

(D) HANDWHEEL WITH ODD NUMBER OF OFFSET SPOKES

FIGURE 9-10-1 Broken-out or partial sections.

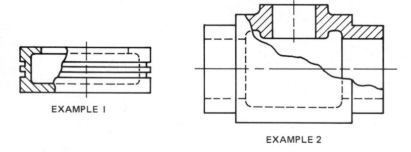

EXAMPLE I

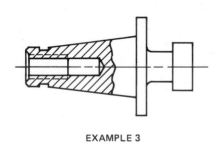

EXAMPLE 2

EXAMPLE 3

FIGURE 10-1-4 Screw thread terms.

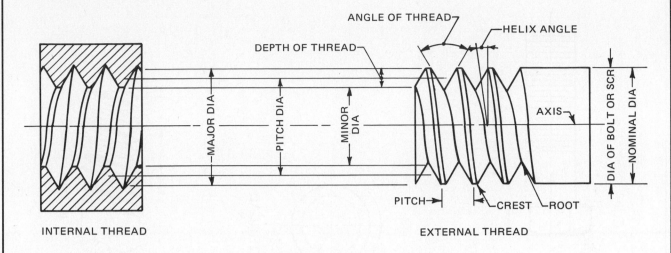

FIGURE 10-1-5 Common thread forms and proportions.

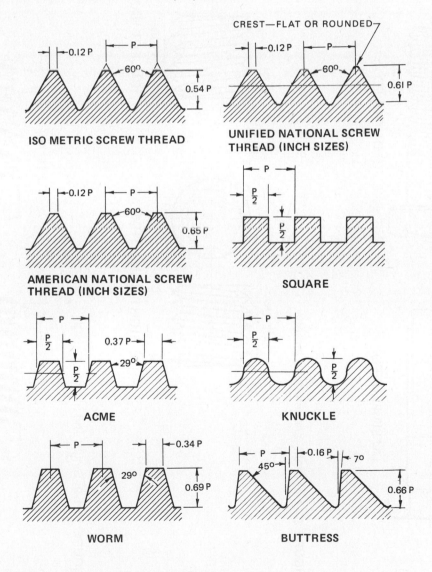

FIGURE 10-1-7 Symbolic thread representation.

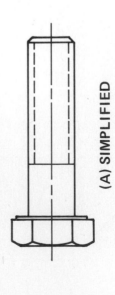

(A) **SIMPLIFIED**

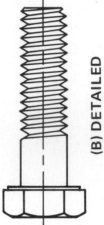

(B) **DETAILED**

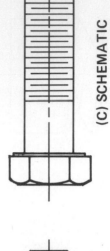

(C) **SCHEMATIC**

FIGURE 10-1-8 Right- and left-hand threads.

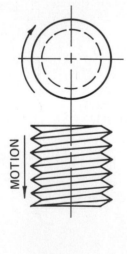

(A) **RIGHT-HAND THREAD**

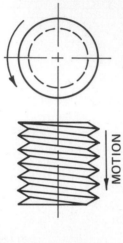

(B) **LEFT-HAND THREAD**

FIGURE 10-1-9 Single and multiple threads.

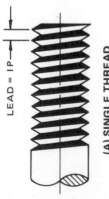

LEAD = IP

(A) **SINGLE THREAD**

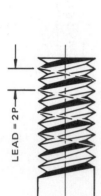

LEAD = 2P

(B) **DOUBLE THREAD**

LEAD = 3P

(C) **TRIPLE THREAD**

FIGURE 10-1-10 Simplified thread representation.

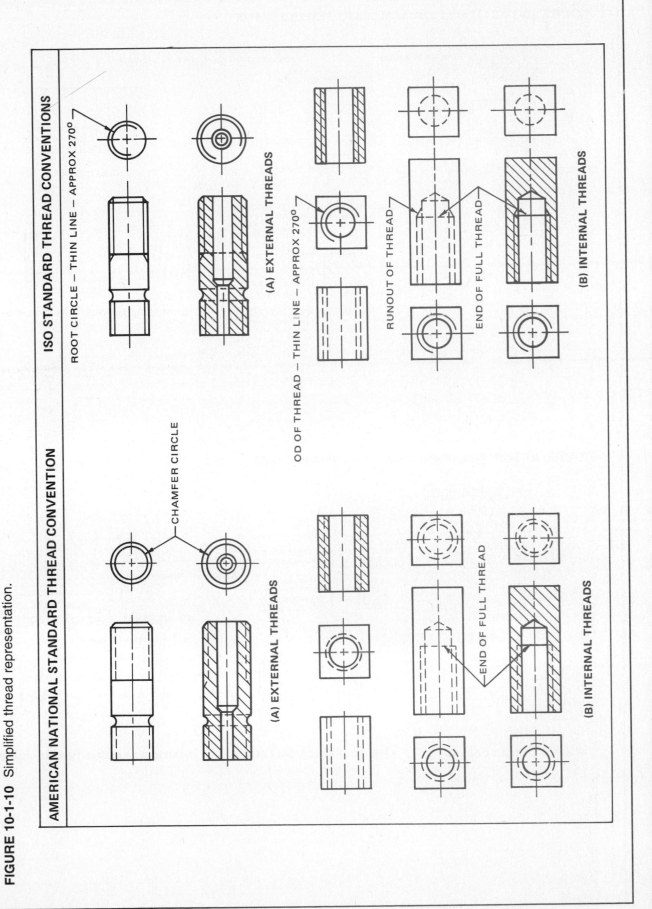

AMERICAN NATIONAL STANDARD THREAD CONVENTION

CHAMFER CIRCLE

(A) EXTERNAL THREADS

END OF FULL THREAD

(B) INTERNAL THREADS

ISO STANDARD THREAD CONVENTIONS

ROOT CIRCLE — THIN LINE — APPROX 270°

(A) EXTERNAL THREADS

OD OF THREAD — THIN LINE — APPROX 270°

RUNOUT OF THREAD

END OF FULL THREAD

(B) INTERNAL THREADS

FIGURE 10-1-12 Thread specifications for inch-size threads.

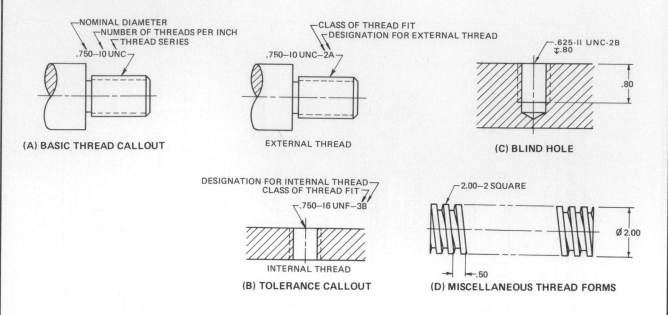

(A) BASIC THREAD CALLOUT

EXTERNAL THREAD

(C) BLIND HOLE

(B) TOLERANCE CALLOUT

(D) MISCELLANEOUS THREAD FORMS

FIGURE 10-1-13 Thread specifications for metric threads.

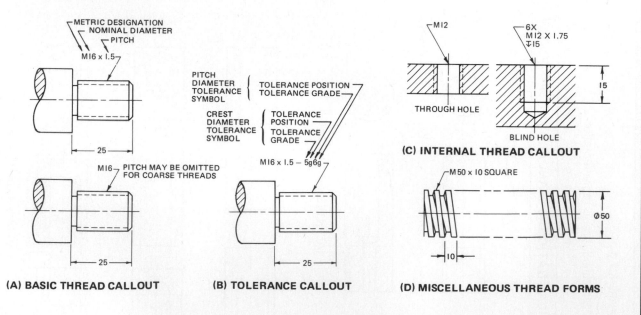

(A) BASIC THREAD CALLOUT

(B) TOLERANCE CALLOUT

(C) INTERNAL THREAD CALLOUT

(D) MISCELLANEOUS THREAD FORMS

CHAPTER 10-TRANSPARENCY 5

FIGURE 10-2-1 Detailed representation of threads.

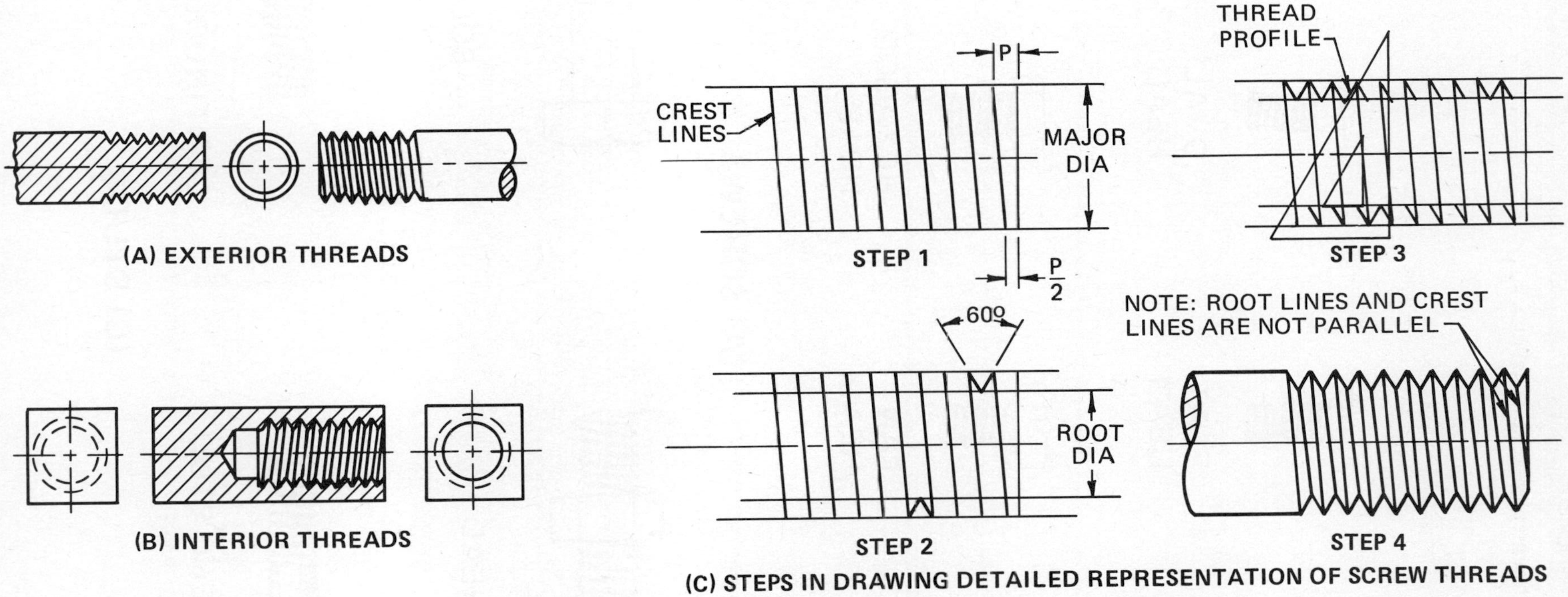

(A) EXTERIOR THREADS
(B) INTERIOR THREADS
CREST LINES
P
MAJOR DIA
STEP 1
P/2
60º
ROOT DIA
STEP 2
THREAD PROFILE
STEP 3
NOTE: ROOT LINES AND CREST LINES ARE NOT PARALLEL
STEP 4
(C) STEPS IN DRAWING DETAILED REPRESENTATION OF SCREW THREADS

FIGURE 10-3-1 Common threaded fasteners.

ROUND
HEAD

FLAT
HEAD

OVAL
HEAD

FILLISTER
HEAD

TRUSS
HEAD

PAN
HEAD

HEXAGON
HEAD

HEXAGON
WASHER
HEAD

(A) SCREWS

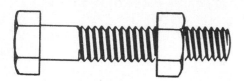
HEX HEAD

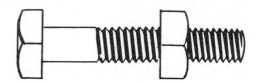
SQUARE HEAD

(B) BOLTS

DOUBLE–END STUD

CONTINUOUS–THREAD STUD

(C) STUDS

FIGURE 10-3-2 Fastener applications.

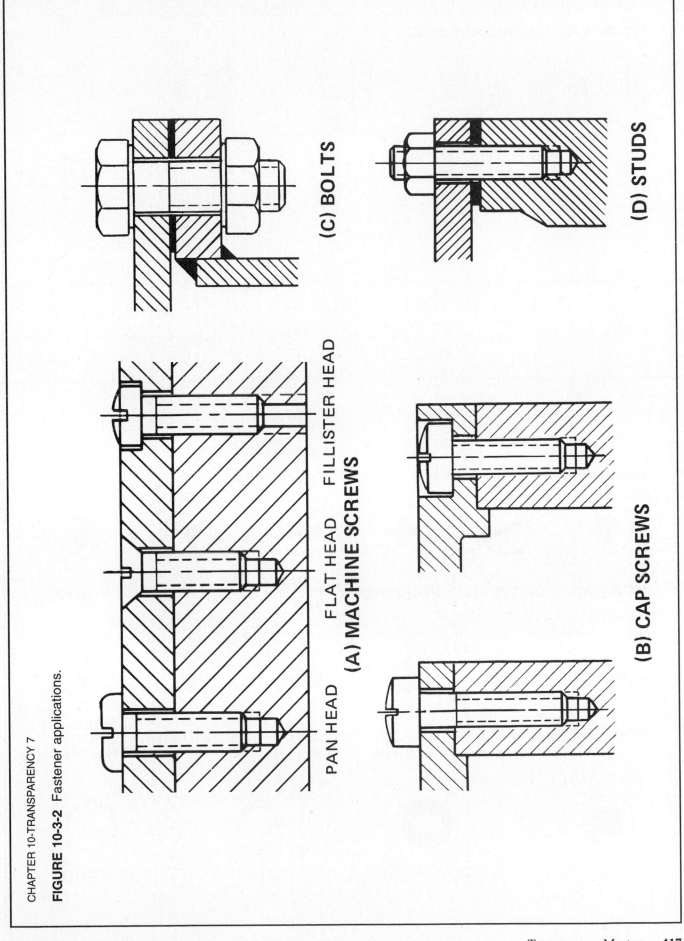

(C) BOLTS

(D) STUDS

PAN HEAD   FLAT HEAD   FILLISTER HEAD

(A) MACHINE SCREWS

(B) CAP SCREWS

FIGURE 10-3-3 Common head styles.

PAN

BINDING

WASHER
(FLANGED)

OVAL

TRUSS

HEX

SQUARE

FLAT

FILLISTER

12-SPLINE
FLANGE

FIGURE 10-3-4 Drive configurations.

HEX CAP

SLOTTED

PHILLIPS®

CLUTCH
TYPE A

TRI-WING®

TORX®

TORQ-SET®

TRIPLE
SQUARE

MULTI-
SPLINE

CLUTCH
TYPE G

POZIDRIV®

REED &
PRINCE
(FREARSON)

SCRULOX®

SLAB HEAD

SQUARE

HEXAGON

FIGURE 10-3-12 Approximate head proportions for hexhead cap screws, bolts, and nuts.

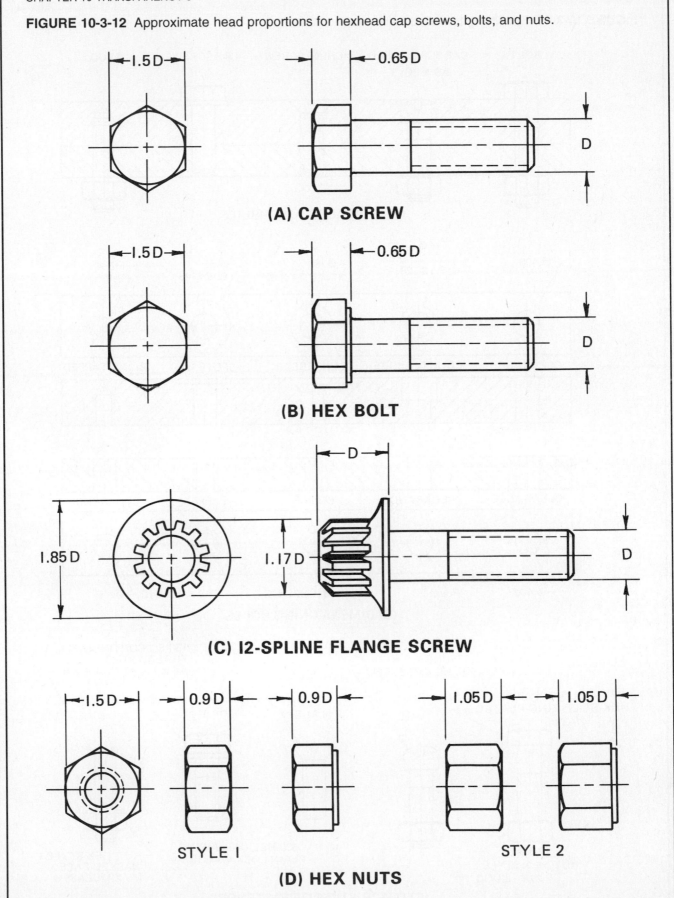

(A) CAP SCREW

(B) HEX BOLT

(C) 12-SPLINE FLANGE SCREW

STYLE 1

STYLE 2

(D) HEX NUTS

FIGURE 10-3-18 Specifying threaded fasteners and holes.

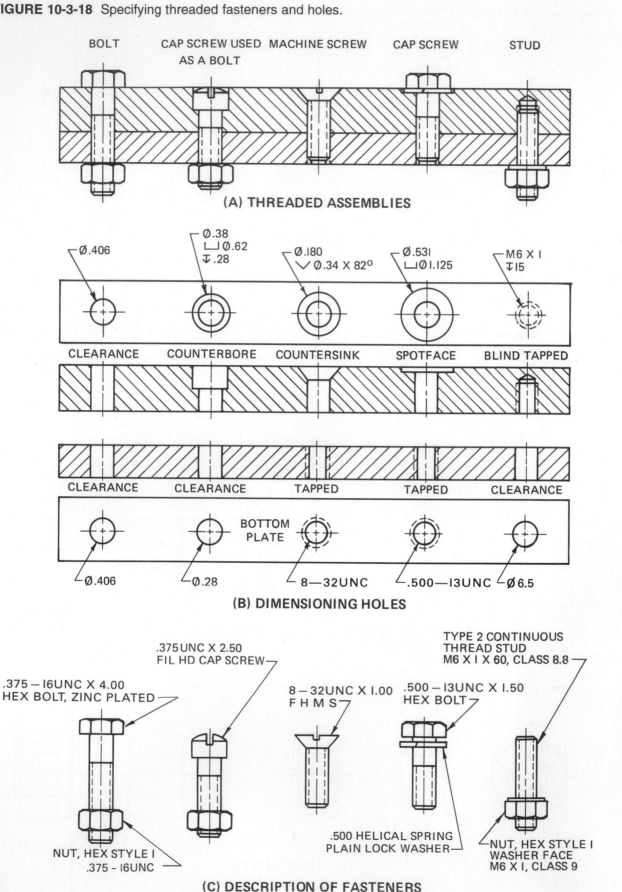

BOLT CAP SCREW USED MACHINE SCREW CAP SCREW STUD
AS A BOLT

(A) THREADED ASSEMBLIES

Ø.406

Ø.38
⌴Ø.62
⊼.28

Ø.180
⌵Ø.34 X 82°

Ø.531
⌴Ø1.125

M6 X 1
⊼15

CLEARANCE COUNTERBORE COUNTERSINK SPOTFACE BLIND TAPPED

CLEARANCE CLEARANCE TAPPED TAPPED CLEARANCE

BOTTOM
PLATE

Ø.406 Ø.28 8—32UNC .500—13UNC Ø6.5

(B) DIMENSIONING HOLES

.375—16UNC X 4.00
HEX BOLT, ZINC PLATED

.375UNC X 2.50
FIL HD CAP SCREW

8—32UNC X 1.00
F H M S

.500—13UNC X 1.50
HEX BOLT

TYPE 2 CONTINUOUS
THREAD STUD
M6 X 1 X 60, CLASS 8.8

.500 HELICAL SPRING
PLAIN LOCK WASHER

NUT, HEX STYLE I
.375—16UNC

NUT, HEX STYLE I
WASHER FACE
M6 X 1, CLASS 9

(C) DESCRIPTION OF FASTENERS

CHAPTER 11-TRANSPARENCY 1

FIGURE 11-1-1 Miscellaneous types of fasteners.

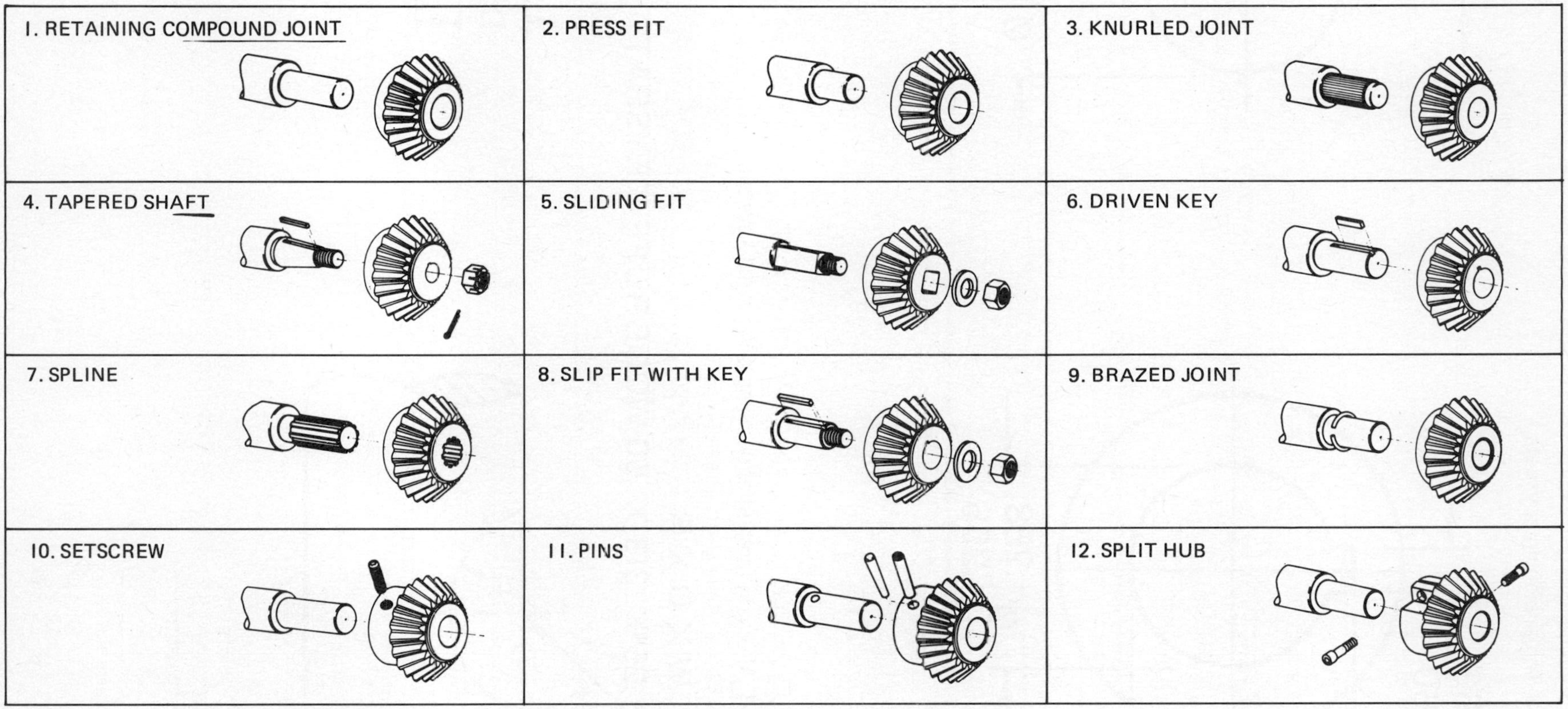
1. RETAINING COMPOUND JOINT
2. PRESS FIT
3. KNURLED JOINT
4. TAPERED SHAFT
5. SLIDING FIT
6. DRIVEN KEY
7. SPLINE
8. SLIP FIT WITH KEY
9. BRAZED JOINT
10. SETSCREW
11. PINS
12. SPLIT HUB

FIGURE 11-1-3 Dimensioning keyseats.

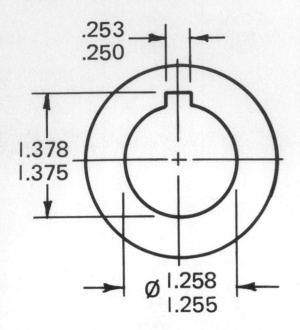

.253
.250

1.378
1.375

Ø 1.258
1.255

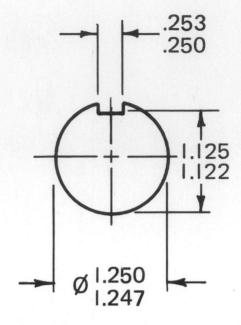

.253
.250

1.125
1.122

Ø 1.250
1.247

FIGURE 11-1-4 Alternate method of detailing a Woodruff keyseat.

W X D X R
.25 X .313 X .50 WOODRUFF KEYSEAT

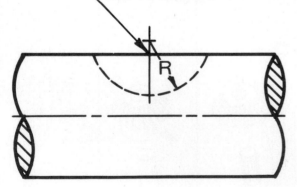

R

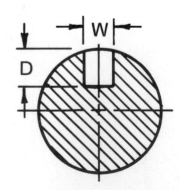

W

D

FIGURE 11-1-7 Callout and representation of splines.

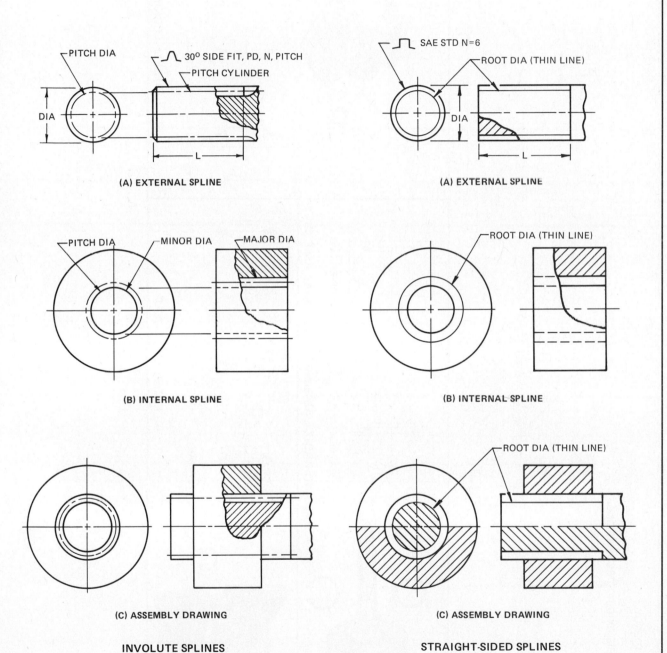

(A) EXTERNAL SPLINE

(A) EXTERNAL SPLINE

(B) INTERNAL SPLINE

(B) INTERNAL SPLINE

(C) ASSEMBLY DRAWING

(C) ASSEMBLY DRAWING

INVOLUTE SPLINES

STRAIGHT-SIDED SPLINES

CHAPTER 11-TRANSPARENCY 4

FIGURE 11-3-1 Retaining ring applications.

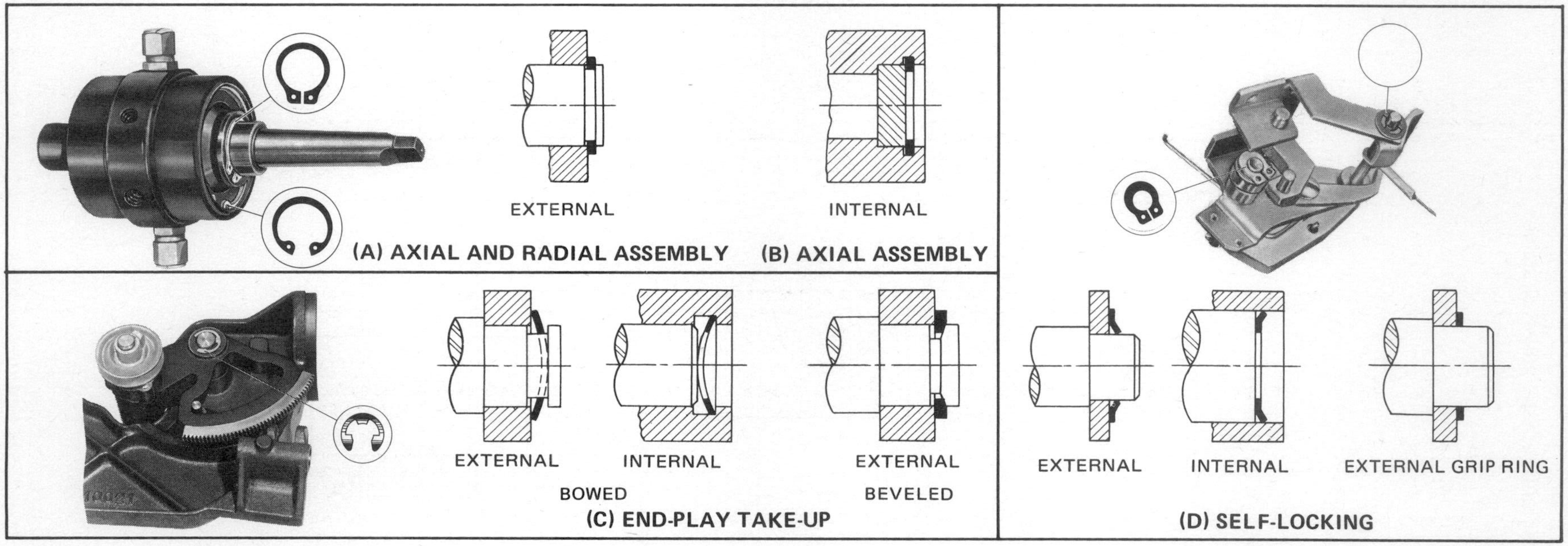
EXTERNAL
INTERNAL
(A) AXIAL AND RADIAL ASSEMBLY
(B) AXIAL ASSEMBLY
EXTERNAL
INTERNAL
BOWED
EXTERNAL
BEVELED
(C) END-PLAY TAKE-UP
EXTERNAL
INTERNAL
EXTERNAL GRIP RING
(D) SELF-LOCKING

CHAPTER 11-TRANSPARENCY 5

FIGURE 11-4-1 Types of springs.

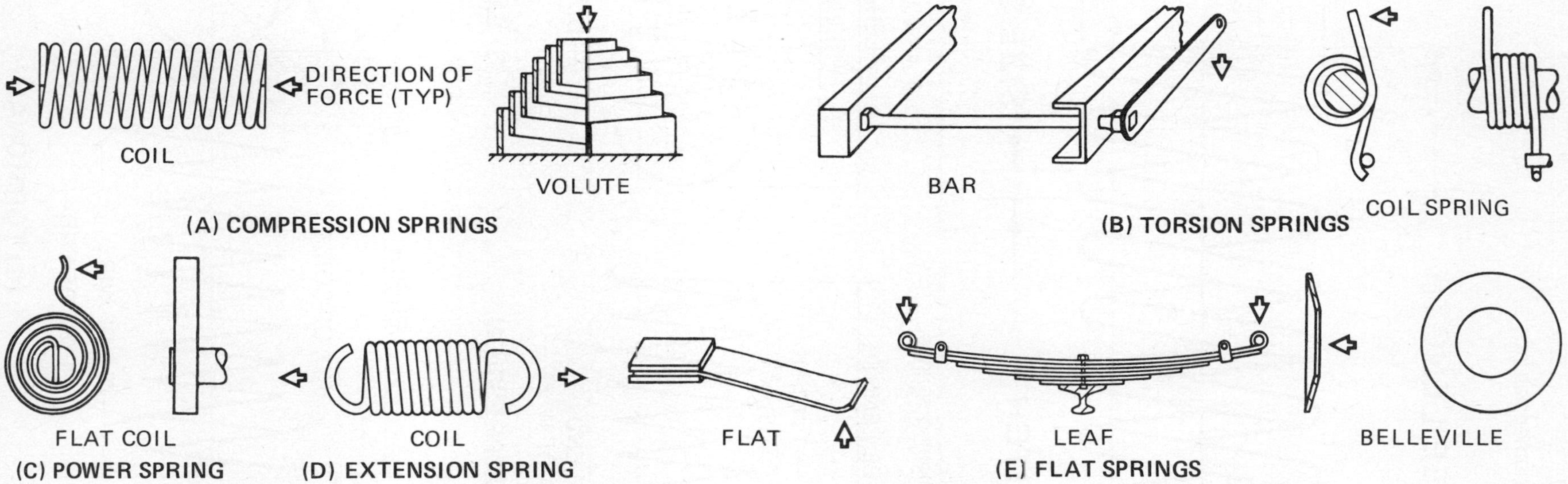
DIRECTION OF
FORCE (TYP)
COIL
VOLUTE
(A) COMPRESSION SPRINGS
BAR
COIL SPRING
(B) TORSION SPRINGS
FLAT COIL
(C) POWER SPRING
COIL
(D) EXTENSION SPRING
FLAT
LEAF
BELLEVILLE
(E) FLAT SPRINGS

FIGURE 11-4-2 Spring nomenclature.

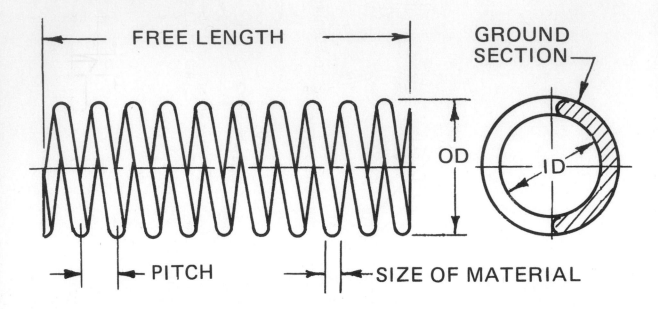

FREE LENGTH

GROUND SECTION

OD

ID

PITCH

SIZE OF MATERIAL

FIGURE 11-4-8 Dimensioning springs.

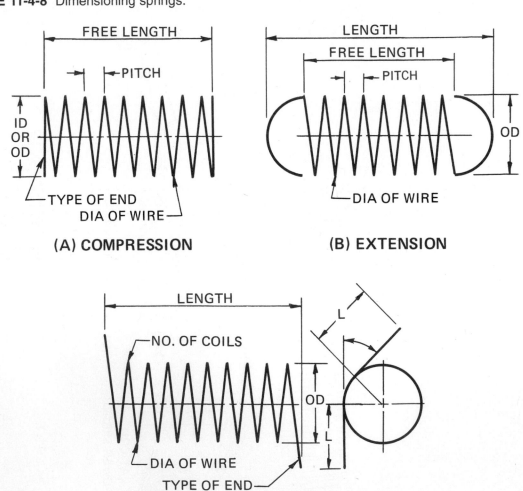

FREE LENGTH

PITCH

ID OR OD

TYPE OF END
DIA OF WIRE

(A) COMPRESSION

LENGTH

FREE LENGTH

PITCH

OD

DIA OF WIRE

(B) EXTENSION

LENGTH

NO. OF COILS

OD

L

L

DIA OF WIRE

TYPE OF END

(C) TORSIONAL

FIGURE 11-5-2 Approximate sizes and types of large rivets .50 in. (12mm) and

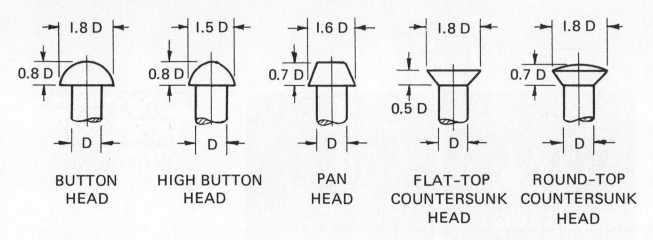

BUTTON HEAD   HIGH BUTTON HEAD   PAN HEAD   FLAT-TOP COUNTERSUNK HEAD   ROUND-TOP COUNTERSUNK HEAD

FIGURE 11-5-3 Conventional rivet symbols.

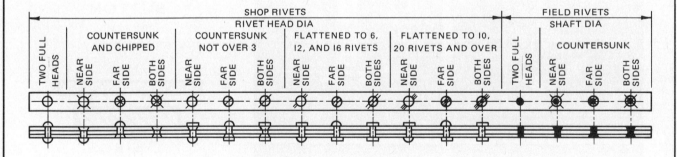

FIGURE 11-6-1 Resistance-welded fasteners.

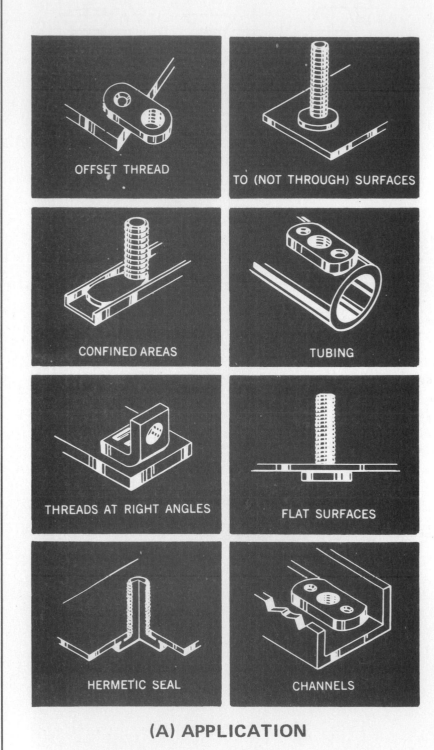

(A) APPLICATION

OFFSET THREAD

TO (NOT THROUGH) SURFACES

CONFINED AREAS

TUBING

THREADS AT RIGHT ANGLES

FLAT SURFACES

HERMETIC SEAL

CHANNELS

SPHERICAL

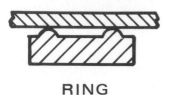

RING

BUTTON

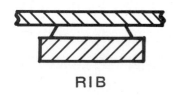

RIB

PYRAMIDAL

(B) WELD PROJECTION

FIGURE 11-7-1 Stresses in bonded joints.

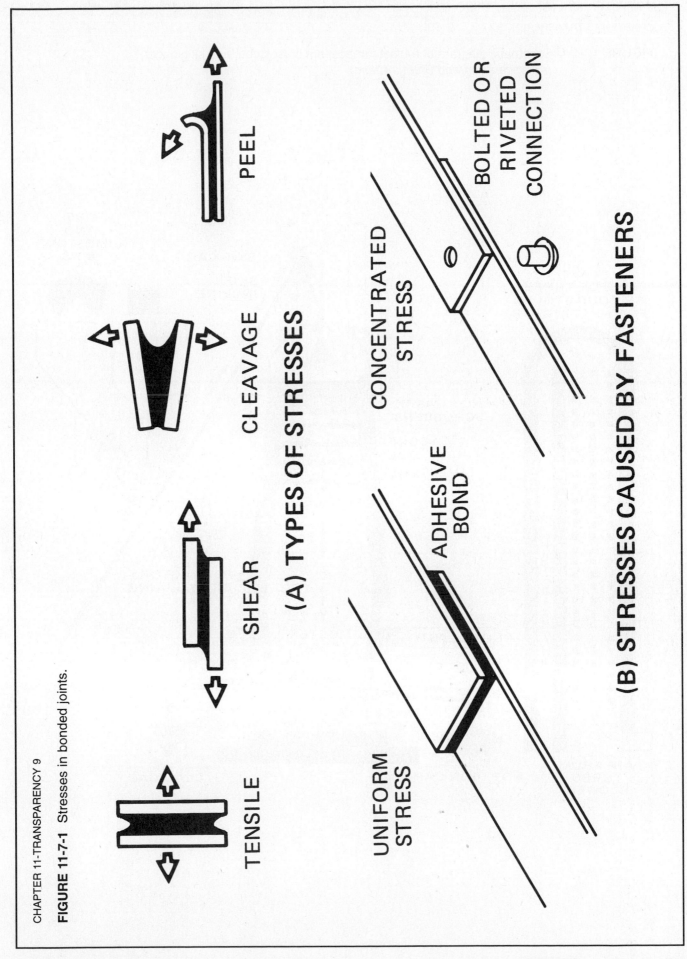

PEEL

CLEAVAGE

SHEAR

TENSILE

(A) TYPES OF STRESSES

BOLTED OR RIVETED CONNECTION

CONCENTRATED STRESS

ADHESIVE BOND

UNIFORM STRESS

(B) STRESSES CAUSED BY FASTENERS

FIGURE 12-1-1 Schematic diagram of a blast furnace, hot blast stone, and skiploader.
(American Iron and Steel Institute)

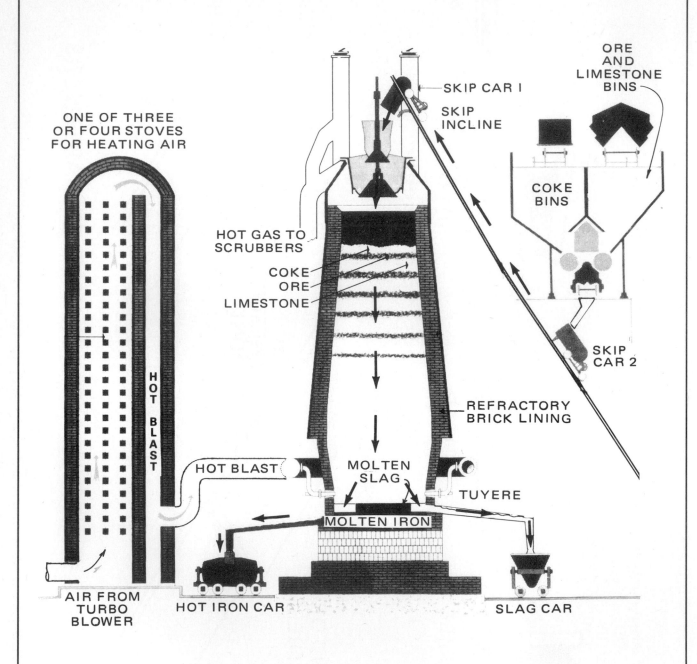

FIGURE 12-2-1 Flowchart for steelmaking.

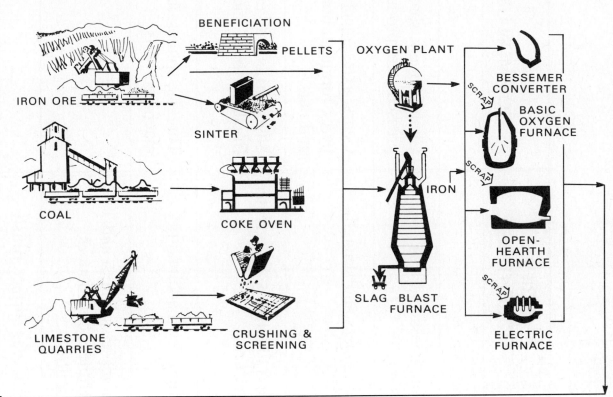

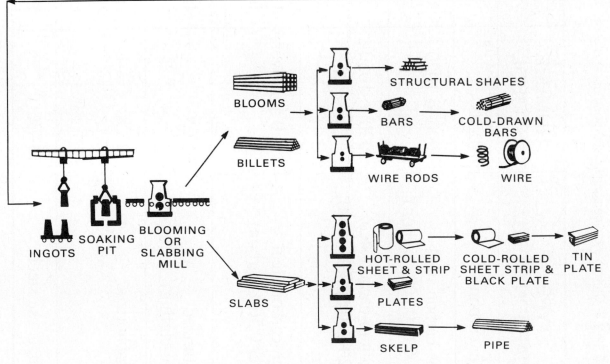

CHAPTER 12-TRANSPARENCY 3

FIGURE 12-2-3 Carbon steel designations, properties, and uses.

TYPE OF CARBON STEEL	NUMBER SYMBOL	PRINCIPAL PROPERTIES	COMMON USES
Plain carbon	10XX		
Low-carbon steel (0.06 to 0.20% carbon)	1006 to 1020	Toughness and less strength	Chains, rivets, shafts, and pressed steel products
Medium-carbon steel (0.20 to 0.50% carbon)	1020 to 1050	Toughness and strength	Gears, axles, machine parts, forgings, bolts, and nuts
High-carbon steel (over 0.50% carbon)	1050 and over	Less toughness and greater hardness	Saws, drills, knives, razors, finishing tools, and music wire
Sulfurized (free-cutting)	11XX	Improves machinability	Threads, splines, and machined parts
Phosphorized	12XX	Increases strength and hardness but reduces ductility	
Manganese steels	13XX	Improves surface finish	

FIGURE 12-2-8 AISI designation system for alloy steel.

TYPE OF STEEL	ALLOY SERIES	APPROXIMATE ALLOY CONTENT (%)	PRINCIPAL PROPERTIES	COMMON USES
Manganese steel	13xx	Mn 1.6–1.9	Improve surface finish	
Molybdenum steels	40xx	Mo 0.15–0.3	High strength	Axles, forgings, gears, cams, mechanical parts
	41xx	Cr 0.4–1.1; Mo 0.08–0.35		
	43xx	Ni 1.65–2; Cr 0.4–0.9; Mo 0.2–0.3		
	44xx	Mo 0.45–0.6		
	46xx	Ni 0.7–2; Mo 0.15–0.3		
	47xx	Ni 0.9–1.2; Cr 0.35–0.55; Mo 0.15–0.4		
	48xx	Ni 3.25–3.75; Mo 0.2–0.3		
Chromium steels	50xx	Cr 0.3–0.5	Hardness, great strength and toughness	Gears, shafts, bearings, springs, connecting rods
	51xx	Cr 0.7–1.15		
	E51100	C 1.0; Cr 0.9–1.15		
	E52100	C 1.0; Cr 0.9–1.15		
Chromium-vanadium steel	61xx	Cr 0.5–1.1; V 0.1–0.15	Hardness and strength	Punches and dies, piston rods, gears, axles
Nickel-chromium-molybdenum steels	86xx	Ni 0.4–0.7; Cr 0.4–0.6; Mo 0.15–0.25	Rust resistance, hardness, and strength	Food containers, surgical equipment
	87xx	Ni 0.4–0.7; Cr 0.4–0.6; Mo 0.2–0.3		
	88xx	Ni 0.4–0.7; Cr 0.4–0.6; Mo 0.3–0.4		
Silicon-manganese steel	92xx	Si 1.8–2.2	Springiness and elasticity	Springs

CHAPTER 12-TRANSPARENCY 5

FIGURES 12-3-1 Common methods of forming metals.

FORMING METHOD ↓ / METAL →	ALUMINUM	COPPER	IRON	LEAD	MAGNESIUM	NICKEL	SILVER, GOLD, PLATINUM	MOLYBDENUM, COPPER TANTALUM, TUNGSTEN	STEEL	TIN	TITANIUM	ZINC
Casting												
Centrifugal	✔	✔	✔			✔			✔			
Continuous	✔	✔	✔	✔					✔			
Ceramic mold	✔	✔	✔		✔	✔			✔			✔
Investment	✔	✔			✔	✔	✔		✔			
Permanent mold	✔	✔	✔	✔	✔	✔			✔	✔		✔
Sand	✔	✔	✔	✔	✔	✔			✔	✔		✔
Shell mold	✔	✔	✔			✔			✔			
Die casting	✔	✔		✔	✔					✔		✔
Cold heading	✔	✔		✔		✔	✔		✔			
Deep drawing	✔	✔			✔	✔			✔		✔	✔
Extruding	✔	✔		✔	✔	✔		✔	✔	✔	✔	
Forging	✔	✔			✔	✔		✔	✔		✔	
Machining	✔	✔	✔		✔	✔	✔	✔	✔		✔	✔
PM compacting	✔	✔	✔			✔	✔	✔	✔		✔	
Stamping and forming	✔	✔			✔	✔	✔	✔	✔		✔	✔

FIGURE 12-3-2 Wrought aluminum alloy designations.

MAJOR ALLOYING ELEMENT	DESIGNATION
Aluminum (99% or more)	1xxx
Copper	2xxx
Manganese	3xxx
Silicon	4xxx
Magnesium	5xxx
Magnesium and silicon	6xxx
Zinc	7xxx
Other elements	8xxx
Unused series	9xxx

CHAPTER 12-TRANSPARENCY 7

Common terms and definitions.

FAMILY OF PLASTICS

THERMOSETTING

Alkyds
Allylics
Amino (Melamine and Urea)
Casein
Epoxy
Phenolics
Polyesters (fiberglass)
Silicones

THERMOPLASTICS

ABS
Acetal Resin
Acrylics
Cellulosics
Fluorocarbons
Nylon
Polycarbonate
Polyethylene
Polystyrene
Polypropylenes
Urethanes
Vinyls

FIGURE 13-1-2 Sequence in preparing a sand casting.

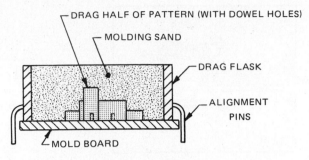

DRAG HALF OF PATTERN (WITH DOWEL HOLES)

MOLDING SAND

DRAG FLASK

ALIGNMENT PINS

MOLD BOARD

(A) STARTING TO MAKE THE SAND MOLD

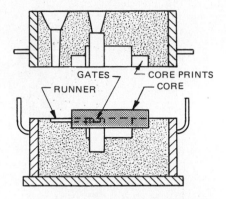

GATES — CORE PRINTS

RUNNER — CORE

(E) PARTING FLASKS TO REMOVE PATTERN AND TO ADD CORE AND RUNNER

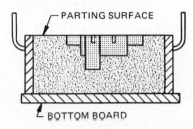

PARTING SURFACE

BOTTOM BOARD

(B) AFTER ROLLING OVER THE DRAG

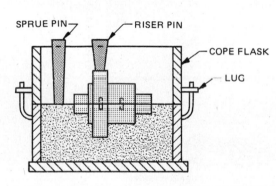

SPRUE PIN — RISER PIN

COPE FLASK

LUG

(C) PREPARING TO RAM MOLDING SAND IN COPE

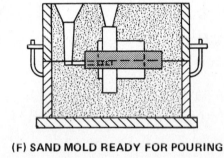

(F) SAND MOLD READY FOR POURING

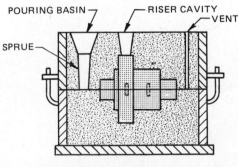

POURING BASIN — RISER CAVITY

VENT

SPRUE

(D) REMOVING RISER AND SPRUE PINS AND ADDING POURING BASIN

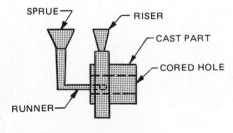

SPRUE — RISER

CAST PART

CORED HOLE

RUNNER

SPRUE, RISER, AND RUNNER TO BE REMOVED FROM CASTING.

(G) CASTING AS REMOVED FROM THE MOLD

CHAPTER 13-TRANSPARENCY 2

FIGURE 13-1-5 Investment mold casting.

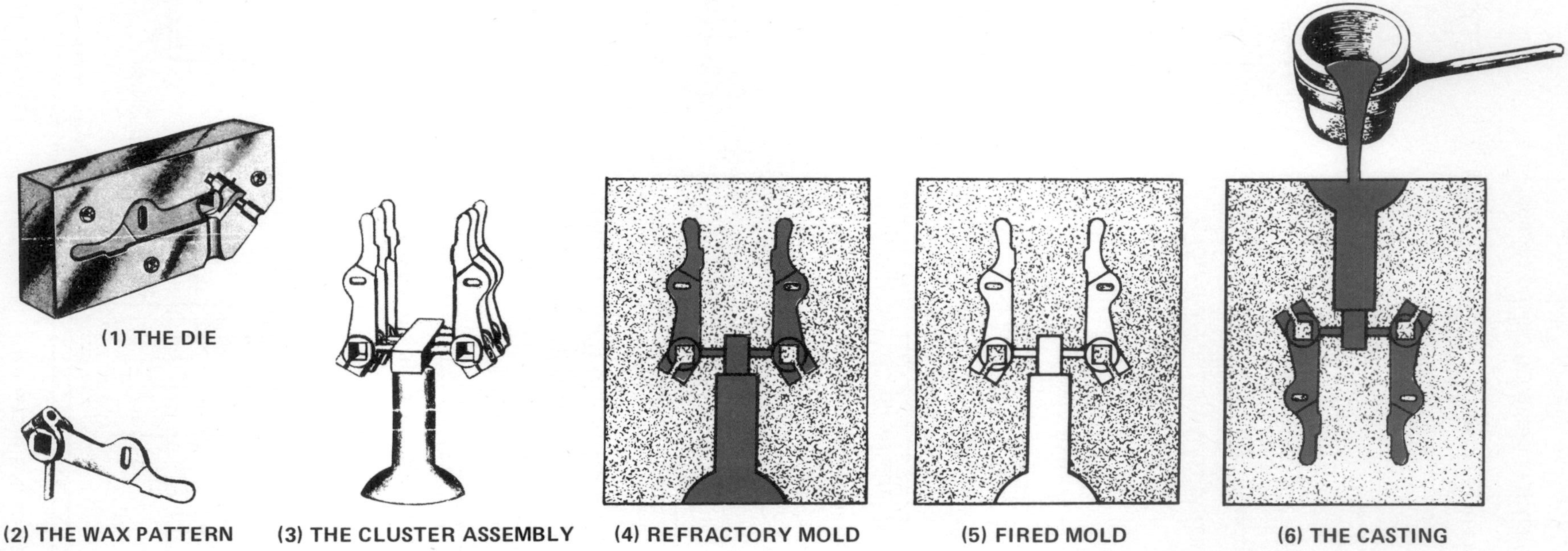
(1) THE DIE
(2) THE WAX PATTERN
(3) THE CLUSTER ASSEMBLY
(4) REFRACTORY MOLD
(5) FIRED MOLD
(6) THE CASTING

FIGURE 13-1-7 General characteristics of casting processes.

PROCESS	METALS CAST	USUAL WEIGHT (MASS) RANGE	MINIMUM PRODUCTION QUANTITIES	RELATIVE SET-UP COST	CASTING DETAIL FEASIBLE	MINIMUM THICKNESS IN. (MM)	DIMENSIONAL TOLERANCES IN. (MM)	SURFACE FINISH, RMS (μIN.)
SAND (Green, Dry, and Core) CO$_2$ Sand	All ferrous and nonferrous	Less than 1 lb. (0.5 kg) to several tons	3, without mechanization	Very low to high depending on mechanization	Fair	.12 to .25 (3 to 6) .10 to .25 (2.5 to 6)	± .03 (0.8) ± .02 (0.5)	350 250
SHELL	All ferrous and nonferrous	0.5 to 30 lb. (0.2 to 15 kg)	50	Moderate to high depending on mechanization	Fair to good	.03 to .10 (0.8 to 2.5)	± .015 (0.4)	200
PLASTER	Al, Mg, Cu, and Zn alloys	Less than 1 lb. to 3000 lb. (0.5 to 1350 kg)	1	Moderate	Excellent	.03 to .08 (0.8 to 2)	± .01 (0.2)	100
INVESTMENT	All ferrous and nonferrous	Less than 1 oz. to 50 lb. (30 g to 25 kg)	25	Moderate	Excellent	0.2 to .06 (0.5 to 1.5)	± .005 (0.1)	80
PERMANENT MOLD Metal Mold Graphite Mold	Nonferrous and cast iron Steel	1 to 40 lb. (0.5 to 20 kg) 5 to 300 lb. (2 to 150 kg)	100 100	Moderate to high	Poor	.18 to .25 (4.5 to 6) .25 (6)	± .02 (0.5) ± .03 (0.8)	200 200
DIE	Sn, Pb, Zn, Al, Mg, and Cu alloys	Less than 1 lb. to 20 lb. (0.5 to 10 kg)	1000	High	Excellent	.05 to .08 (1.2 to 2)	± .002 (0.5)	60

* Values listed are primarily for aluminum alloys, but data applies generally to other metals also.

‡ Depends on surface area. Double if dimension is across parting line.

FIGURE 13-1-16 Cast-part drawings. (A & B)

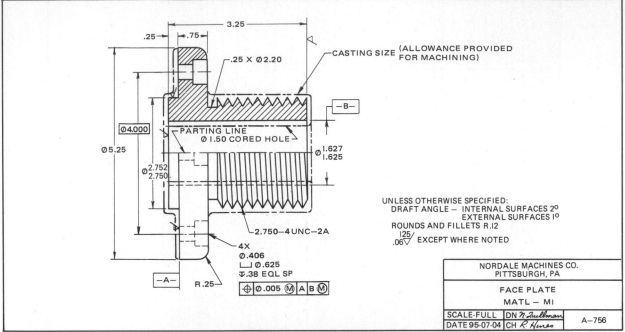

(A) WORKING DRAWING OF A CAST PART

UNLESS OTHERWISE SPECIFIED:
DRAFT ANGLE — INTERNAL SURFACES 2°
EXTERNAL SURFACES 1°
ROUNDS AND FILLETS R.12
.06 ∇ 125 EXCEPT WHERE NOTED

NORDALE MACHINES CO.
PITTSBURGH, PA

FACE PLATE
MATL — MI

| SCALE-FULL | DN N. Zullman | A—756 |
| DATE 95-07-04 | CH R. Hines | |

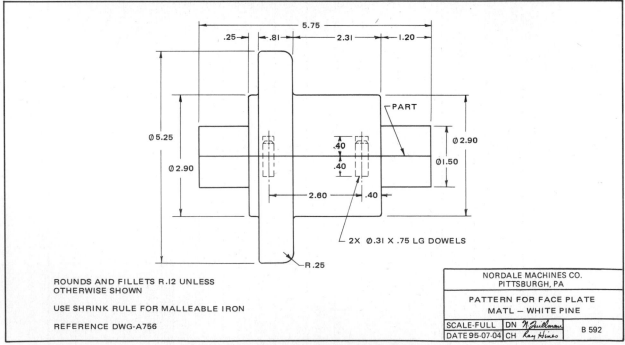

(B) PATTERN DRAWING FOR THE CAST PART SHOWN IN (A)

ROUNDS AND FILLETS R.12 UNLESS OTHERWISE SHOWN

USE SHRINK RULE FOR MALLEABLE IRON

REFERENCE DWG-A756

NORDALE MACHINES CO.
PITTSBURGH, PA

PATTERN FOR FACE PLATE
MATL — WHITE PINE

| SCALE-FULL | DN N. Zullman | B 592 |
| DATE 95-07-04 | CH Ray Hines | |

FIGURE 13-1-17 Guide to machining and tolerance allowance in inches for casting.

CASTING ALLOY	DIMENSIONS WITHIN THIS RANGE	CASTING ALLOWANCE	STANDARD DRAWING TOLERANCE (±)
Cast Iron, Aluminum, Bronze, Etc. Sand Castings	Up to 8.00	.06	.03
	8.00 to 16.00	.09	.06
	16.00 to 24.00	.12	.07
	24.00 to 32.00	.18	.09
	Over 32.00	.25	.12
Pearlitic, Malleable, and Steel Sand Castings	Up to 8.00	.06	.03
	8.00 to 16.00	.09	.06
	16.00 to 24.00	.18	.09
	Over 24.00	.25	.12
Permanent and Semipermanent Mold Castings	Up to 12.00	.06	.03
	12.00 to 24.00	.09	.06
	Over 24.00	.18	.09
Plaster Mold Castings	Up to 8.00	.03	.02
	8.00 to 12.00	.06	.03
	Over 12.00	.10	.06

FIGURE 13-1-19 Casting datums.

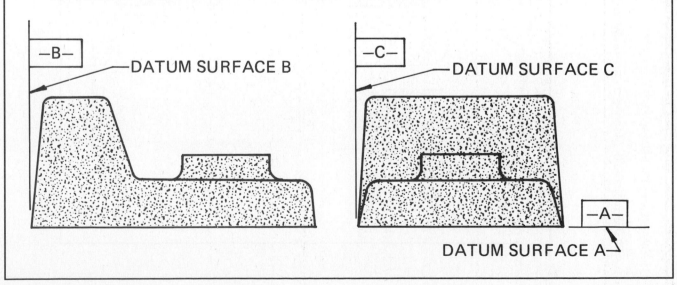

FIGURE 13-1-20 Machined cast drawing illustrating datum lines, setup points and surface finish.

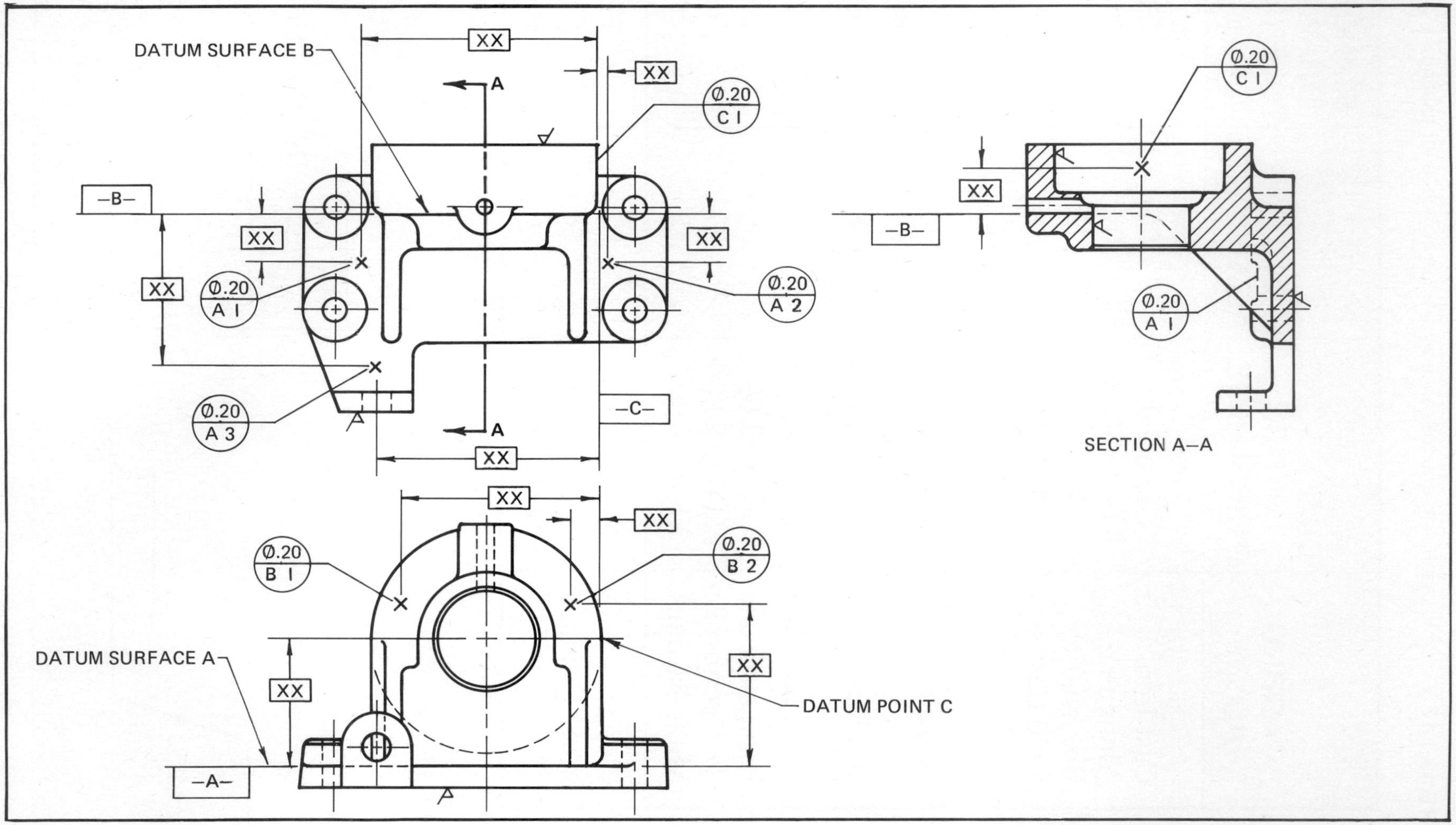

SECTION A–A

DATUM SURFACE B

DATUM POINT C

DATUM SURFACE A

FIGURE 13-2-3 Forging dies.

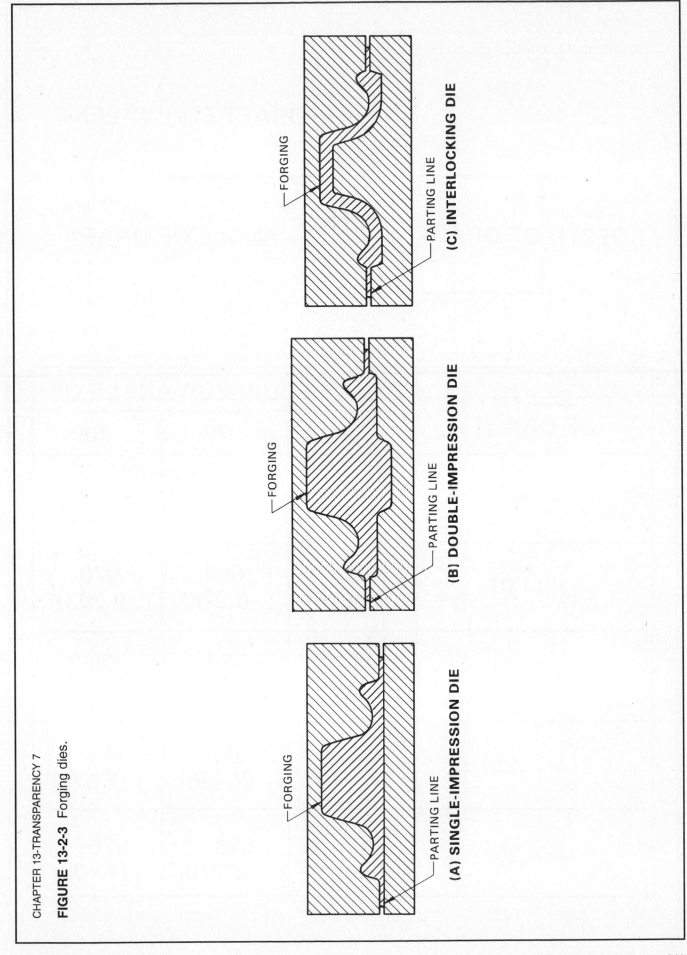

FORGING

PARTING LINE

(A) SINGLE-IMPRESSION DIE

FORGING

PARTING LINE

(B) DOUBLE-IMPRESSION DIE

FORGING

PARTING LINE

(C) INTERLOCKING DIE

FIGURE 13-2-8 Die draft equivalent.

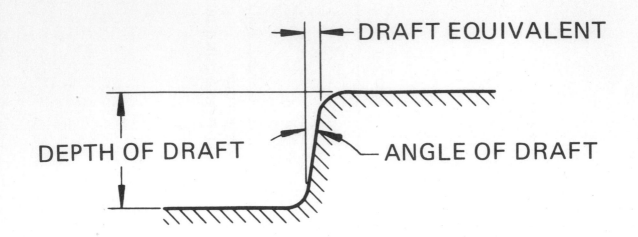

DEPTH OF DRAFT	DRAFT EQUIV FOR ANGLE OF		
	5°	7°	10°
.20 (5)	.018 (0.437)	.024 (0.614)	.035 (0.882)
.40 (10)	.035 (0.875)	.050 (1.228)	.070 (1.763)
.60 (15)	.052 (1.312)	.074 (1.842)	.106 (2.645)
.80 (20)	.070 (1.750)	.100 (2.456)	.140 (3.527)
1.00 (25)	.088 (2.187)	.123 (3.070)	.176 (4.408)

FIGURE 13-2-10 Forged-part drawings.

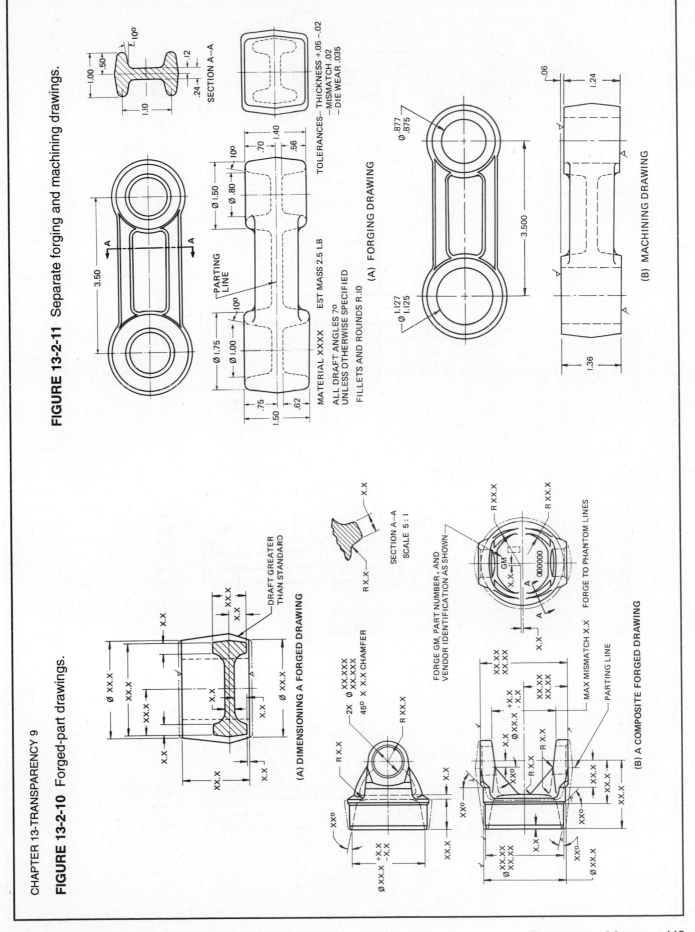

(A) DIMENSIONING A FORGED DRAWING

SECTION A–A
SCALE 5 : 1

(B) A COMPOSITE FORGED DRAWING

FIGURE 13-2-11 Separate forging and machining drawings.

SECTION A–A

TOLERANCES— THICKNESS +.05 –.02
— MISMATCH .02
— DIE WEAR .035

PARTING LINE

EST MASS 2.5 LB

MATERIAL XXXX

ALL DRAFT ANGLES 7°
UNLESS OTHERWISE SPECIFIED
FILLETS AND ROUNDS R.10

(A) FORGING DRAWING

(B) MACHINING DRAWING

FIGURE 13-3-1 Compacting sequence for powder metallurgy.

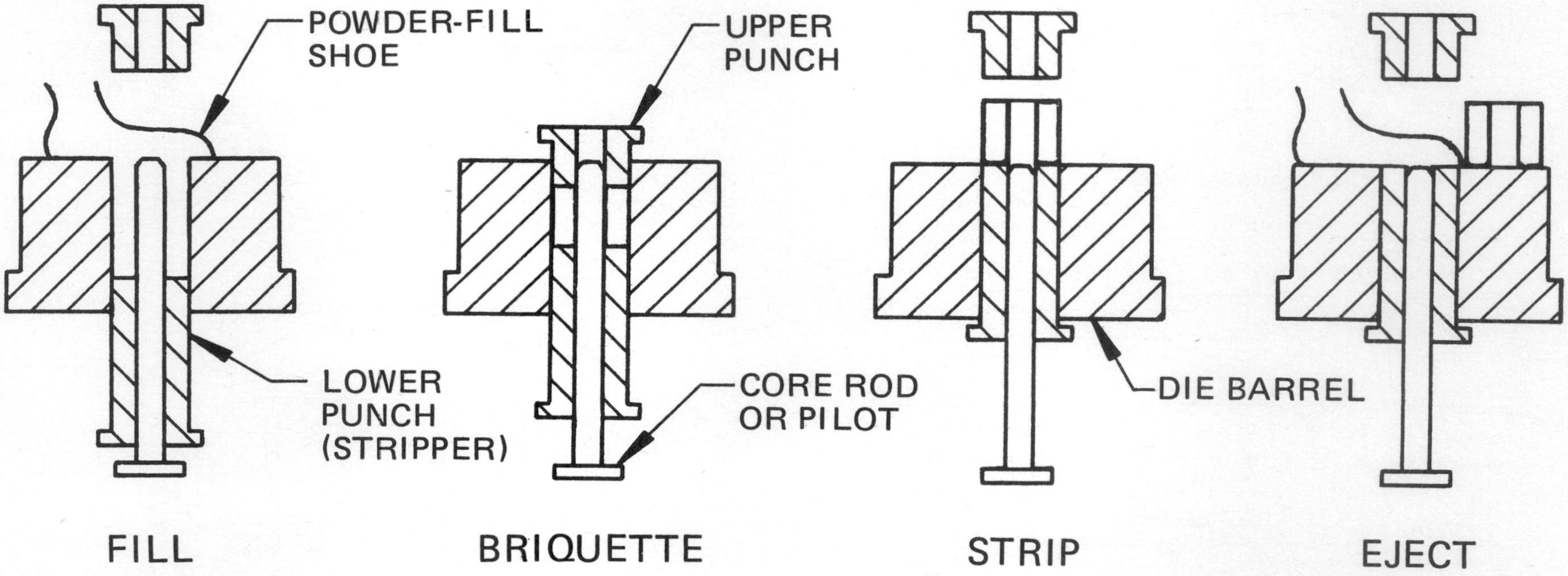

POWDER-FILL
SHOE

UPPER
PUNCH

LOWER
PUNCH
(STRIPPER)

CORE ROD
OR PILOT

DIE BARREL

FILL

BRIQUETTE

STRIP

EJECT

FIGURE 14-2-5 Detail assembly drawing of a sawhorse.

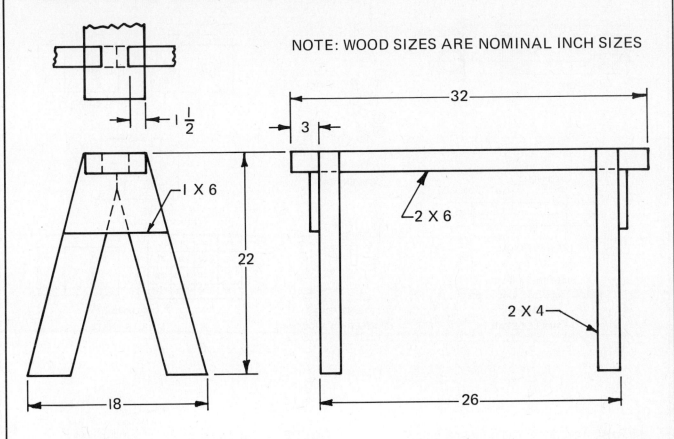

NOTE: WOOD SIZES ARE NOMINAL INCH SIZES

FIGURE 14-2-6 Selecting the most suitable type of projection.

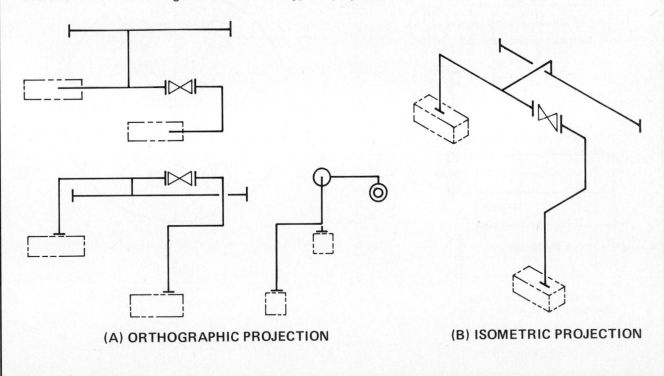

(A) ORTHOGRAPHIC PROJECTION   **(B) ISOMETRIC PROJECTION**

FIGURE 14-2-7 Comparision between conventional and simplified representation.

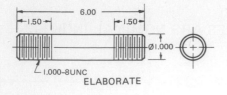

ELABORATE

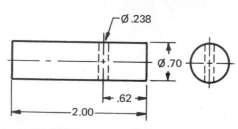

CONVENTIONAL

Ø1.000 STUD 6.00 LG

THREAD ENDS 1.000–8UNC X 1.50LG

SIMPLIFIED

(A) SIMPLE DETAIL

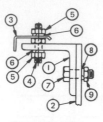

CONVENTIONAL

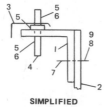

SIMPLIFIED

(B) ASSEMBLY DRAWING

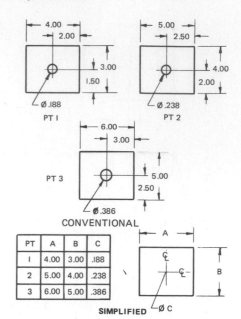

PT 1   PT 2

PT 3

CONVENTIONAL

PT	A	B	C
1	4.00	3.00	.188
2	5.00	4.00	.238
3	6.00	5.00	.386

SIMPLIFIED

(C) SIMILAR PARTS

FIGURE 14-2-8 Simplified representation for detailed parts.

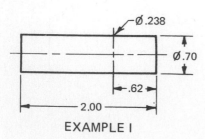

CONVENTIONAL DRAWING

Ø.238

Ø.70

.62

2.00

EXAMPLE 1

NOTE PT 2 Ø.70 X 2.00LG
Ø.238 HOLE – .62 FROM END

PART DESCRIBED BY A NOTE

EXAMPLE 2

FIGURE 14-2-9 Identification of similar-size holes.

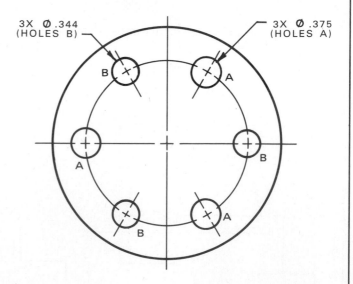

3X Ø .344
(HOLES B)

3X Ø .375
(HOLES A)

CHAPTER 14-TRANSPARENCY 3

FIGURE 14-3-1 A simple detail drawing.

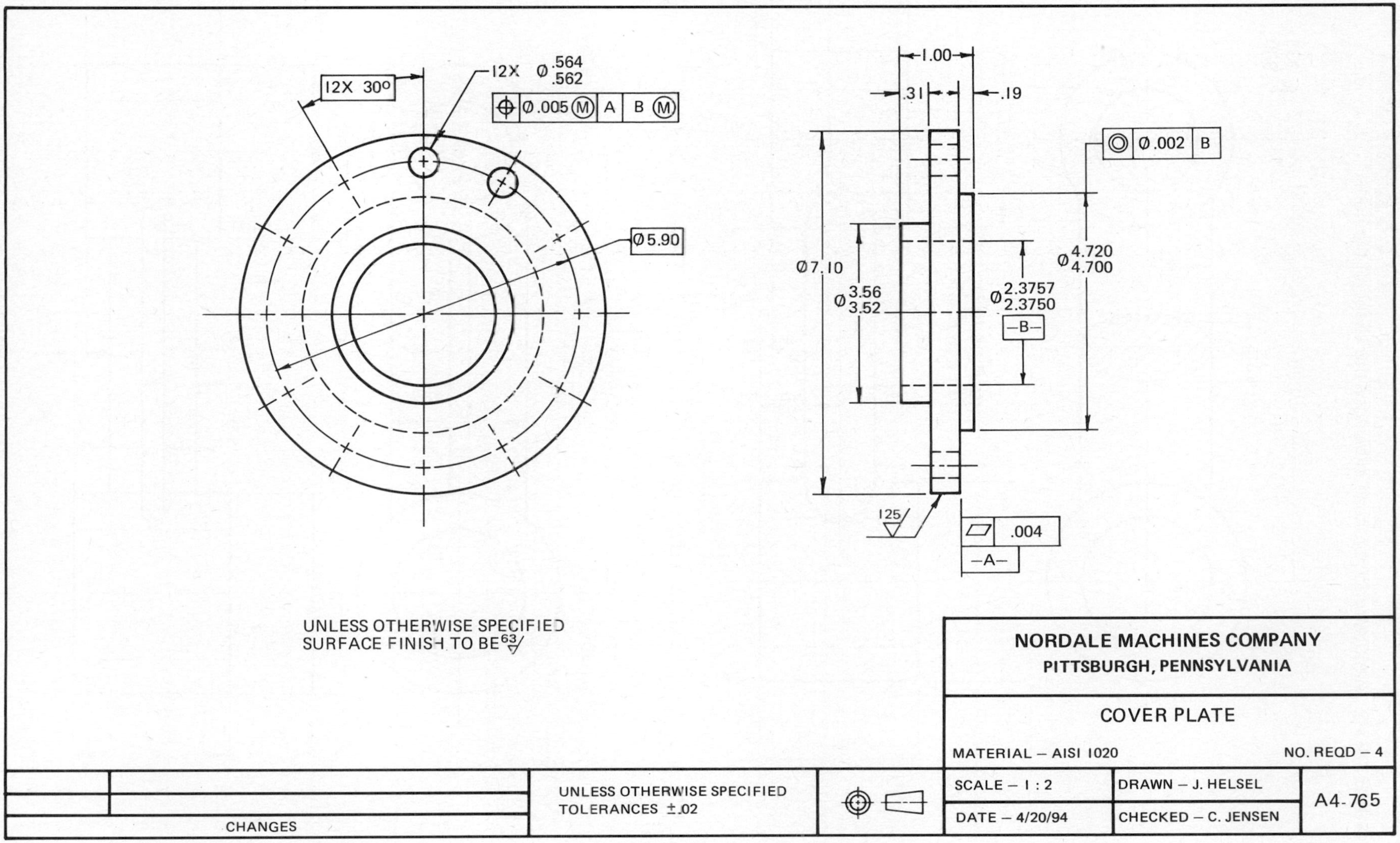

12X 30°
12X Ø .564 .562
⌖ Ø.005 Ⓜ A B Ⓜ
Ø5.90
1.00
.31
.19
◎ Ø.002 B
Ø7.10
Ø 3.56 3.52
Ø 2.3757 2.3750
-B-
Ø 4.720 4.700
125
⏥ .004
-A-
UNLESS OTHERWISE SPECIFIED
SURFACE FINISH TO BE 63
NORDALE MACHINES COMPANY
PITTSBURGH, PENNSYLVANIA
COVER PLATE
MATERIAL – AISI 1020
NO. REQD – 4
SCALE – 1 : 2
DRAWN – J. HELSEL
A4-765
DATE – 4/20/94
CHECKED – C. JENSEN
UNLESS OTHERWISE SPECIFIED
TOLERANCES ±.02
CHANGES

FIGURE 14-3-2 Manufacturing process influences the shape of the parts.

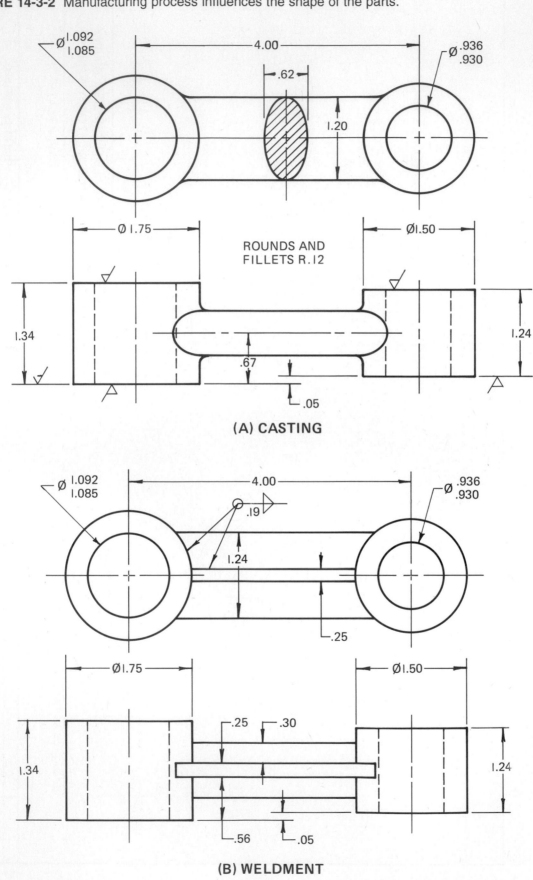

(A) CASTING

(B) WELDMENT

FIGURE 14-3-2 Manufacturing process influences the shape of the parts (continued).

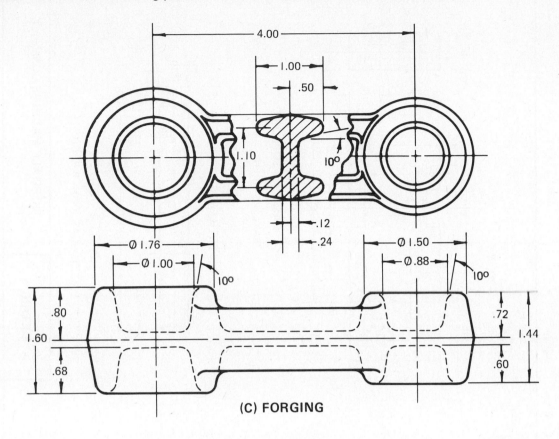

(C) FORGING

UNLESS OTHERWISE SPECIFIED FINISH IS $^{32}\!\!/$
TOLERANCE ON DIMENSIONS ±.02

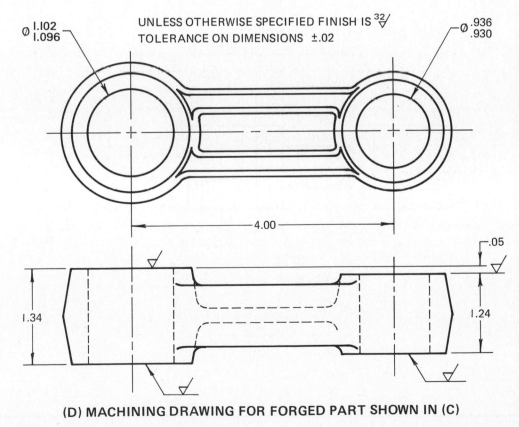

(D) MACHINING DRAWING FOR FORGED PART SHOWN IN (C)

FIGURE 14-4-1 Detail drawing containing many details on one drawing.

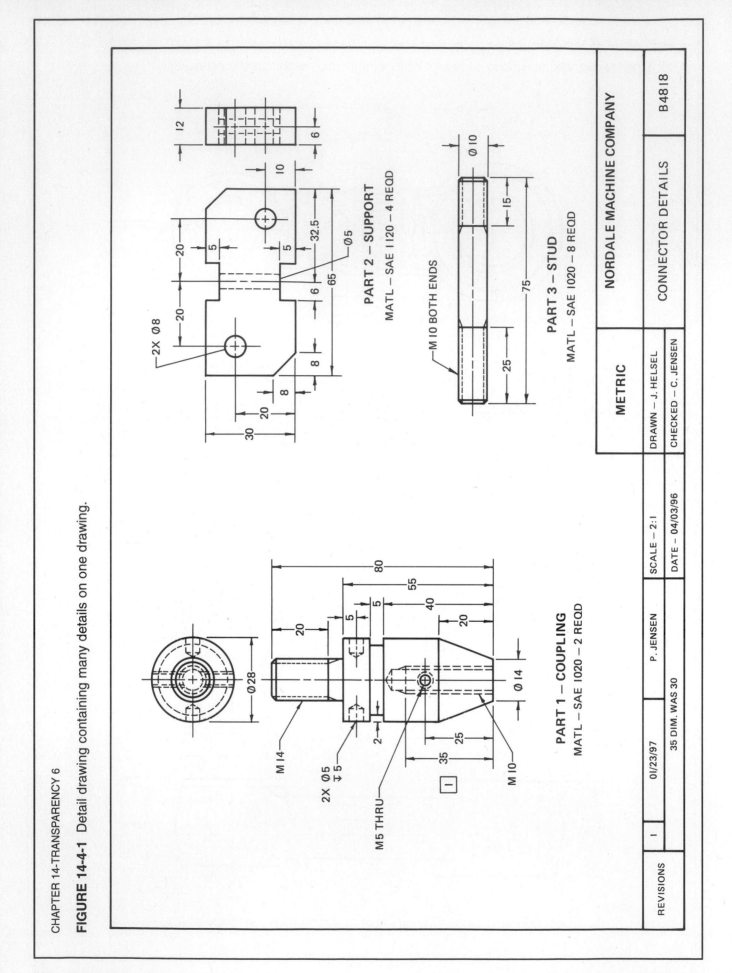

PART 2 – SUPPORT
MATL – SAE 1120 – 4 REQD

PART 3 – STUD
MATL – SAE 1020 – 8 REQD

M 10 BOTH ENDS

PART 1 – COUPLING
MATL – SAE 1020 – 2 REQD

NORDALE MACHINE COMPANY		
CONNECTOR DETAILS		B4818
METRIC	DRAWN – J. HELSEL	
	CHECKED – C. JENSEN	
P. JENSEN	SCALE – 2:1	
	DATE – 04/03/96	
01/23/97		
REVISIONS	35 DIM. WAS 30	

FIGURE 14-6-5 Item list.

QTY	ITEM	MATL	DESCRIPTION	PT NO.
I	BASE	GI	PATTERN # A3154	I
I	CAP	GI	PATTERN # B7156	2
I	SUPPORT	AISI-1212	.38 X 2.00 X 4.38	3
I	BRACE	AISI-1212	.25 X 1.00 X 2.00	4
I	COVER	AISI-1035	.1345 (#10 GA USS) X 6.00 X 7.50	5
I	SHAFT	AISI-1212	Ø1.00 X 6.50	6
2	BEARINGS	SKF	RADIAL BALL # 6200Z	7
2	RETAINING CLIP	TRUARC	N5000-725	8
I	KEY	STL	WOODRUFF # 608	9
I	SETSCREW	CUP POINT	HEX SOCKET .25UNC X 1.50	10
4	BOLT—HEX HD—REG	SEMI-FIN	.38UNC X 1.50LG	11
4	NUT—REG HEX	STL	.38UNC	12
4	LOCK WASHER—SPRING	STL	.38 – MED	13
				14

NOTE: PARTS 7 TO 13 ARE PURCHASED ITEMS.

(A) TYPICAL ITEM LIST.

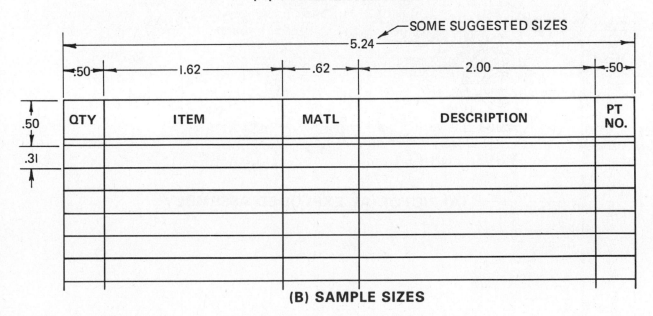

(B) SAMPLE SIZES

FIGURE 14-7-1 (A) Exploded assembly drawings.

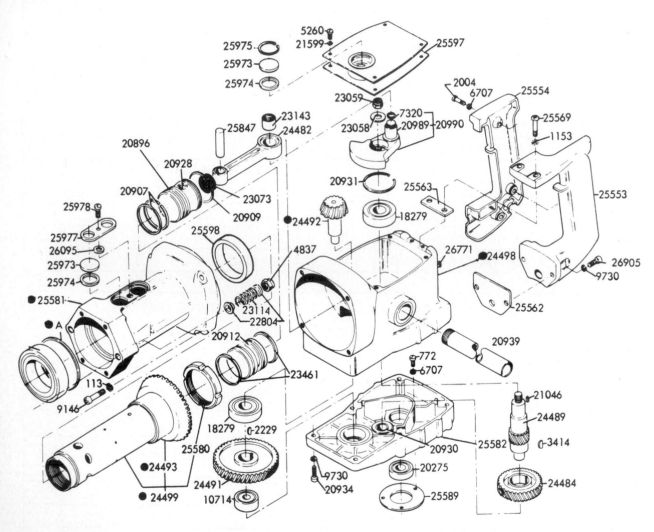

(A) PICTORIAL EXPLODED ASSEMBLY

FIGURE 14-8-1 Detail assembly drawing.

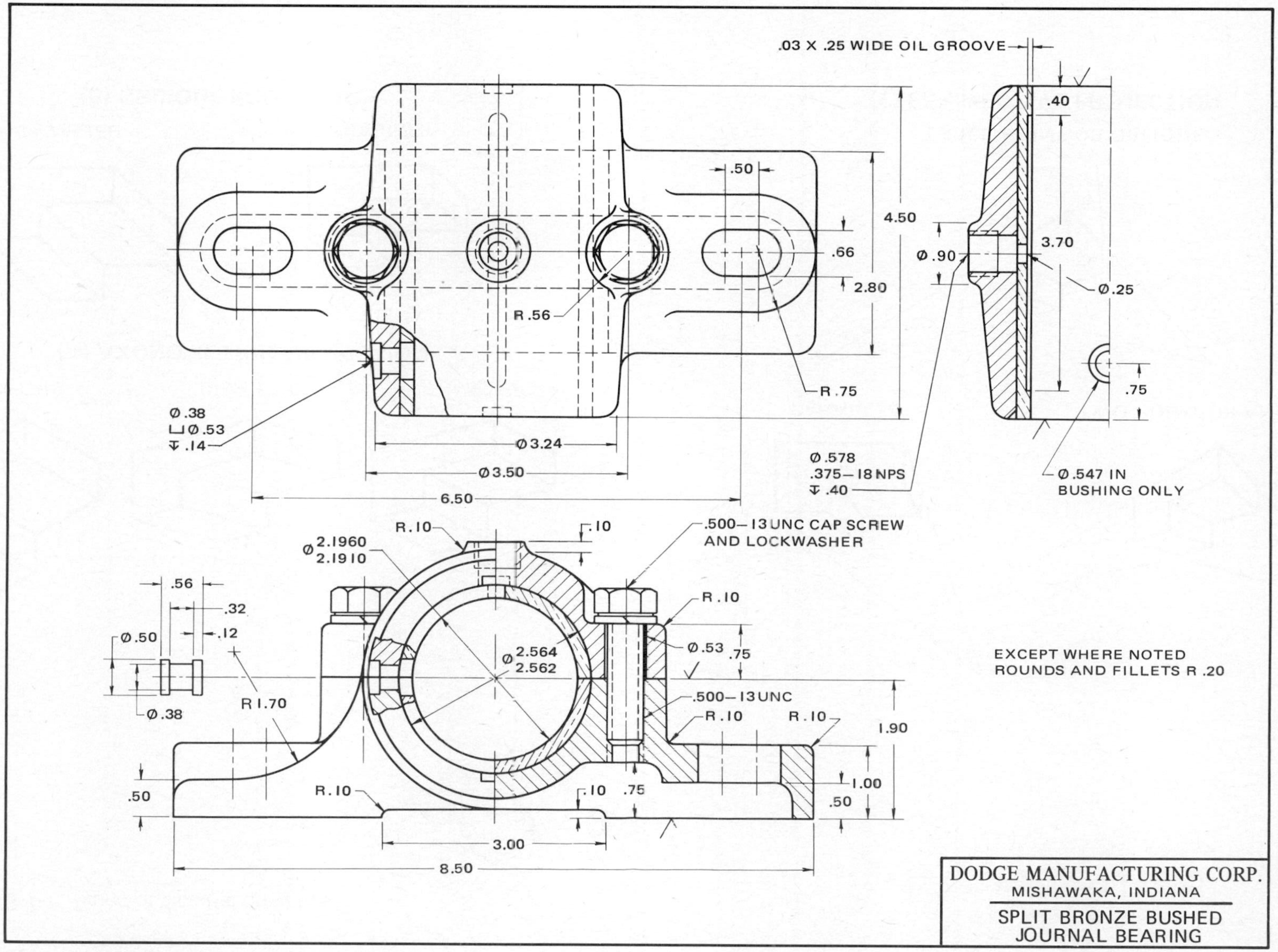

.03 X .25 WIDE OIL GROOVE

Ø.25

.75

.40

3.70

Ø.547 IN
BUSHING ONLY

Ø.90

DODGE MANUFACTURING CORP.
MISHAWAKA, INDIANA
SPLIT BRONZE BUSHED
JOURNAL BEARING

EXCEPT WHERE NOTED
ROUNDS AND FILLETS R .20

4.50

2.80

.66

.50

R .75

Ø.578
.375—18 NPS
.40

.500—13 UNC CAP SCREW
AND LOCKWASHER

R .56

Ø3.24

Ø3.50

6.50

Ø.38
Ø.53
.14

1.90

1.00

.50

R .10

Ø.53 .75

.500—13 UNC

R .10

R .10

.10

.75

.10

Ø2.564
2.562

3.00

8.50

Ø2.1960
2.1910

R .10

R 1.70

R .10

.50

.32

.12

.56

Ø.50

Ø.38

FIGURE 15-1-1 Types of pictorial drawings.

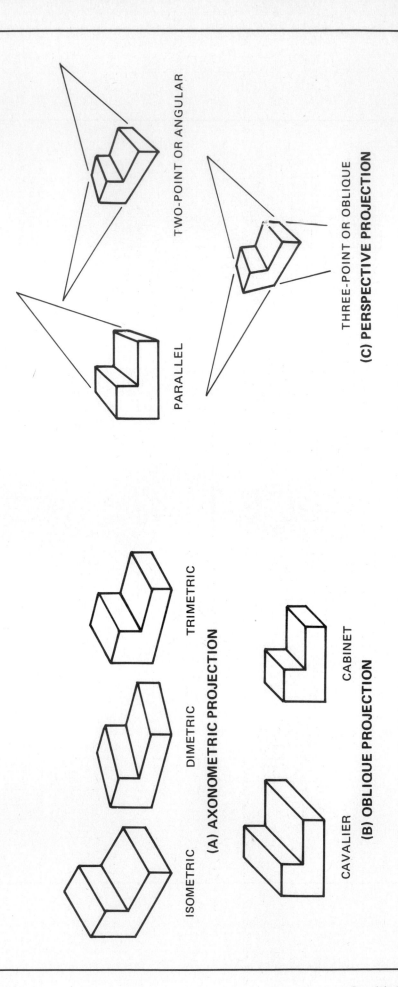

ISOMETRIC

DIMETRIC

TRIMETRIC

(A) AXONOMETRIC PROJECTION

CAVALIER

CABINET

(B) OBLIQUE PROJECTION

PARALLEL

TWO-POINT OR ANGULAR

THREE-POINT OR OBLIQUE

(C) PERSPECTIVE PROJECTION

FIGURE 15-1-3 Types of axonometric drawings.

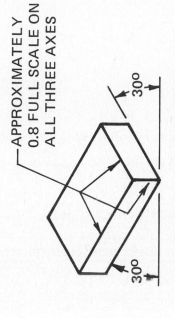

APPROXIMATELY 0.8 FULL SCALE ON ALL THREE AXES

30°

30°

(A) ISOMETRIC PROJECTION

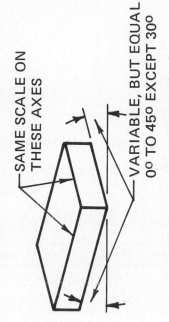

SAME SCALE ON THESE AXES

VARIABLE, BUT EQUAL 0° TO 45° EXCEPT 30°

(B) DIMETRIC PROJECTION

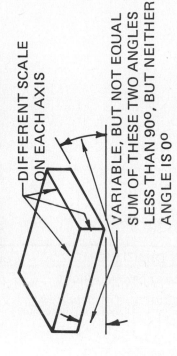

DIFFERENT SCALE ON EACH AXIS

VARIABLE, BUT NOT EQUAL SUM OF THESE TWO ANGLES LESS THAN 90°, BUT NEITHER ANGLE IS 0°

(C) TRIMETRIC PROJECTION

FIGURE 15-1-2 Kinds of projections.

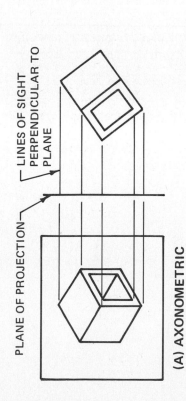

PLANE OF PROJECTION

LINES OF SIGHT PERPENDICULAR TO PLANE

(A) AXONOMETRIC

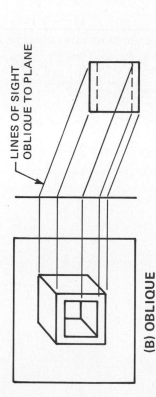

LINES OF SIGHT OBLIQUE TO PLANE

(B) OBLIQUE

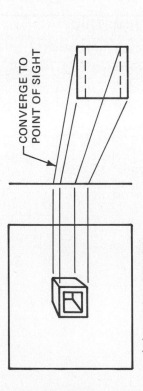

CONVERGE TO POINT OF SIGHT

(C) PERSPECTIVE

FRONT VIEW

SIDE VIEW

FIGURE 15-1-4 Axonometric projection. *(Graphic Standard Instrument Co.)*

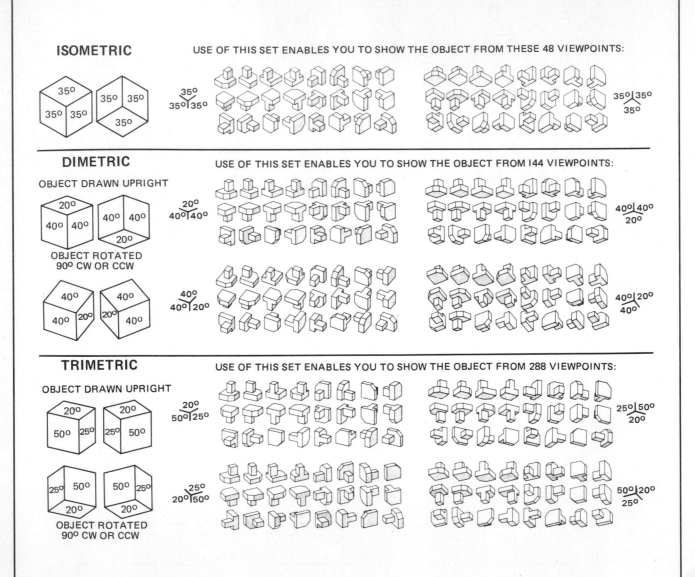

ISOMETRIC
USE OF THIS SET ENABLES YOU TO SHOW THE OBJECT FROM THESE 48 VIEWPOINTS:

DIMETRIC
USE OF THIS SET ENABLES YOU TO SHOW THE OBJECT FROM 144 VIEWPOINTS:

OBJECT DRAWN UPRIGHT

OBJECT ROTATED
90° CW OR CCW

TRIMETRIC
USE OF THIS SET ENABLES YOU TO SHOW THE OBJECT FROM 288 VIEWPOINTS:

OBJECT DRAWN UPRIGHT

OBJECT ROTATED
90° CW OR CCW

FIGURE 15-1-6 Isometric axes and projection.

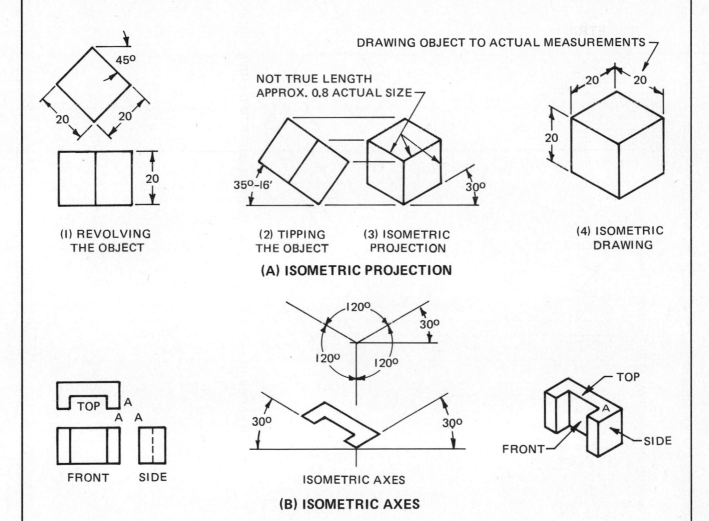

DRAWING OBJECT TO ACTUAL MEASUREMENTS

45°

20   20

20

(I) REVOLVING
THE OBJECT

NOT TRUE LENGTH
APPROX. 0.8 ACTUAL SIZE

35°-16′   30°

(2) TIPPING
THE OBJECT

(3) ISOMETRIC
PROJECTION

(A) ISOMETRIC PROJECTION

20   20

20

(4) ISOMETRIC
DRAWING

120°   30°

120°   120°

30°   30°

ISOMETRIC AXES

(B) ISOMETRIC AXES

TOP
A A A

FRONT   SIDE

TOP
A
FRONT   SIDE

CHAPTER 15-TRANSPARENCY 5

FIGURE 15-1-9 Sequence in drawing an object having nonisometric lines.

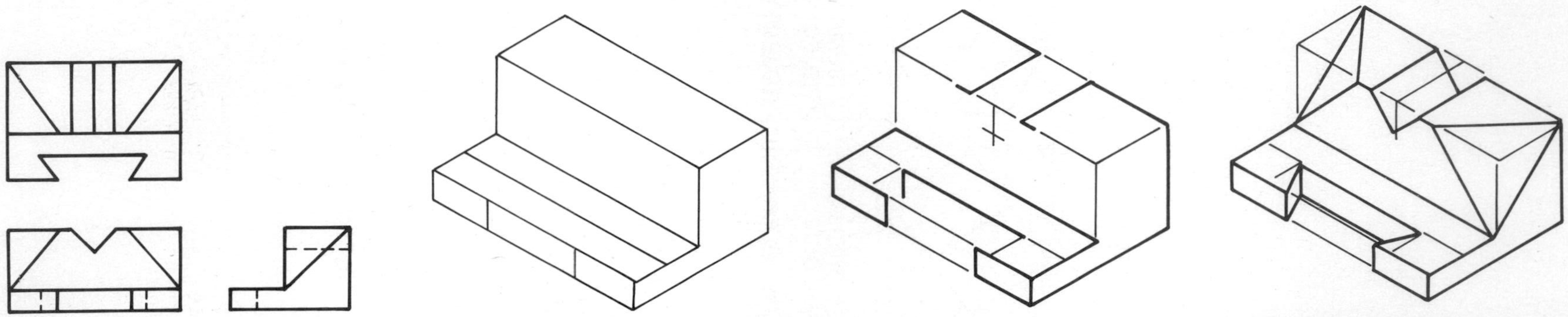

(A) PART (B) BLOCK IN FEATURES (C) DARKEN ISOMETRIC LINES (D) COMPLETE NONISOMETRIC LINES

FIGURE 15-2-2 Sequence in drawing isometric circles.

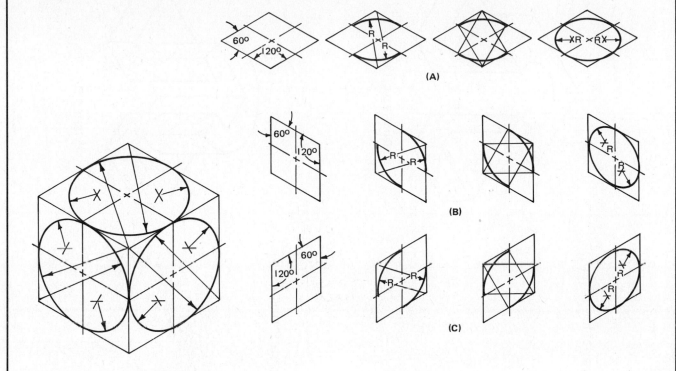

FIGURE 15-2-4 Drawing isometric arcs and circles.

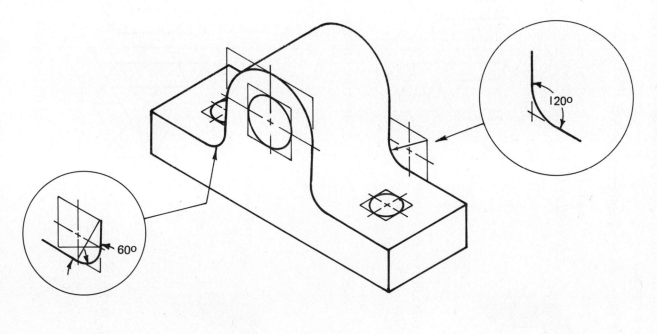

FIGURE 15-3-2 Examples of isometric half-sections.

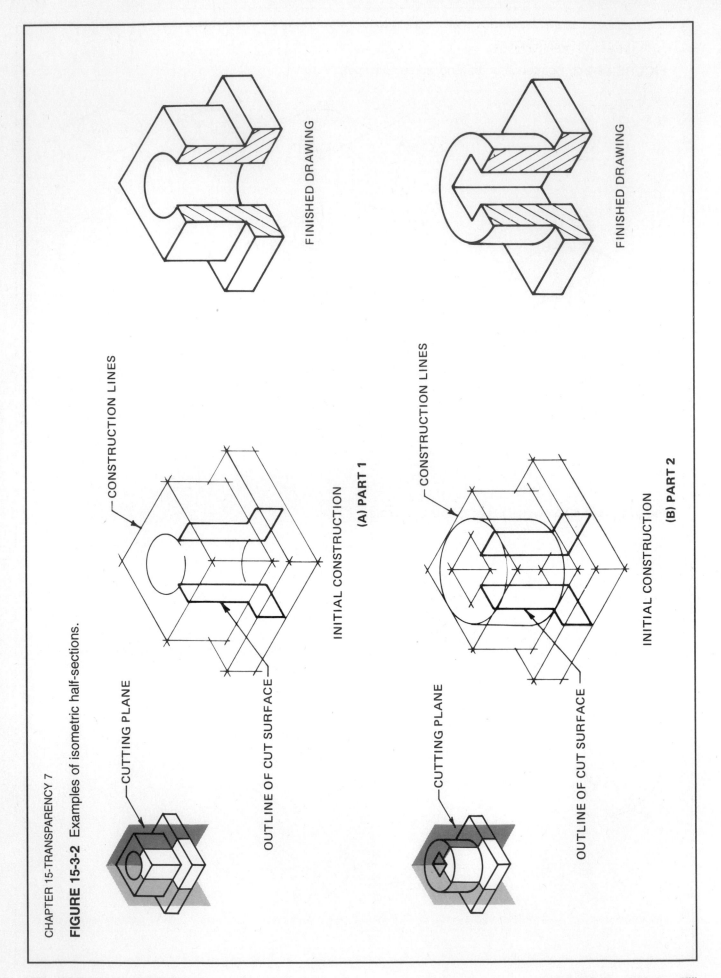

CONSTRUCTION LINES

CUTTING PLANE

OUTLINE OF CUT SURFACE

INITIAL CONSTRUCTION

FINISHED DRAWING

(A) PART 1

CONSTRUCTION LINES

CUTTING PLANE

OUTLINE OF CUT SURFACE

INITIAL CONSTRUCTION

FINISHED DRAWING

(B) PART 2

FIGURE 15-4-3 Types of oblique projection.

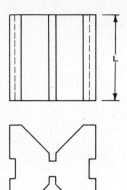

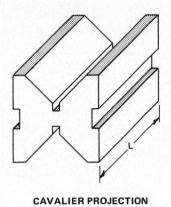

CAVALIER PROJECTION

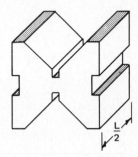

CABINET PROJECTION

FIGURE 15-4-4 Oblique construction by the box method.

(A)

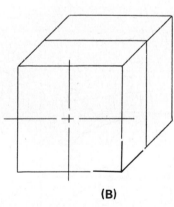

(B)

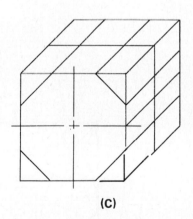

(C)

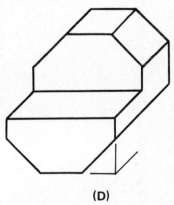

(D)

FIGURE 15-5-3 Approximate ellipse construction for oblique drawings with 45° axis.

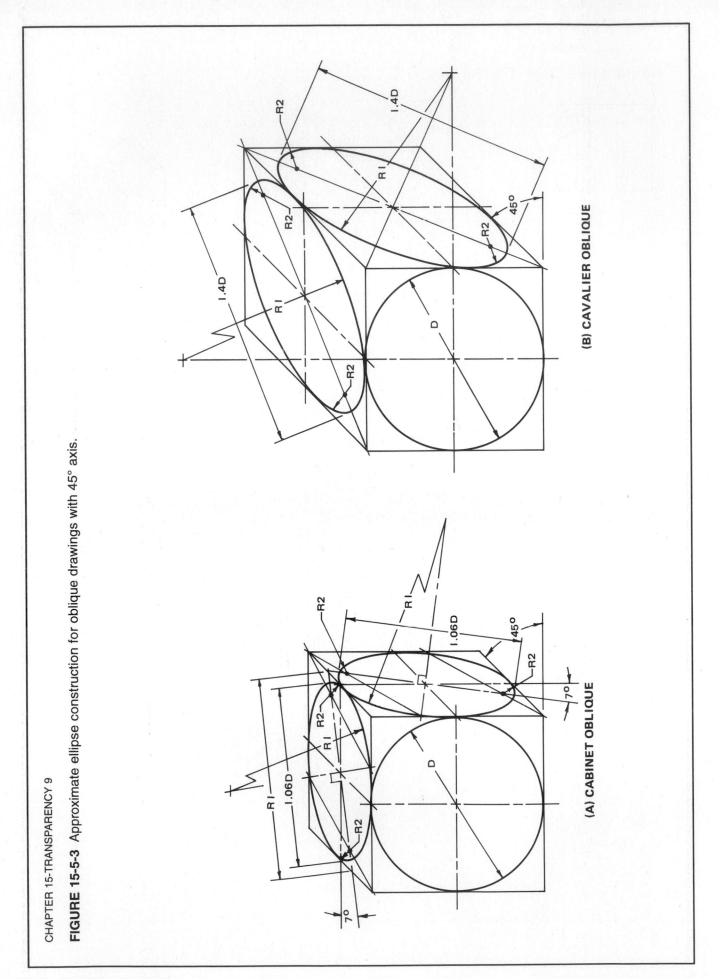

(A) CABINET OBLIQUE

(B) CAVALIER OBLIQUE

FIGURE 15-6-4 Types of perspective drawings.

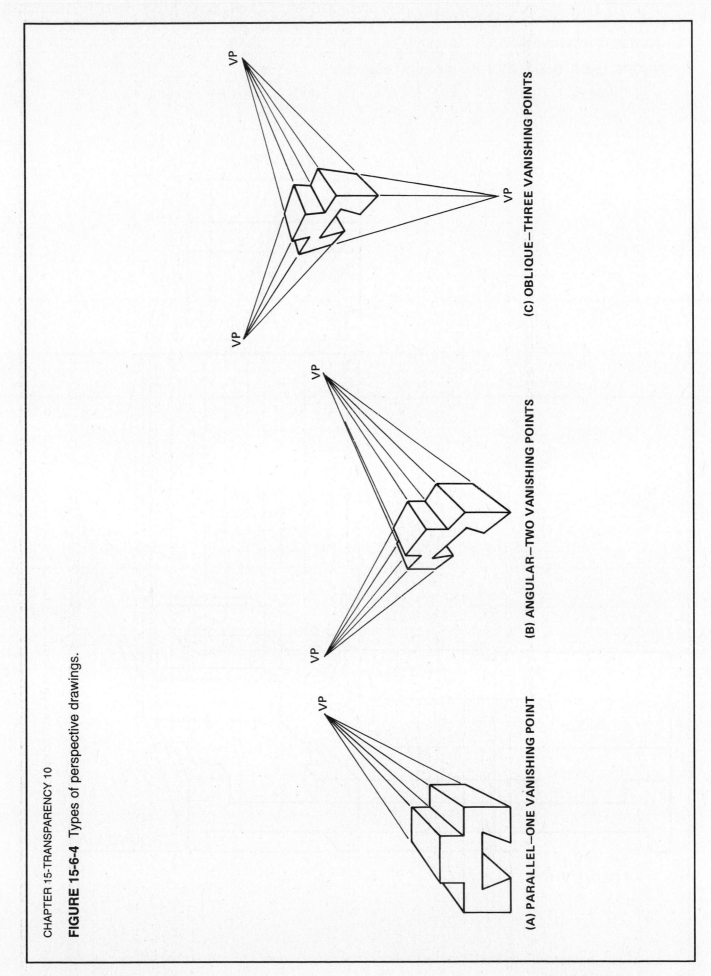

(A) PARALLEL—ONE VANISHING POINT

(B) ANGULAR—TWO VANISHING POINTS

(C) OBLIQUE—THREE VANISHING POINTS

FIGURE 15-6-6 Construction of a one-point perspective.

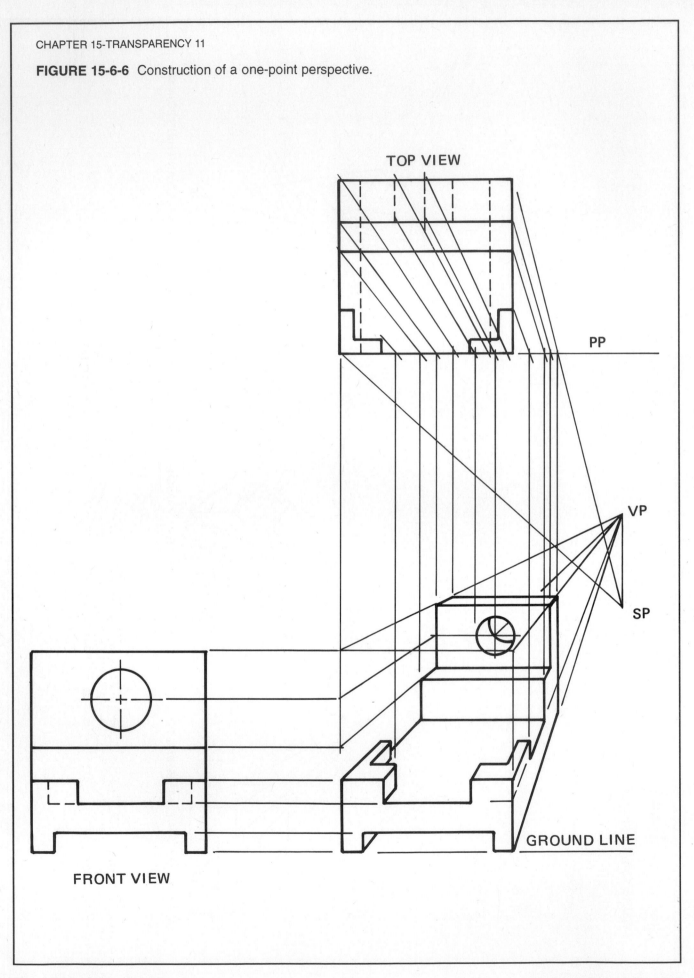

TOP VIEW

PP

VP

SP

GROUND LINE

FRONT VIEW

FIGURE 15-7-2 Angular-perspective drawing of a prism.

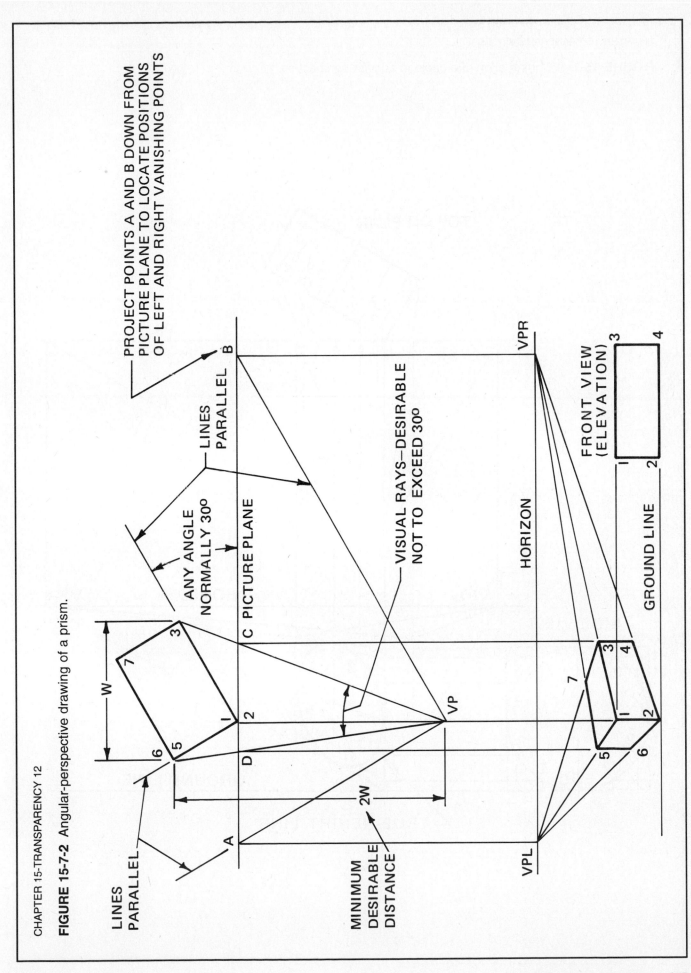

PROJECT POINTS A AND B DOWN FROM
PICTURE PLANE TO LOCATE POSITIONS
OF LEFT AND RIGHT VANISHING POINTS

LINES PARALLEL

ANY ANGLE NORMALLY 30°

VISUAL RAYS—DESIRABLE NOT TO EXCEED 30°

PICTURE PLANE

LINES PARALLEL

MINIMUM DESIRABLE DISTANCE

HORIZON

VPR

VPL

VP

FRONT VIEW (ELEVATION)

GROUND LINE

FIGURE 15-7-4 Construction of a circle in angular perspective.

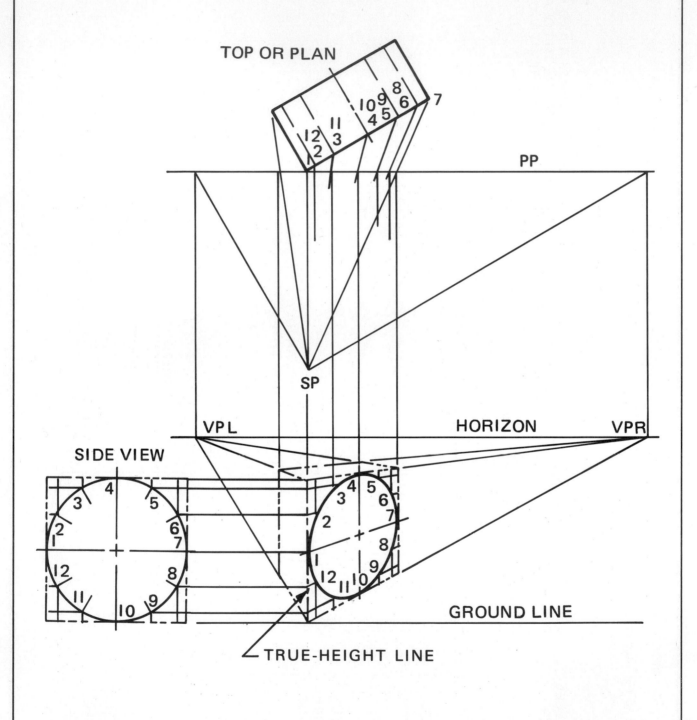

FIGURE 15-7-5 Horizon lines.

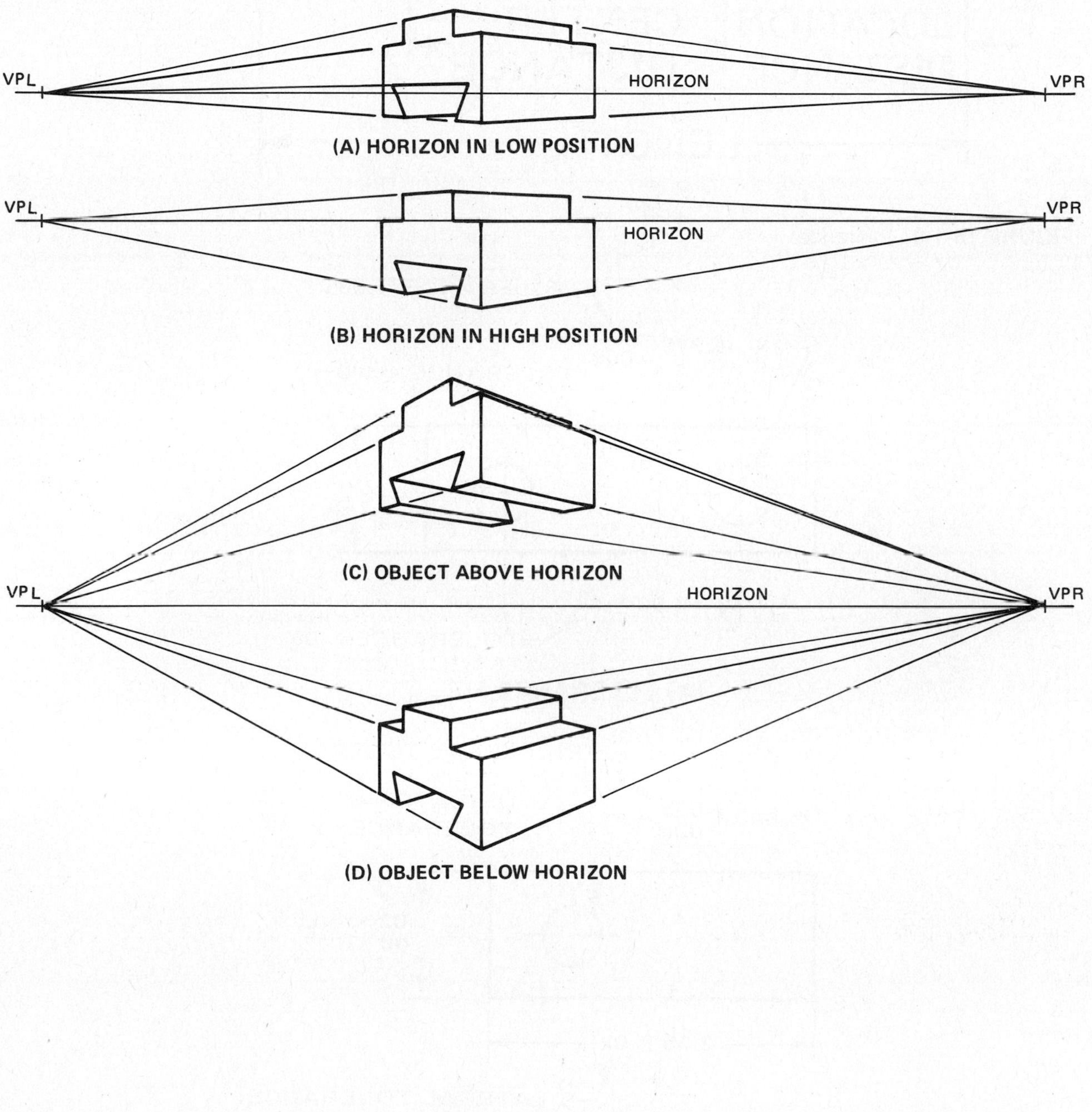

VPL

HORIZON

VPR

(A) HORIZON IN LOW POSITION

VPL

HORIZON

VPR

(B) HORIZON IN HIGH POSITION

(C) OBJECT ABOVE HORIZON

VPL

HORIZON

VPR

(D) OBJECT BELOW HORIZON

FIGURE 16-1-2 Dimension of a part.

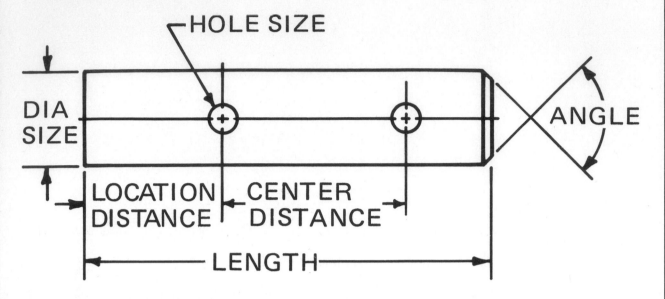

FIGURE 16-1-3 Tolerance.

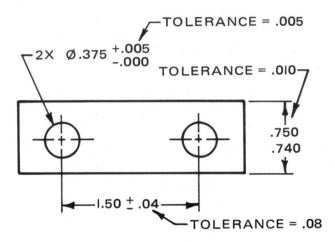

(A) TOLERANCE SIZE

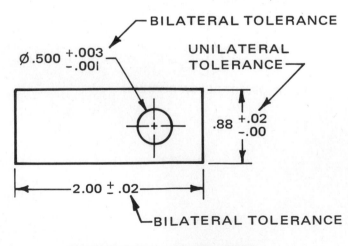

(B) TYPE OF TOLERANCE

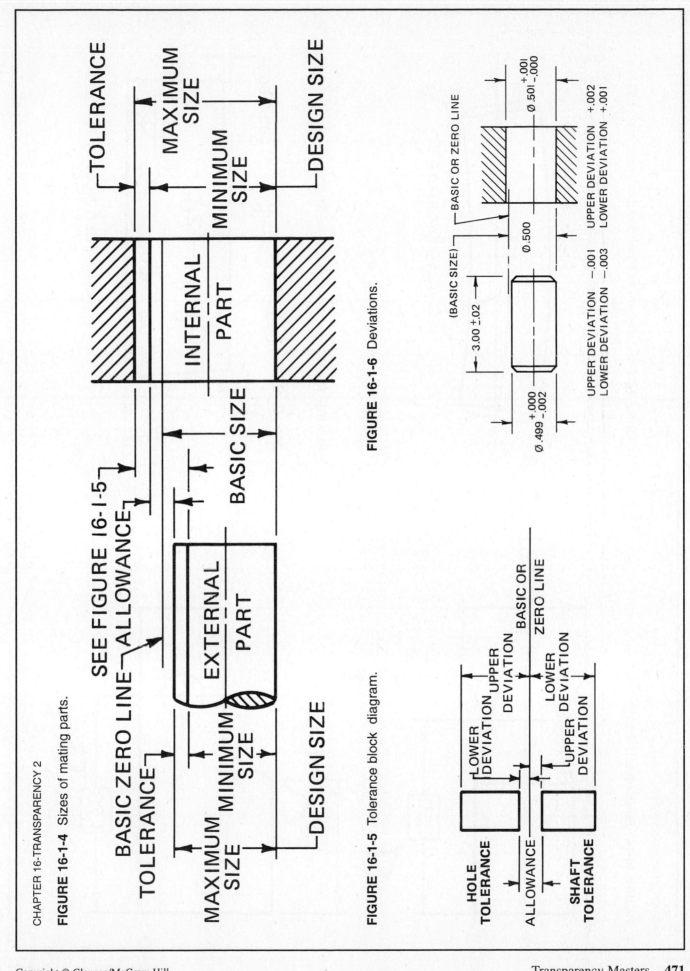

FIGURE 16-1-4 Sizes of mating parts.

FIGURE 16-1-5 Tolerance block diagram.

FIGURE 16-1-6 Deviations.

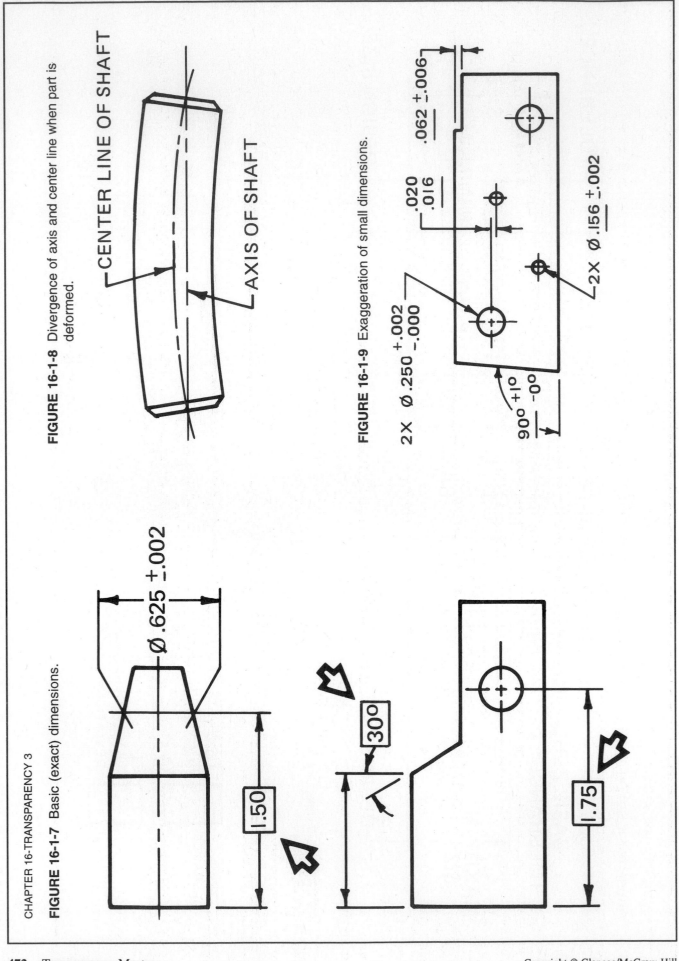

FIGURE 16-1-7 Basic (exact) dimensions.

FIGURE 16-1-8 Divergence of axis and center line when part is deformed.

FIGURE 16-1-9 Exaggeration of small dimensions.

FIGURE 16-1-10 Point-to-point dimensions when datum are not used.

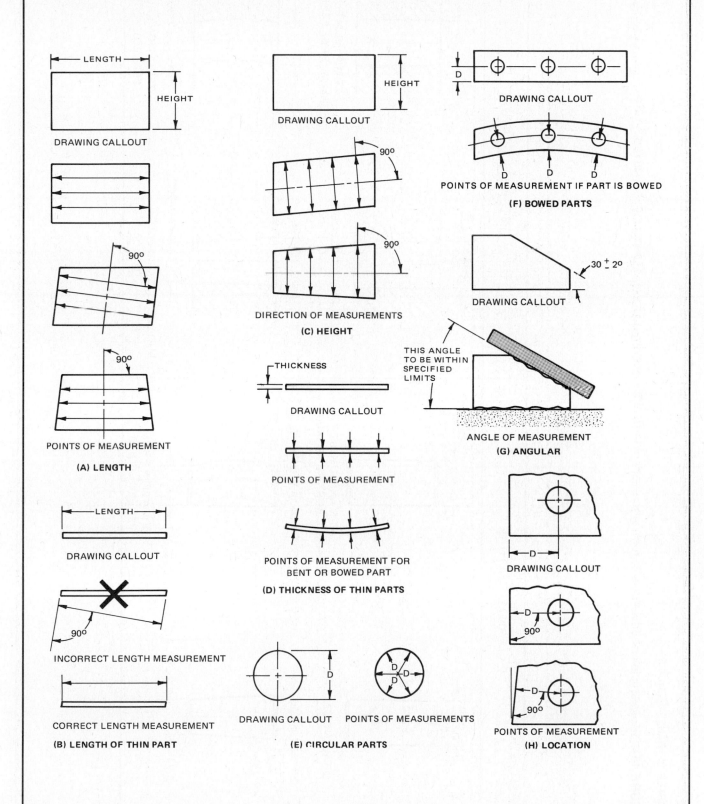

FIGURE 16-1-14 Examples of deviations of form when perfect form at the maximum material condition is required.

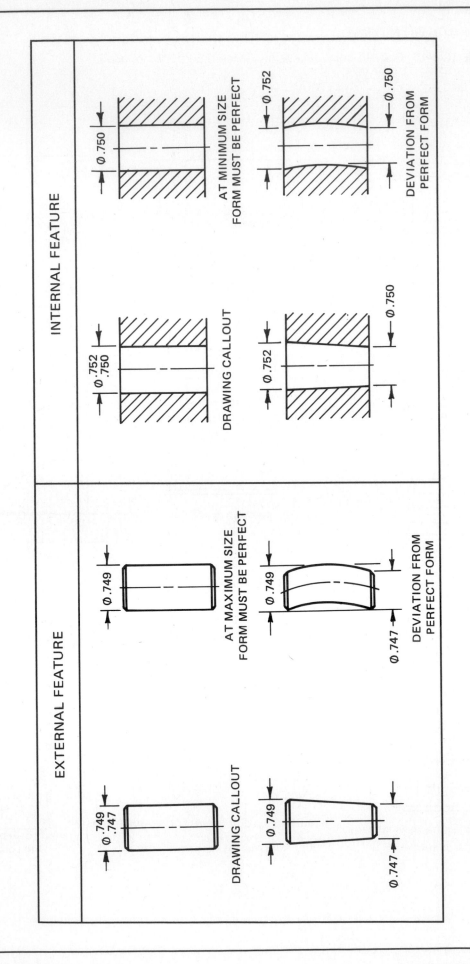

FIGURE 16-2-2 Geometric characteristic symbols.

FEATURE	TYPE OF TOLERANCE	CHARACTERISTIC	SYMBOL	SEE UNIT
INDIVIDUAL FEATURES	FORM	STRAIGHTNESS	—	16-2, 16-5
		FLATNESS	▱	16-3
		CIRCULARITY (ROUNDNESS)	◯	16-12
		CYLINDRICITY	⌭	
INDIVIDUAL OR RELATED FEATURES	PROFILE	PROFILE OF A LINE	⌒	16-13
		PROFILE OF A SURFACE	⌓	
RELATED FEATURES	ORIENTATION	ANGULARITY	∠	16-7, 16-8
		PERPENDICULARITY	⊥	
		PARALLELISM	//	
	LOCATION	POSITION	⊕	16-9
		CONCENTRICITY	◎	16-14
	RUNOUT	CIRCULAR RUNOUT	*↗	16-14
		TOTAL RUNOUT	*↗↗	
SUPPLEMENTARY SYMBOLS		MAXIMUM MATERIAL CONDITION	Ⓜ	16-4
		REGARDLESS OF FEATURE SIZE	Ⓢ	
		LEAST MATERIAL CONDITION	Ⓛ	
		PROJECTED TOLERANCE ZONE	Ⓟ	16-9
		BASIC DIMENSION	☐XX☐	16-9, 16-11
		DATUM FEATURE	—A—	16-6
		DATUM TARGET	Ø.50 / A2	16-11

* MAY BE FILLED IN

FIGURE 16-2-4 Application of feature control frame.

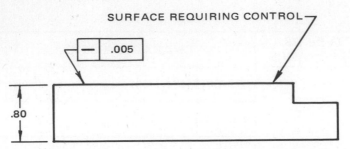

SURFACE REQUIRING CONTROL

— | .005

.80

RUNNING A LEADER FROM THE FRAME TO THE FEATURE

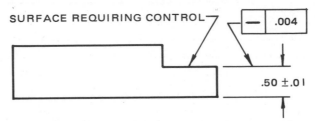

SURFACE REQUIRING CONTROL

— | .004

.50 ±.01

ATTACHED TO AN EXTENSION LINE USING A LEADER

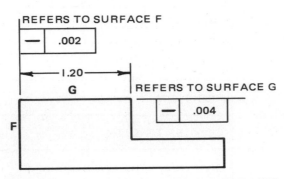

REFERS TO SURFACE F

— | .002

1.20

G

REFERS TO SURFACE G

— | .004

F

ATTACHED DIRECTLY TO AN EXTENSION LINE

(A) CONTROL OF SURFACE OR SURFACE ELEMENTS

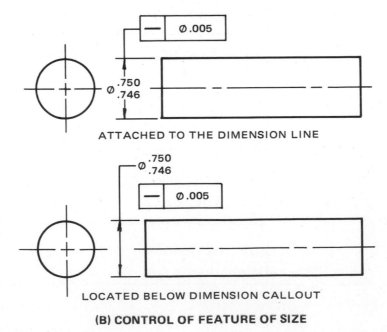

— | Ø.005

Ø .750 / .746

ATTACHED TO THE DIMENSION LINE

Ø .750 / .746

— | Ø.005

LOCATED BELOW DIMENSION CALLOUT

(B) CONTROL OF FEATURE OF SIZE

FIGURE 16-4-5 Effect of location.

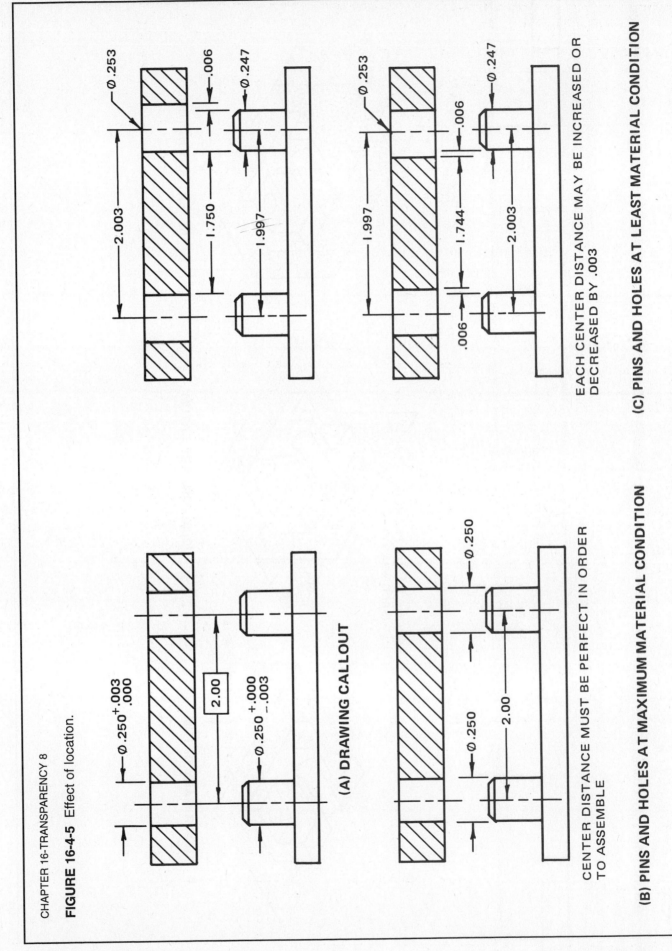

(A) DRAWING CALLOUT

(B) PINS AND HOLES AT MAXIMUM MATERIAL CONDITION

CENTER DISTANCE MUST BE PERFECT IN ORDER TO ASSEMBLE

(C) PINS AND HOLES AT LEAST MATERIAL CONDITION

EACH CENTER DISTANCE MAY BE INCREASED OR DECREASED BY .003

FIGURE 16-5-2 The datum planes.

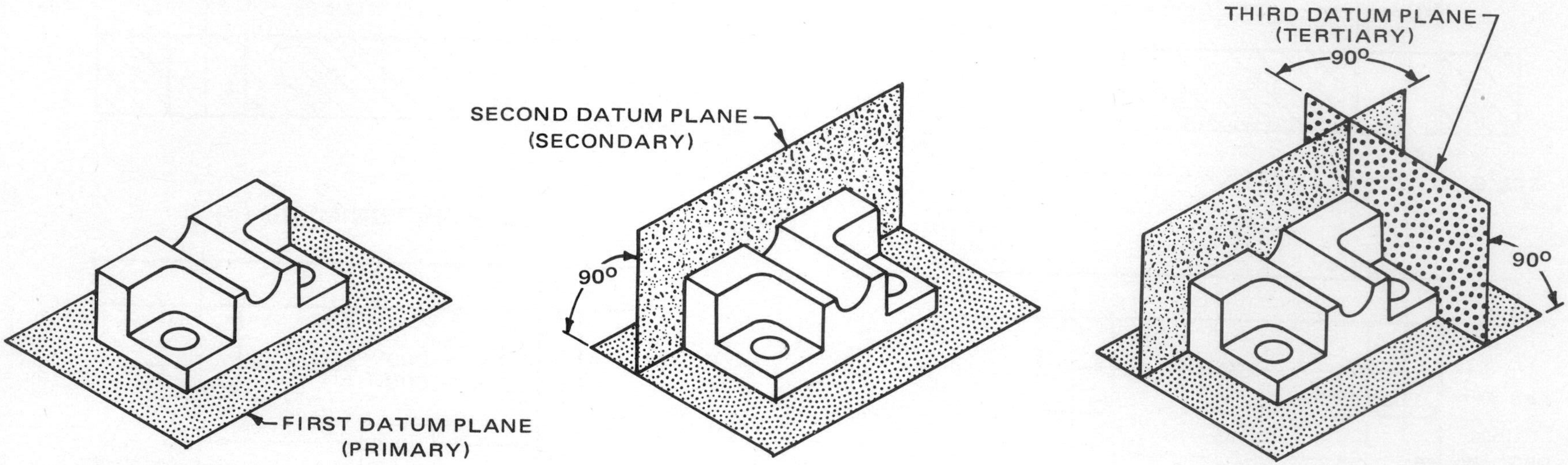

(A) PRIMARY DATUM

FIRST DATUM PLANE (PRIMARY)

(B) SECONDARY DATUM

SECOND DATUM PLANE (SECONDARY)

90°

(C) TERTIARY DATUM

THIRD DATUM PLANE (TERTIARY)

90°

90°

FIGURE 16-5-3 Three-plane datum system.

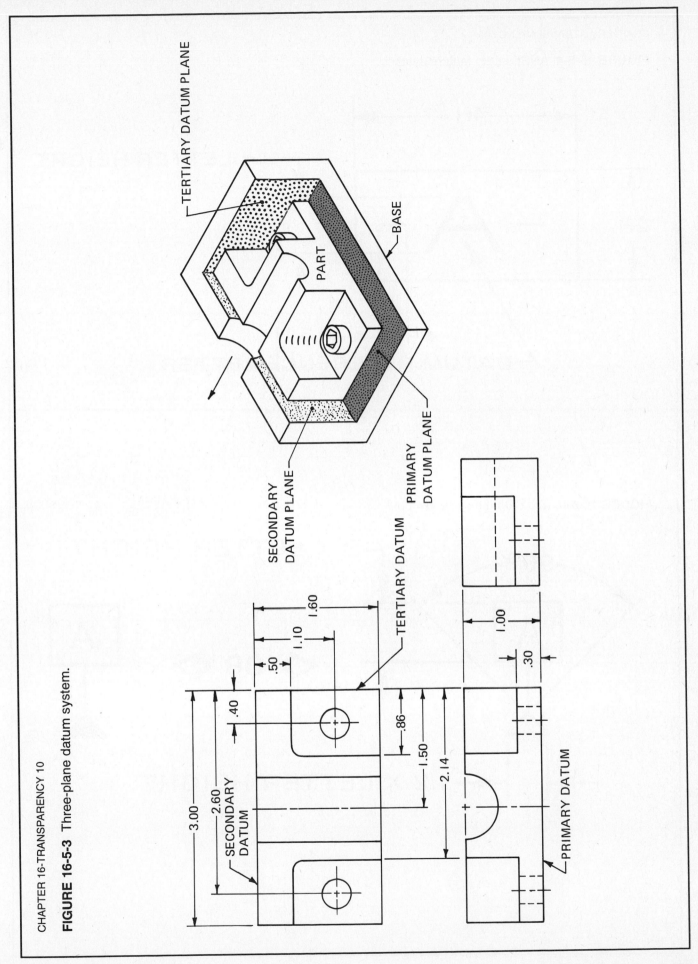

FIGURE 16-5-4 ANSI datum feature symbol.

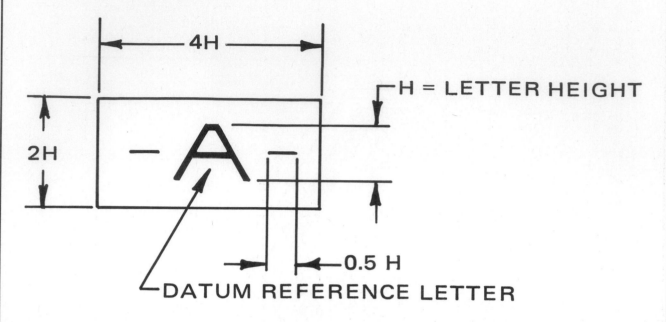

FIGURE 16-5-6 ISO datum feature symbol.

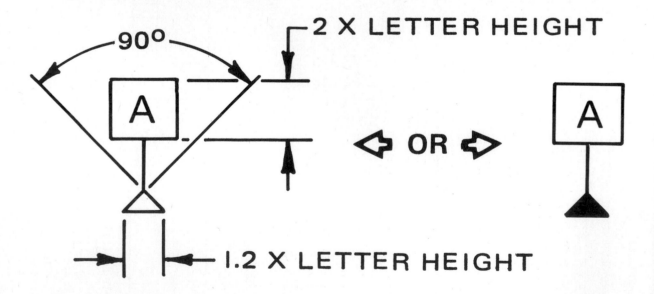

FIGURE 16-6-2 Orientation tolerancing for flat surfaces.

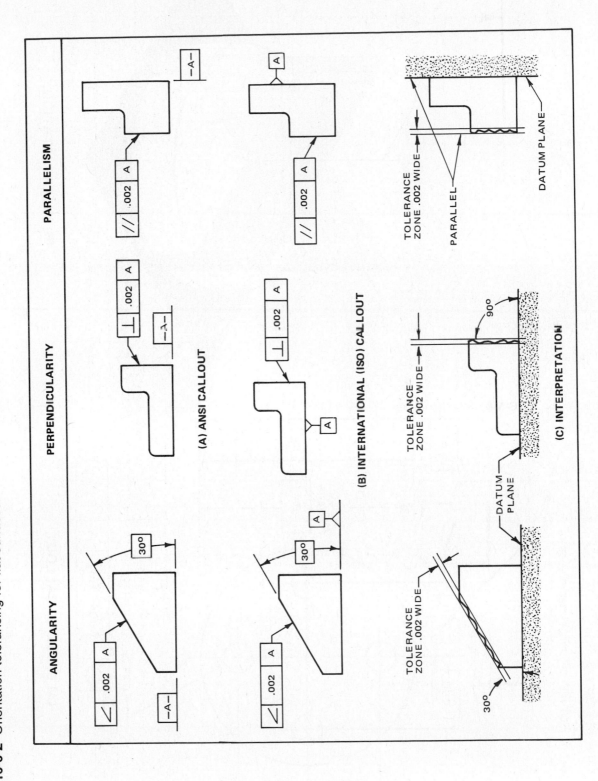

FIGURE 16-7-1 Part with cylindrical datum feature.

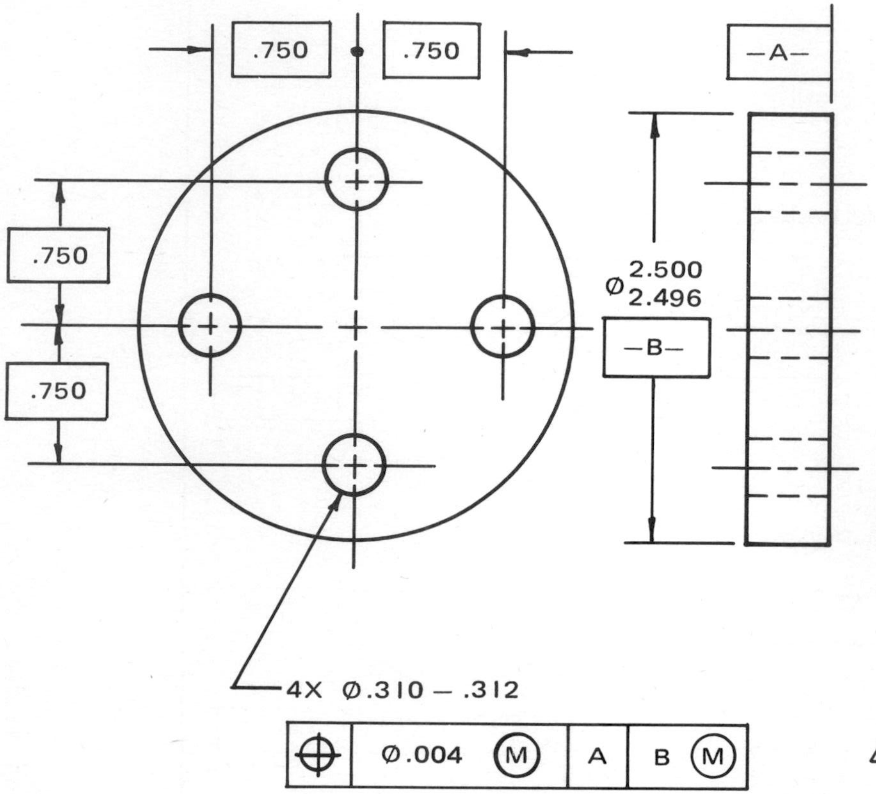

DATUM AXIS B

X

Y

FIRST DATUM PLANE

(B) INTERPRETATION

—A—

$\varnothing$ 2.500
$\varnothing$ 2.496

—B—

.750

.750

.750

.750

.750

4X $\varnothing$ .310 − .312

| $\oplus$ | $\varnothing$.004 Ⓜ | A | B Ⓜ |

(A) DRAWING CALLOUT

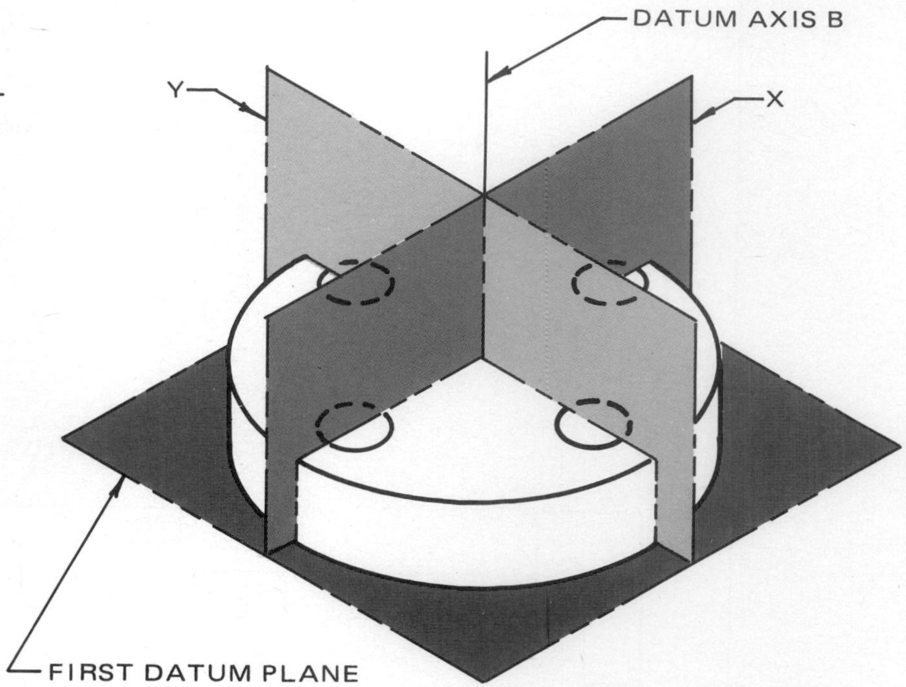

FIGURE 16-7-9 Secondary and tertiary datum features—MMC.

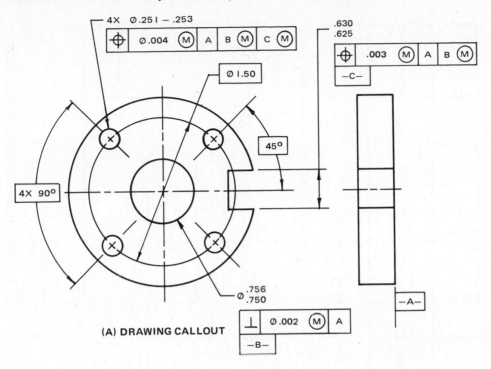

(A) DRAWING CALLOUT

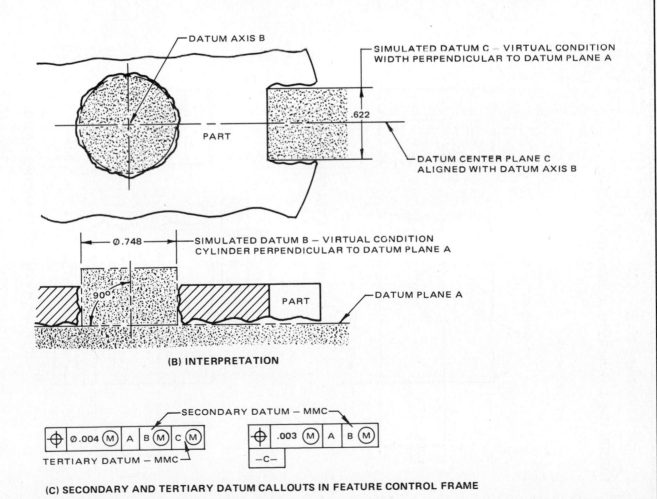

(B) INTERPRETATION

SECONDARY DATUM — MMC

TERTIARY DATUM — MMC

(C) SECONDARY AND TERTIARY DATUM CALLOUTS IN FEATURE CONTROL FRAME

FIGURE 16-9-1 Comparison of tolerancing methods.

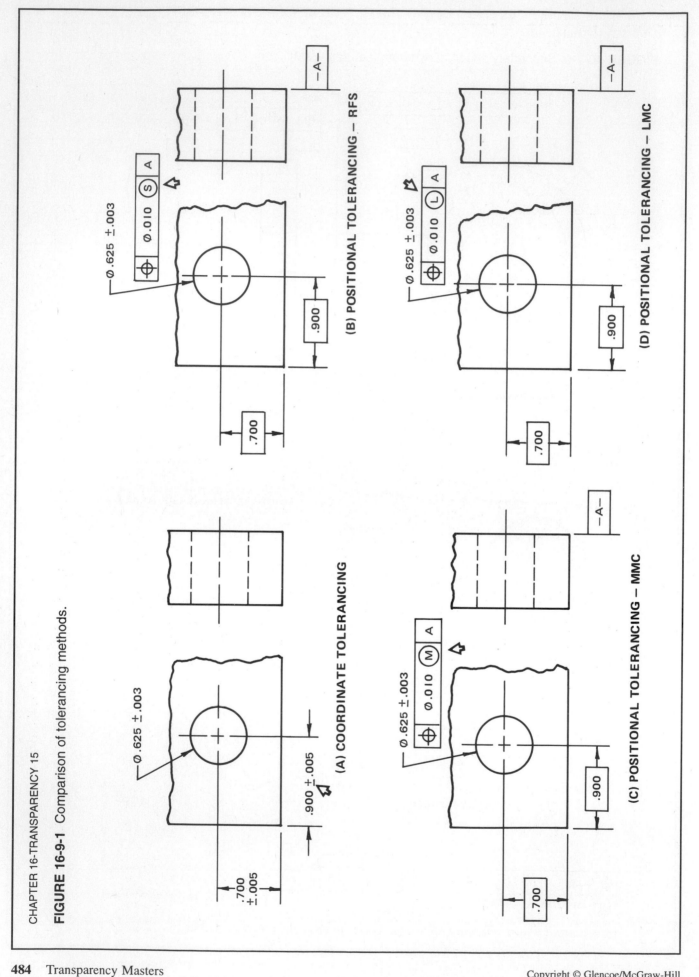

(A) COORDINATE TOLERANCING

(B) POSITIONAL TOLERANCING – RFS

(C) POSITIONAL TOLERANCING – MMC

(D) POSITIONAL TOLERANCING – LMC

FIGURE 16-9-2 Tolerance zones for coordinate tolerancing.

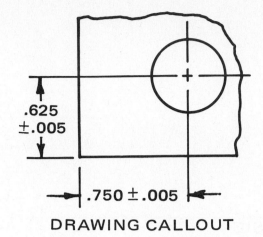

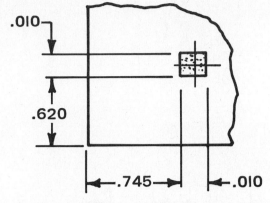

DRAWING CALLOUT · TOLERANCE ZONE AT SURFACE

(A) EQUAL TOLERANCES

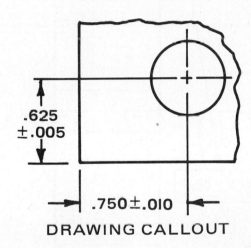

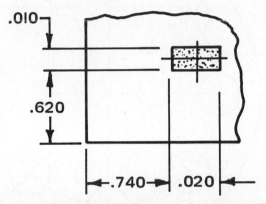

DRAWING CALLOUT · TOLERANCE ZONE AT SURFACE

(B) UNEQUAL TOLERANCES

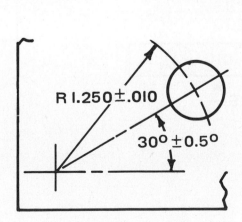

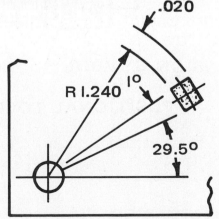

DRAWING CALLOUT · TOLERANCE ZONE AT SURFACE

(C) POLAR TOLERANCES

FIGURE 16-9-7 Identifying basic dimensions.

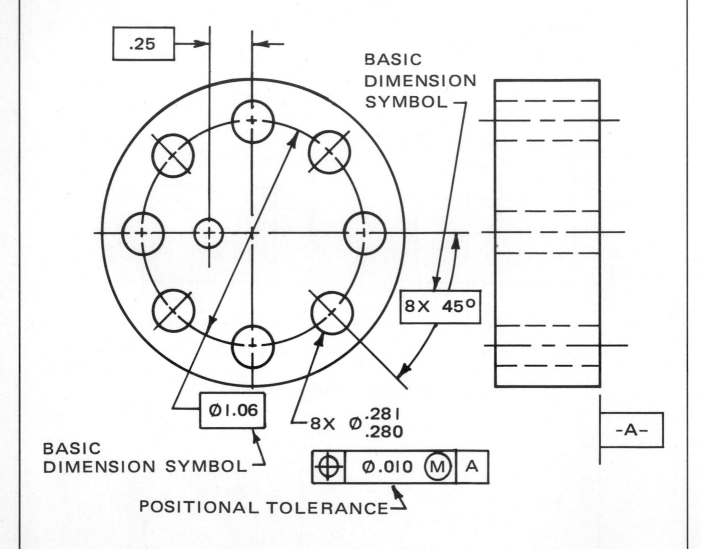

FIGURE 16-10-1 Illustrating how a fastener can interfere with a mating part.

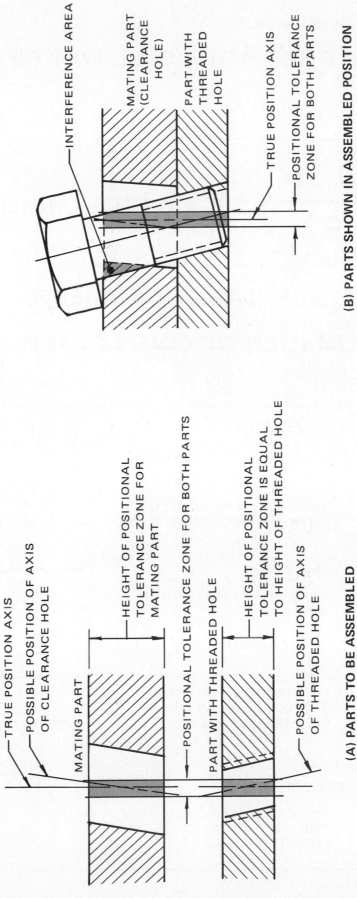

INTERFERENCE AREA

MATING PART (CLEARANCE HOLE)

PART WITH THREADED HOLE

TRUE POSITION AXIS

POSITIONAL TOLERANCE ZONE FOR BOTH PARTS

(B) PARTS SHOWN IN ASSEMBLED POSITION

TRUE POSITION AXIS

POSSIBLE POSITION OF AXIS OF CLEARANCE HOLE

HEIGHT OF POSITIONAL TOLERANCE ZONE FOR MATING PART

MATING PART

POSITIONAL TOLERANCE ZONE FOR BOTH PARTS

PART WITH THREADED HOLE

HEIGHT OF POSITIONAL TOLERANCE ZONE IS EQUAL TO HEIGHT OF THREADED HOLE

POSSIBLE POSITION OF AXIS OF THREADED HOLE

(A) PARTS TO BE ASSEMBLED

FIGURE 16-11-1 Datum target symbol.

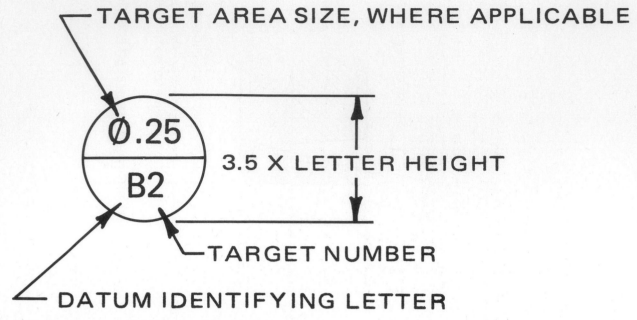

TARGET AREA SIZE, WHERE APPLICABLE

3.5 X LETTER HEIGHT

TARGET NUMBER

DATUM IDENTIFYING LETTER

FIGURE 16-11-2 Identification of datum targets.

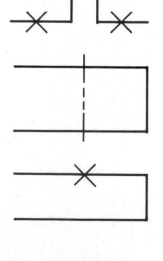

TARGET POINT

A CROSS ON THE SURFACE

OR

DATUM POINT LOCATED ON ADJACENT VIEWS

TARGET LINE

A PHANTOM LINE ON THE SURFACE

AND/OR

A CROSS MAY BE ADDED ON THE PROFILE (WHERE THE LINE APPEARS AS A POINT ON THE SURFACE)

TARGET AREA

A SECTION-LINED AREA ON THE SURFACE ENCLOSED BY PHANTOM LINES

FIGURE 16-11-13 Part with a surface and three target lines used as datum features.

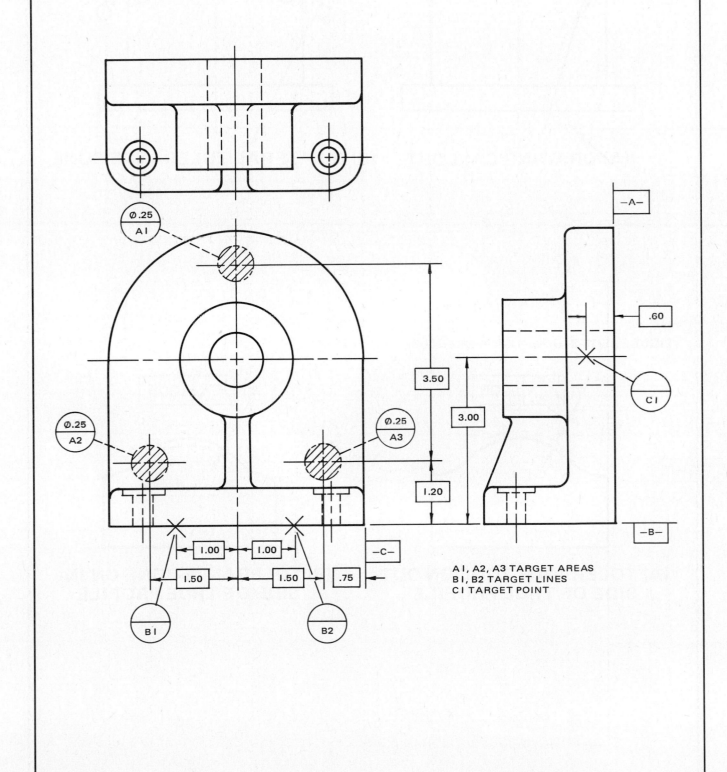

A1, A2, A3 TARGET AREAS
B1, B2 TARGET LINES
C1 TARGET POINT

FIGURE 16-13-2 Simple profile with a bilateral profile tolerance zone.

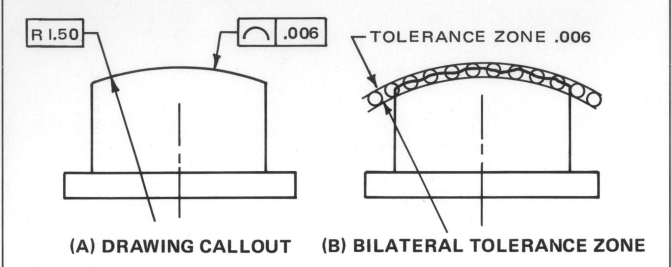

(A) DRAWING CALLOUT   **(B) BILATERAL TOLERANCE ZONE**

FIGURE 16-13-3 Unilateral tolerance zones.

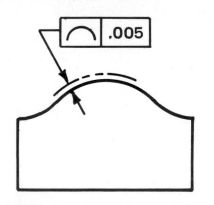 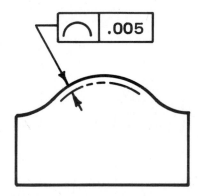

(A) TOLERANCE ZONE ON OUT-SIDE OF TRUE PROFILE   **(B) TOLERANCE ZONE ON IN-SIDE OF TRUE PROFILE**

FIGURE 16-13-16 Profile-of-a-surface tolerance controls form and size of cam profile.

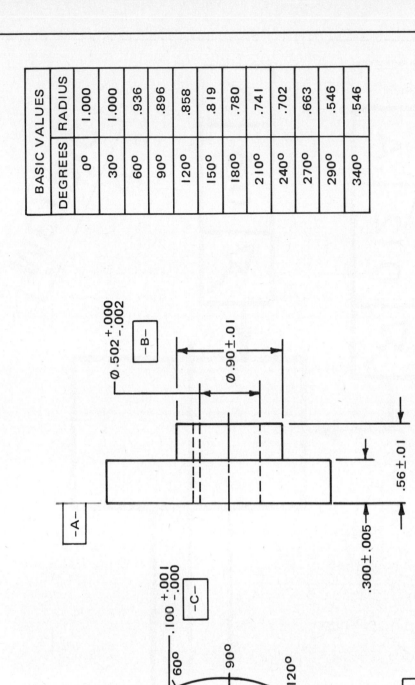

BASIC VALUES	
DEGREES	RADIUS
0°	1.000
30°	1.000
60°	.936
90°	.896
120°	.858
150°	.819
180°	.780
210°	.741
240°	.702
270°	.663
290°	.546
340°	.546

NOTE: VALUES SHOWN IN THE CHART ARE BASIC.

FIGURE 16-14-19 Cylindrical datum feature.

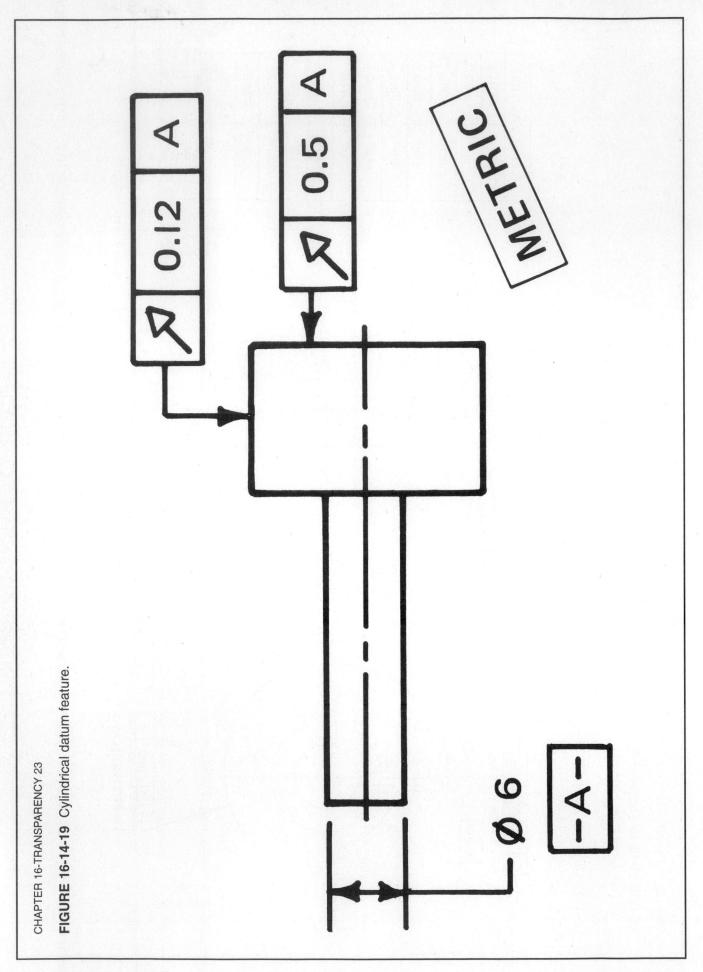

METRIC

FIGURE 16-15-3 Positional tolerancing applied to slots at MMC.

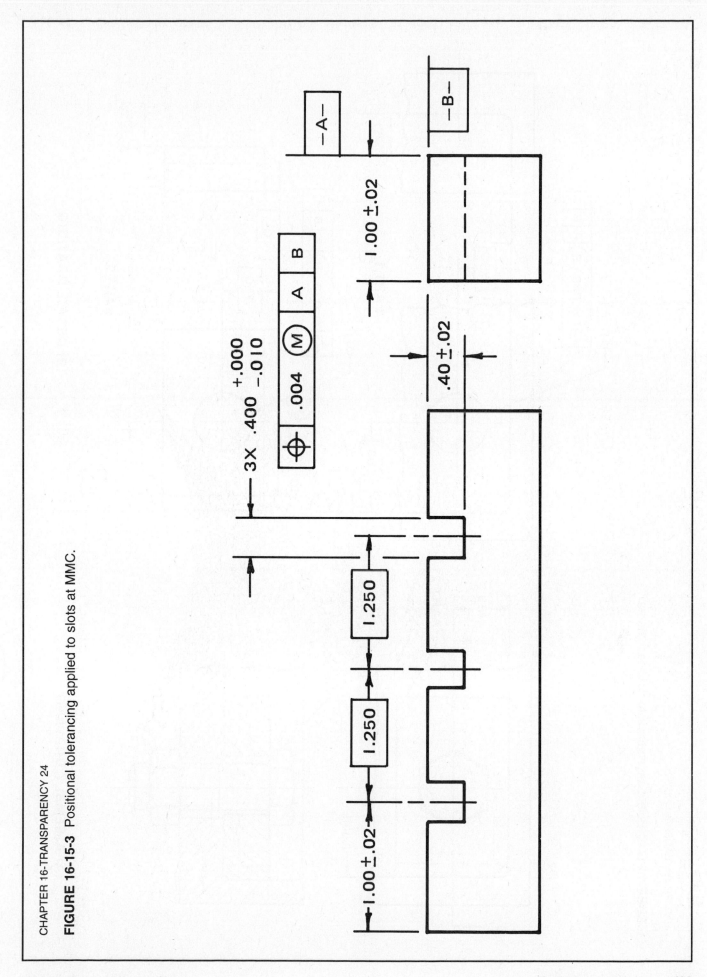

FIGURE 16-17-3 Bolted assembly with floating fastener.

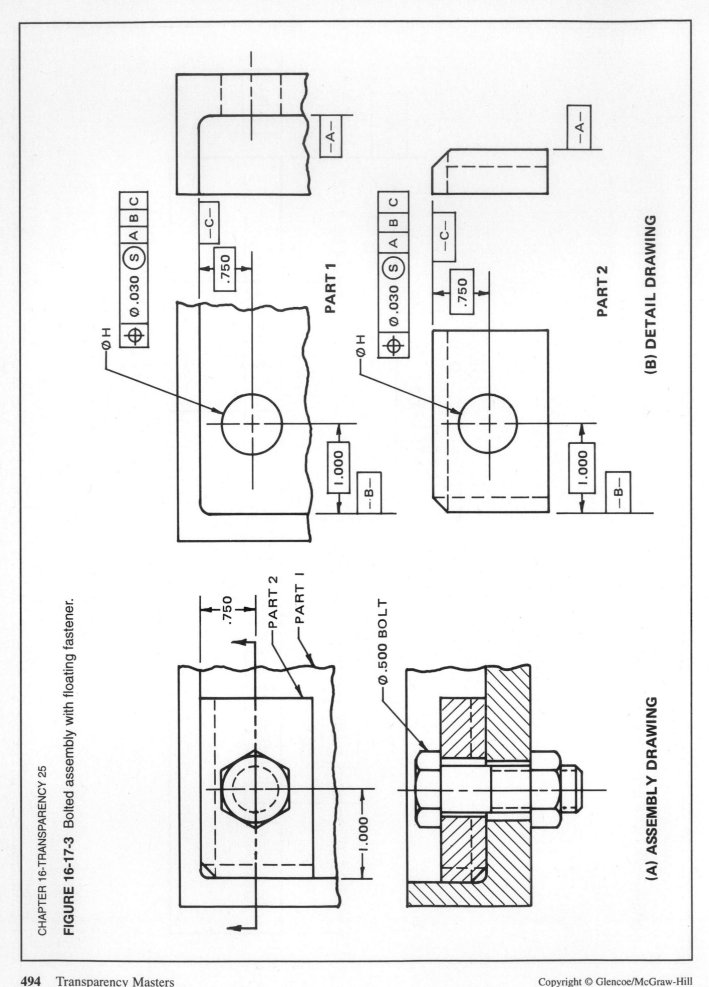

PART 1

PART 2

(B) DETAIL DRAWING

(A) ASSEMBLY DRAWING

CHAPTER 17-TRANSPARENCY 1

FIGURE 17-1-2 Integrated computer-aided engineering (CAE) systems.

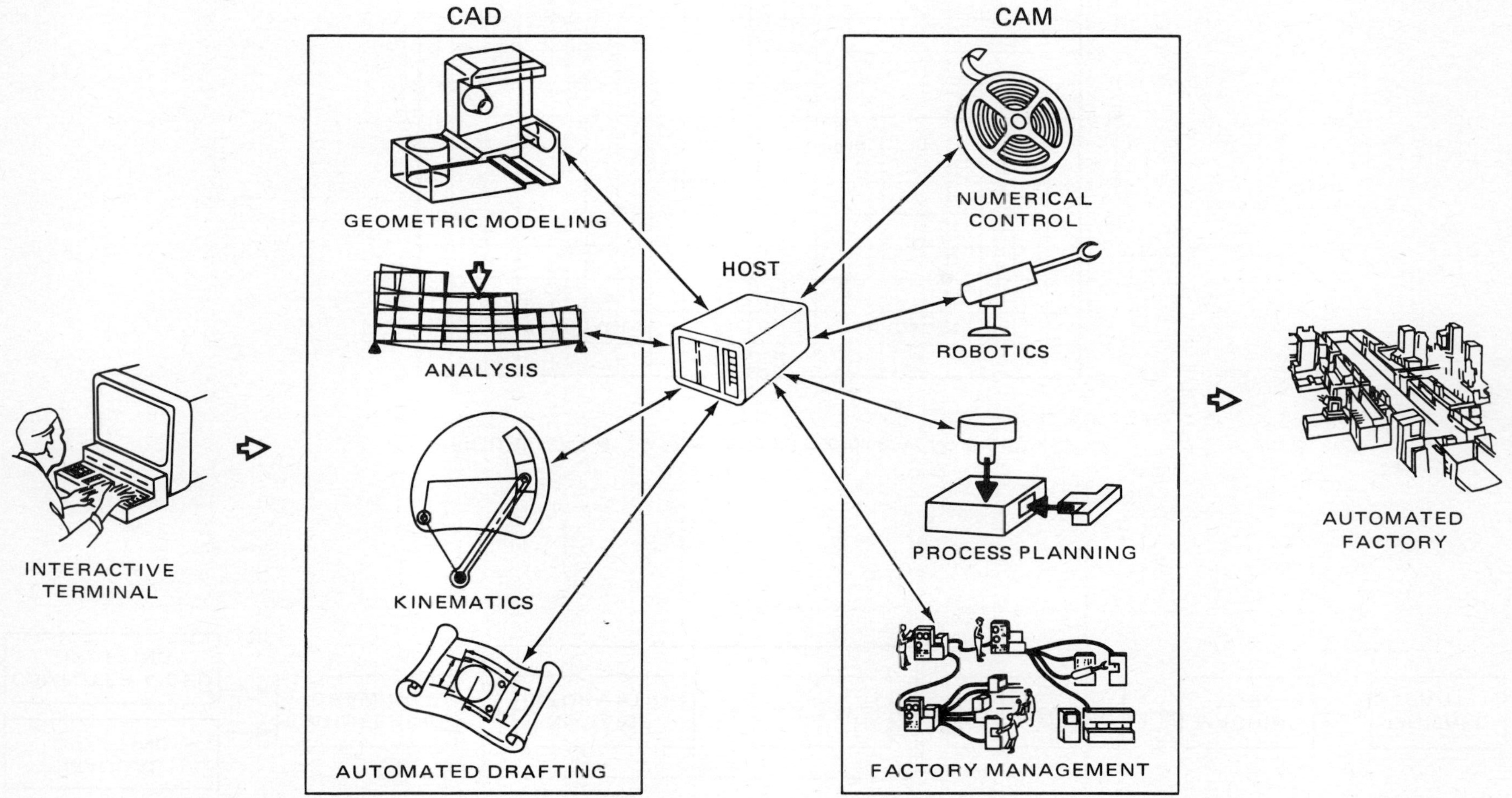
CAD
CAM
GEOMETRIC MODELING
ANALYSIS
KINEMATICS
AUTOMATED DRAFTING
HOST
NUMERICAL CONTROL
ROBOTICS
PROCESS PLANNING
FACTORY MANAGEMENT
INTERACTIVE TERMINAL
AUTOMATED FACTORY

FIGURE 17-1-3 Computer numerical control sequence.

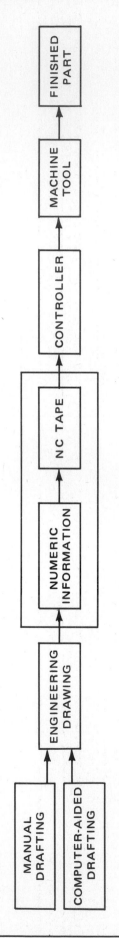

FIGURE 17-1-4 Two-dimensional coordinates (X and Y).

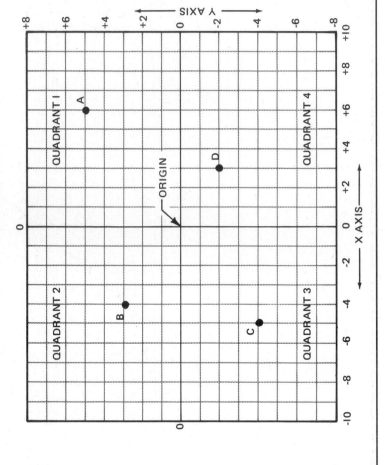

FIGURE 17-1-5 Positioning the work.

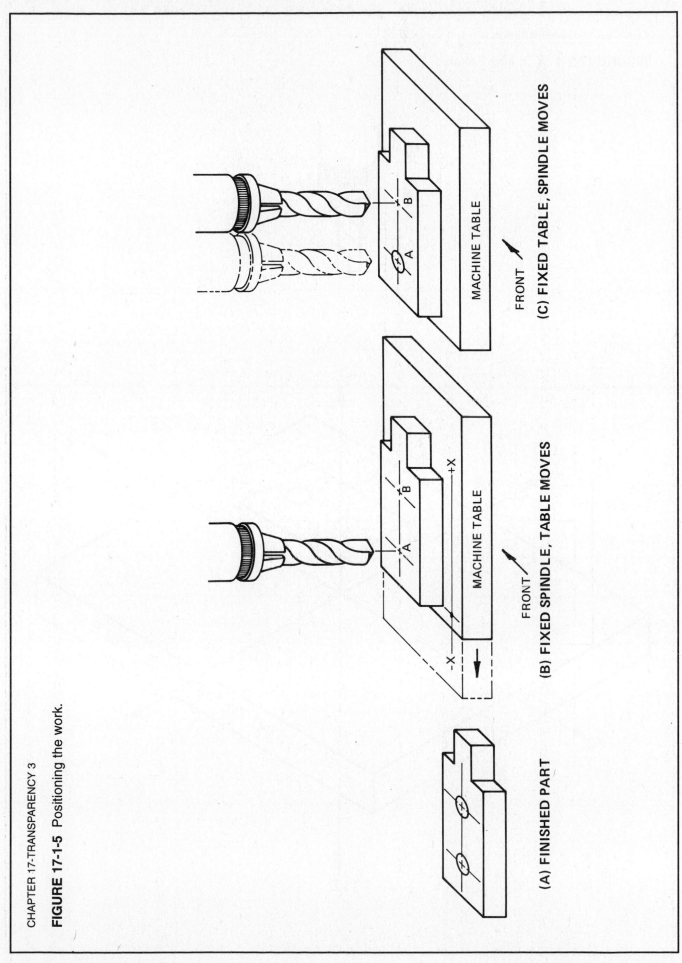

(A) FINISHED PART

(B) FIXED SPINDLE, TABLE MOVES

(C) FIXED TABLE, SPINDLE MOVES

MACHINE TABLE

FRONT

FIGURE 17-2-1 *X, Y,* and *Z* axes.

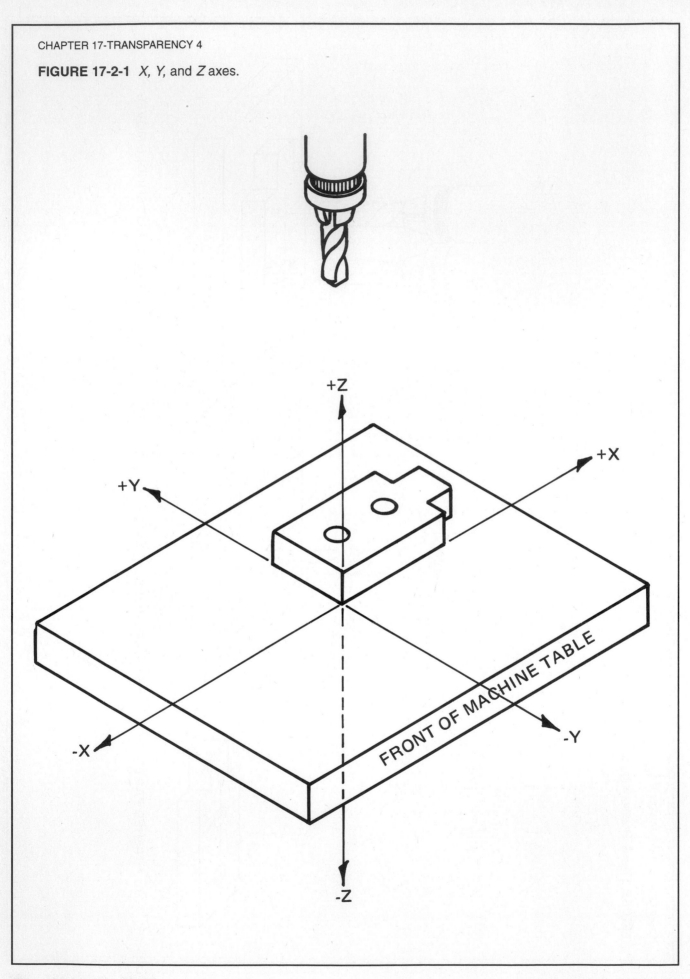

FIGURE 17-2-4 Calculating *Z* distance.

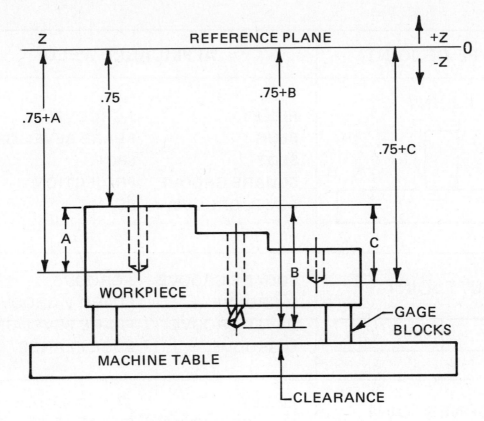

FIGURE 17-2-5 Determining gage block height.

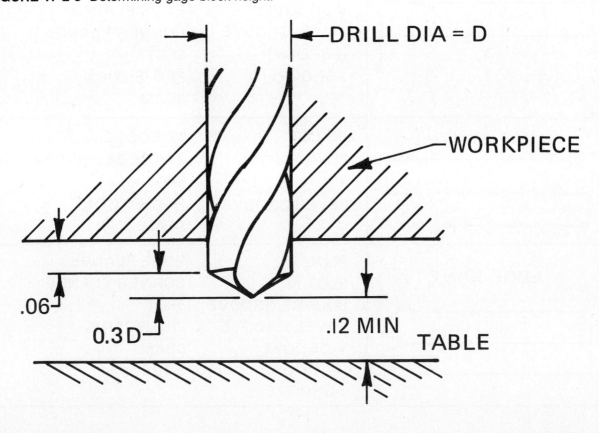

FIGURE 18-1-4 Basic welding joints.

TYPE OF JOINT	APPLICABLE WELDS	
T-JOINT 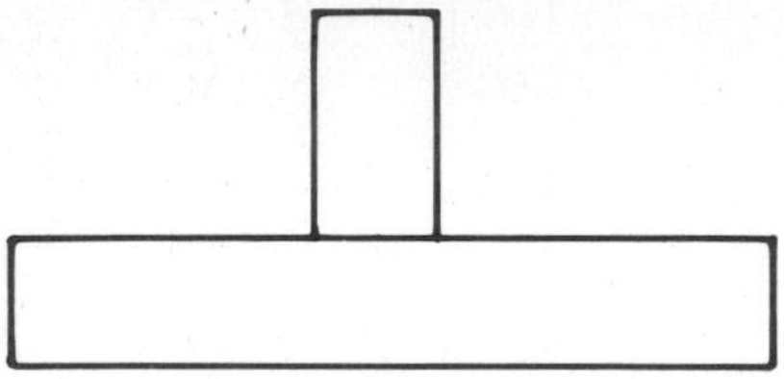	FILLET PLUG SLOT SQUARE GROOVE BEVEL GROOVE	J-GROOVE FLARE BEVEL GROOVE SPOT PROJECTION SEAM
BUTT JOINT 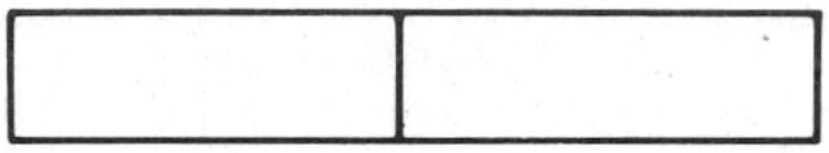	SQUARE GROOVE V-GROOVE BEVEL GROOVE U-GROOVE	J-GROOVE FLARE V-GROOVE FLARE BEVEL GROOVE EDGE FLANGE
CORNER JOINT 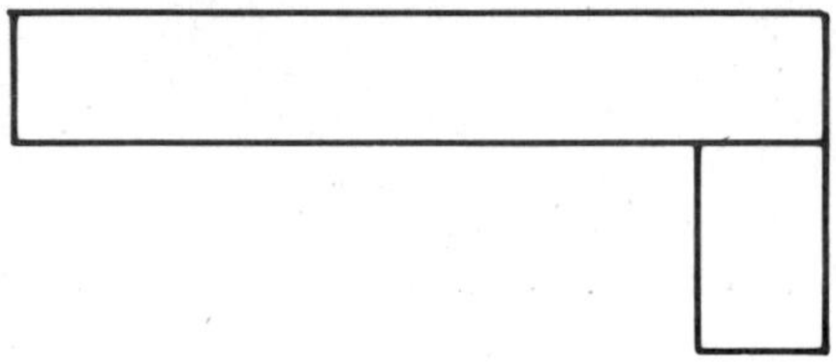	FILLET SQUARE GROOVE V-GROOVE BEVEL GROOVE U-GROOVE J-GROOVE	FLARE V-GROOVE FLARE BEVEL GROOVE EDGE FLANGE CORNER FLANGE SPOT PROJECTION SEAM
LAP JOINT 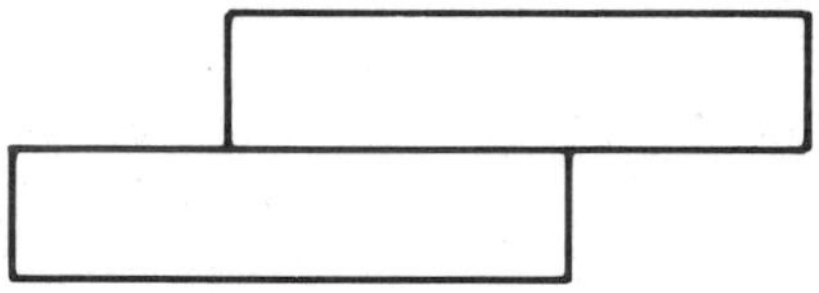	FILLET PLUG SLOT BEVEL GROOVE	J-GROOVE FLARE BEVEL GROOVE SPOT PROJECTION SEAM
EDGE JOINT 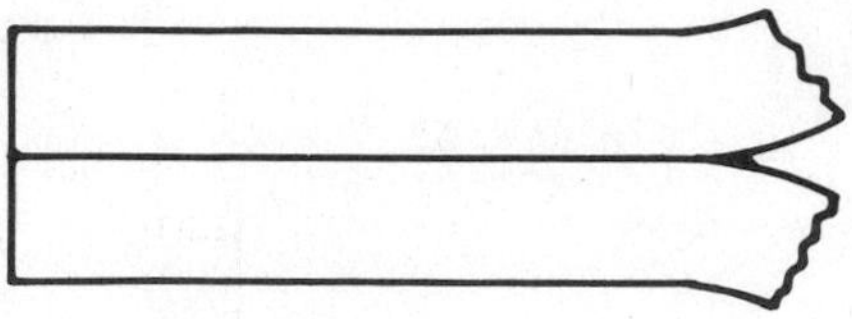	PLUG SLOT SQUARE GROOVE BEVEL GROOVE V-GROOVE U-GROOVE J-GROOVE	EDGE FLANGE CORNER FLANGE SPOT PROJECTION SEAM EDGE

FIGURE 18-2-1 Weld terminology.

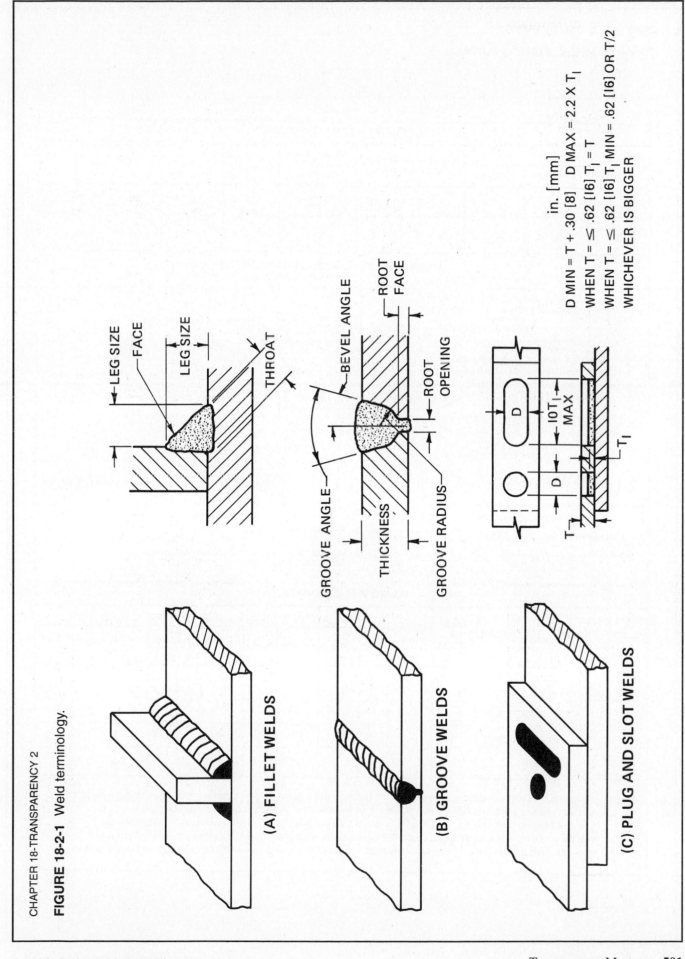

(A) FILLET WELDS

(B) GROOVE WELDS

(C) PLUG AND SLOT WELDS

in. [mm]

D MIN = T + .30 [8]   D MAX = 2.2 X T_l

WHEN T = ≤ .62 [16] T_l = T

WHEN T = ≤ .62 [16] T_l MIN = .62 [16] OR T/2

WHICHEVER IS BIGGER

FIGURE 18-2-2 Welding symbols.

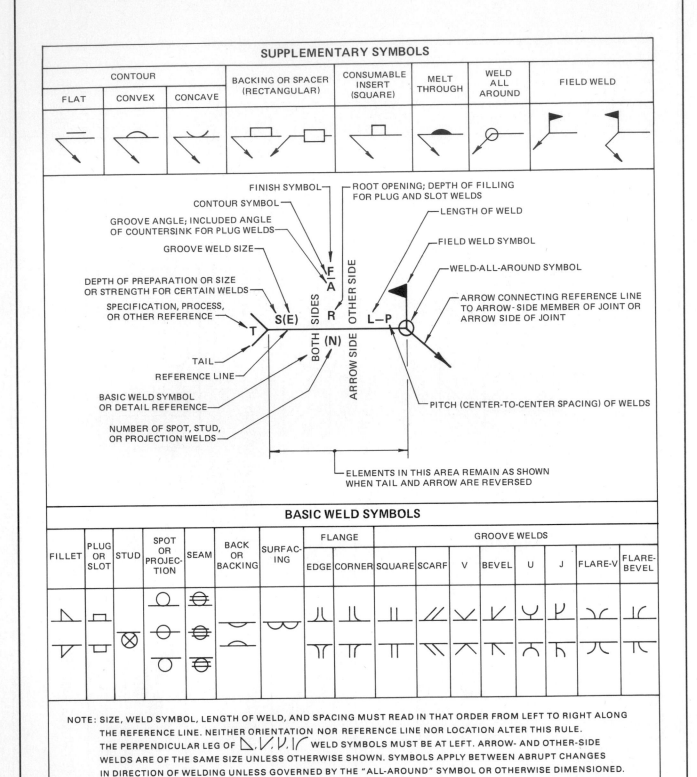

FIGURE 18-2-3 Fillet and groove welds.

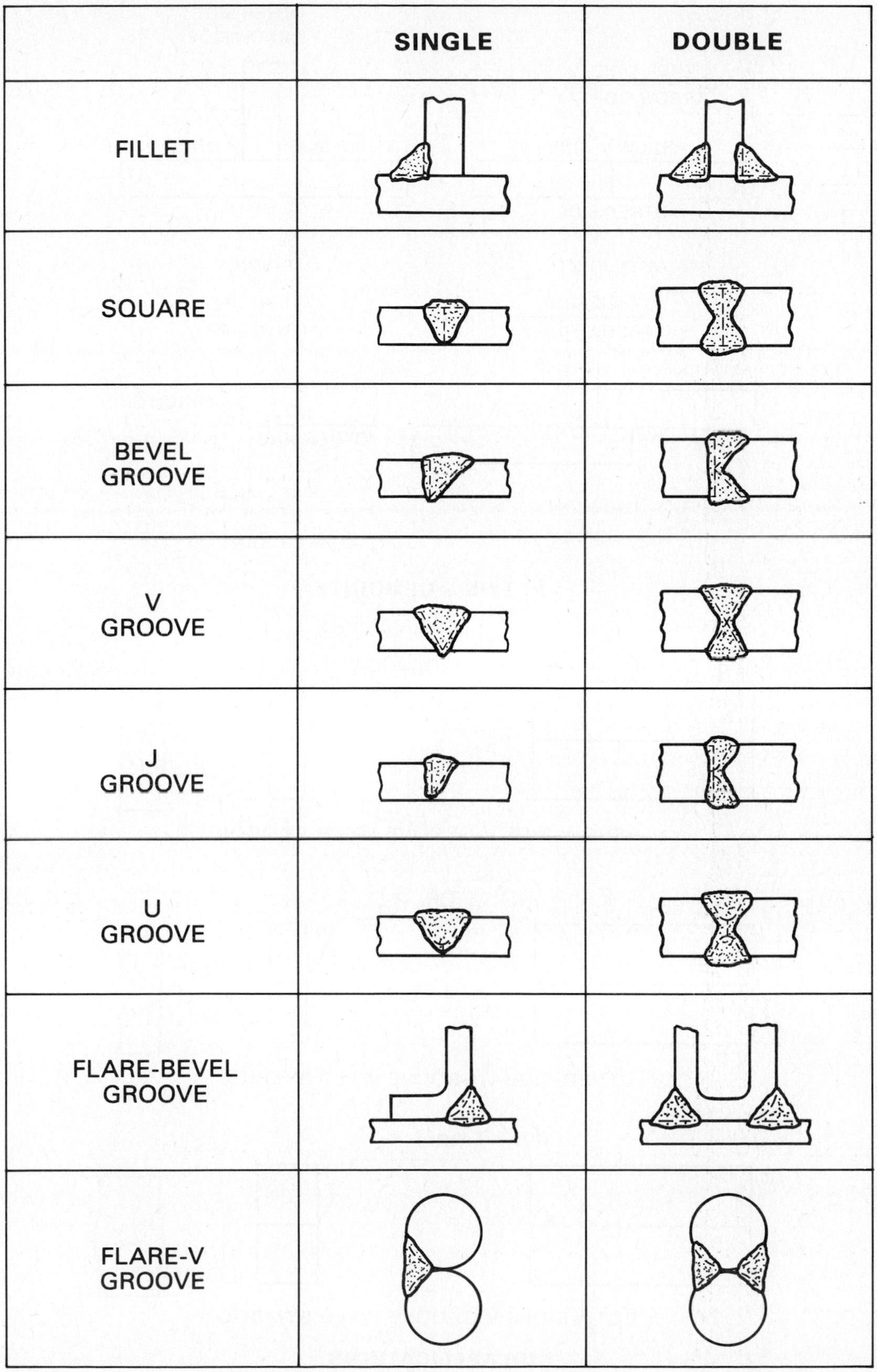

	SINGLE	DOUBLE
FILLET		
SQUARE		
BEVEL GROOVE		
V GROOVE		
J GROOVE		
U GROOVE		
FLARE-BEVEL GROOVE		
FLARE-V GROOVE		

FIGURE 18-2-7 Arrow side and other side of joint.

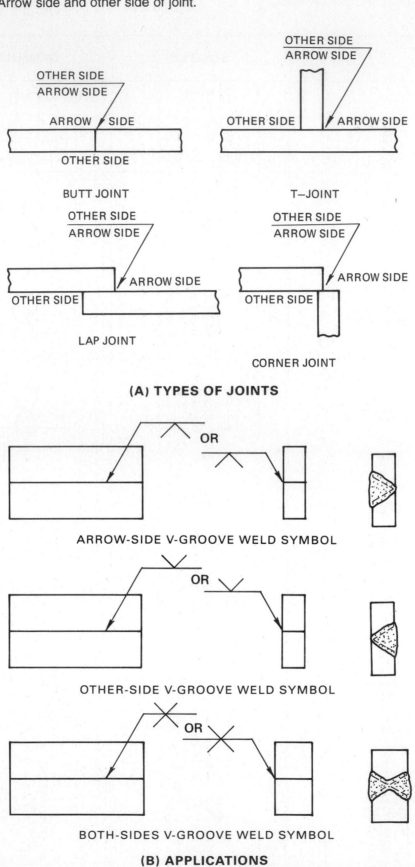

BUTT JOINT

T—JOINT

LAP JOINT

CORNER JOINT

(A) TYPES OF JOINTS

ARROW-SIDE V-GROOVE WELD SYMBOL

OTHER-SIDE V-GROOVE WELD SYMBOL

BOTH-SIDES V-GROOVE WELD SYMBOL

(B) APPLICATIONS

FIGURE 18-2-11 Combined welding symbols.

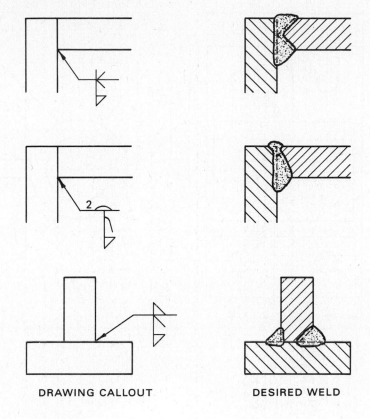

DRAWING CALLOUT DESIRED WELD

FIGURE 18-2-12 Finishing of welds.

FLAT   CONVEX   CONCAVE

(A) CONTOUR SYMBOLS

G   C   M
GRINDING CHIPPING MACHINING

R   H
ROLLING   HAMMERING

(B) POSTWELD FINISHING SYMBOLS

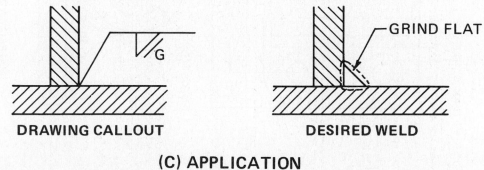

GRIND FLAT

DRAWING CALLOUT   DESIRED WELD

(C) APPLICATION

FIGURE 18-3-5 Comparison of a cast shaft support with a welded steel shaft support.

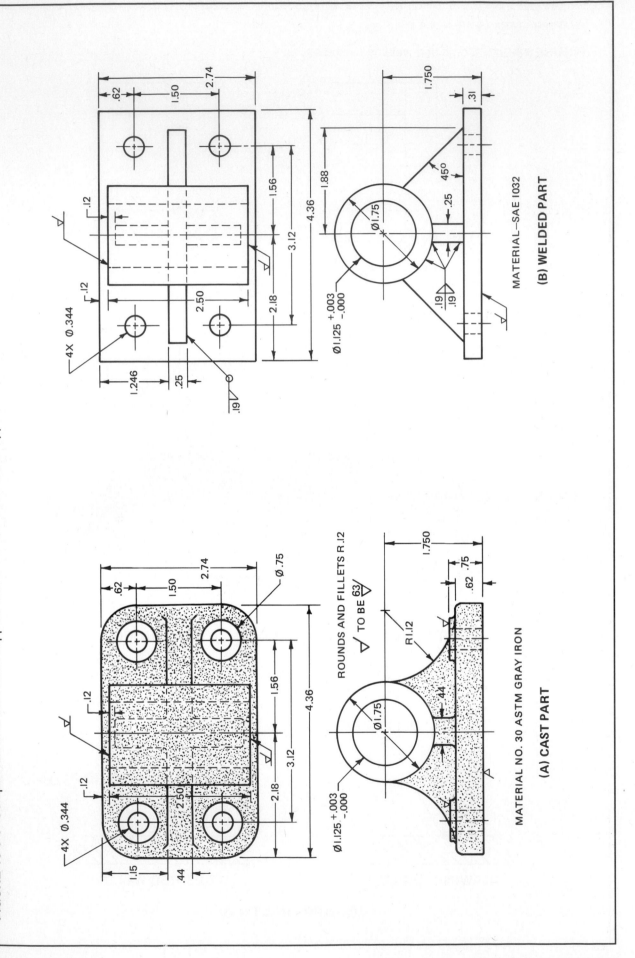

MATERIAL—SAE 1032

(B) WELDED PART

ROUNDS AND FILLETS R.12

MATERIAL NO. 30 ASTM GRAY IRON

(A) CAST PART

FIGURE 18-4-6 Application of back and backing weld symbols.

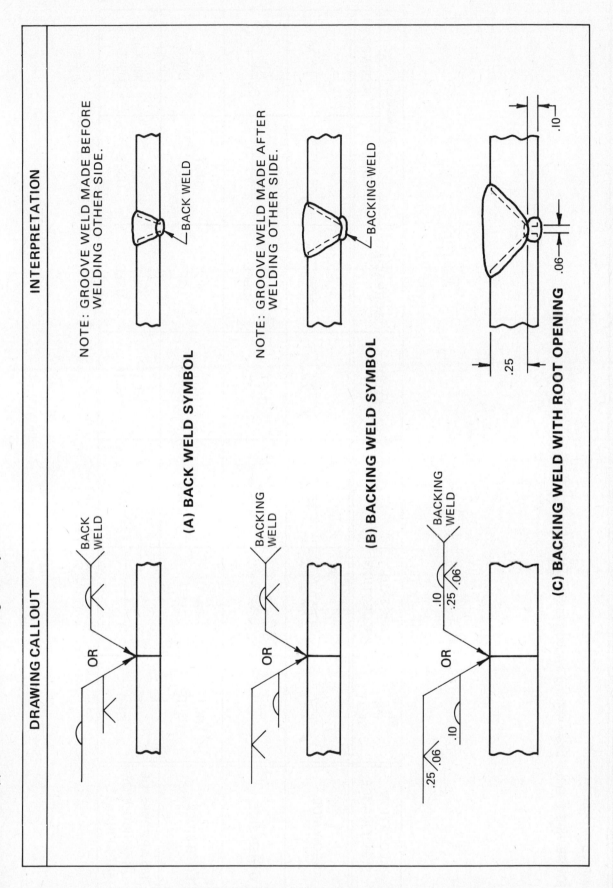

FIGURE 18-5-1 Other basic welding symbols and their location significance.

LOCATION SIGNIFICANCE	PLUG OR SLOT	SPOT OR PROJECTION	SEAM	STUD	SURFACING	FLANGE EDGE	FLANGE CORNER
ARROW SIDE	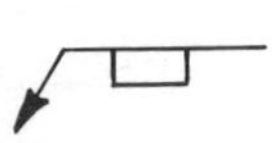	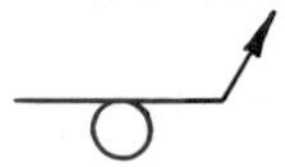		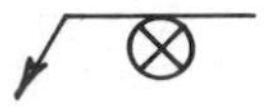		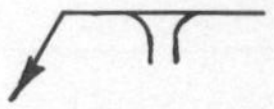	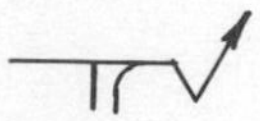
OTHER SIDE	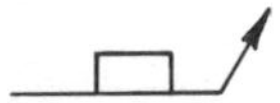	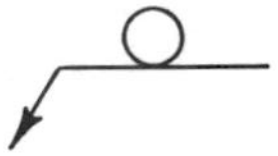	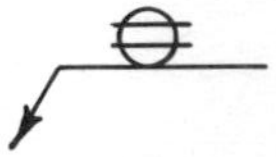	NOT USED	NOT USED	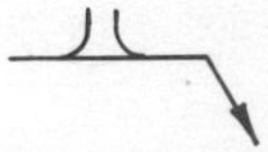	
BOTH SIDES	NOT USED	NOT USED	NOT USED	NOT USED	NOT USED	NOT USED	NOT USED
NO ARROW-SIDE OR OTHER-SIDE SIGNIFICANCE	NOT USED	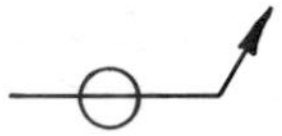		NOT USED	NOT USED	NOT USED	NOT USED

FIGURE 18-5-3 Application of slot weld symbols.

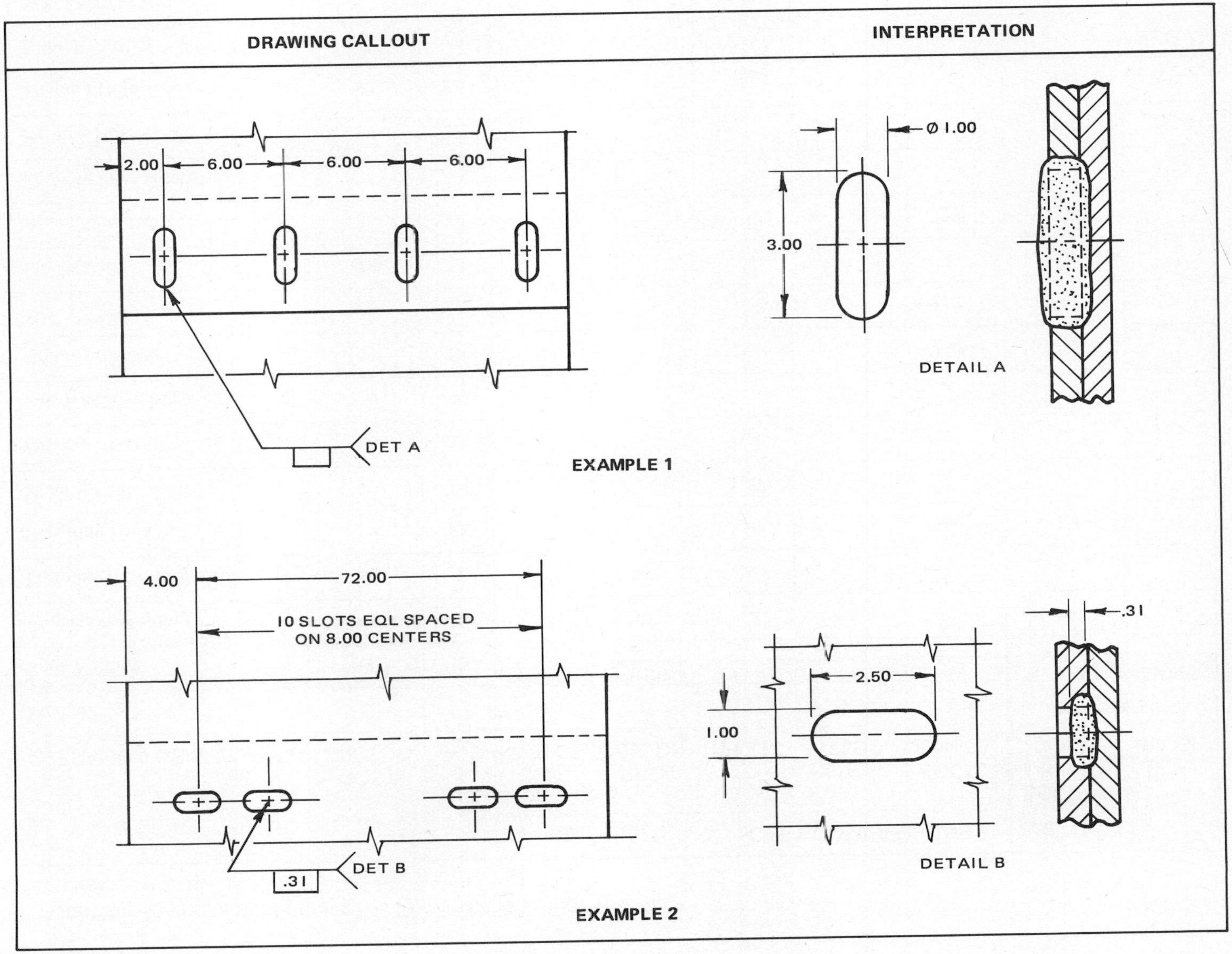

INTERPRETATION

DRAWING CALLOUT

DETAIL A

∅ 1.00

3.00

EXAMPLE 1

6.00 6.00 6.00 2.00

DET A

DETAIL B

.31

2.50

1.00

EXAMPLE 2

72.00

10 SLOTS EQL SPACED
ON 8.00 CENTERS

4.00

DET B

.31

FIGURE 19-2-1 Assembly methods cost analysis with reference to Fig. 19-2-2.

COST COMPARISONS

OPERATION OR MATERIAL	COST FACTOR	QTY & POS	SPOT WELDS	PROJ WELDS	RIVETS	ARC WELDS	BOLTS & NUTS	BOLTS & TAPPING PLATE	BLIND RIVETS
Method			1	2	3	4	5	6	7
Spot Welding	100	3B	300						
Projection Welding	100	3B		300				200	
Forming Weld Projection	89	3B		267					
Punching Hole	89	4A			356		356	356	356
Rivet	70	2A			140				
Driving Rivets	96	2A			192				192
Arc Welding (1 in.)	250	3C				750			
Bolt	115	2A					230	230	
Nut	106	2A					212		
Lockwasher	18	2A					36	36	
Assembling Bolts	136	2A					272	272	
Tapping Plate (Matl)	321	1A						321	
Drilling Hole	89	2A						178	
Tapping Hole	89	2A						178	
Blind Rivet	742	2A							1484
TOTAL COST			300	567	688	750	1106	1771	2032

The above table is for illustrative purposes only and its application should be adjusted to costs prevailing at the time of its use. Cost comparisons are based on spot welding as Unit 100. The table is not intended to indicate that the least costly method is the best; function and strength of assembly must also be considered.

FIGURE 20-1-1 Flat-belt drives.

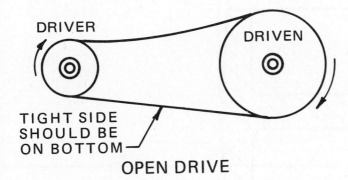

DRIVER

DRIVEN

TIGHT SIDE
SHOULD BE
ON BOTTOM

OPEN DRIVE

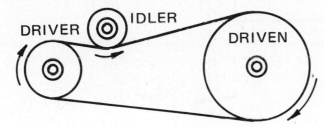

DRIVER IDLER

DRIVEN

OPEN DRIVE WITH IDLER

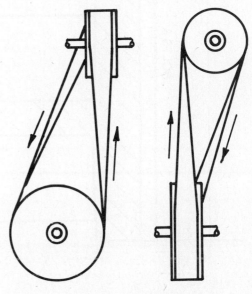

QUARTER-TWIST DRIVE

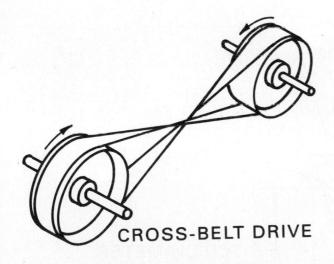

CROSS-BELT DRIVE

(A) PARALLEL SHAFTS

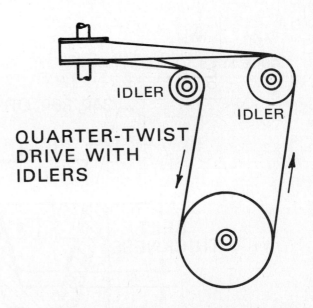

IDLER

IDLER

**QUARTER-TWIST
DRIVE WITH
IDLERS**

(B) PERPENDICULAR SHAFTS

FIGURE 20-1-3 Crown on pulley.

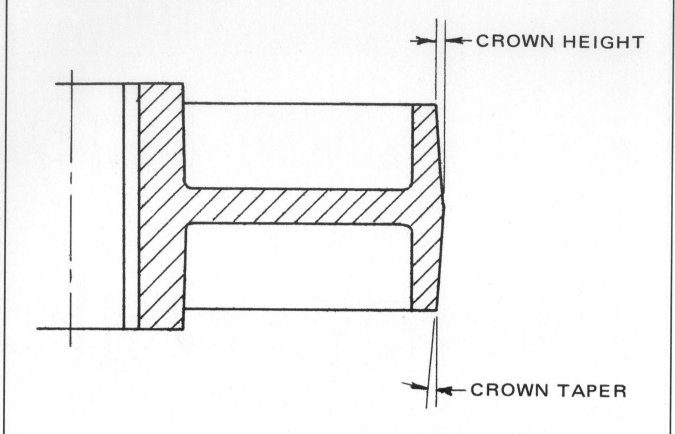

FIGURE 20-1-5 V-belt and pulley.

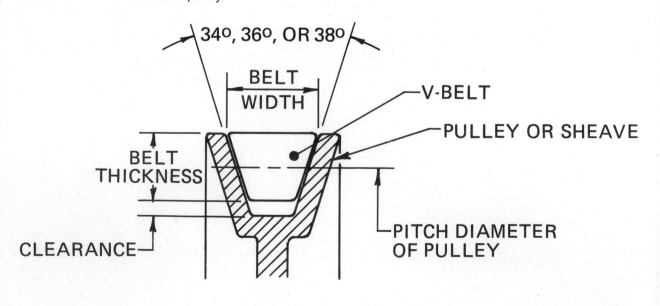

FIGURE 20-1-10 Location of idler pulleys.

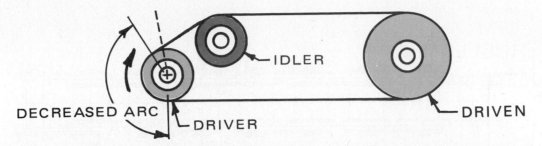

(A) INSIDE IDLER PULLEY, AT LEAST AS LARGE AS THE SMALL SHEAVE, ON THE SLACK SIDE OF THE DRIVE

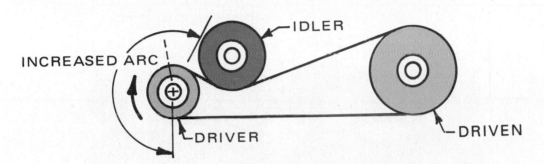

(B) OUTSIDE IDLER PULLEY, AT LEAST 1.3 LARGER THAN THE SMALL SHEAVE

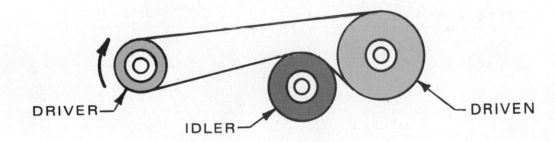

(C) OUTSIDE IDLER PULLEY ON THE TIGHT SIDE OF THE DRIVE

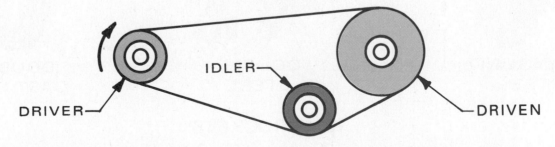

(D) INSIDE IDLER PULLEY ON THE TIGHT SIDE OF THE DRIVE

FIGURE 20-2-3 Roller chain terminology and sprockets.

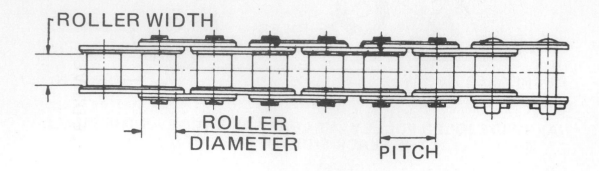

ROLLER WIDTH

ROLLER DIAMETER

PITCH

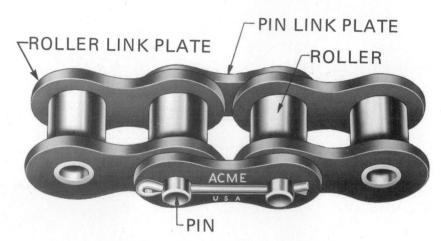

ROLLER LINK PLATE

PIN LINK PLATE

ROLLER

ACME

U S A

PIN

(A) CHAIN TERMINOLOGY

SINGLE STEEL

DOUBLE STEEL

DOUBLE CAST IRON

(B) SPROCKETS

FIGURE 20-2-7 (A) Chain drives.

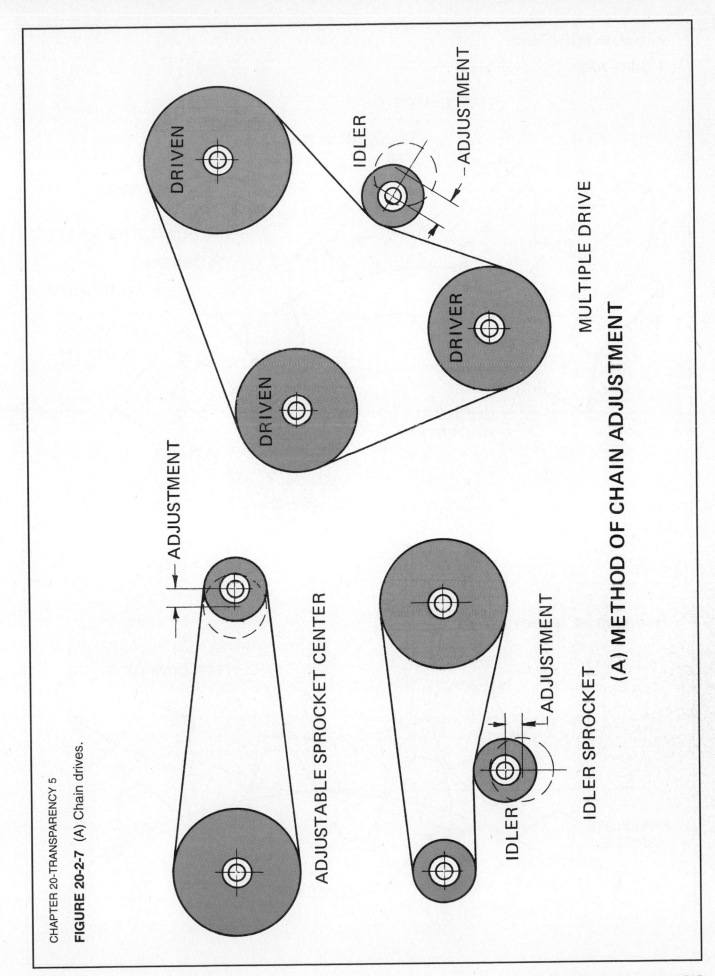

ADJUSTMENT

DRIVEN

DRIVEN

ADJUSTABLE SPROCKET CENTER

IDLER

ADJUSTMENT

ADJUSTMENT

DRIVER

MULTIPLE DRIVE

IDLER

ADJUSTMENT

IDLER SPROCKET

(A) METHOD OF CHAIN ADJUSTMENT

FIGURE 20-3-3 Gear-teeth terms.

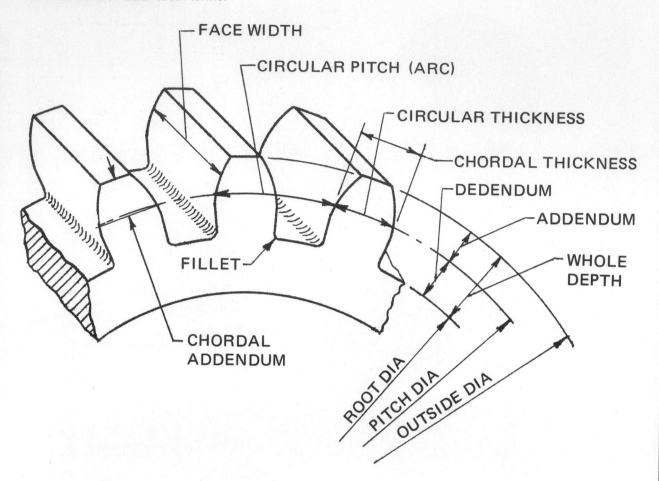

FIGURE 20-3-4 Meshing of gear teeth.

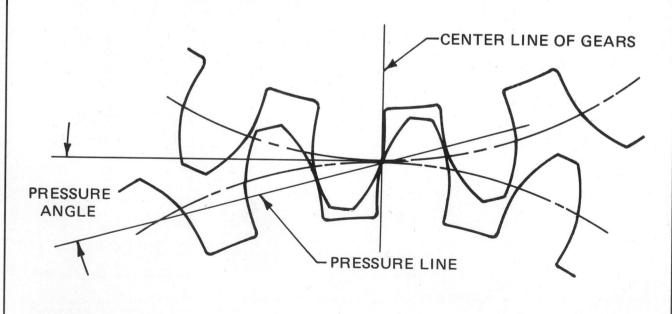

CHAPTER 20-TRANSPARENCY 7

FIGURE 20-3-8 Working drawing of a spur gear.

.32
Ø2.00
Ø2.20
Ø5.00
Ø6.400
1.50
1.00
ROUNDS AND FILLETS R.10
Ø 1.002
1.000
.25 X .12
KEYSEAT
.76
.56

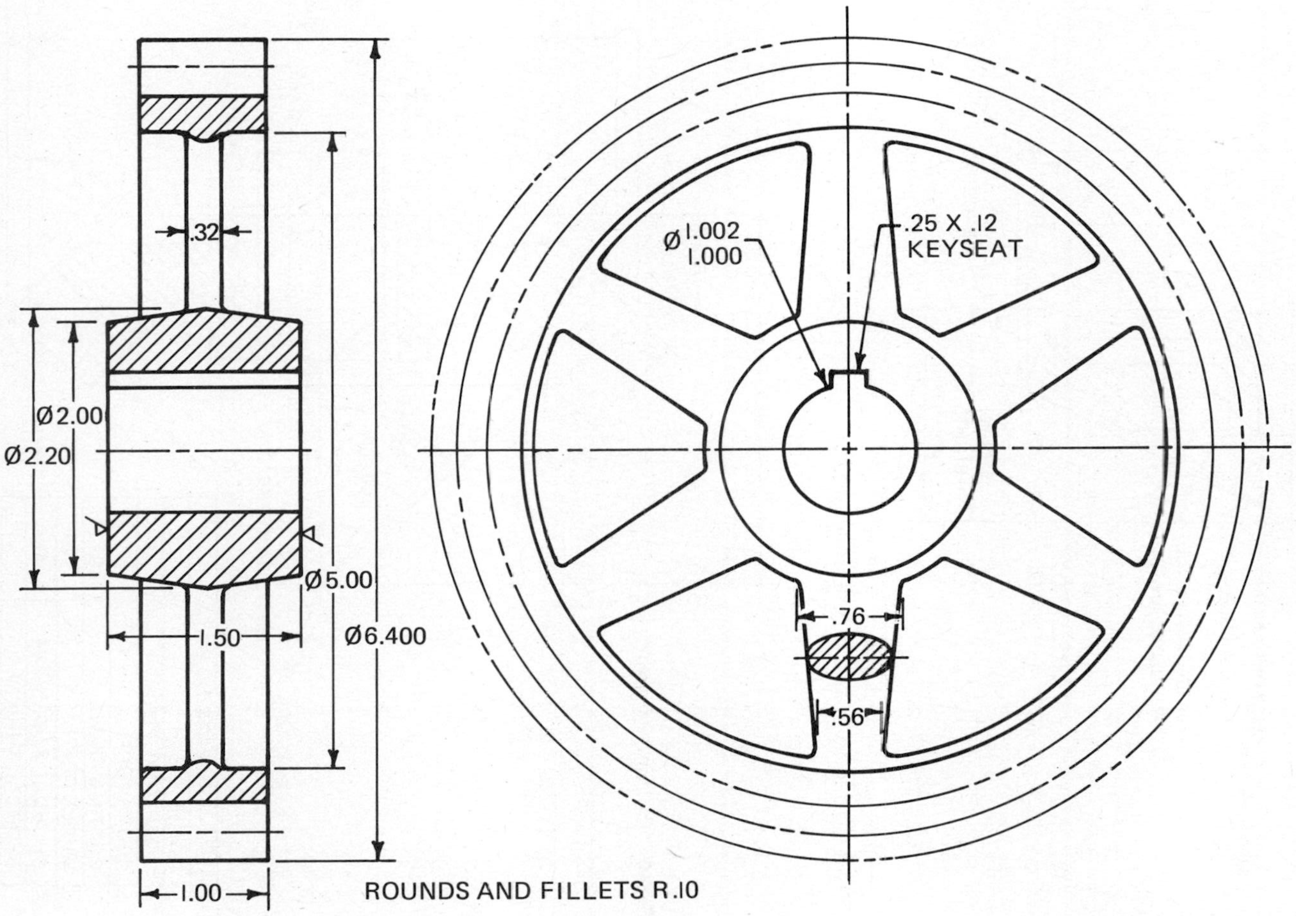

CUTTING DATA	
NUMBER OF TEETH	30
PITCH DIAMETER	6.000
DIAMETRAL PITCH	5
PRESSURE ANGLE	25°
WHOLE DEPTH	.431
CHORDAL ADDENDUM	.204
CHORDAL THICKNESS	.300
CIRCULAR THICKNESS	.314
WORKING DEPTH	.400

FIGURE 20-6-2 Bevel gear nomenclature.

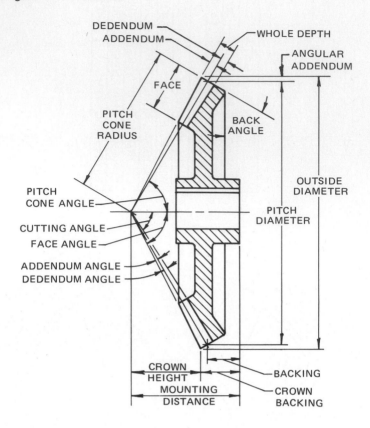

FIGURE 20-6-3 Working drawing of a bevel gear.

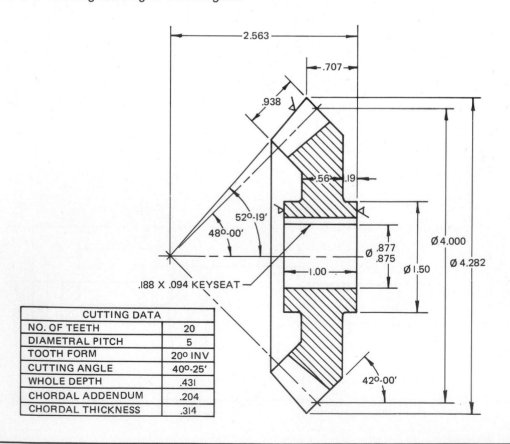

CUTTING DATA	
NO. OF TEETH	20
DIAMETRAL PITCH	5
TOOTH FORM	20° INV
CUTTING ANGLE	40°-25'
WHOLE DEPTH	.431
CHORDAL ADDENDUM	.204
CHORDAL THICKNESS	.314

FIGURE 20-6-4 Bevel gear assembly or display drawing.

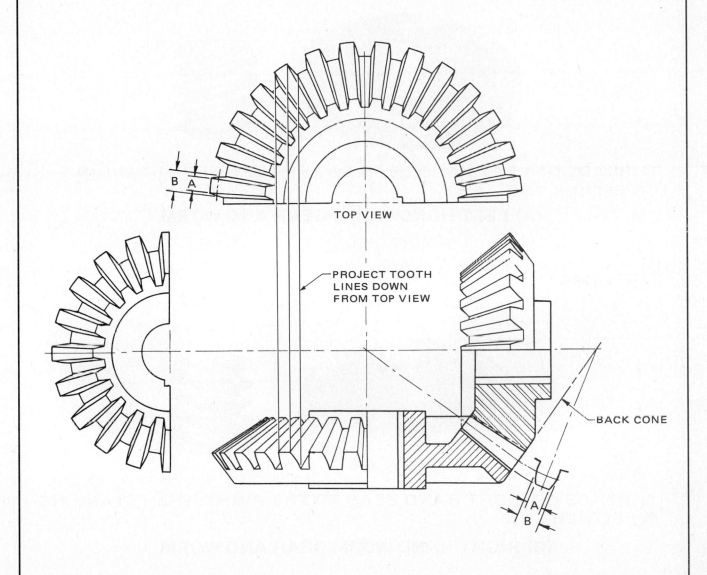

TOP VIEW

PROJECT TOOTH
LINES DOWN
FROM TOP VIEW

BACK CONE

B A

A
B

FIGURE 20-7-4 Identifying right- and left-hand worms and worm gears.

THREADS OF LEFT HAND LEAN TO THE LEFT WHEN STANDING ON EITHER END

(A) LEFT-HAND WORM GEAR AND WORM

THREADS OF RIGHT HAND LEAN TO THE RIGHT WHEN STANDING ON EITHER END

(B) RIGHT-HAND WORM GEAR AND WORM

FIGURE 20-7-6 Working drawing of a worm and worm gear.

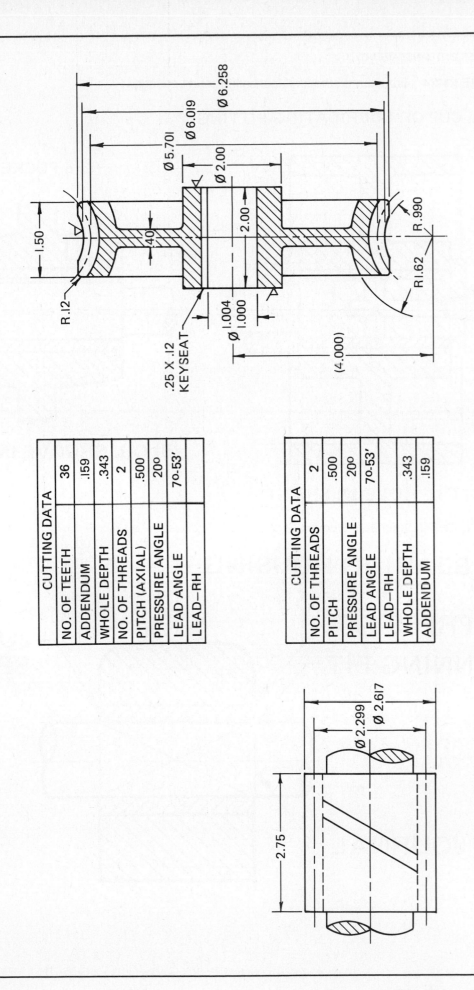

CUTTING DATA	
NO. OF TEETH	36
ADDENDUM	.159
WHOLE DEPTH	.343
NO. OF THREADS	2
PITCH (AXIAL)	.500
PRESSURE ANGLE	20º
LEAD ANGLE	7º-53'
LEAD—RH	

CUTTING DATA	
NO. OF THREADS	2
PITCH	.500
PRESSURE ANGLE	20º
LEAD ANGLE	7º-53'
LEAD—RH	
WHOLE DEPTH	.343
ADDENDUM	.159

FIGURE 21-2-1 Common methods of lubricating plain bearings.

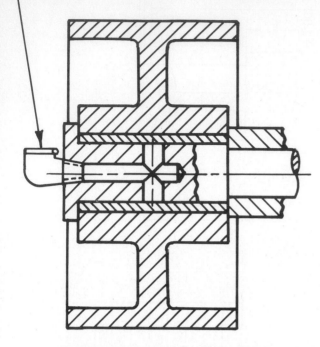

OIL CUP OR LUBRICATING FITTING

(A) OIL HOLE IN SHAFT

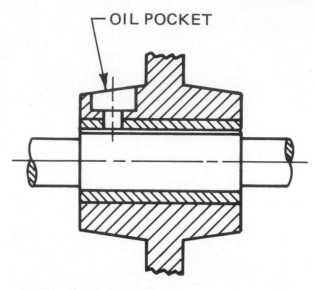

OIL POCKET

(B) OIL GROOVE IN BEARING

FIGURE 21-2-2 Journal or sleeve bearing.

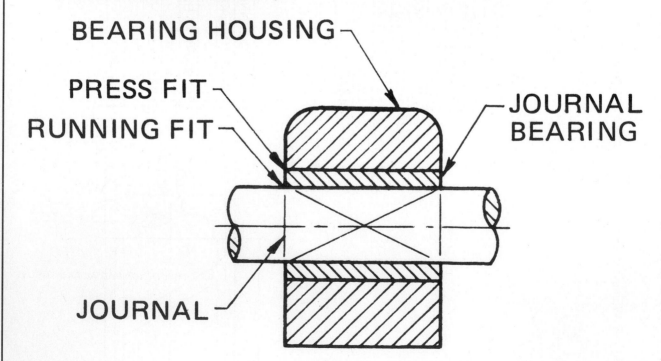

BEARING HOUSING

PRESS FIT

RUNNING FIT

JOURNAL BEARING

JOURNAL

FIGURE 21-3-1 Antifriction-bearing nomenclature. *(SKF Co.)*

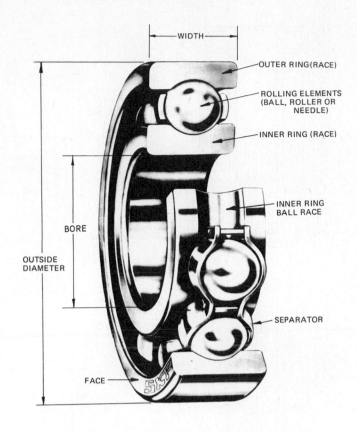

FIGURE 21-3-2 Types of bearing loads.

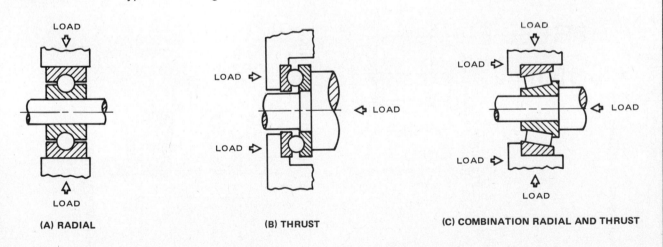

CHAPTER 21-TRANSPARENCY 3

FIGURE 21-3-8 Axial mounting of inner rings.

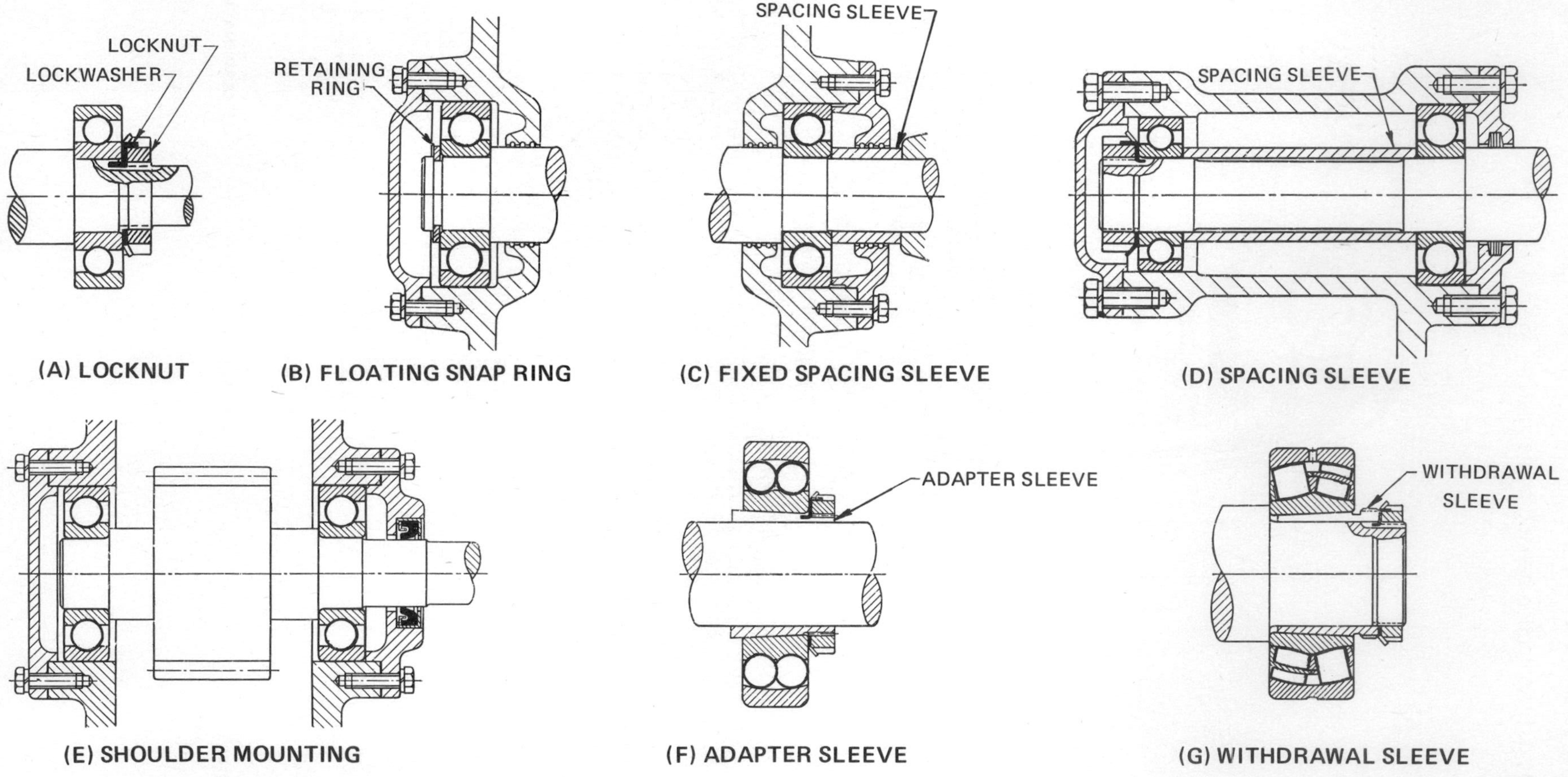
LOCKNUT
LOCKWASHER
(A) LOCKNUT
RETAINING RING
(B) FLOATING SNAP RING
SPACING SLEEVE
(C) FIXED SPACING SLEEVE
SPACING SLEEVE
(D) SPACING SLEEVE
(E) SHOULDER MOUNTING
ADAPTER SLEEVE
(F) ADAPTER SLEEVE
WITHDRAWAL SLEEVE
(G) WITHDRAWAL SLEEVE

CHAPTER 21-TRANSPARENCY 4

FIGURE 21-3-9 Outer ring mountings.

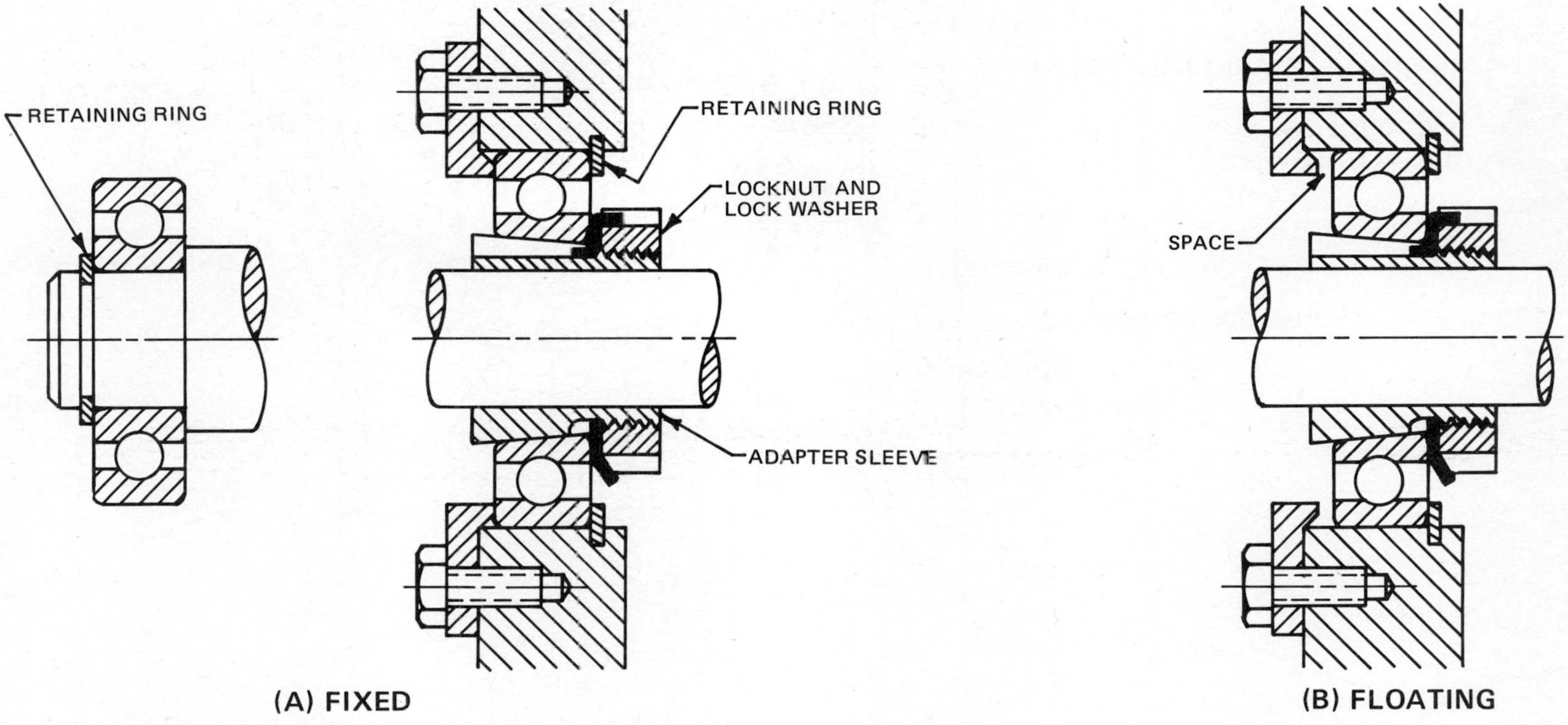
RETAINING RING
RETAINING RING
LOCKNUT AND
LOCK WASHER
ADAPTER SLEEVE
SPACE
(A) FIXED
(B) FLOATING

FIGURE 21-5-4 Metal-cased seal nomenclature.

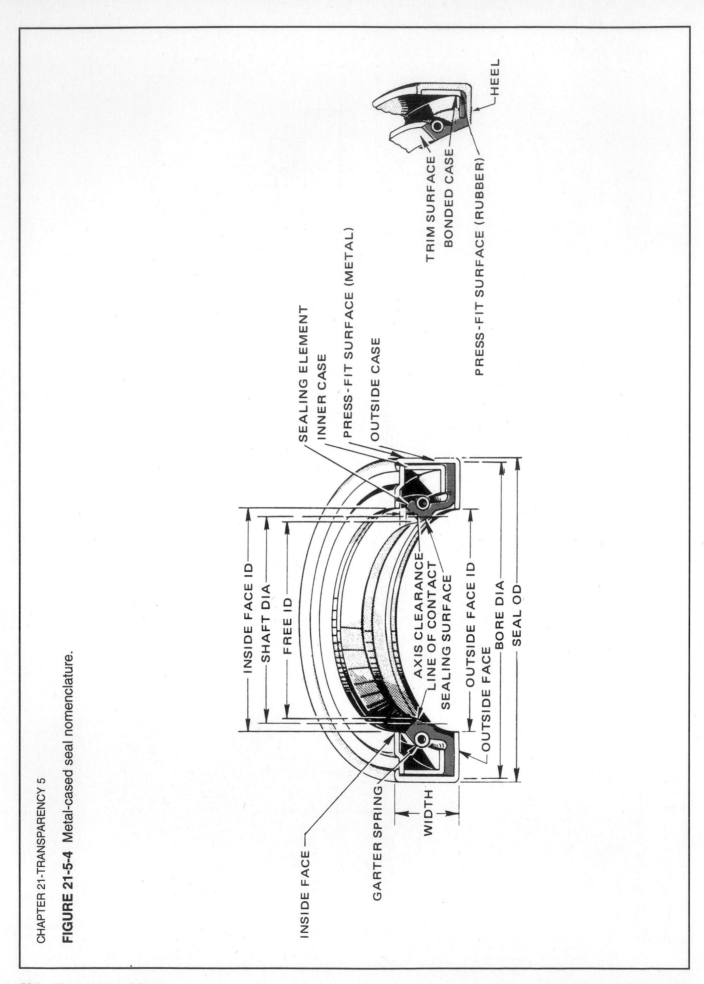

FIGURE 21-5-7 Basic end-face seal design.

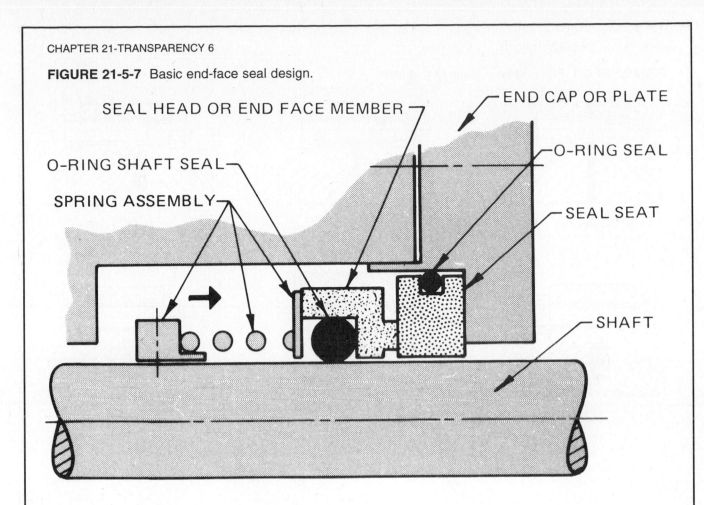

FIGURE 21-5-8 Shaft seal configurations.

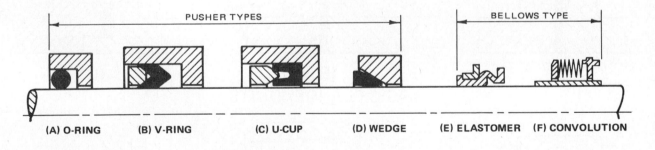

FIGURE 21-6-1 Flanged-type static O-ring seal

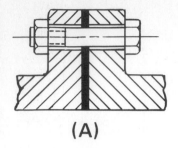

(A)

(B)

(C)

BASIC FLANGE JOINTS

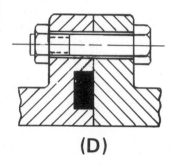

(D)

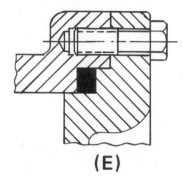

(E)

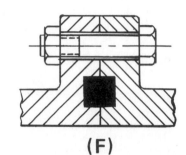

(F)

METAL-TO-METAL JOINTS

FIGURE 21-6-2 Flat gasket joints.

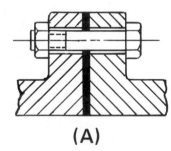

(A)

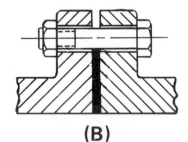

(B)

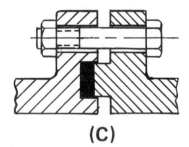

(C)

BASIC FLANGE JOINTS

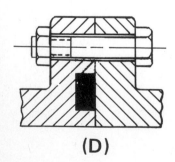

(D)

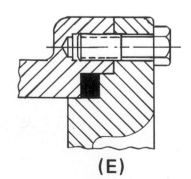

(E)

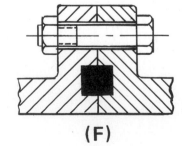

(F)

METAL-TO-METAL JOINTS

FIGURE 22-1-3 Cam nomenclature.

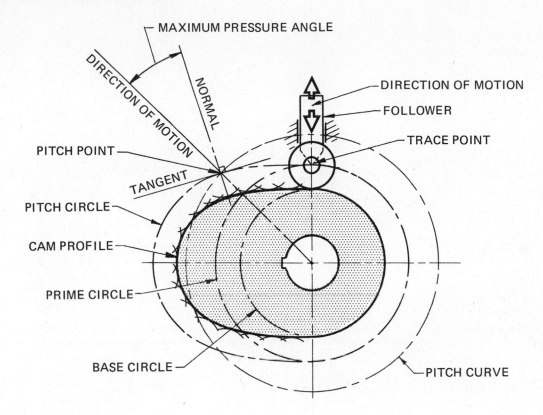

FIGURE 22-1-4 Cam displacement diagram.

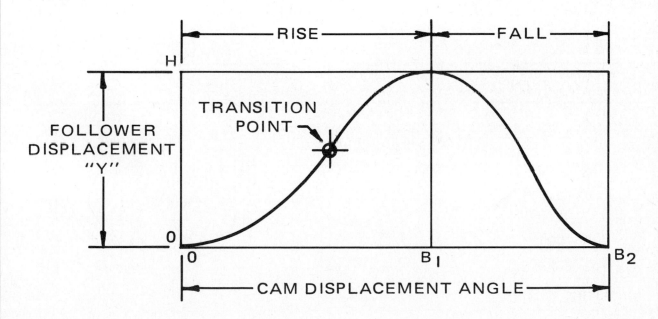

FIGURE 22-1-5 Types of cam followers.

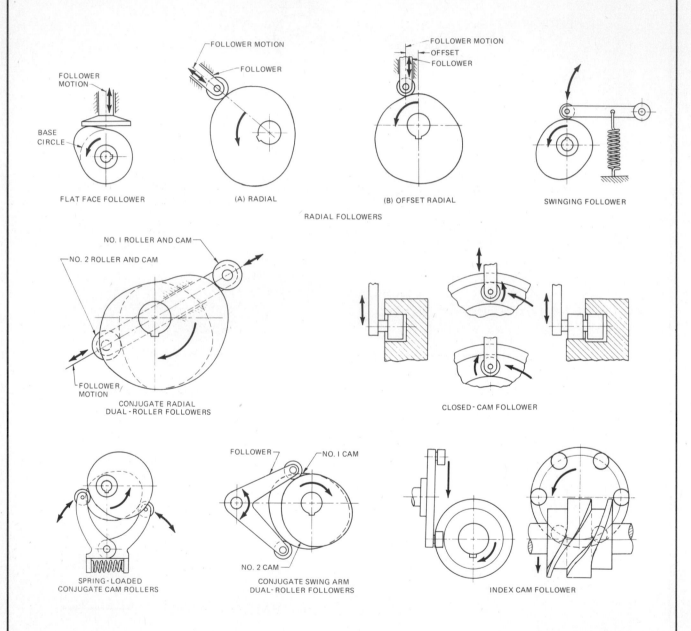

FIGURE 22-1-8 Cam motions.

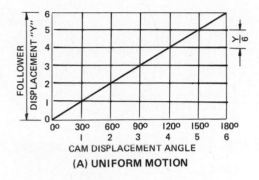

(A) UNIFORM MOTION

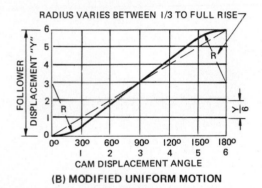

(B) MODIFIED UNIFORM MOTION

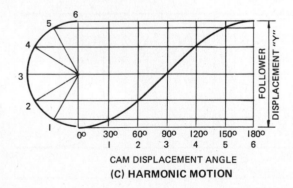

(C) HARMONIC MOTION

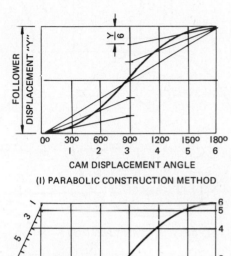

(I) PARABOLIC CONSTRUCTION METHOD

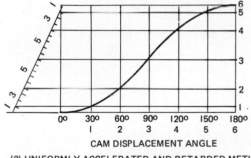

(2) UNIFORMLY ACCELERATED AND RETARDED METHOD

(D) PARABOLIC MOTION

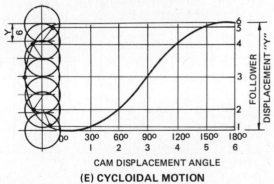

(E) CYCLOIDAL MOTION

FIGURE 22-1-9 Eccentric plate cam.

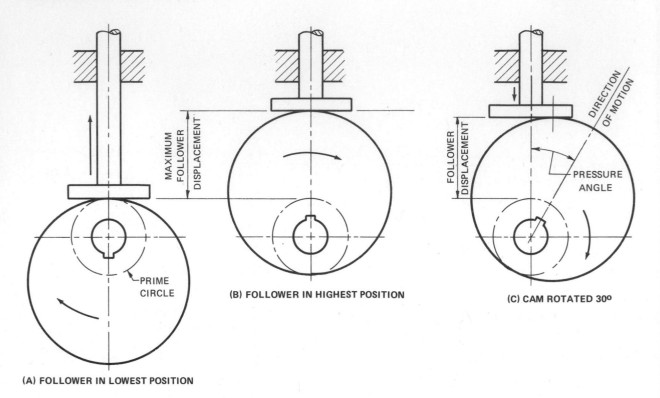

(A) FOLLOWER IN LOWEST POSITION

(B) FOLLOWER IN HIGHEST POSITION

(C) CAM ROTATED 30°

FIGURE 22-1-10 Cam displacement diagram.

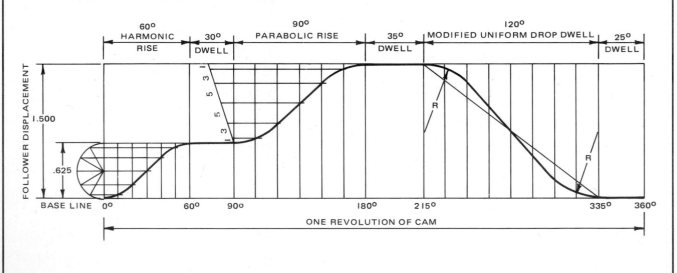

FIGURE 22-2-1 Eccentric plate cam.

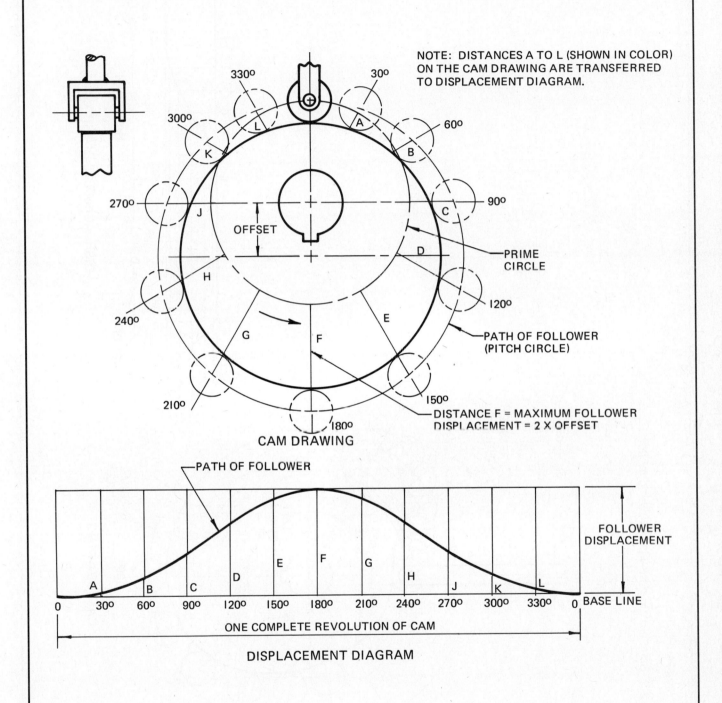

NOTE: DISTANCES A TO L (SHOWN IN COLOR) ON THE CAM DRAWING ARE TRANSFERRED TO DISPLACEMENT DIAGRAM.

CAM DRAWING

DISTANCE F = MAXIMUM FOLLOWER DISPLACEMENT = 2 X OFFSET

PATH OF FOLLOWER

ONE COMPLETE REVOLUTION OF CAM

DISPLACEMENT DIAGRAM

FIGURE 22-2-2 Simple plate cam with harmonic motion.

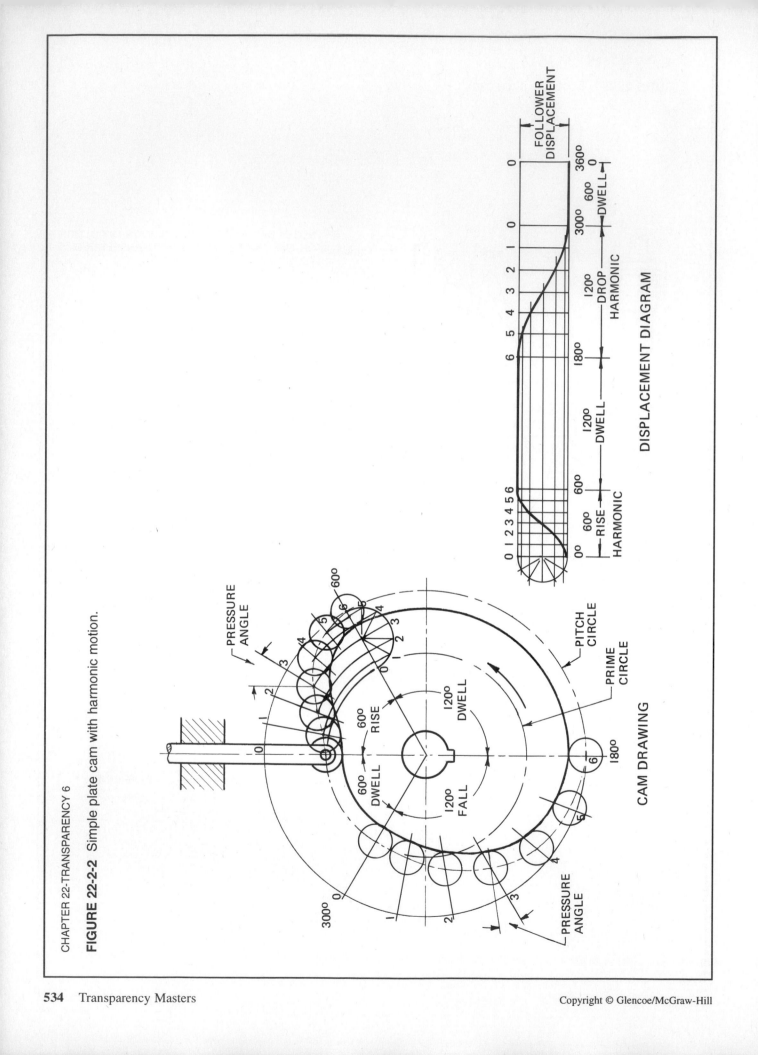

DISPLACEMENT DIAGRAM

CAM DRAWING

CHAPTER 22-TRANSPARENCY 7

FIGURE 22-2-6 Dimensioning point *A* on the cam profile.

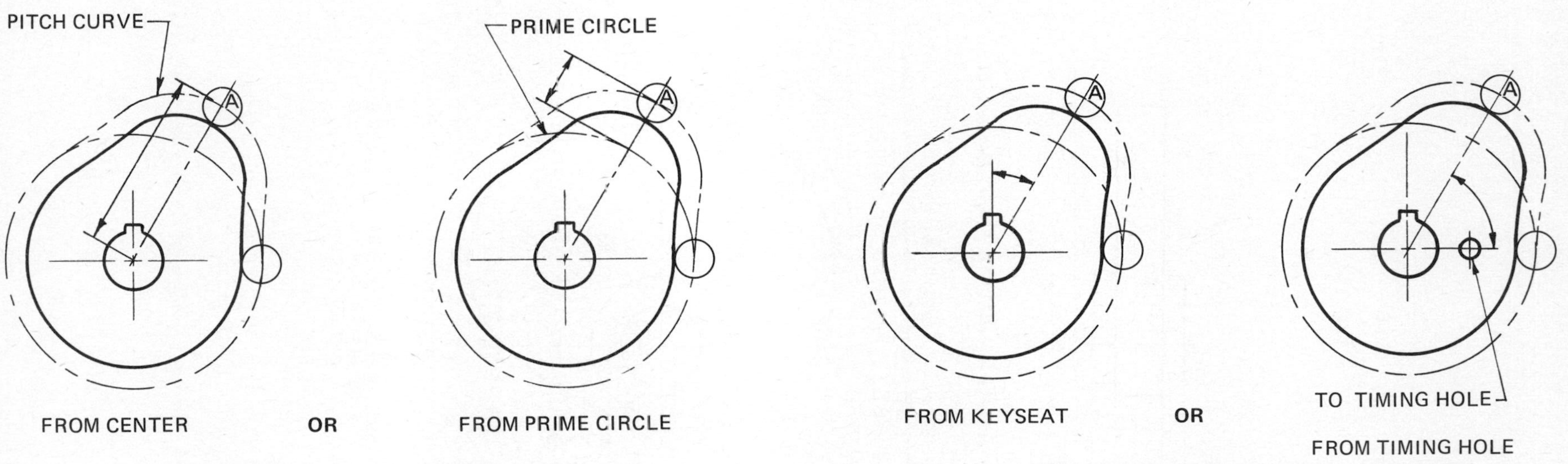
PITCH CURVE
PRIME CIRCLE
A
A
A
A
FROM CENTER
OR
FROM PRIME CIRCLE
FROM KEYSEAT
OR
TO TIMING HOLE
FROM TIMING HOLE

(A) DIMENSIONING RADIAL DISPLACEMENT **(B) DIMENSIONING ANGULAR DISPLACEMENT**

FIGURE 22-2-9 Plate cam drawing.

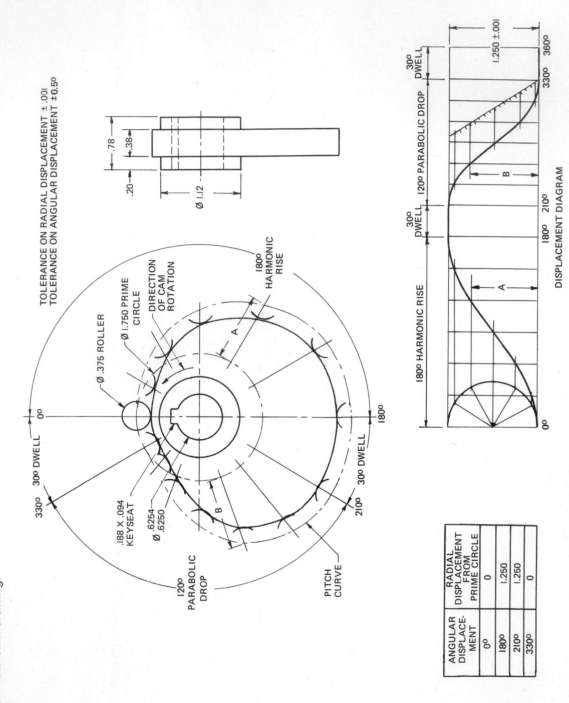

TOLERANCE ON RADIAL DISPLACEMENT ± .001
TOLERANCE ON ANGULAR DISPLACEMENT ± 0.5°

.375 ROLLER
Ø 1.750 PRIME CIRCLE
DIRECTION OF CAM ROTATION
180° HARMONIC RISE
.188 X .094 KEYSEAT
.6254
Ø .6250
120° PARABOLIC DROP
PITCH CURVE
330°
30° DWELL
0°
180°
210°
30° DWELL

1.250 ± .001
30° DWELL
120° PARABOLIC DROP
30° DWELL
180° HARMONIC RISE
DISPLACEMENT DIAGRAM
A
B
0°
180°
210°
330°
360°

ANGULAR DISPLACE-MENT	RADIAL DISPLACEMENT FROM PRIME CIRCLE
0°	0
180°	1.250
210°	1.250
330°	0

NOTE: ANGULAR AND RADIAL DISPLACEMENT DIMENSIONS FOR MOTIONS SUPPLIED BY CAM MANUFACTURER

.78
.38
.20
Ø 1.12

FIGURE 22-4-2 Drum cam drawing.

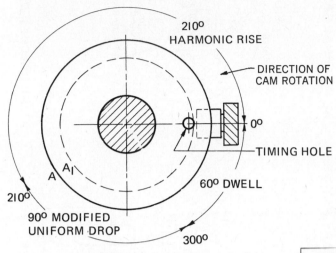

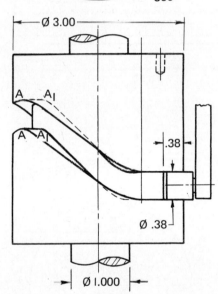

ANGULAR DISPLACEMENT FROM TIMING HOLE	DISPLACEMENT FROM BASE LINE
0º	0
210º	1.250
300º	0

TOLERANCE ON RADIAL DISPLACEMENT FROM BASELINE   ±.001

TOLERANCE ON ANGULAR DISPLACEMENT FROM BASELINE ±0.5º

NOTE: ANGULAR DISPLACEMENT AND DISPLACE-
MENT FROM BASELINE SUPPLIED BY CAM
MANUFACTURER.

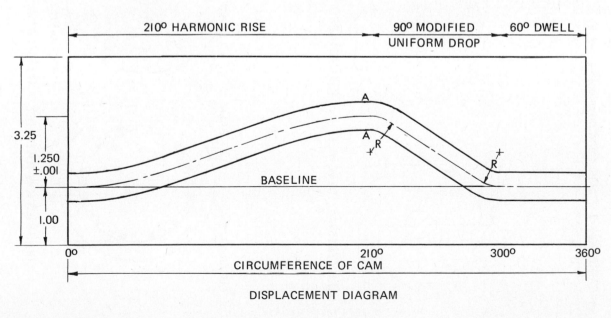

DISPLACEMENT DIAGRAM

FIGURE 22-7-1 Ratchet and pawl applications.

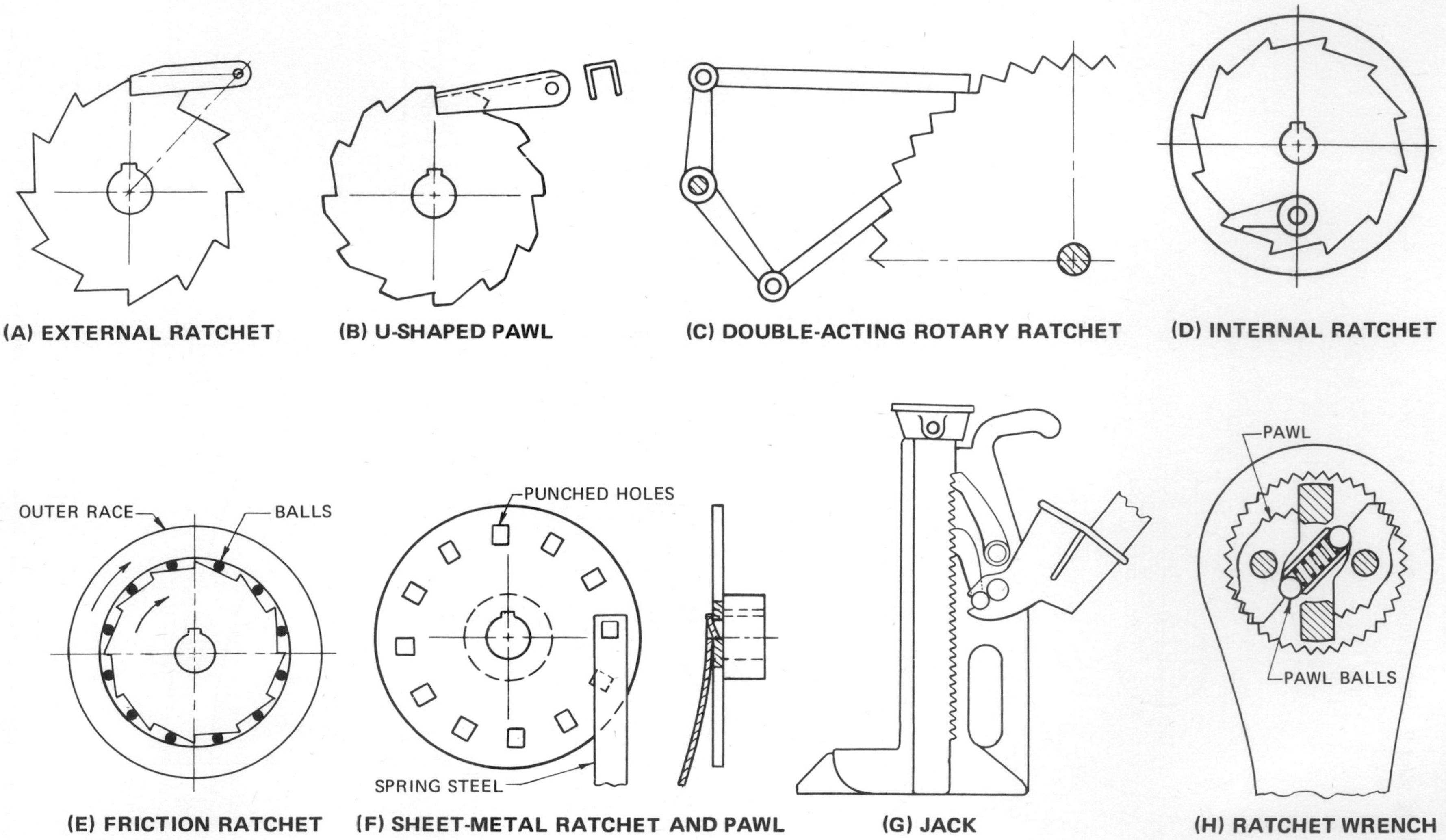
(A) EXTERNAL RATCHET
(B) U-SHAPED PAWL
(C) DOUBLE-ACTING ROTARY RATCHET
(D) INTERNAL RATCHET
OUTER RACE
BALLS
PUNCHED HOLES
SPRING STEEL
PAWL
PAWL BALLS
(E) FRICTION RATCHET
(F) SHEET-METAL RATCHET AND PAWL
(G) JACK
(H) RATCHET WRENCH

FIGURE 23-1-1 Development of a rectangular box.

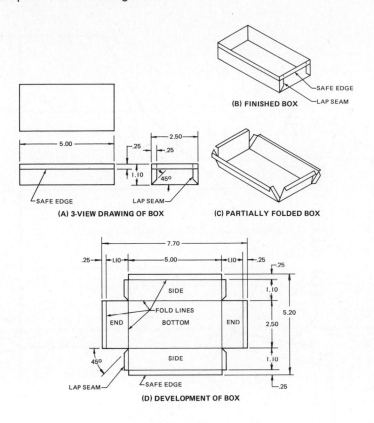

(B) FINISHED BOX

(A) 3-VIEW DRAWING OF BOX

(C) PARTIALLY FOLDED BOX

(D) DEVELOPMENT OF BOX

FIGURE 23-1-2 Development drawing with a complete set of folding and assembly instructions.

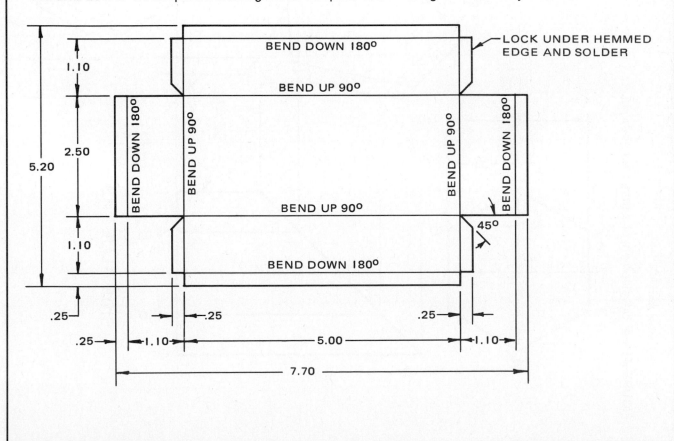

FIGURE 23-2-4 Development of a truncated hexagon.

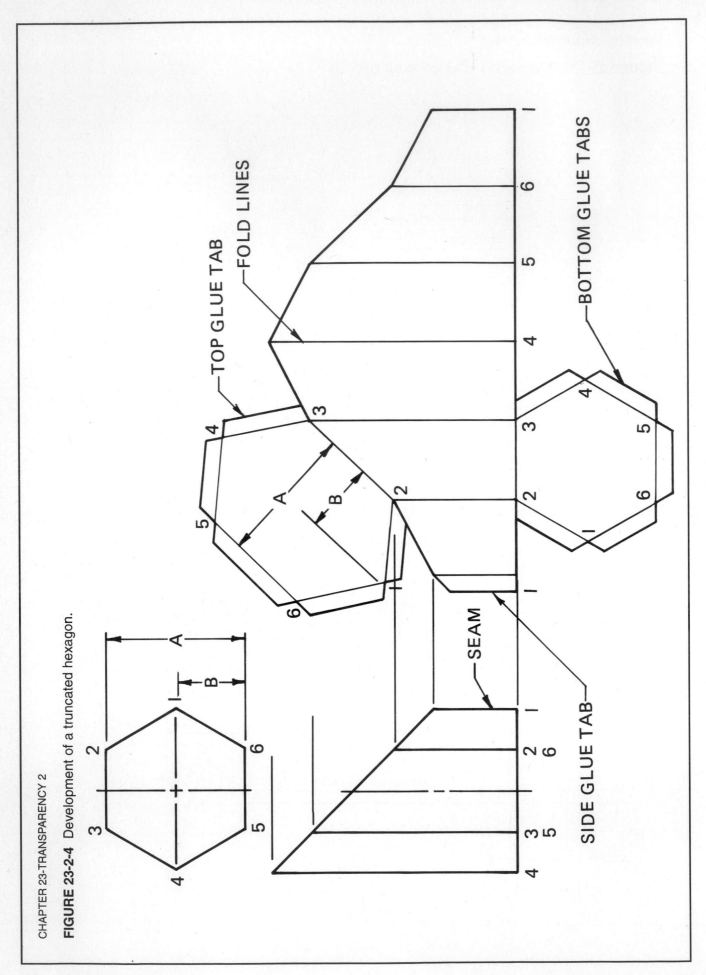

TOP GLUE TAB

FOLD LINES

BOTTOM GLUE TABS

SEAM

SIDE GLUE TAB

FIGURE 23-3-1 Development of a right pyramid with true length of edge lines shown.

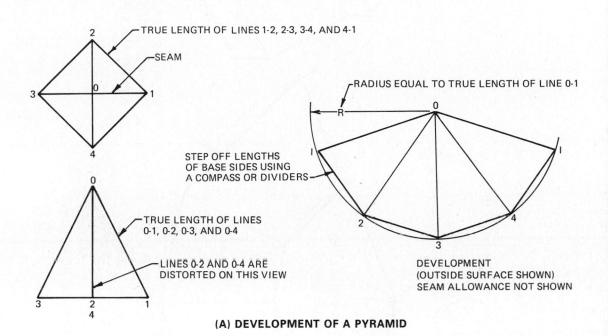

(A) DEVELOPMENT OF A PYRAMID

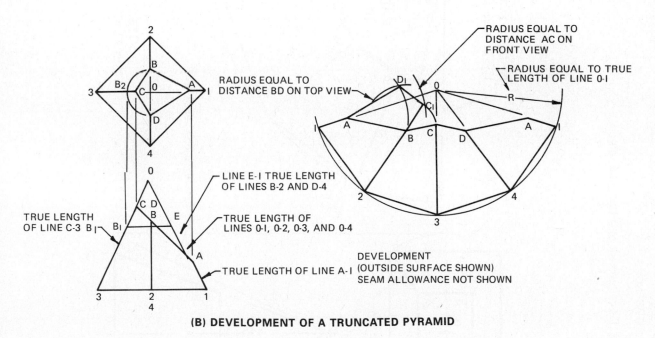

(B) DEVELOPMENT OF A TRUNCATED PYRAMID

FIGURE 23-3-4 Development of a transition piece.

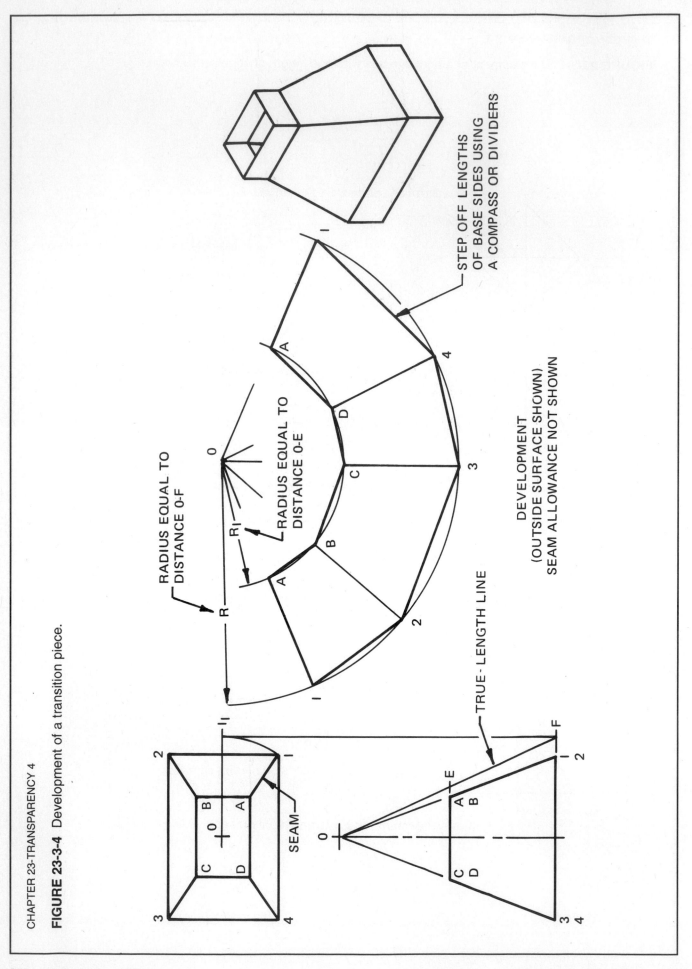

CHAPTER 23-TRANSPARENCY 5

FIGURE 23-4-3 Development of a cylinder with the top and bottom truncated.

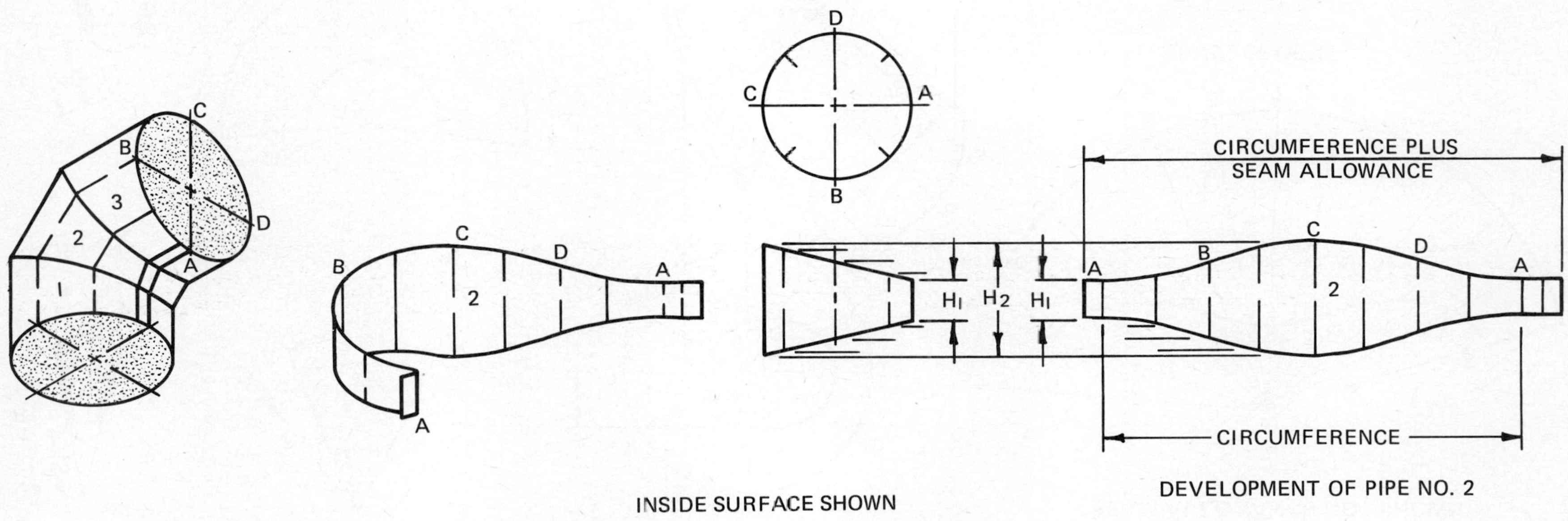

FIGURE 23-5-2 Development of a truncated cone.

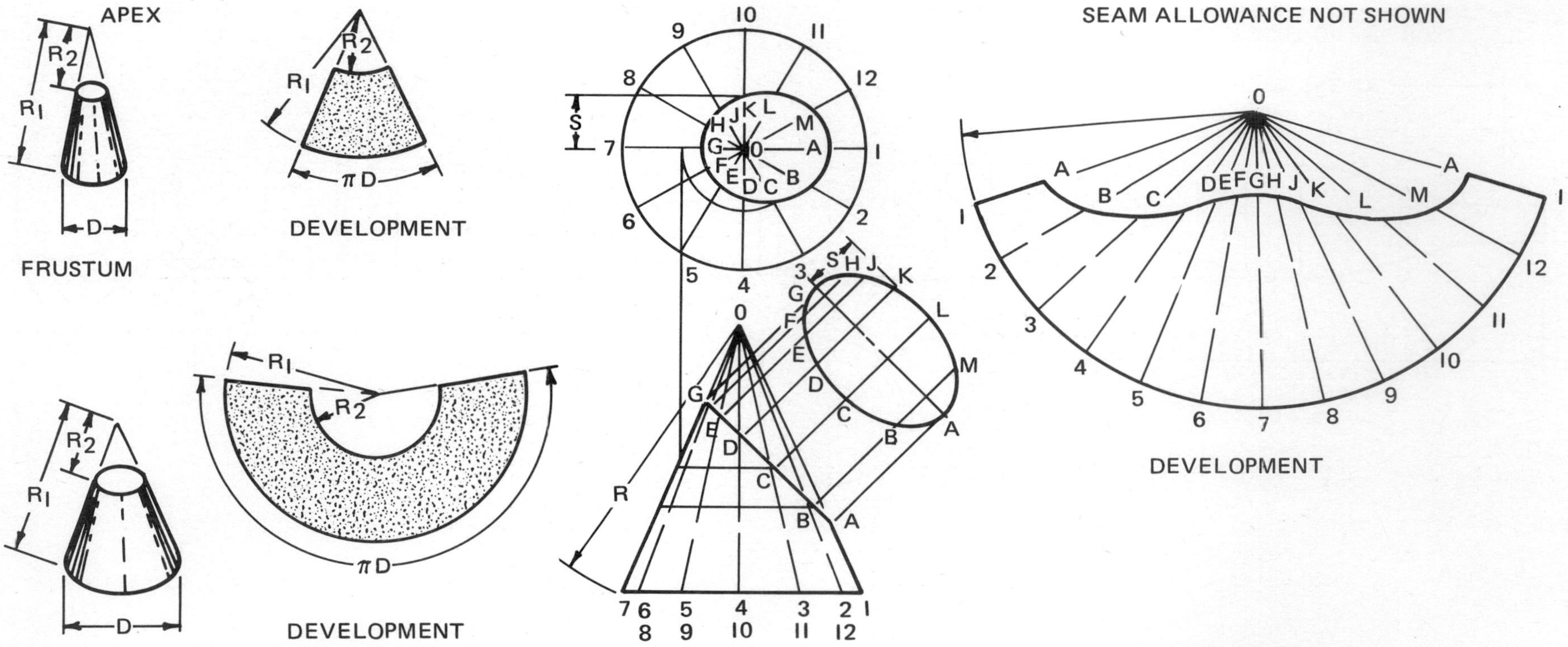

SEAM ALLOWANCE NOT SHOWN

DEVELOPMENT

(B) DEVELOPMENT PROCEDURE

APEX

DEVELOPMENT

FRUSTUM

DEVELOPMENT

(A) PROPORTION OF HEIGHT TO BASE

FIGURE 23-6-4 Development of an offset transition piece—rectangular to round.

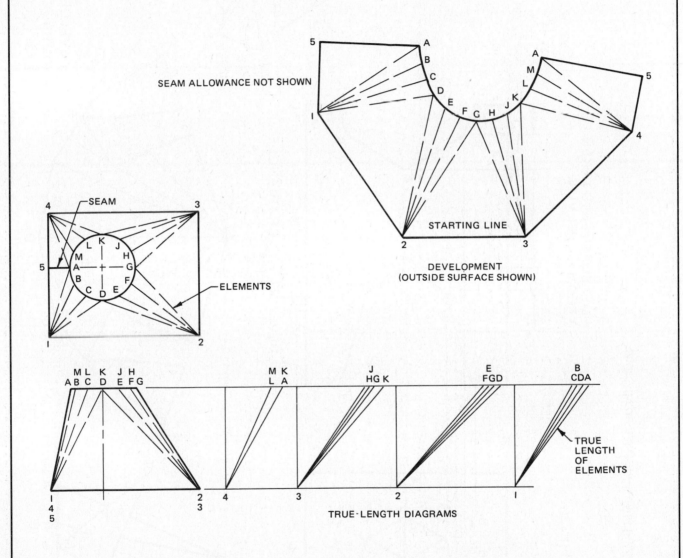

FIGURE 23-8-3 Intersecting prisms—triangle and pyramid.

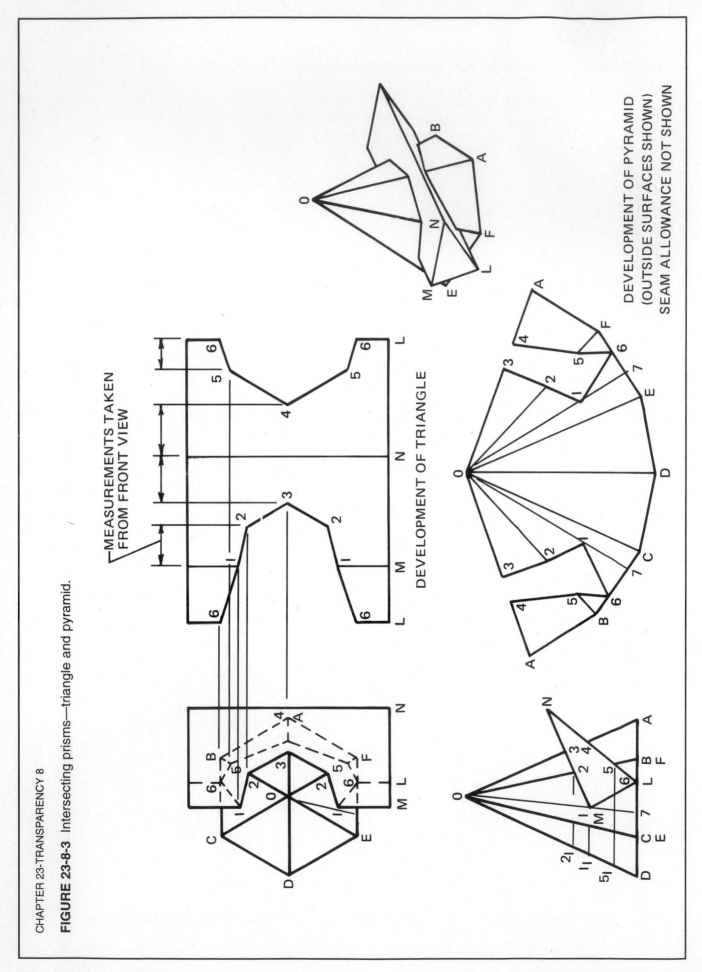

MEASUREMENTS TAKEN FROM FRONT VIEW

DEVELOPMENT OF TRIANGLE

DEVELOPMENT OF PYRAMID
(OUTSIDE SURFACES SHOWN)
SEAM ALLOWANCE NOT SHOWN

FIGURE 23-9-1 Plotting lines of intersection and making development drawings for a 45° reducing tee.

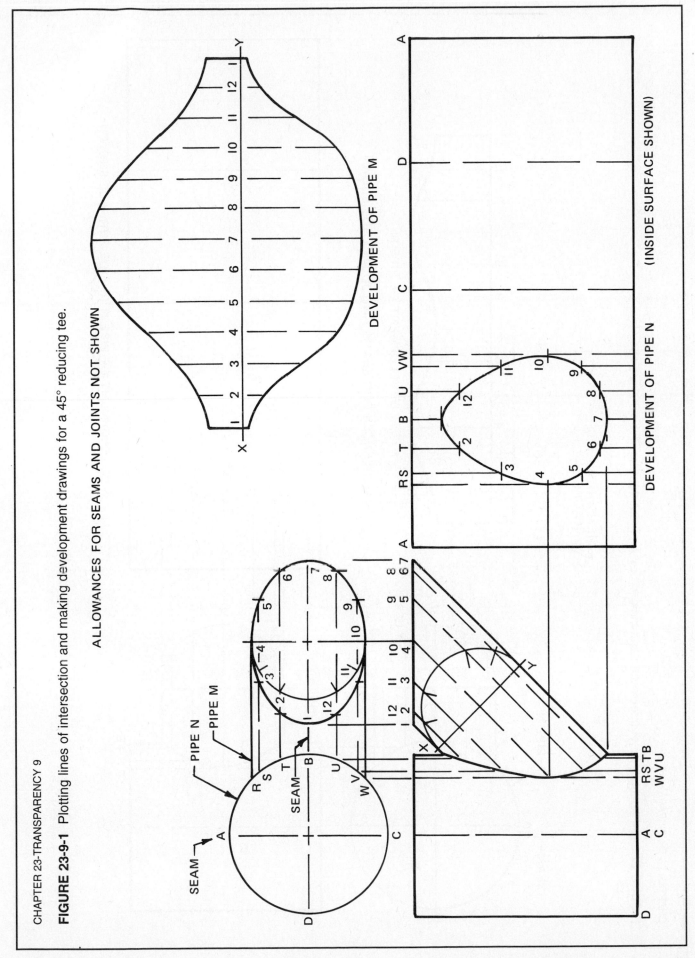

ALLOWANCES FOR SEAMS AND JOINTS NOT SHOWN

FIGURE 23-9-2 Plotting lines of intersection and making development drawings for a 90° reducing tee.

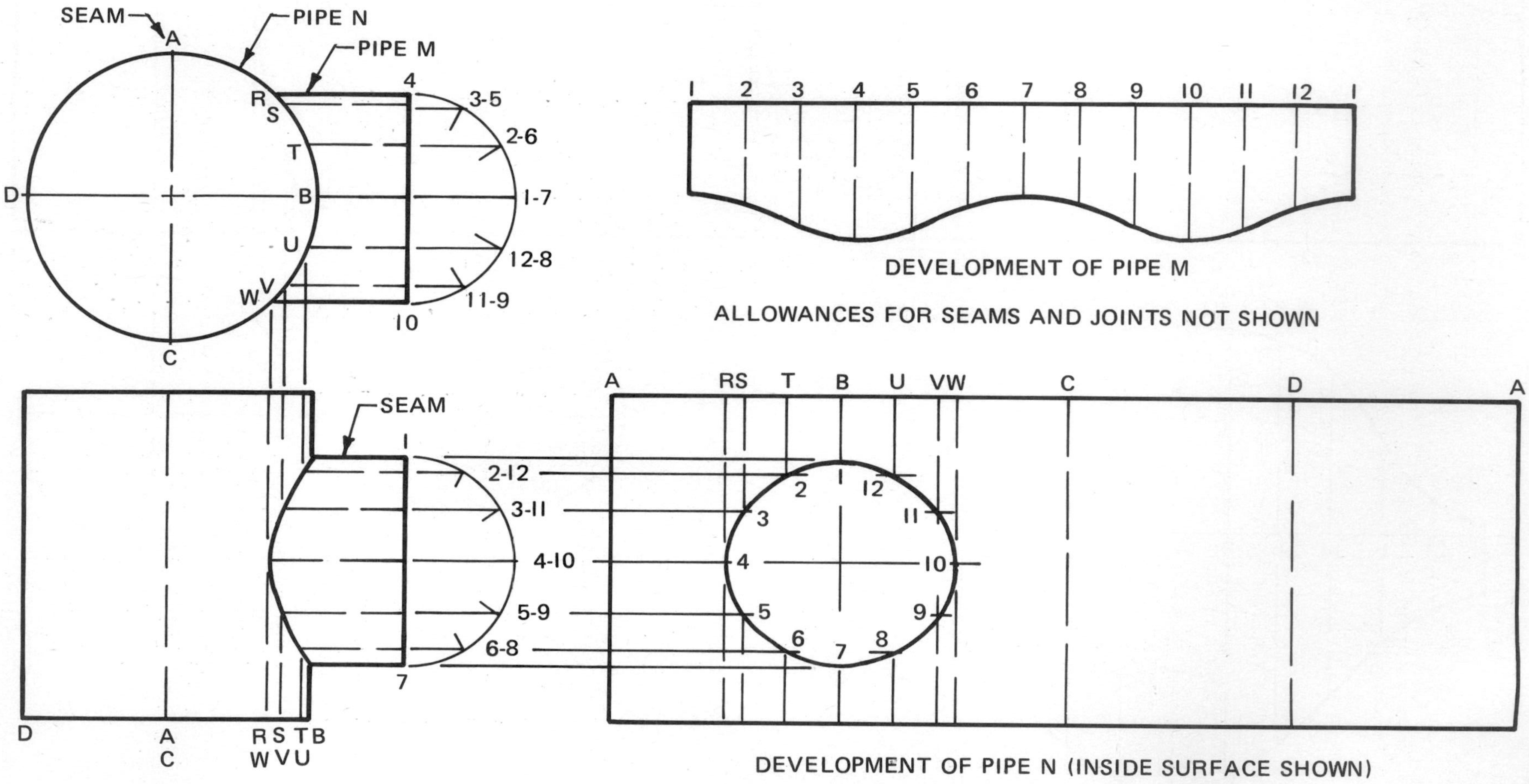
SEAM
PIPE N
PIPE M
A
R
S
T
B
U
V
W
D
C
4
3-5
2-6
1-7
12-8
11-9
10
I 2 3 4 5 6 7 8 9 10 11 12 I
DEVELOPMENT OF PIPE M
ALLOWANCES FOR SEAMS AND JOINTS NOT SHOWN
SEAM
2-12
3-11
4-10
5-9
6-8
7
A RS T B U VW C D A
D
A C
R S T B
W V U
DEVELOPMENT OF PIPE N (INSIDE SURFACE SHOWN)

FIGURE 24-1-2 Common types of pipe joints.

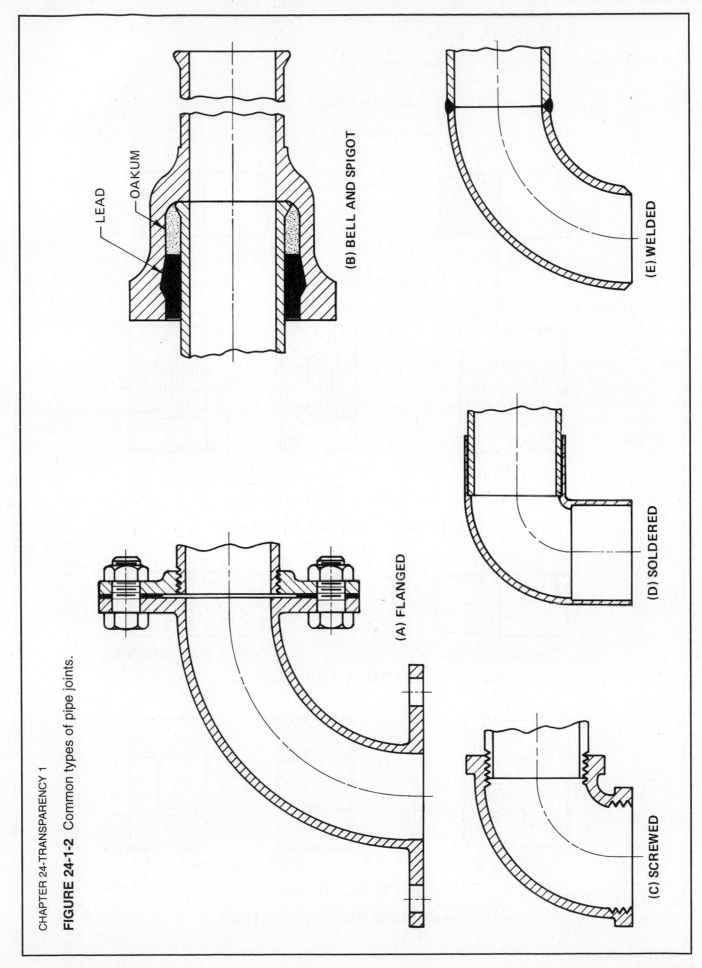

LEAD

OAKUM

(B) BELL AND SPIGOT

(E) WELDED

(A) FLANGED

(D) SOLDERED

(C) SCREWED

FIGURE 24-1-6 Screwed fitting, *(Crane Canada Ltd.)*

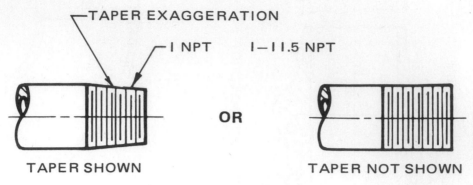

TAPER EXAGGERATION

I NPT   I—I I.5 NPT

OR

TAPER SHOWN   TAPER NOT SHOWN

EXTERNAL THREAD

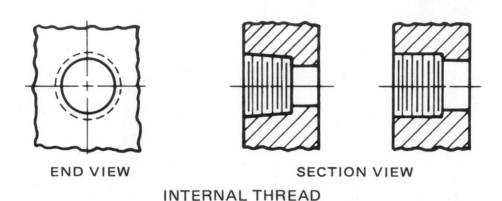

END VIEW   SECTION VIEW

INTERNAL THREAD

(A) SCHEMATIC REPRESENTATION

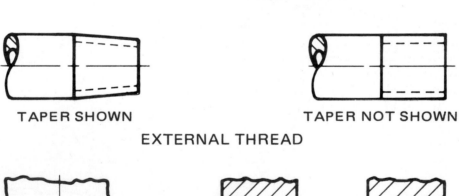

TAPER SHOWN   TAPER NOT SHOWN

EXTERNAL THREAD

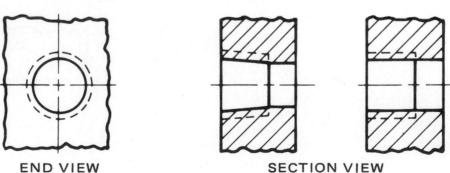

END VIEW   SECTION VIEW

INTERNAL THREAD

(B) SIMPLIFIED REPRESENTATION

FIGURE 24-1-11 Pipe drawing symbols.

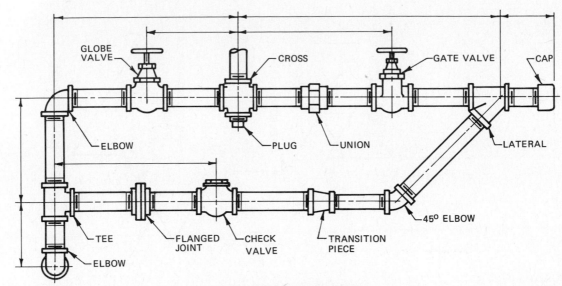

(A) DOUBLE-LINE DRAWING

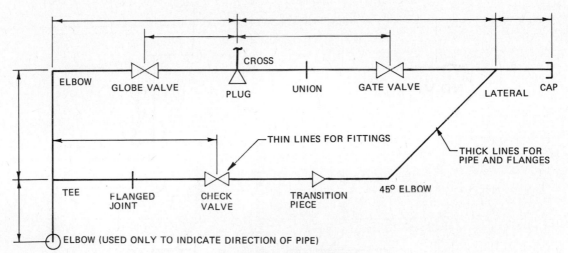

(B) SINGLE-LINE DRAWING

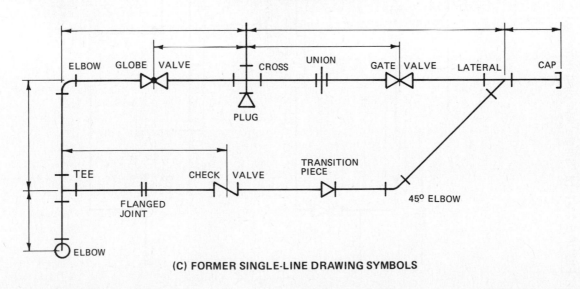

(C) FORMER SINGLE-LINE DRAWING SYMBOLS

FIGURE 24-1-12 Single-line pipe drawings.

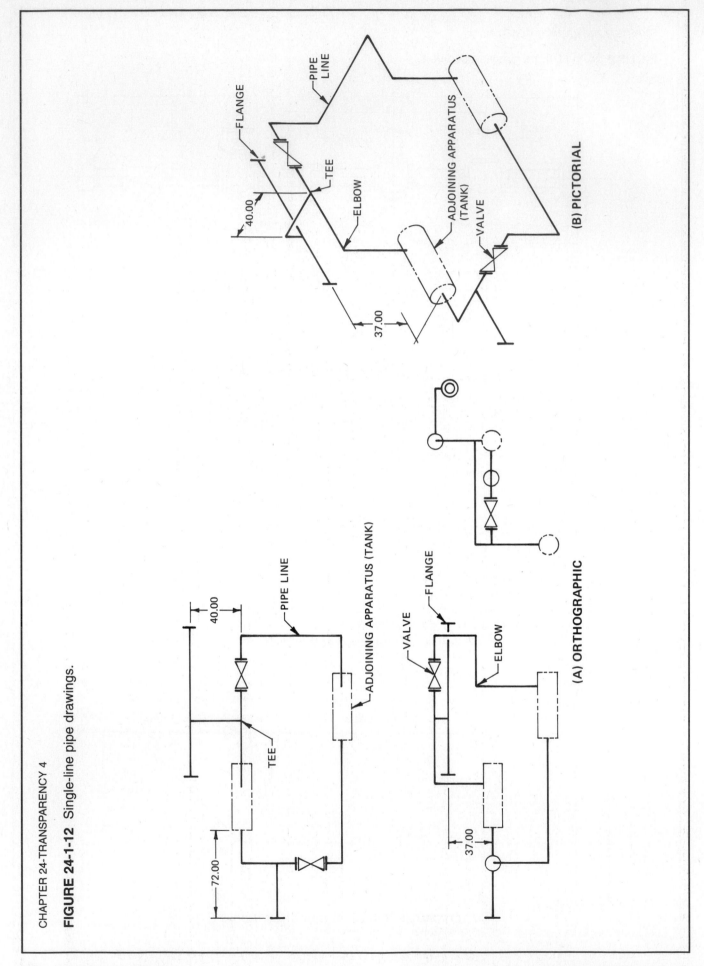

(B) PICTORIAL

(A) ORTHOGRAPHIC

FIGURE 24-2-2 Coordinate axes for pipe drawings.

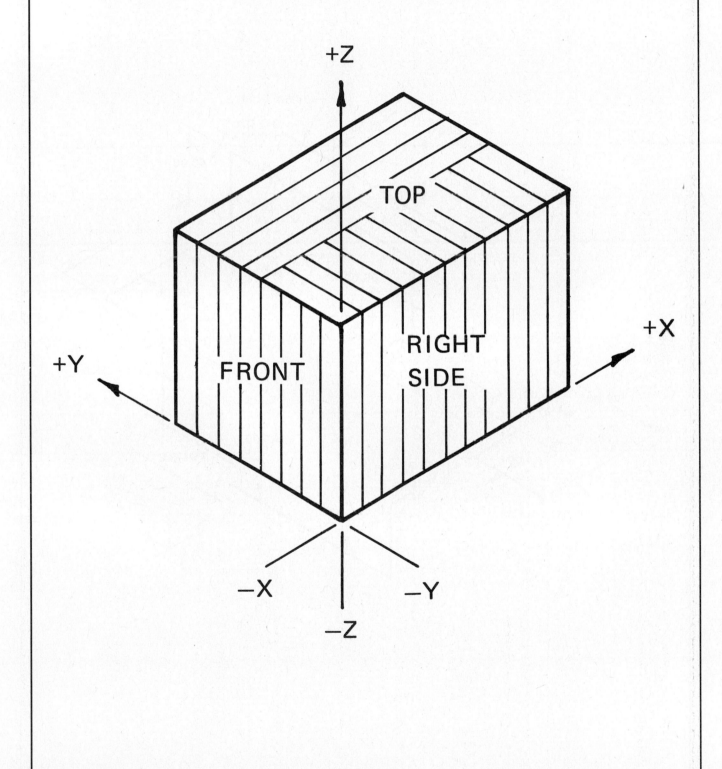

FIGURE 24-2-6 Unidirectional dimensioning.

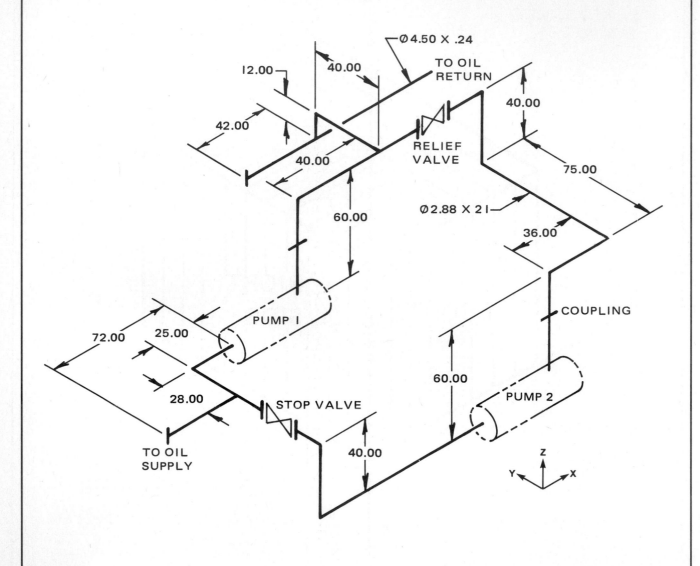

FIGURE 24-3-3 Specifying slope of pipes.

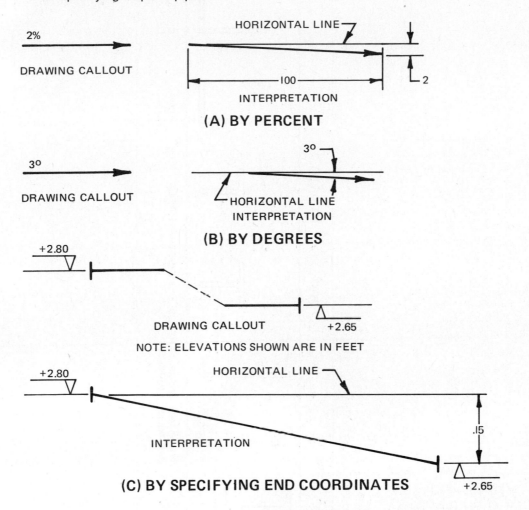

(A) BY PERCENT

(B) BY DEGREES

NOTE: ELEVATIONS SHOWN ARE IN FEET

(C) BY SPECIFYING END COORDINATES

FIGURE 24-3-4 Supports and hangers.

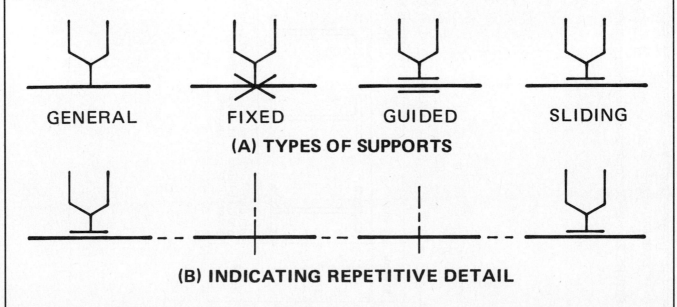

GENERAL FIXED GUIDED SLIDING

(A) TYPES OF SUPPORTS

(B) INDICATING REPETITIVE DETAIL

FIGURE 25-1-2 Common structural steel shapes.

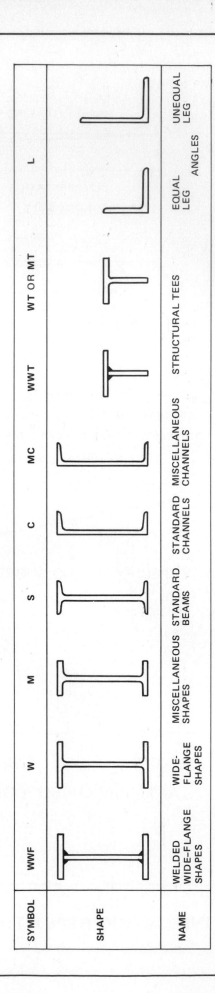

SYMBOL	WWF	W	M	S	C	MC	WWT	WT OR MT	L	
SHAPE										
NAME	WELDED WIDE-FLANGE SHAPES	WIDE-FLANGE SHAPES	MISCELLANEOUS SHAPES	STANDARD BEAMS	STANDARD CHANNELS	MISCELLANEOUS CHANNELS	STRUCTURAL TEES		EQUAL LEG	UNEQUAL LEG
									ANGLES	

FIGURE 25-1-6 Structural steel terms.

1. Anchors or hangers for open-web steel joists
2. Anchors for structural steel
3. Bases of steel and iron for steel or iron columns
4. Beams, purlins, girts
5. Bearing plates for structural steel
6. Bracing for steel members or frames
7. Brackets attached to the steel frame
8. Columns, concrete-filled pipe, and struts
9. Conveyor structural steel framework
10. Steel joists, open-web steel joists, bracing, and accessories supplied with joists
11. Separators, angles, tees, clips, and other detail fittings
12. Floor and roof plates (raised pattern or plain (connected to steel frame)
13. Girders
14. Rivets and bolts
15. Headers or trimmers for support of open-web steel joists where such headers or trimmers frame into structural steel members
16. Light-gage cold-formed steel used to support floor and roofs
17. Lintels shown on the framing plans or otherwise scheduled
18. Shelf angles

FIGURE 25-1-15 Dimensioning structural drawings.

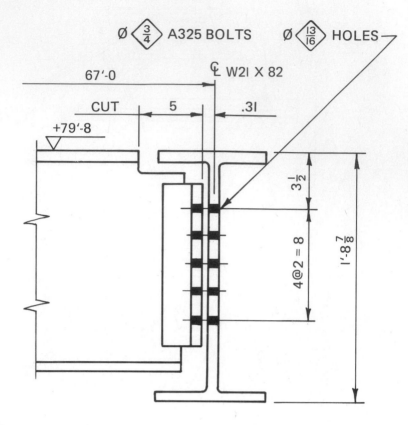

FIGURE 25-1-16 Bolt symbols.

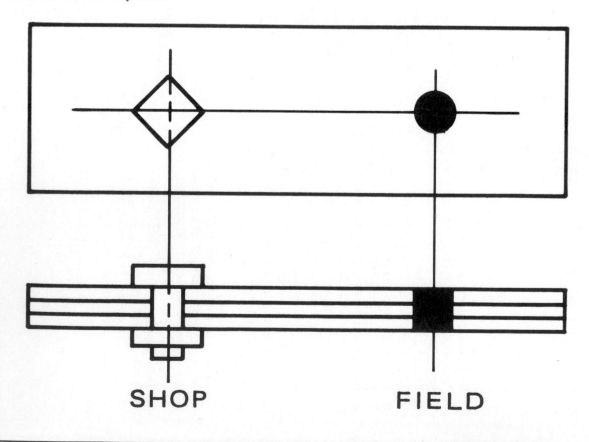

FIGURE 25-2-5 Detail of W 18 x 60 beam connections.

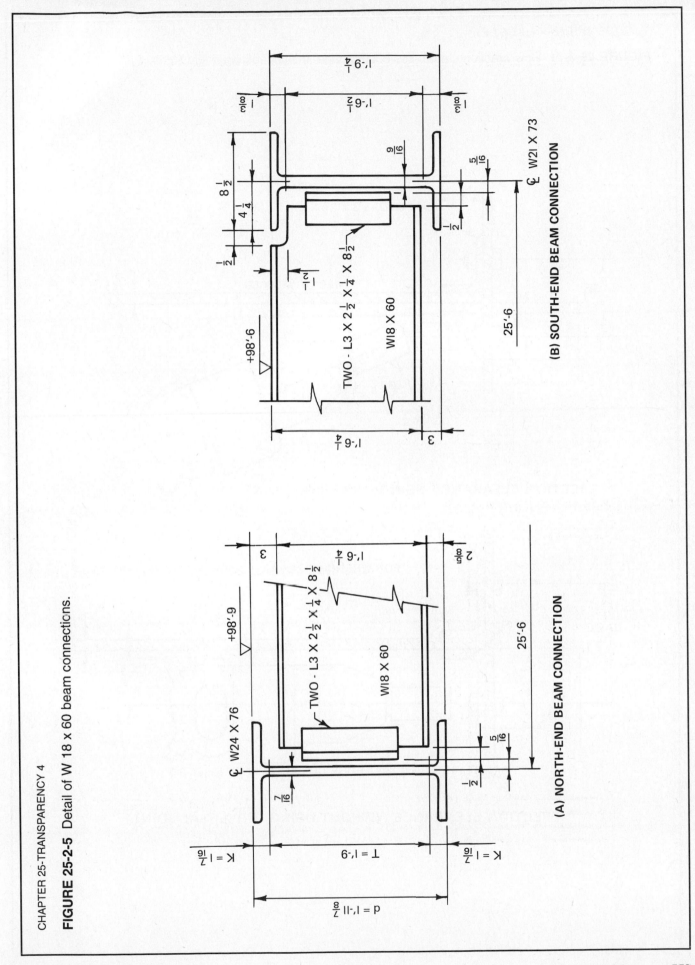

(A) NORTH-END BEAM CONNECTION

(B) SOUTH-END BEAM CONNECTION

FIGURE 25-3-11 How erection clearances control gage and connecting angle sizes.

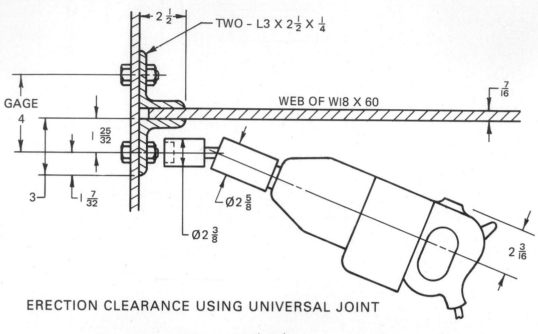

ERECTION CLEARANCE USING UNIVERSAL JOINT

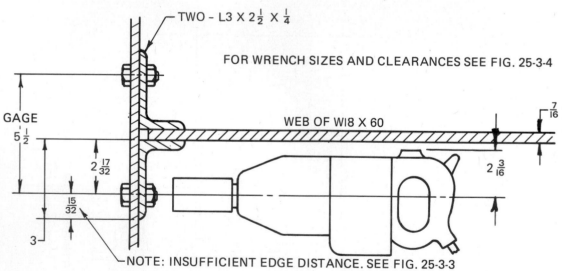

ERECTION CLEARANCE WITHOUT USING UNIVERSAL JOINT

FIGURE 25-3-13 Detail of north end of W 18 x 60 beam from partial design drawing, Fig. 25-3-7.

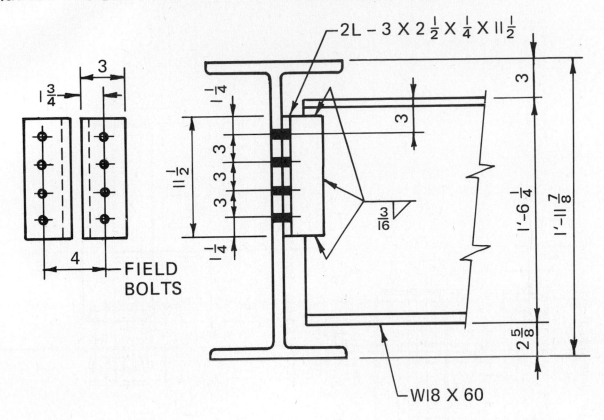

FIGURE 25-3-14 Complete beam detail of W 18 x 60 from partial design drawing, Fig. 25-3-7.

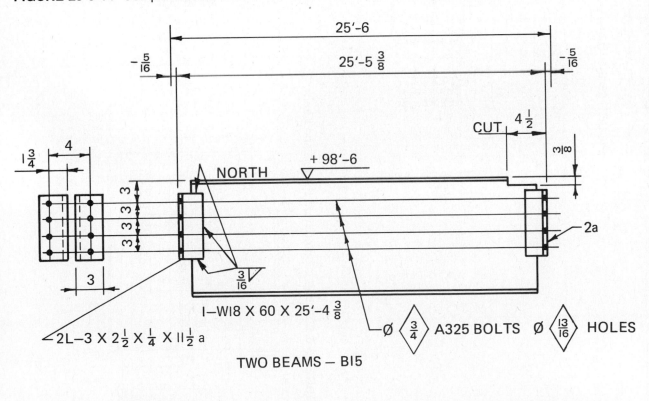

FIGURE 25-6-1 Dimensioning to center line of channel webs.

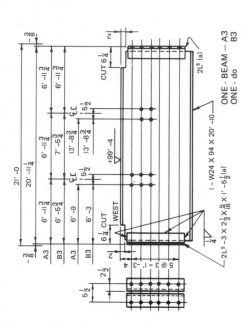

FIGURE 25-6-2 Dimensioning from the left end beam.

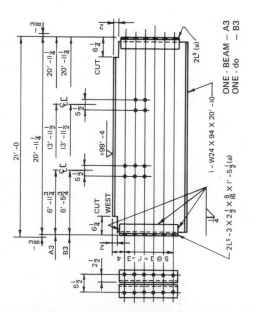

FIGURE 25-6-3 Dimensioning to the back of channels.

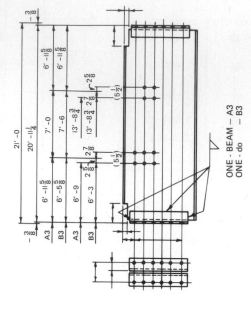

FIGURE 26-1-3 Simple plate jig.

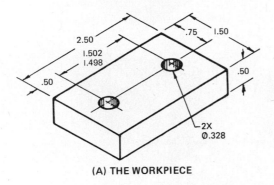

(A) THE WORKPIECE

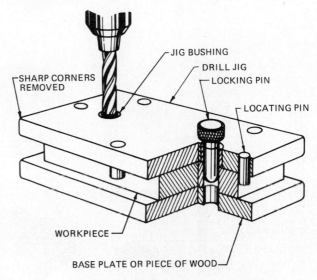

(C) ADD LOCKING PIN BEFORE STARTING SECOND HOLE

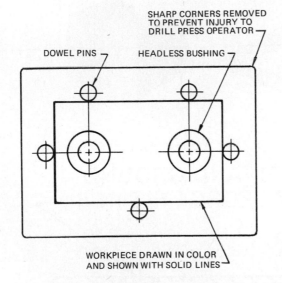

(B) PLACE JIG OVER WORKPIECE AND DRILLING FIRST HOLE

FIGURE 26-2-1 Machined sections for construction of jigs and fixtures. *(Standard Parts Co.)*

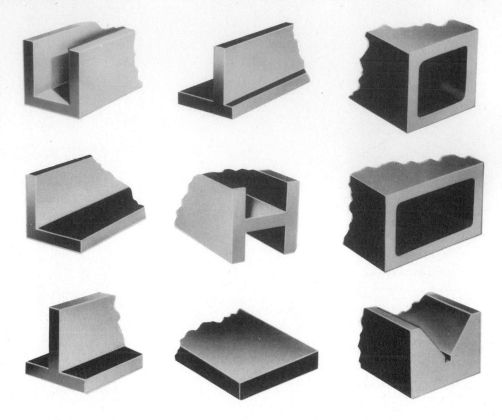

FIGURE 26-2-3 Dowel pins and cap screws.

DOWEL PINS—HARDENED AND GROUND. CLOSE SLIP FIT IN A, FORCE FIT IN B

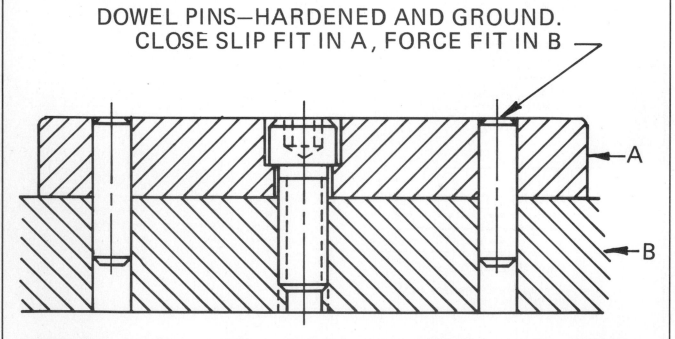

DOWEL PINS — USED TO ALIGN PARTS. MINIMUM OF 2 SOCKET SCREWS USED TO HOLD PARTS TOGETHER.

FIGURE 26-2-10 Common clamping devices.

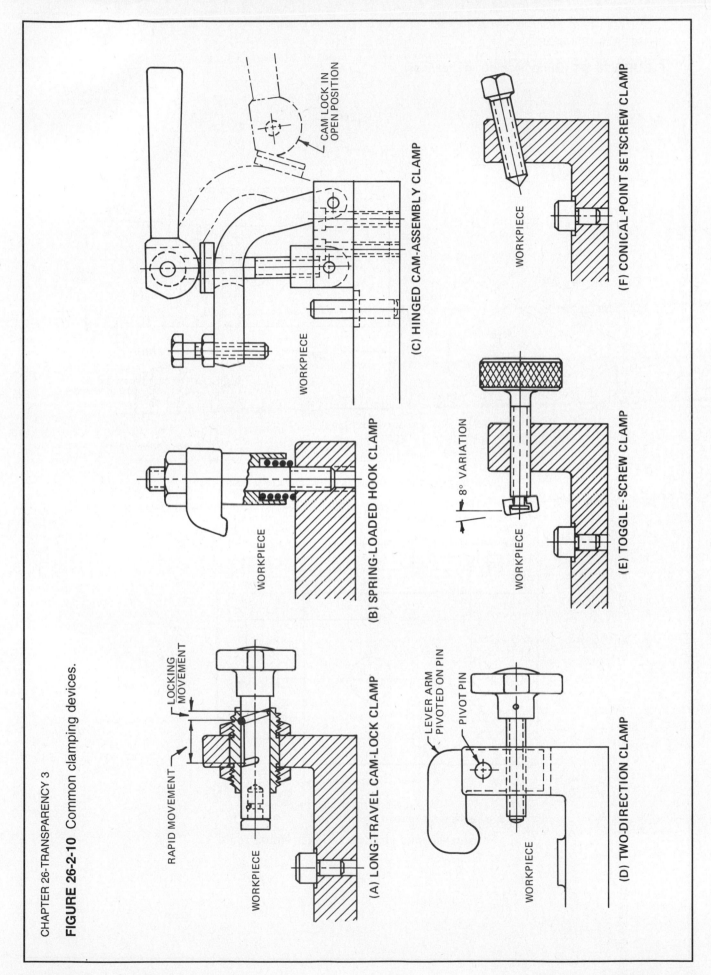

(A) LONG-TRAVEL CAM-LOCK CLAMP

LOCKING MOVEMENT

RAPID MOVEMENT

WORKPIECE

(B) SPRING-LOADED HOOK CLAMP

WORKPIECE

(C) HINGED CAM-ASSEMBLY CLAMP

CAM LOCK IN OPEN POSITION

WORKPIECE

(D) TWO-DIRECTION CLAMP

LEVER ARM PIVOTED ON PIN

PIVOT PIN

WORKPIECE

(E) TOGGLE-SCREW CLAMP

8° VARIATION

WORKPIECE

(F) CONICAL-POINT SETSCREW CLAMP

WORKPIECE

FIGURE 26-3-1 Dimensioning jig drawings.

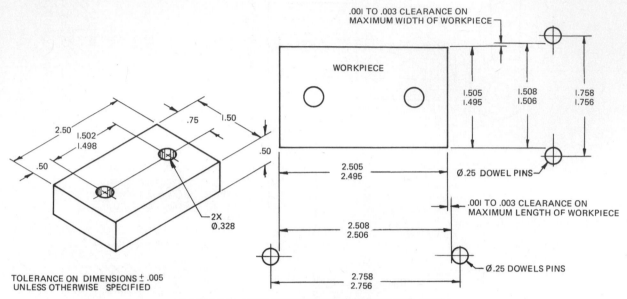

.001 TO .003 CLEARANCE ON
MAXIMUM WIDTH OF WORKPIECE

WORKPIECE

1.505
1.495

1.508
1.506

1.758
1.756

2.505
2.495

Ø.25 DOWEL PINS

.001 TO .003 CLEARANCE ON
MAXIMUM LENGTH OF WORKPIECE

2.508
2.506

2.758
2.756

Ø.25 DOWELS PINS

2.50
1.502
1.498
.50
.75
1.50
.50

2X
Ø.328

TOLERANCE ON DIMENSIONS ± .005
UNLESS OTHERWISE SPECIFIED

(A) CALCULATING DISTANCES BETWEEN DOWEL PINS

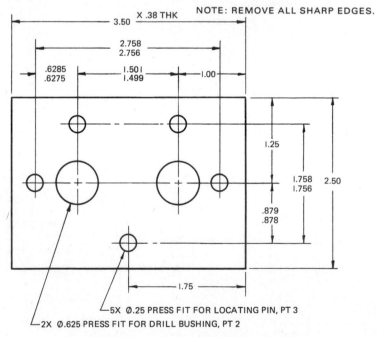

NOTE: REMOVE ALL SHARP EDGES.

X .38 THK

3.50

2.758
2.756

.6285
.6275

1.501
1.499

1.00

1.25

1.758
1.756

2.50

.879
.878

1.75

5X Ø.25 PRESS FIT FOR LOCATING PIN, PT 3

2X Ø.625 PRESS FIT FOR DRILL BUSHING, PT 2

(B) DIMENSIONING JIG PLATE SHOWN IN FIGURE 26-1-3

FIGURE 27-2-1 Partial schematic diagram of a receiver.

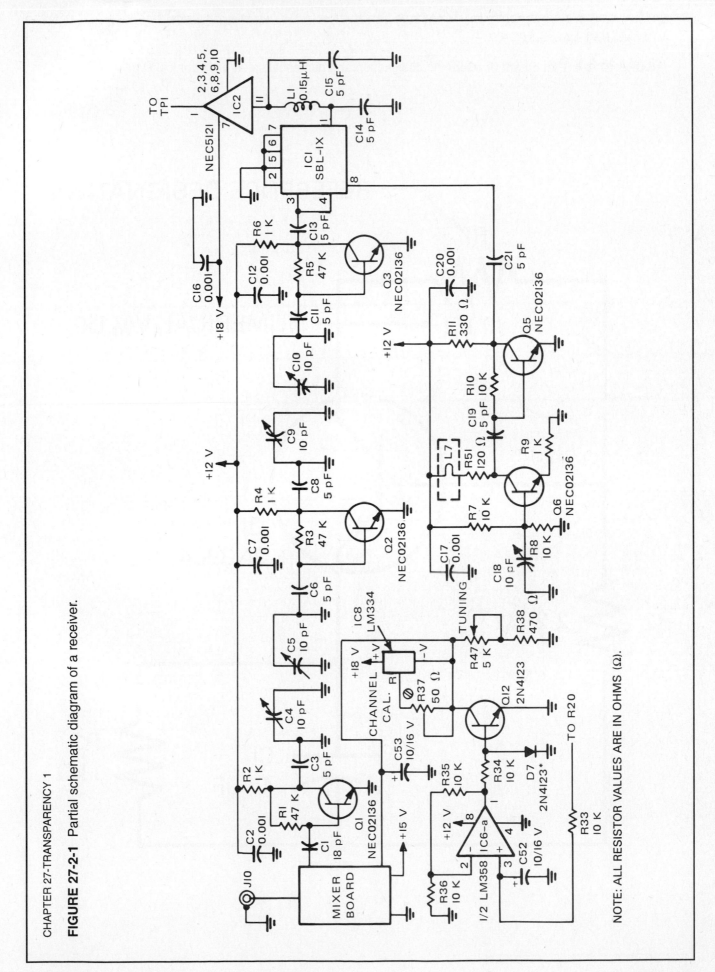

NOTE: ALL RESISTOR VALUES ARE IN OHMS (Ω).

FIGURE 27-2-4 Placement of reference designation and numerical value on graphical symbols.

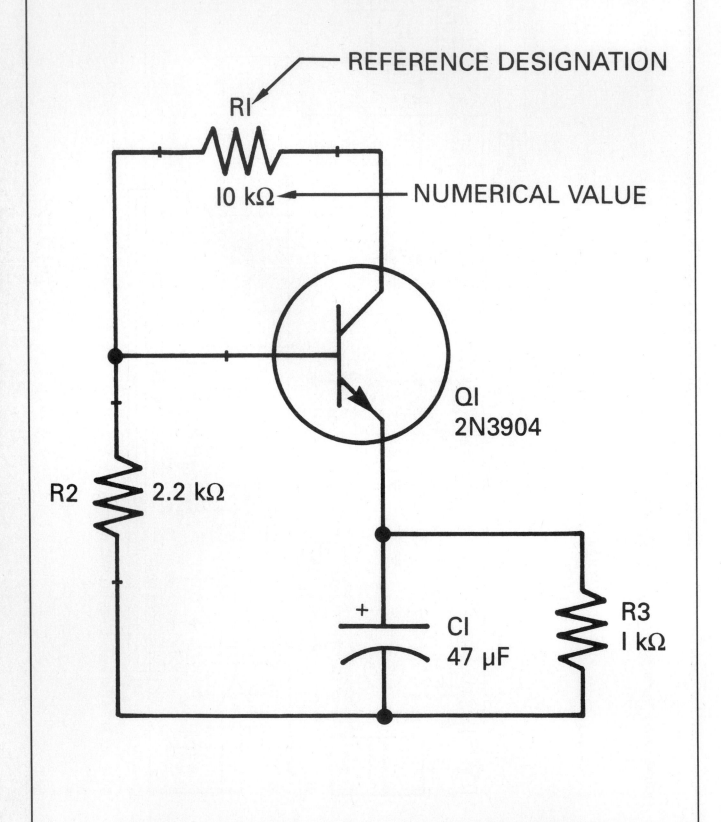

FIGURE 27-3-1 Point-to-point connection diagram of a boat's electrical system.

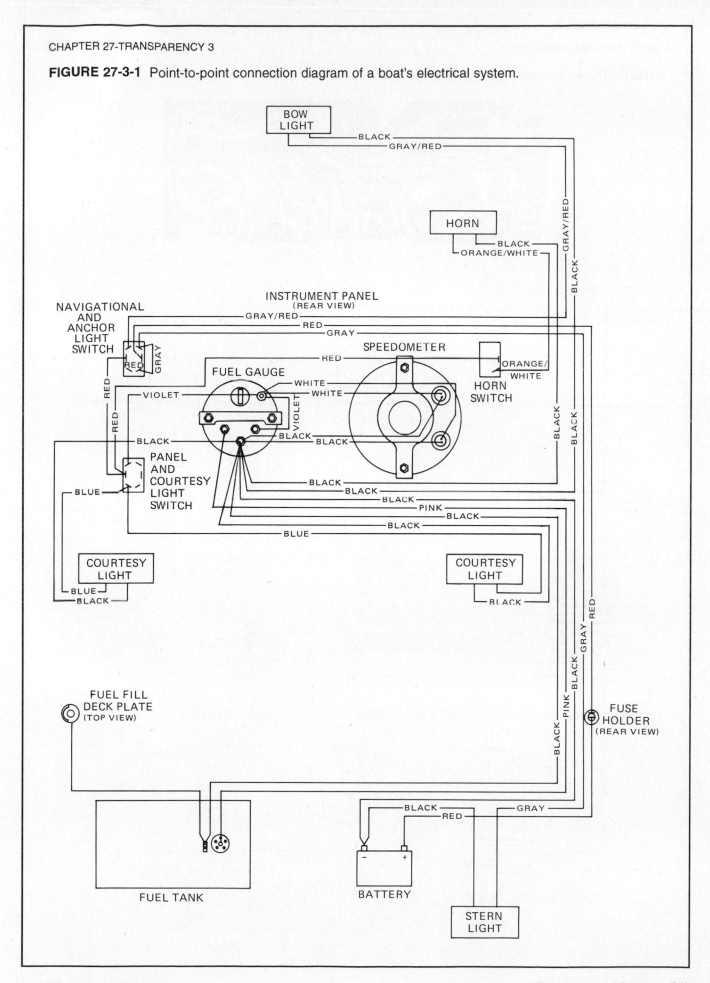

FIGURE 27-4-1 Printed circuit used on the amplifier in Fig. 27-4-2. *(General Electric)*

FIGURE 27-4-2 Schematic diagram for a simple amplifier. *(General Electric)*

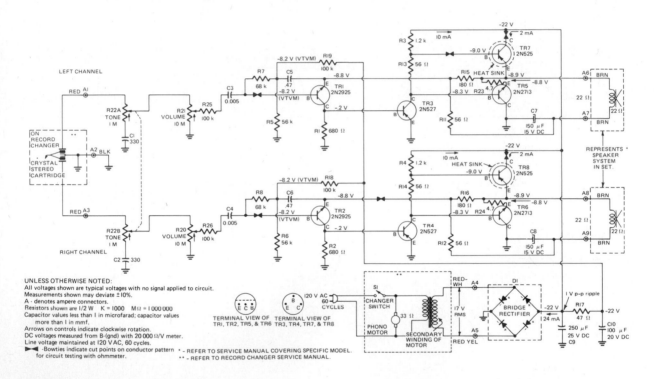

FIGURE 27-4-3 Layout of components for amplifier shown Fig. 27-4-2. *(General Electric)*

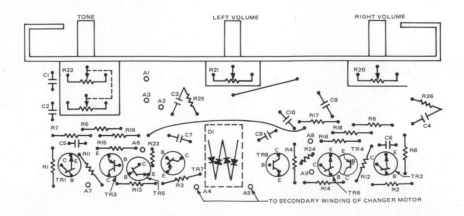

FIGURE 27-5-1 Block diagram of a radio-phonograph combination.

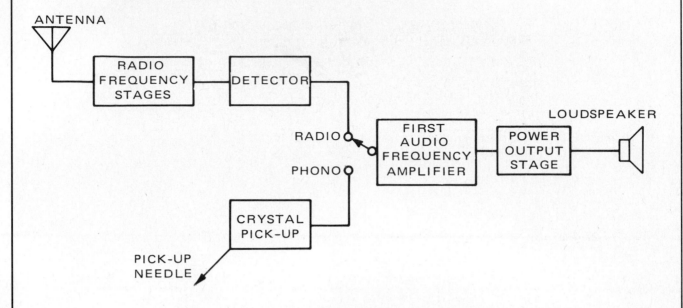

FIGURE 27-5-2 Logic flow diagram.

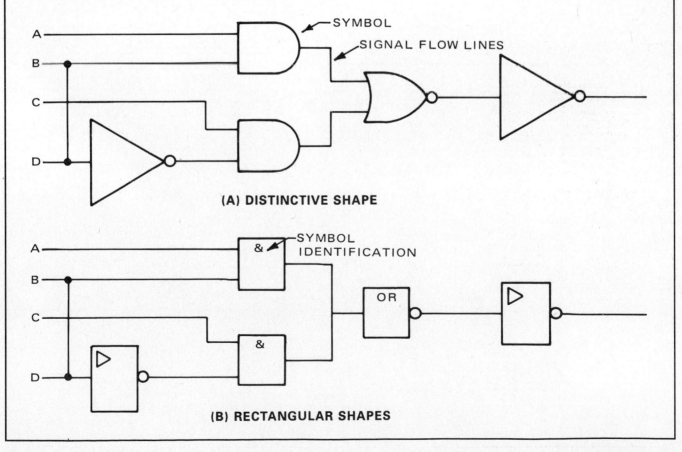

(A) DISTINCTIVE SHAPE

(B) RECTANGULAR SHAPES